The chemical philosopher, J. B. van Helmont, illumined by the light shed by chemical knowledge (in contrast to Galen, Avicenna, and Theophrastus). From J. B. van Helmont, *Aufgang der Artzney-Kunst* (Sulzbach, 1683).

THE CHEMICAL PHILOSOPHY

Paracelsian Science and Medicine in
the Sixteenth and Seventeenth Centuries

TWO VOLUMES BOUND AS ONE

ALLEN G. DEBUS

Morris Fishbein Professor Emeritus of the
History of Science and Medicine
The University of Chicago

DOVER PUBLICATIONS, INC.
Mineola, New York

To Bruni

Copyright

Copyright © 1977 by Allen G. Debus
All rights reserved under Pan American and International Copyright Conventions.

Published in the United Kingdom by David & Charles, Brunel House, Forde Close, Newton Abbot, Devon TQ12 4PU.

Bibliographical Note

This Dover edition, first published in 2002, is an unabridged republication, with some minor corrections, of the work first published as *The Chemical Philosophy: Paracelsian Science and Medicine in the Sixteenth and Seventeenth Centuries*, Volume I and Volume II, by Science History Publications, a division of Neale Watson Academic Publications, Inc., New York, in 1977. A new Preface and a list of errata have been prepared especially for this edition by the author.

Dover Publications wishes to thank Neale W. Watson, of Watson Publishing International, for his kind assistance in making this new edition possible.

Library of Congress Cataloging-in-Publication Data

Debus, Allen G.
    The chemical philosophy / Allen G. Debus.
       p. cm.
    Originally published: New York : Science History Publications, 1977.
    "A new preface and a list of errata have been prepared especially for this edition by the author"–T.p. verso.
    Includes bibliographical references and index.
    ISBN 0-486-42175-9 (pbk.)
    1. Chemistry–History. 2. Paracelsus, 1493–1541. 3. Medicine–History–16th century. 4. Medicine–History–17th century. I. Title.

QD14 .D43 2002
540'.9'031–dc21

2002017457

Manufactured in the United States of America
Dover Publications, Inc., 31 East 2nd Street, Mineola, N.Y. 11501

CONTENTS

Volume I

LIST OF PLATES	vii
PREFACE TO DOVER EDITION	ix
PREFACE	xxi

1 CHEMISTRY AND NATURE IN THE RENAISSANCE   1

The Chemical Heritage: Alchemy in Antiquity, 3/ Islamic Alchemy, 7/ The Latin Alchemy of the West, 11/ Medieval Medical Chemistry: The Analysis of Spa Waters, 14/ The Chemically Prepared Medicines, 19/ Renaissance Factors: The Educational Problem, 25/ The Hermetic Revival and the Study of Nature, 30/ Magic, Mathematics, and Nature, 35/ Paracelsus: The Man, 45/ The Paracelsian System, 51/ Conclusion, 60

2 THE CHEMICAL PHILOSOPHY   63

The Paracelsian Universe, 66/ Paracelsian "Mathematics," 73/ Chemistry and the New Science, 76/ Chemical Theory and the Elements, 78/ Chemistry and the Geocosm, 84/ The Microcosm and Medical Theory, 96/ The Chemical Analysis of Spa Waters, 109/ The New Medicines, 112/ The *Basilica chymica* of Oswald Crollius (1609), 117/ Conclusion, 124

3 THE PARACELSIAN DEBATES   127

Synthesis and Reaction: The Work of Severinus (1571) and Erastus (1572-1574), 128/ The Search for Common Ground: Albertus Wimpenaeus (1569) and Guinter von Andernach (1571), 135/ French Paracelsism in the Late Sixteenth Century, 145/ The Paris Confrontation (1603), 159/ The English Solution, 173/ The College of Physicians and the *Pharmacopoeia* (1618), 182/ The *Agreement and Disagreement* of Daniel Sennert (1619), 191/ Conclusion, 200

4 THE SYNTHESIS OF ROBERT FLUDD   205

Robert Fludd and the Rosicrucian Problem (1617), 206/ The Fluddean Philosophy, 224/ The Initial Reaction in England (1618-1623), 253/ Fludd and Kepler (1619-1623), 256/ Fludd and the French Mechanists (1623-1633), 260/ Fludd and the Weapon-Salve Controversy (1631-1638), 279/ Conclusion, 290

Volume II

5 THE BROKEN CHAIN: THE HELMONTIAN RESTATEMENT OF THE CHEMICAL PHILOSOPHY   295

Van Helmont: The Early Years, 297/ The Tract on the Weapon-Salve (1621), 303/ The Letters to Mersenne (1630-1631), 306/ Van Helmont's Final Years (1631-1644), 310/ The Helmontian Philosophy of Nature: Mathematics and Motion, 311/ The Elements and the Principles, 317/ Chemistry as the Key to Nature, 322/ Quantification: A New Chemical Tool, 327/ The Vacuum and the *Magnal*, 329/ A Model for the Geocosm, 334/ The Mineral Kingdom, 339/ The Chemical Geocosm of Edward Jorden (1631), 344/ Helmontian Medicine: The Divine Office of the Physician, 357/ The Theory of Disease, 359/ Tartaric Disease, 362/ Chemical Inquiries: The Search for the Vital Spirit, 365/ A New Concept of Digestion, 368/ The Chemical Remedies, 371/ A Challenge for the Future, 376

6 THE CHEMICAL PHILOSOPHY IN TRANSITION: NATURE, EDUCATION, AND STATE 381

Educational Reform: Background, 382/ John Webster and the *Academiarum examen* (1654), 393/ The *Vindiciae academiarum* of John Wilkins and Seth Ward (1654), 400/ Thomas Hall's *Whip for Webster* (1654), 406/ Chemistry and the State: The Agricultural Problem, 410/ Agricultural Chemistry in Seventeenth-Century England, 420/ Chemistry and Economic Policy: Johann Rudolph Glauber, 425/ *The Prosperity of Germany* (1656-1661), 434/ Conclusion, 441

7 THE CHEMICAL PHILOSOPHY IN TRANSITION: TOWARD A NEW CHEMISTRY AND MEDICINE 447

Chemistry in Mid-Century: Lefévre (1660) and Rhumelius (1648), 448/ Geocosmic Considerations: F. M. van Helmont (1685) and John Webster (1671), 455/ J. J. Becher's *Physica subterranea* (1669), 458/ G. E. Stahl and Chemical Tradition, 463/ The Chemical Corpuscularians: Walter Charleton and the Chemical Philosophy, 469/ The "Helmontian" Robert Boyle, 473/ The Analysis Problem, 484/ The Nitro-Aerial Particles in Mid-Century, 492/ The Acid-Alkali Theory after van Helmont, 499/ Chemistry and Late-Seventeenth-Century Medicine: The Chemical Medicine of Noah Biggs (1651), 502/ Chemistry and the London College of Physicians, 507/ Chemistry and the Blood, 512/ The Chemical Medicine of Willis and Sylvius, 519/ A Newtonian Postscript, 531/ Conclusion, 536

8 POSTSCRIPT 539

The Chemical Philosophy in Retrospect, 540/ Acceptance and Rejection: The Question of Influence, 543/ Aftermath, 547

BIBLIOGRAPHY 555

INDEX 597

ERRATA 607

LIST OF PLATES

Volume I

The chemical philosopher, J. B. van Helmont, illumined by the light shed by chemical knowledge (in contrast to Galen, Avicenna, and Theophrastus).
frontispiece

I	The use of an enclosed analytical balance (1652)	39
II	Paracelsus at the age of forty-seven (Hirschvogel)	49
III	The true chemical philosopher worshipping the Creator through prayer and the study of nature in the chemical laboratory (H. Khunrath, 1609)	67
IV	The chemical philosopher following in the footsteps of nature (M. Maier, 1618)	71
V	Michael Sendivogius (1556-1636/1646)	85
VI	Volcanoes depicted as the result of the earthly central fire (A. Kircher, 1678)	91
VII	The macrocosm and the microcosm (R. Fludd, 1617)	97
VIII	An eighteenth-century chemist exhibiting the phenomenon of palingenesis (E. Sibly, 1792)	101
IX	Examples of sympathetic action in nature (*Theatrum sympatheticum auctum*, 1662)	105
X	A comparison of the human body and the proper distillation equipment for urine: an example of Paracelsian "chemical anatomy" (Gerhard Dorn, ascribed to Paracelsus, 1577)	113
XI	Johannes Wimpenaeus at the age of thirty (1570)	137
XII	Joannes Guinter of Andernach (c.1505)	141
XIII	Joseph Duchesne (Quercetanus) (c.1544-1609)	151
XIV	Andreas Libavius (c.1540-1616)	171
XV	Sir Theodore Turquet de Mayerne (1573-1655)	187
XVI	Daniel Sennert (1572-1637)	193
XVII	Robert Fludd (1574-1637)	209
XVIII	The universe of Robert Fludd (1617)	221
XIX	The harmony of the universe illustrated in terms of Fludd's monochord (1617)	233
XX	Robert Fludd's thermoscope and its relationship to the cosmos (1631)	237

XXI	Robert Fludd's use of mechanical analogies to indicate the most probable location of the source of motion in the universe (1617)	241
XXII	Man besieged in his castle of health (1631)	251
XXIII	David de Planis Campy (1589-c. 1644)	263

Volume II

XXIV	J.B. van Helmont (1579-1644)	299
XXV	The chemical philosopher as imitator and interpreter of natural phenomena a(1682)	323
XXVI	Robert Fludd's illustration of the candle experiment (1631)	335
XXVII	Johann Rudolph Glauber (1603-1670)	427
XXVIII	Johann Joachim Becher (1635-1682)	443
XXIX	An allegorical representation of a vital universe having a valid relationship between its macrocosmic and geocosmic phenomena (Becher, 1738)	465
XXX	Robert Boyle (1627-1691)	475
XXXI	The macrocosm and the microcosm (T. Schütze, 1654)	515
XXXII	Thomas Willis (1621-1675)	521
XXXIII	Franciscus de le Boë Sylvius (1614-1672)	527

PREFACE TO THE DOVER EDITION

Forty years ago the Scientific Revolution was generally interpreted as a progression from Copernicus to Kepler and Galileo, culminating in the *Principia Mathematica* of Isaac Newton (1687). Emphasis was placed on the acceptance of the heliocentric universe, the development of the mechanical philosophy, and an experimental approach to the understanding of nature. This was largely a positivistic approach to the subject that focused on the exact sciences and relegated to the sidelines important contemporary developments in chemistry, biology and medicine. The events leading to the discovery of the circulation of the blood by William Harvey in 1628 are a notable exception, included because Harvey seemed to fit the model of scientific progress that was desired by mid-twentieth century historians.

As a graduate student in the history of science trained originally in chemistry I was struck by the extensive chemical literature of the sixteenth and seventeenth centuries. The authors of these works frequently wrote of their "chemical philosophy," an approach to nature which they thought to be the proper replacement for the works of the ancient philosophers and physicians which were still entrenched at the universities. For them it was not mathematics or the laws of motion which were to be the basis of a new understanding of nature, but rather an observational chemical investigation of man and the great world about us. The early sixteenth-century figure Paracelsus inspired many toward this end. His disciples sought to replace the teachings of the Aristotelians and Galenists, and in the course of the seventeenth century they found themselves in conflict not only with the followers of the ancients, but with the more recent mechanists as well.

In writing the present book my goal was to describe the views of these early modern chemists and to follow their debates with others. *The Chemical Philosophy* was completed after a year at the Institute for Advanced Study at Princeton and published in 1977. The book was awarded the Pfizer Prize of the History of Science Society in 1978, but it was printed in a relatively small edition that was exhausted in a few years. Since then it has been translated into Japanese, but the original edition has become difficult to locate on the used book market. For this reason a reprint has seemed appropriate.

Influential in shaping my views were Walter Pagel's *Paracelsus: An Introduction to Philosophical Medicine in the Era of the Renaissance* (1958), which not only presented a fresh account of the work of Paracelsus, but also delved deeply into his sources and influence on later figures. Also important were Frances Yates' many works, but especially her *Giordano Bruno and the Hermetic Tradition* (1964), which for many scholars, myself included, showed the widespread influence of the *Corpus Hermeticum*.

After the manuscript of *The Chemical Philosophy* was already in the hands of the publisher several books appeared on related topics. Owen Hannaway showed the wide spectrum of views held by chemical philosophers in his comparison of Oswald Croll and Andreas Libavius in *The Chemists and the Word* (1975) and Betty Jo Teeter Dobbs began her investigation of Isaac Newton's alchemy in *The Foundations of Newton's Alchemy* (1975).

During the past quarter century a number of scholars have written studies that have contributed to our understanding of the role of chemistry in the period of the Scientific Revolution. Only a few of these can be mentioned here. My emphasis will be on books although a number of articles will be included. Among general studies special attention should be given to Walter Pagel's posthumous *The Smiling Spleen* (1984). Sir Hugh Trevor-Roper surveyed the Paracelsian movement in a paper with this title in his *Renaissance Essays* (1985). I summarized the views and debates of the Paracelsians in a series of four lectures presented at the Istituto Filosofici in Naples in 1990 which were translated into Italian and published as *Paracelso e la Tradizione Paracelsiana* (1996). I concentrated on the medical aspects of their work in "La médecine chimique" in *De la Renaissance aux Lumières* (1997). Earlier I had prepared a more general account of Renaissance science and medicine based on my lectures at the University of Chicago, which attempted to integrate the chemical reaction against ancient tradition and the views of the mechanists (*Man and Nature in the Renaissance*, 1978).

A number of international meetings were held in 1993 to mark the 500th anniversary of the birth of Paracelsus. Volumes of papers presented at some of these meetings have been published, among them *Beiträge des symposiums zum 500. Geburtstag von Theophrastus Bombastus von Hohenheim, genannt Paracelsus (1493–1541)* (1995); *Un esame critico del pensiero di Paracelso, collocato nella sua dimensione storica* (1994); Gerhart Harrer, ed., *500 Jahre Paracelsus: Nachlese zum Jubiläum kongress* (1995); and Ole Peter Grell, ed., *Paracelsus, The Man and His Reputation, His Ideas and Their Transformation* (1998). Another collection devoted largely to Paracelsian and early chemical studies is *Reading the Book of Nature: The Other Side of the Scientific Revolution*, edited by Allen G. Debus and Michael T. Walton (1997).

A number of regional studies have appeared. Jole Shackelford's "Paracelsianism in Denmark and Norway in the Sixteenth and Seventeenth Centuries" (doctoral dissertation, 1989) is an important discussion of the Scandinavian scene relating religion and Paracelsian thought. José María López Piñero has published frequently on the relation of Paracelsian thought to the rise of modern science in Spain as in his *La Introducción de la Ciencia Moderna en España* (1969) and his *El "Dialogus" (1589) del paracelsista Llorenç Coçar y la cátedra de medicamentos químicos de la Universidad de Valencia (1591)* (1977). Recently Mar Rey Bueno has added to our knowledge of late seventeenth century Spanish iatrochemistry in her *El Hechizado: Medicina, alquimia y superstición en el corte de Carlos II* (1998). I have discussed the relation of Spanish and Portuguese iatrochemistry in "Chemistry and Iatrochemistry in Early Eighteenth Century Portugal: A Spanish Connection" in *História e Desenvolvimento da Ciência em Portugal até ao Século XX* (1986) and in "Paracelsus and the Delayed Scientific Revolution in Spain: A Legacy of Philip II" in *Reading the Book of Nature*.

A number of studies of Italian Paracelsianism have been prepared by Giancarlo Zanier including *Medicina e Filosofia tra '500 e '600* (1983). Another work that adds to our knowledge of the Italian scene is William Eamon's *Science and the Secrets of Nature: Books of Secrets in Medieval and Early Modern Culture* (1994). And going beyond the books of secrets, there is a study of the relation of early modern astrology, Paracelsianism and magic on medicine by Wolf-Dieter Müller-Jahncke, *Astrologisch-Magische Theorie und Praxis in der Heilkunde der Frühen Neuzeit* (1985).

An essential work on early modern French medicine is Laurence Brockliss and Colin Jones, *The Medical World of Early Modern France* (1997) which covers medical debates in France as well as the system of medical education. Focused more on the French iatrochemists is my *The French Paracelsians: The Chemical Challenge to Medical and Scientific Tradition in Early Modern France* (1991).

There are a number of works dealing with specific figures as well.

Zbigniew Szydlo has concentrated his research on Michael Sendivogius, the subject of his Ph.D. dissertation at University College, London, "The Life and Work of Michael Sendivogius (1566–1636)" (1991). Szydlo has had a special interest in the origins of the aerial niter theory. Leonhard Thurneisser, a significant figure for his Paracelsian views and for his chemically oriented herbal, is the subject of a monograph by Peter Morys, *Medizin und Pharmazie in der Kosmologie Leonhard Thurneisser zum Thurn (1531–1596)* (1982). And Bernard Joly has turned to the work of the seventeenth century French figure, Pierre-Jean Fabre in his *Rationalité de l'Alchimie au XVII Siècle* (1992). This work includes a transcription and translation of, and commentary on, Fabre's *Manuscriptum ad Federicum*.

Bruce T. Moran has discussed the work of Johannes Hartmann and the early academic teaching of chemistry in *Chemical Pharmacy Enters the University: Johannes Hartmann and the Didactic Care of Chymiatria in the Early Seventeenth Century* (1991). He has expanded this research in *The Alchemical World of the German Court: Occult Philosophy and Chemical Medicine in the Circle of Moritz of Hessen (1572–1632)* (1991). I have discussed the introduction of chemistry to the European academic world in "Chemistry and the Universities in the Seventeenth Century," in *Academiae Analecta: Klasse der Wetenschappen* (1986).

William H. Huffman has reexamined the life and career of Robert Fludd in *Robert Fludd and the End of the Renaissance* (1988). This is the first such study since J. B. Craven's *Doctor Robert Fludd* was published in 1902. I have published Fludd's manuscript at Trinity College in *Robert Fludd and His Philosophicall Key: Being a Transcription of the Manuscript at Trinity College, Cambridge* (1979). Van Helmont has been the subject of Walter Pagel's *Joan Baptista Van Helmont: Reformer of Science and Medicine* (1981) and several monographs: Paulo Alves Porto, *Van Helmont e o Conceito de Gas: Quimica e Medicina no Século XVII* (1995); Berthold Heinecke, *Wissenschaft und Mystik bei J. B. van Helmont (1579–1644)* (1996); and Guido Giglioni, *Immaginazione e Malattia: Saggio su Jan Baptiste van Helmont* (2000). A significant work emphasizing Johann Joachim Becher's social and commercial views has appeared in Pamela H. Smith's *The Business of Alchemy: Science and Culture in the Holy Roman Empire* (1994). However, there remains much to be done on Becher's chemical view of nature.

In the past decade much attention has been given to the work of Robert Boyle. An edition of his works including new material and variant readings is being edited by Michael Hunter and Edward B. Davis. The first seven volumes have already appeared from the press of Pickering and Chatto, 1999. A number of new books and articles on Boyle have appeared as well. Among the books are those of Rose-Mary Sargent, *The Diffident Naturalist: Robert Boyle and the Philosophy of Experiment* (1995); Barbara Beguin Kaplan, *Divulging of Useful Truths in Physick: the Medical Agenda of Robert Boyle* (1993); and Lawrence M. Principe, *The Aspiring Adept: Robert Boyle and His Alchemical Quest* (1998). The emergent view of Boyle as a chemist rather than a mechanical philosopher who succeeded in making chemistry a physical science is summarized by Antonio Clericuzio in *Elements, Principles and Corpuscles: A Study of Atomism and Chemistry in the Seventeenth Century* (2000). Significant articles by William Newman, Antonio Clericuzio and others have been gathered by Michael Hunter in *Robert Boyle Reconsidered* (1994). The social and intellectual connections between the circle of chemists that surrounded Johan Hartlib and nurtured Starkey, Boyle and other Helmontians who shaped late seventeenth century chemical discourse is described in John T. Young, *Faith, Medical Alchemy and Natural Philosophy: Johann Moriaen, Reformed Intelligencer, and the Hartlib Circle* (1998).

Recent research has fully established the widespread interest and acceptance of alchemy by otherwise "respectable" scientists of the seventeenth and eighteenth centuries. In addition to the work of Principe on Boyle's alchemy already mentioned, Betty Jo Teeter Dobbs continued her research on Isaac Newton in *The Janus Faces of Genius: The Role of Alchemy in Newton's Thought* (1991). William R. Newman's *Gehennical Fire: The Lives of George Starkey, an American Alchemist in the Scientific Revolution* (1994) sheds light not only on Starkey, but on Boyle and Newton as well. Both Hermann Boerhaave and Georg Ernst Stahl are now known to have seriously considered the transmutation of metals and one can point to a large number of alchemical works well into the eighteenth century, both newly composed and traditional reprints.

The continued interest in alchemy has resulted in a stream of scholarly studies, many of which may be found in journals such as *Ambix, Chrysopoeia,* and *Cauda pavonis.* Among the many books on alchemy published recently I will mention only a few. An important collection will be found in *Alchemy Revisited: Proceedings of the International Conference on the History of Alchemy at the University of Groningen 17–19 April 1989* edited by Z. R. W. M. von Martels

(1990) which covers a broad spectrum of research by over thirty scholars. William R. Newman's exhaustive study, *The Summa Perfectionis of Pseudo-Geber: A Critical Edition, Translation and Study* (1991) is essential reading for anyone interested in this medieval text that remained authoritative into the eighteenth century. And on a far different note, Charles Nicholl's *The Chemical Theatre* (1980) shows the impact of chemistry and alchemy on literature and popular culture primarily in the English Renaissance. A seemingly never-ending profusion of editions of alchemical illustrations has appeared–perhaps too many. However, I have found the collection of Stanislas Klossowski de Rola, *The Golden Game: Alchemical Engravings of the Seventeenth Century* (1988) a particularly handsome volume, even though one may disagree with some of the author's interpretations.

My own work since the publication of *The Chemical Philosophy* has moved forward in time to concentrate more on the late seventeenth and early eighteenth centuries. Much of this research has appeared in the form of articles although several books have already been mentioned. To these I will add only *Chemistry and Medical Debate: van Helmont to Boerhaave* (2001). Here I discuss the continuing debates relating to chemistry in medicine into the eighteenth century. Following the work of van Helmont, Thomas Willis and Franciscus de la Boë Sylvius, these physicians developed theories of acid-alkali dualism and fermentation to explain physiological processes. These chemical explanations were anathema to those physicians influenced by the new mechanical philosophy who sought truth in mechanistic analogies. This debate can be followed well into the eighteenth century and was reflected in the works of authors as important as Hermann Boerhaave and Georg Ernst Stahl both of whom rejected traditional iatrochemical explanations and limited chemistry in medicine to its practical applications. And yet, Boerhaave looked upon van Helmont as one of his favorite authors and and Stahl still reflected an animistic physiology. And at the same time both devoted much time to alchemical pursuits.

In short, there has been an ever-growing interest in the development and influence of the chemical philosophy in the period of the Scientific Revolution. Surely, in part this has been due to an increasing interest in contextual history, an approach fostered by Walter Pagel. It is now generally acknowledged that (a) Paracelsus presented an open challenge to ancient texts still widely accepted academically in the sixteenth century and that he called for an observational study of nature; (b) for Paracelsus and his followers chemistry played a major role in the understanding of nature and man through analogies and laboratory investigation; and (c) the new chemically prepared medicines favored by the Paracelsians led to the academic acceptance of chemistry in the course of the seventeenth century. These early modern chemists clearly played a major role in the Scientific Revolution through their rejection of the ancients and their observational approach to nature. And indeed, if we are to understand the Chemical Revolution of the late eighteenth century of Lavoisier and his colleagues, we must also understand the fundamental changes in chemistry and iatrochemistry made by the chemical philosophers in the previous two centuries. Given the continually expanding interest in this subject it is my hope that the readers of this volume will still find something of value in its pages.

Allen G. Debus
Deerfield, Illinois, August 2001

References

Alves Porto, Paulo. *Van Helmont e o Conceito de Gas: Quimica e Medicina no Século XVII.* São Paulo: Edus, 1995.

Brockliss, Laurence and Colin Jones. *The Medical World of Early Modern France.* Oxford: Clarendon Press, 1997.

Clericuzio, Antonio. *Elements, Principles and Corpuscles: A Study of Atomism and Chemistry in the Seventeenth Century.* Dordrecht: Kluwer, 2000.

Convegno Internazionale su Paracelso. *Un esame critico del pensiero di Paracelso, collocato nella sua dimensione storica.* Roma 17–18 dicembre 1993, Goethe Institut. Roma: Edizioni Paracelso, 1994.

Craven, J. B. *Doctor Robert Fludd (Robertus de Fluctibus). The English Rosicrucian. Life and Writings.* London: William Peace & Sons, 1902.

Debus, Allen G. *The Chemical Philosophy.* 2 vols. New York: Science History Publications, 1977.

——. *The Chemical Philosophy* (Japanese). Tokyo: Heibonsha, 1999.

——. "Chemistry and Iatrochemistry in Early Eighteenth Century

Portugal: A Spanish Connection." In *História e Desenvolvimento da Ciência em Portugal até ao Século XX*, organizaçao do Instituto de Altos Estudos da Academia das Ciências de Lisboa. Colóquio sobre a História e Desenvolvimento da Ciência em Portugal, 1985. 2 vols. Lisboa: Academia das Ciências de Lisboa, 1986.

——. *Chemistry and Medical Debate: van Helmont to Boerhaave.* Canton, Mass.: Science History Publications, 2001.

——. "Chemistry and the Universities in the Seventeenth Century." In *Academiae Analecta: Klasse der Wetenschappen*, 48 (1986), 1-33.

——. *The French Paracelsians: The Chemical Challenge to Medical and Scientific Tradition in Early Modern France.* Cambridge: Cambridge University Press, 1991.

——. "La médecine chimique." In *De la Renaissance aux Lumières.* Vol. 2 of *Histoire de la pensée médicale en Occident*, edited by Mirko D. Grmek and Bernardino Fantini. Paris: Éditions du Seuil, 1997.

——. *Man and Nature in the Renaissance.* Cambridge: Cambridge University Press, 1978.

——. *Paracelso e la Tradizione Pacelsiana.* Napoli: Città del Sole, 1996.

——. "Paracelsus and the Delayed Scientific Revolution in Spain: A Legacy of Philip II." In *Reading the Book of Nature: The Other Side of the Scientific Revolution*, edited by Allen G. Debus and Michael T. Walton. Sixteenth Century Essays and Studies, vol. XLI. Kirksville, Mo.: Thomas Jefferson University Press, 1997.

Debus, Allen G. and Michael T. Walton, eds. *Reading the Book of Nature: The Other Side of the Scientific Revolution.* Sixteenth Century Essays and Studies, vol. XLI. Kirksville, Mo.: Thomas Jefferson University Press, 1997.

Dobbs, Betty Jo Teeter. *The Foundations of Newton's Alchemy, or "The Hunting of the Greene Lyon."* Cambridge: Cambridge University Press, 1975.

——. *The Janus Faces of Genius: The Role of Alchemy in Newton's Thought.* Cambridge: Cambridge University Press, 1991.

Eamon, William. *Science and the Secrets of Nature: Books of Secrets in Medieval and Early Modern Culture.* Princeton: Princeton University Press, 1994.

Fludd, Robert. *Robert Fludd and His Philosophicall Key: Being a Transcription of the Manuscript at Trinity College, Cambridge.*

With an Introduction by Allen G. Debus. New York: Science History Publications, 1979.

Giglioni, Guido. *Immaginazione e Malattia: Saggio su Jan Baptiste van Helmont.* Milano: Franco Angeli, 2000.

Grell, Ole Peter, ed. *Paracelsus, The Man and His Reputation, His Ideas and Their Transformation.* Leiden, Boston, Köln: E. J. Brill, 1998.

Hannaway, Owen. *The Chemists and the Word.* Baltimore: The Johns Hopkins University Press, 1975.

Harrer, Gerhart, ed. *500 Jahre Paracelsus: Nachlese zum Jubiläum kongress.* Salzburger Beiträge zur Paracelsusforschung, 28. Wien: Osterreichischer Kunst- und Kulturverlag, 1995.

Heinecke, Berthold. *Wissenschaft und Mystik bei J. B. van Helmont (1579-1644).* Bern, Berlin, et al.: Peter Land, 1996.

Hunter, Michael, ed. *Robert Boyle Reconsidered.* Cambridge: Cambridge University Press, 1994.

Hunter, Michael and Edward B. Davis, eds. *The Works of Robert Boyle.* 14 vols. London: Pickering and Chatto, 1999-2000.

Huffman, William H. *Robert Fludd and the End of the Renaissance.* London and New York: Routledge, 1988.

Joly, Bernard. *Rationalité de l'Alchime au XVII Siècle.* Paris: J. Vrin, 1992.

Kaplan, Barbara Beguin. *Divulging of Useful Truths in Physick: The Medical Agenda of Robert Boyle.* Baltimore: The Johns Hopkins University Press, 1993.

Klossowski de Rola, Stanislas. *The Golden Game: Alchemical Engravings of the Seventeenth Century.* New York: George Braziller, 1988.

Lopez Piñero, José Maria. *El "Dialogus" (1589) del paracelsista Llorenç Coçar y la cátedra de medicamentos químicos de la Universidad de Valencia (1591).* Cuadernos Valencianos de Historia de la Medicina y de las Ciencia XX, Serie B (Textos Clásicos). Valencia: Cátedra e Instituto de Historia de la Medicina, 1977.

———. *La Introducción de la Ciencia Moderna en España.* Barcelona: Ediciones Ariel, 1969.

Moran, Bruce T. *The Alchemical World of the German Court: Occult Philosophy and Chemical Medicine in the Circle of Moritz of Hessen (1572-1632).* Sudhoffs Archiv, Beiheft 29. Stuttgart: Franz Steiner Verlag, 1991.

———. *Chemical Pharmacy Enters the University: Johannes Hartmann and the Didactic Care of Chymiatria in the Early Seventeenth Century.* Madison, Wisc.: American Institute of the History of Pharmacy, 1991.

Morys, Peter. *Medizin und Pharmazie in der Kosmologie Leonhard Thurneisser zum Thurn (1531–1596).* Abhandlungen zum Geschichte der Medizin und der Naturwissenschaften, Heft 43. Husum: Matthiesen Verlag Ingwert Paulsen Jr., 1982.

Müller-Jahncke, Wolf-Dieter. *Astrologisch-Magische Theorie und Praxis in der Heilkunde der Frühen Neuzeit.* Sudhoffs Archiv, Beiheft 25. Stuttgart: Franz Steiner, 1985.

Newman, William R. *Gehennical Fire: The Lives of George Starkey, an American Alchemist in the Scientific Revolution.* Cambridge, Mass.: Harvard University Press, 1994.

———. *The Summa Perfectionis of Pseudo-Geber: A Critical Edition, Translation and Study.* Leiden: E. J. Brill, 1991.

Nicholl, Charles. *The Chemical Theatre.* London, Boston and Henley: Routledge and Kegan Paul, 1980.

Pagel, Walter. *Joan Baptista Van Helmont: Reformer of Science and Medicine.* Cambridge: Cambridge University Press, 1981.

———. *Paracelsus: An Introduction to Philosophical Medicine in the Era of the Renaissance.* Basel: S. Karger, 1958; reprinted with "Addenda and Errata," Basel et al.: S. Karger, 1982.

———. *The Smiling Spleen: Paracelsianism in Storm and Stress.* Basel et al.: S. Karger, 1984.

Principe, Lawrence M. *The Aspiring Adept: Robert Boyle and His Alchemical Quest.* Princeton: Princeton University Press, 1998.

Rey Bueno, Mar. *El Hechizado: Medicina, alquimia y superstición en el corte de Carlos II.* Madrid: Ediciones Corona Borealis, 1998.

Sargent, Rose-Mary. *The Diffident Naturalist: Robert Boyle and the Philosophy of Experiment.* Chicago: University of Chicago Press, 1995.

Shackelford, Jole. "Paracelsianism in Denmark and Norway in the Sixteenth and Seventeenth Centuries." Ph.D. diss., University of Wisconsin–Madison, 1989.

Smith, Pamela H. *The Business of Alchemy: Science and Culture in the Holy Roman Empire.* Cambridge: Cambridge University Press, 1994.

Szydlo, Zbigniew. "The Life and Work of Michael Sendivogius (1566–1636)." Ph.D. diss., University College, London, 1991.

Trevor-Roper, Hugh. "The Paracelsian Movement." In *Renaissance Essays.* London: Secker and Warburg, 1985.

Von Martels, Z. R. W. M., ed. *Alchemy Revisited: Proceedings of the International Conference on the History of Alchemy at the University of Groningen 17-19 April 1989.* Leiden et al.: E. J. Brill, 1990.

Yates, Frances. *Giordano Bruno and the Hermetic Tradition.* Chicago: University of Chicago Press, 1964.

Young, John T. *Faith, Medical Alchemy and Natural Philosophy: Johann Moriaen, Reformed Intelligencer, and the Hartlib Circle.* Aldershot: Hampshire Ashgate, 1998.

Zanier, Giancarlo. *Medicina e Filosofia tra '500 e '600.* Milan: Franco Angeli Editore, 1983.

Zimmerman, Volker. "Das Werk–die Rezeption." In *Beiträge des symposiums zum 500. Geburtstag von Theophrastus Bombastus von Hohenheim, genannt Paracelsus (1493–1541) an Universität Basel am 3. Und 4. Dezember 1993.* Stuttgart: Franz Steiner, 1995.

PREFACE

The student of sixteenth- and seventeenth-century science and medicine is acutely aware of the distrust of ancient authority that was then widespread—a distrust coupled with a conscious search for some new key to nature. There were many attempts to fulfill this goal, but few seemed more promising than that proposed by the followers of Paracelsus. Building upon a mixture of ancient, medieval, and Renaissance sources, they claimed to have found their key to a truly Christian interpretation of nature in chemistry. For them both macrocosmic and microcosmic events would be revealed through the personal observations of the chemist and the Divine Grace of the Lord. These, they argued, were far better guides to knowledge than the texts of antiquity still being taught in the universities.

There is little doubt that this chemical philosophy attracted much attention in the late sixteenth and seventeenth centuries, yet there has been little attempt by historians to describe the breadth of this approach to nature and medicine or to record its vicissitudes. It is my purpose in the present study to rectify this at least partially. An effort has been made to discuss some of the major concepts of interest to the chemical philosophers as well as the conflicts they engaged in with their contemporaries. The chronological period extends roughly from the death of Paracelsus in 1541 to the publication of Robert Boyle's *Sceptical Chymist* one hundred and twenty years later, but the reader will frequently find himself encountering people and ideas from times both earlier and later.

In an attempt to make a very large subject manageable, it has been necessary to be selective. Thus, although a great number of chemical philosophers are referred to in these pages, there are many others with an equal claim for inclusion who are not here. I hope, though, that the most important themes have been mentioned. There are other legitimate topics for discussion whose many facets could not be fully explored. One is the relationship of Aristotle and Galen to Renaissance science and medicine. While many of the concepts of the Paracelsian chemists were derived from an Aristotelian and Galenic tradition that was still very much alive, here the main emphasis is upon the sharp differences the Paracelsians saw between their own views and those of the ancients.

Similarly, recent research points to the importance of a socioeconomic interaction in the rise of modern science. It would be hard to deny this as a significant area of study, but I remain convinced that research of this sort in the future must be based upon a more thorough understanding of the scientific and philosophical background than has often been the case. Thus, with the exception of the sixth chapter (on the educational, agricultural, and economic implications of the chemical philosophy) the present work may be primarily characterized as "internalist" rather than "externalist." But although this is true, I have not attempted to confine myself to the "positive" contributions made by the chemical philosophers. Rather, my goal has been to give a balanced picture of their system of nature as a whole—and to present this from their point of view rather than our own.

An abstract of my plan for the first part of this book appeared in the form of two essays published in 1973, but actually the present pages are based upon research that was initiated at Harvard University in 1956. This research has resulted in a number of publications, including the book *The English Paracelsians* in 1965. All of this earlier work has been drawn upon throughout the writing of the present volume; indeed, parts of earlier papers have been incorporated. In such cases the footnotes and the bibliography should lead the reader to the original. Still, this is a book with a purpose that differs significantly from that of my earlier studies. It is a synthesis that has required a great deal of additional research, and as a result much new material is appearing in print for the first time.

Over the years I have become deeply indebted to many people, institutions, and foundations. I wish to thank the Morris Fishbein Center for the Study of the History of Science and Medicine at the University of Chicago for long-time support in so many ways that it would be impossible to be both specific and comprehensive. I would also add that the award of a Guggenheim Fellowship (1966-1967) and many years of support from the National Institutes of Health (1962-1970: USPHS GM-09855 and LM-00046) and the

Preface

National Science Foundation (GS-29189, GS-37063) were essential for the preparation of fundamental segments of this research. More recently, the University of Chicago in cooperation with the National Endowment for the Humanities (Grant H5426) made it possible for me to devote the year 1972-1973 to writing the first draft of the manuscript.

If financial support has been essential, no less important has been the aid I have received from countless librarians in this country and abroad. The staff members at the Library of the School of Historical Studies at the Institute for Advanced Study, at the Cambridge University Library, and the British Museum have always simplified problems which occasionally seemed quite difficult. Other colleges and university librarians have gone far beyond the call of duty, and some officials at continental libraries have responded to my letters by having unique Paracelsian publications rapidly microfilmed. Dr. F. N. L. Poynter of the Wellcome Historical Medical Library (now retired) has taken a personal interest in my research since my first visit to London as a graduate student in 1959. And here I must also add Mr. Robert Rosenthal, the Director of the Department of Special Collections at the University of Chicago Libraries. His keen eye and balanced judgment have been of great aid in tracking down numerous elusive texts over the years.

During the year 1966-1967 and the summer sessions of 1969 and 1971 it was my pleasure to be an Overseas Fellow at Churchill College, Cambridge. I would like to express my gratitude to the two masters and the many fellows of the College I have known for their encouragement, but especially I wish to thank Dr. M. A. Hoskin, who has always shown an interest in the problems I have sought to unravel. Similarly I wish to thank the director and the members of the Institute for Advanced Study for the opportunity to spend a precious year in semi-seclusion, at a time when I was ready to prepare the first draft of the present manuscript.

To my many friends in England—Professor P. M. Rattansi and Dr. W. A. Smeaton at University College, London, Dr. Charles Webster at Oxford University, Dr. Frances A. Yates at the Warburg Institute, and in years past Professors Douglas McKie and J. R. Partington—I gladly express my debt for long hours of illuminating discussion. Dr. David M. Knight kindly read the concluding chapter of this manuscript. In this country Professor Marshall Clagett, of the Institute for Advanced Study, and my colleagues Dr. Lester S. King and Dr. Jerome J. Bylebyl have frequently borne the burden of my involvement with the texts of the Paracelsians. The influence of my training with Professor I. B. Cohen at Harvard University forms an indispensable part of all my research.

Professor Walter Pagel deserves a special note of gratitude. As the doyen of Paracelsian scholars his vast studies have time and time again served as the solid bedrock for my own interpretations. No less important has been his approach to the history of science and medicine, especially his insistence that the historian must study early scientific and medical thought as a whole rather than separate out only those passages which appear to be "modern." In addition, his interest in the progress of this research may make the word "collaborator" as appropriate as the word "friend."

Finally, there remain some others whose efforts were no less important for the completion of this manuscript. My secretary, Mrs. Helen Little, has willingly accepted the task of typing well over one thousand pages of copy in between other, often more pressing, duties connected with the Morris Fishbein Center for the Study of the History of Science and Medicine. My publisher, Mr. Neale Watson, has always shown an enthusiasm for my work. I have always been grateful for this, but I am also in his debt for obtaining Mrs. Diana D. Menkes, Assistant Editor of *Isis,* to help guide this volume through the press. It was my pleasure to work with Mrs. Menkes once before, and I feel that on both occasions the final volumes have benefited greatly from her editorial wisdom.

My gratitude is also extended to a group long deceased—the small army of dedicated sixteenth- and seventeenth-century translators of the iatrochemical texts. Some of the difficulties found in using these translations will be noted in the Bibliography, but in general these men—Nicholas Culpeper, John Chandler, Richard Russell, and many others—were conscientious in their effort to present accurate versions of the works they chose to make available in English. Although many passages I have translated myself, I have attempted when possible to present those contemporary translations, for they provide something of the intense flavor of the period that would be difficult if not impossible to imitate.

Last I must thank my family: my father, both for aid and understanding, and my wife and children for having put up with the difficulties associated with their frequent transplantation into different locations, schools, and indeed different lives because of the constant quest for rare texts and free time for my research.

Deerfield, Illinois   *Allen G. Debus*
January 1, 1974

Volume I

1
CHEMISTRY AND NATURE IN THE RENAISSANCE

> The great virtues that lie hidden in nature would never have been revealed if alchemy had not uncovered them and made them visible. Take a tree, for example; a man sees it in the winter, but he does not know what it is, he does not know what it conceals within itself, until summer comes and discloses the buds, the flowers, the fruit. . . . Similarly the virtues in things remain concealed to man, unless the alchemists disclose them, as the summer reveals the nature of the tree.
>
> Paracelsus, *Paragranum* (1530)*

THE SIXTEENTH and seventeenth centuries were the time of the scientific revolution, the period which saw the achievements of Copernicus, Vesalius, Kepler, Galileo, Harvey, and Newton. The impact of their work was to fundamentally change our understanding of nature and the universe. The developments of science acted to alter man's comfortable relationship with his Creator while at the same time furthering the quest for the betterment of life on earth through the belief in man's power over nature.

The history of the scientific revolution is normally presented in terms of those advances in the physical sciences that led to the Newtonian synthesis and in the progression in the medical sciences from Vesalius to Harvey and the circulation of the blood. This positivistic approach is of the utmost importance in establishing the major avenues of scientific development, but far too often its practitioners have ignored the complexity of this extremely rich period. In recent years new research has pointed to seemingly

---

*From the lengthy discussion of alchemy as the third column of medicine in Paracelsus, *Opera Bücher und Schrifften*, edited by Johann Huser, 2 vols. (Strasbourg: L. Zetzner, 1616), Vol. I, p. 222. The Latin translation is that by Fridericus Bitiskius in the *Opera omnia medico-chemico-chirurgica*, 3 vols. (Geneva: Ioan. Antonij, & Samuelis De Tournes, 1658), Vol. I, p. 210. The rather free English translation cited here is that of Norbert Guterman in Paracelsus, *Selected Writings*, ed. Jolande Jacobi (New York: Bollingen Series XXVIII, Pantheon Books, 1951), p. 218. This is from the German, but a complete translation of the

"nonmodern" or even "anti-modern" themes persisting in the scientific literature until an unexpectedly late period. Even though this is true, there are those who argue that the seventeenth-century Hermetic, alchemical, and magical literature may be safely ignored. Nowhere here, they say, will we find anything to compare with Kepler's laws of planetary motion, Harvey's work on the circulation, Boyle's investigation of the vacuum, or Newton's laws of motion. These historians claim that we should not devote our time to the study of "lesser" men and outmoded concepts.

But can we close our eyes to the persistent interest of seventeenth-century savants in natural magic and astrology? Is it really possible to ignore the fascination with alchemy that is found in the work of Tycho Brahe, Robert Boyle, and Isaac Newton? It seems more likely that if we are to understand these men, and many of their contemporaries, we cannot limit ourselves to their positive achievements alone. To properly assess them—and the rise of modern science—we must make some attempt to examine the totality of their work, and we must investigate some fields of study that were then—but are no longer—respectable.

In any such program of research it would be proper to point to chemistry as a subject of special interest. Problems raised by chemical hypotheses and broad mystical alchemical speculations occur frequently in the sixteenth- and seventeenth-century scientific and medical literature. Furthermore, in the course of a century and a half, a change occurred in chemistry that rivalled in magnitude the alterations that were seen in astronomy and in the physics of motion. The scientific revolution need not be ascribed completely to the advent of a "modern" mechanical philosophy of nature.

A description of the state of chemistry at the opening of the sixteenth century would have to include a number of fields that seemingly had little relationship to one another. To be sure, some of the mining and metallurgical techniques of that period employed chemical processes of great significance. But the chemical technology of the mines was far removed from the medical chemistry of the period. Here a medieval tradition already

---

chapter on alchemy in the *Paragranum* from the Latin exists in *The Hermetic and Alchemical Writings of Paracelsus*, trans. Arthur Edward Waite, 2 vols. (London: James Elliott, 1894), Vol. II, pp. 148-164 (156).

suggested that chemically distilled substances might well be more potent than the more common materia medica described in herbals. A third field was that of the alchemists—men who spoke of a mystical approach to nature and wrote on the transmutation of the metals. But their works were written for the "initiated"; there was no rush to have them set in print. Lynn Thorndike has pointed to the fact that alchemical texts were not published prior to 1500—and thereafter were printed only slowly.[1]

How very different was the chemical scene a century and a half later. By that time the works of the chemists had become the center of intense interest and debate. Traditional transmutatory alchemy was a subject of more concern than it had ever been. Yet this interest was dwarfed by the truly impressive number of physicians and natural philosophers who sought to establish a new science and a new medicine upon chemistry and chemical analogies. Not only was this "chemical philosophy" meant to replace the works of Aristotle, Galen, and their followers, it was to become the basis for suggested economic and educational reforms that were seemingly in line with a new age of religious truth.

It is the story of this chemical philosophy that forms the central theme of the present book. To some extent an attempt has been made to show its relationship to more familiar areas of science and medicine, but the primary object is to develop and illustrate this approach to nature as it was viewed by those who believed in it.

The Chemical Heritage: Alchemy in Antiquity

The study of Renaissance chemistry and its relationship to medicine and natural philosophy is complicated by the fact that alchemy was a fusion of many different concepts derived from the Near and the Far East as well as the Latin West.[2] While we always point to a special interest in the study of matter and its changes—especially transmutation—there are other themes which also recur

[1] Lynn Thorndike, "Alchemy During the First Half of the Sixteenth Century," *Ambix*, 2 (1938), 26-37.

[2] The present brief account of ancient and medieval alchemy is based upon my article "Alchemy" in the *Dictionary of the History of Ideas*, 5 vols. (New York: Charles Scribner's Sons, 1973, 1974), Vol. I, pp. 27-34. Discussions of the extensive literature on alchemy will be found on pp. 33-34 and in Allen G. Debus,

frequently. One of these is the deep-seated belief that alchemy had a special role to play in medicine both through pharmacy and through the more esoteric search to prolong human life. No less important was the persistent conviction that alchemical study was fundamental for a true understanding of nature because of its reliance on observation and experience rather than logic and argument. The alchemist believed in a unified nature, and he expressed this belief most frequently through the macrocosm-microcosm analogy. There was little doubt in his mind that man's direct connection with the greater world would permit him—through its study—to reach a more profound knowledge of the Creator.

There is no one source for these beliefs, but there is general agreement that the search for the origins of alchemy must be conducted not in philosophical works alone but also in surviving texts that illustrate the practical craft traditions of antiquity. In fact, we do find that the oldest surviving works of metal craftsmen combine an emphasis on the change in the appearance of metals with the acceptance of a vitalistic view of nature—a view that included the belief that metals live and grow within the earth in a fashion analogous to the growth of the human fetus. It was to become basic to alchemical thought that the operator might hasten the natural process of metallic growth in his laboratory and thus bring about perfection in far less time than that required by nature.

This ancient craft tradition may be traced in a number of different texts widely separated in time. Babylonian tablets of the thirteenth century B.C. describe the production of "silver" from a copper-bronze mixture.[3] Here elements of ritual are already present and the processes themselves impose secrecy upon the operator. The Leiden and Stockholm papyri, although dating from the third century A.D., would appear to form part of this same tradition. Again there are directions for the imitation of the noble metals, while on occasion further information is offered to make it possible to double the quantities. Even more interesting is the

---

"The Significance of the History of Early Chemistry," *Cahiers d'Histoire Mondiale*, 9 (1965), 39-58. Mircea Eliade's important *The Forge and the Crucible* (New York: Harper, 1962) contains a good bibliography on alchemy, which he has updated in "The Forge and the Crucible: A Postscript," *History of Religion*, 8 (1968), 74-88.

[3] A. Leo Oppenheim, "Mesopotamia in the Early History of Alchemy," *Revue d'Assyriologie et d'Archéologie Orientale*, 60 (1966), 29-45.

Physica et mystica of Bolos of Mendes (perhaps as early as 200 B.C.).[4] Here practical directions were combined with mystical passages which foreshadow the Hellenistic texts of late antiquity.

Alexandrian alchemy was based on Greek philosophy as well as on the practical tradition of the metal workers. The early comparisons of man and nature found in the pre-Socratics and in Plato's *Timaeus* fostered an interest in the relationship of the macrocosm and the microcosm, while systems of intermediary beings and the *pneuma* were employed by the Stoics, the neo-Platonists, and other philosophical sects in antiquity to provide connecting links between the two worlds. The doctrine of the two worlds continued to play a significant role until well into the seventeenth century, while the *pneuma* had some similarity with seventeenth-century views of the aerial niter, or saltpeter.

Also important for the development of alchemical thought was the problem of the Creation. Philosophers interested in the Creation and nature were inevitably drawn to the question of the origin of the elements and the possibility of a *prima materia*. The views of the pre-Socratics on the prime matter were to be the springboard from which later authors launched their own concepts. Thus Aristotle conveniently summarized the views of his predecessors prior to refuting them in his *Metaphysics*. Inherent in his own thought was the need for both matter and form. Accepting the four elements of Empedocles (earth, air, water, fire), he postulated their origin from the *prima materia* by the pairing of qualities (humid, dry, hot, cold). And, since the paired qualities might be altered, it seemed theoretically possible to the later alchemists to transmute one substance into another.

The genesis of the elements also forms an important section of Plato's *Timaeus*, where the subject is developed mathematically; but to alchemical authors of late antiquity who were influenced by neo-Platonic, Gnostic, and Christian sources, the accounts found

[4]This work is discussed in considerable depth by Robert P. Multhauf in *The Origins of Chemistry* (London: Oldbourne, 1966), pp. 92-101. The standard source for Greek alchemy is the *Catalogue des manuscrits alchimiques Grecs*, ed. J. Bidez, F. Cumont, J.L. Heiberg, O. Lagercrantz, *et al.*, 8 vols. (Brussels, 1924-1932). Earlier, but still useful is the *Collection des anciens alchimistes Grecs* by M.P.E. Berthelot, 3 vols. (Paris, 1887-1888; reprint London: Holland Press, 1963). See also Jack Lindsay, *The Origins of Alchemy in Graeco-Roman Egypt* (London: Muller, 1970).

in Genesis and the *Pymander* attributed to Hermes Trismegistus were no less significant. Surely the alchemical literature was stamped with a Creation-element theme throughout its existence. In the sixteenth and seventeenth centuries chemists and physicians were still focusing on the subject of the elements as they established or attacked any system of nature.

With the possible exception of Bolos Democritos, the earliest true alchemical texts in Greek date from the end of the third century A.D. These are connected with the early practical tradition as well as current philosophical and religious thought. Works ascribed to Mary the Jewess, Cleopatra, and Zosimos offered detailed descriptions of laboratory equipment and procedures which indicate that there was even at that time a strong emphasis on distillation and sublimation (an emphasis not to be appreciably altered for over a millennium). In addition, these Alexandrian texts are openly concerned with transmutation. Here color change was stressed as a guide to progress—from black to white to yellow to red and violet.[5] This sequence was clearly associated with the change from a chaotic and undefined primal matter to metallic perfection.

Although practical recipes form part of these third- and fourth-century texts, there is also present a pronounced interest in mysticism and the belief in a secret tradition. Allegorical dream sequences abound, and spirits are important in the transformation of matter. Some scientific information can surely be extracted from these alchemical codices, but it is difficult to separate this material from the dominating religious aura. For example, the Great Work, the transmutation of metals, was seen as analogous to death, purgatory, and resurrection. It was this aspect of alchemical thought that stands out most clearly in the later Greek texts. Byzantine authors such as Stephanos (c. 610-641) offered their readers prayers, invocations, and allegorical descriptions, but there is little indication that they had any personal contact with the laboratory. Before the tenth century the basic texts had been codified, and few new works were composed in Greek after that time.

[5] An interpretation of this color change in terms of modern chemistry forms the main theme of *Alchemy, Child of Greek Philosophy* (New York: Columbia University Press, 1934) by Arthur John Hopkins.

Islamic Alchemy

There is little question about the importance of Greek sources in the development of Islamic alchemy.[6] Traditionally Prince Khālid ibn Yazīd (d. 704) was the first Muslim convert to alchemy, and it is significant that his teacher was said to have been one Morienos, a pupil of Adfar of Alexandria. It is unlikely that this story is historically true, but the strong Alexandrian influence on Islamic alchemy is confirmed by the frequent references to Byzantine authors and the widespread use of Greek philosophical concepts. At learned centers throughout the Near East Arabic translations were made not only of the works of major authors such as Aristotle, Galen, and Ptolemy, but also of the alchemical texts of Bolos Democritos, Stephanos, and Zosimos. Among these centers was the old Sassanian academy at Jundī-Shāpūr, while texts from farther East were translated by a group of Sabians at Harran.

The ascription of alchemical texts to earlier historical, religious, or mythological figures was common among the Alexandrian authors and was a custom continued by their Muslim successors. The short alchemical classic the "Emerald Table" was said to have been written by Hermes Trismegistus, but the earliest surviving version is an early-ninth-century Arabic text ascribed to the first-century (A.D.) magician Apollonios of Tyana. Similarly, the *Turba philosophorum*, which exists only in Latin and is in the form of a dialogue between the Greek philosophers of antiquity, has been shown to have been composed in Arabic early in the tenth century.[7]

Islamic alchemy is characterized by both the practical and the mystical elements seen in the earlier Greek texts. There are frequent warnings that the information being revealed is for the in-

---

[6] A collection of Arabic and Syriac texts will be found in M.P.E. Berthelot, *La chimie au moyen age*, 3 vols. (Paris, 1893; reprint Osnabruck: Otto Zeller, Amsterdam: Philo Press, 1967). This work should be supplemented with the numerous studies of Julius Ruska on all aspects of Islamic alchemy and the intensive study of Paul Kraus, *Jābir ibn Hayyān* (2 vols., Cairo, 1942-1943).

[7] Martin Plessner, "The Place of the *Turba Philosophorum* in the Development of Alchemy," *Isis*, **45** (1954), 331-338; "The *Turba Philosophorum*," *Ambix*, **7** (1959), 159-163. An English translation by Arthur Edward Waite exists as *Alchemy, The Turba Philosophorum or Assembly of the Sages* (London: George Redway, 1896).

itiated alone, and there is a continued use of the allegorical approach which had become common in late Greek works. The religious nature of the art is emphasized, and the predominant vitalism favored by alchemical authors is seen in discussions of the generation of metals and in the sexual interpretation of fundamental stages of the Great Work. As in the Alexandrian texts the progress of the operator could be followed through the now standard sequence of color changes. The concept of the philosopher's stone is also well developed in the Arabic literature. This stone, which allegedly provided a substance that brought about the rapid transmutation of base metals to gold, derived from the earlier concept of special elixirs which could cure illnesses in man and which analogously could perfect—or cure—imperfect metals in inanimate nature.

Aristotelian element theory was commonly employed by Islamic authors, but in addition the Arabic works ascribed to the eighth-century scholar Jābir ibn Hayyān (c.721-815) employed a sulfur-mercury theory to describe metals. This concept suggests that all metals are composed of different proportions of a sophic sulfur and a sophic mercury. Authors were in general agreement that these two substances had a resemblance to common sulfur and mercury, but it was also asserted that they were much purer than anything that could be produced in the laboratory. A quantitative relationship between the two was implied, but this reflected the number mysticism favored by the neo-Pythagoreans and Eastern mystics. Although the sulfur-mercury theory appeared first in these Arabic texts, it seems to have been a modification of an earlier Aristotelian concept of two exhalations within the earth that lead to the formation of minerals and metals.[8]

[8] I have discussed the conflicting views on the origin of the sulfur-mercury theory in *The English Paracelsians* (London: Oldbourne, 1965; New York: Franklin Watts, 1966), p. 45, n. 35. It is most commonly described as an extension of the Aristotelian view of smoky and watery vapors within the earth being transformed into stones and minerals on the one hand and metals on the other (*Meteorologica*, Bk. 4). According to this interpretation the Islamic authors suggested that the dry and smoky exhalation first is changed to sulfur while the watery exhalation is converted to mercury; a combination of these results in the metals. If pure and combined in the proper proportion, the result is gold. A different interpretation has been suggested by R. Hooykaas, who argues that this dichotomy actually derives from the emphasis on active and passive "spirits" pursued by Stoic

Also characteristic of the Arabic literature is a special interest in medical chemistry. This is in distinct contrast to the Greek tradition. Although Pliny and Dioscorides had described mineral substances of medicinal value, Hellenistic alchemical texts did not display any real concern with pharmaceutical chemistry. Rather, this interest seems to have originated in the alchemy of China and India.[9] As early as the eighth century B.C. there was a belief in physical immortality in China, which was later to become closely

authors. Here see R. Hooykaas, "Die Elementenlehre des Paracelsus," *Janus*, **39** (1935), 175-188; "Chemical Trichotomy before Paracelsus?," *Archives Internationales d'Histoire des Sciences*, **28** (1949), 1063-1074. Walter Pagel has discussed the background to the Paracelsian principles in his *Paracelsus. An Introduction to Philosophical Medicine in the Era of the Renaissance* (Basel/New York: S. Karger, 1958). Here (pp. 270 ff.) he discusses the trichotomy of sulfur, arsenic, and mercury in the *Summa* of Geber (fourteenth century, ascribed to Jābir), pointing out that the substances are described as having both corporeal and spiritual natures. He examines the medieval sources in still greater detail in his "Paracelsus: Traditionalism and Medieval Sources" in *Medicine, Science and Culture: Historical Essays in Honor of Owsei Temkin*, ed. Lloyd G. Stevenson and Robert P. Multhauf (Baltimore: Johns Hopkins Press, 1968), pp. 51-76 (59-60). Here Pagel discusses the *faex* or *terra* used with sulfur and mercury in the medieval *Septem tractatus Hermetici*.
[9]Recent editions of Chinese alchemical texts include *Alchemy, Medicine and Religion in the China of A.D. 320. The Nei P'ien of Ko Hung*, trans. and ed. James R. Ware (Cambridge, Mass.: M.I.T. Press, 1966), and Nathan Sivin, *Chinese Alchemy: Preliminary Studies* (Cambridge, Mass.: Harvard University Press, 1968). A useful review of Chinese alchemy based on the literature through the early 1950s will be found in Henry M. Leicester, *The Historical Background of Chemistry* (New York: John Wiley, 1956), pp. 52-61, but the subject will surely be deeply affected by the fifth volume (on chemistry and chemical technology) of Joseph Needham's *Science and Civilisation in China*. The second part of this volume, *Spagyrical Discovery and Invention: Magisteries of Gold and Immortality*, was published in Cambridge at the University Press in 1974. Two significant studies include Joseph Needham, "Artisans et alchimistes en Chine et dans le monde hellénistique," *La Pensée*, No. 152 (1970), 2-24 and Lu Gwei-Djen, Joseph Needham, and Dorothy Needham, "The Coming of Ardent Water," *Ambix*, **19** (1972), 69-112. The relation of both Eastern and Western alchemy to views on the lengthening of life has been discussed by Gerald J. Gruman, "A History of Ideas About the Prolongation of Life. The Evolution of Prolongevity Hypotheses to 1800," *Transactions of the American Philosophical Society*, N.S. **56** (Part 9) (Philadelphia: American Philosophical Society, 1966). An important reprint is the *History of Chemistry in Ancient and Medieval India*, incorporating the *History of Hindu Chemistry* by Achaiya Profulla Chandra Ray, ed. P. Ray (Calcutta: Indian Chemical Society, 1956).

associated with Taoist thought. The classic *Nei P'ien* of Ko Hung (c. 320 A.D.) contains sections on the transmutation of metals and on elixirs of life. Chinese alchemy paralleled Alexandrian alchemy in its frequent reference to the macrocosm-microcosm analogy as well as in the development of both esoteric and exoteric approaches to the subject. Thus, while the Chinese alchemist sought a potable gold and various chemically prepared drugs in his quest for longevity and immortality, the texts also indicate a real interest in alchemy as the search for the inner perfection of the soul. The alchemy of India was similar to that of China. Here the Hatha Yoga texts (post-eighth century) explain that the alchemist undergoes the experience of an initiatory death which is followed by a resurrection in the course of the work. In metals the result is seen in the perfection of gold; in the alchemist himself there is induced a similar separation of spirit from gross matter, resulting in a perfected person with an infinitely prolonged youth.

The medical importance of alchemy was accepted by Islamic authors later than concepts more clearly labeled Hellenistic. Among the first was the physician al-Razi (Rhazes) (860-925), whose work is decidedly practical in nature. Although he accepted the truth of transmutation and discussed elixirs of varying powers, in the *Book of the Secret of Secrets* al-Razi spoke at length of chemical equipment and the laboratory operations requisite for the chemist. He described a large number of laboratory reagents which he classified as animal, mineral, vegetable, and "derivative." Chemical texts continued to employ the first three categories as a basic scheme until well into the eighteenth century. Al-Razi's interest in medicine and practical chemistry influenced later Islamic and Western work in medical chemistry. Ibn Sina (Avicenna) (980-1037) and Abu Mansur Muwaffak (late tenth century) indicated a special interest in chemically prepared substances of pharmaceutical value.

The Latin Alchemy of the West

Western alchemy was directly dependent upon Arabic sources.[10] As Islamic scholars had sought alchemical texts to translate in the eighth century, so their Latin counterparts sought similar works four centuries later. A very early Latin translation of this genre is the story of Prince Khālid and Morienos. Traditionally this was completed by Robert of Chester on February 11, 1144, just one year after he had translated the Koran and a year prior to his completion of the translation of the *Algebra* of al-Khwārizmī. Recent scholarship has questioned both this date and the name of the translator, but there is little doubt that this text was considered by later alchemists to be the first in the Latin tradition.[11] And although there is little evidence to indicate that many Arabic alchemical texts were translated in their entirety, there is no doubt that knowledge of this "new science" spread rapidly.

There are frequent references to alchemy in the work of Thomas Aquinas (c. 1225-1274), while from the commentaries of Aristotle written by Albertus Magnus (c. 1193-1280) it is clear that the subject was of great interest to thirteenth-century scholars. Albertus knew the work of Avicenna and commented on the fact that this Islamic scholar had, in different texts ascribed to him,

---

[10]Bibliographies of alchemical texts date from an early period, but the two standard lists are J. Ferguson, *Bibliotheca chemica*, 2 vols. (Glasgow: James Maclehose, 1906) and Denis I. Duveen, *Bibliotheca alchemica et chemica* (London: Dawsons, 1949). Basic collected editions of the Latin alchemical texts include the 6-volume *Theatrum chemicum*, published by Lazarus Zetzner (Strasbourg, 1659-1661), and the 2-volume *Bibliotheca chemica curiosa*, ed. Jean Jacques Manget (Geneva: Chouet, G. De Tournes, et al., 1702). The most extensive German collection is the *Deutsches theatrum chemicum*, prepared by Friedrich Roth-Scholtz, 3 vols. (Nuremburg, 1728-1732). The standard French collection is the *Bibliothèque des philosophes chimiques*, prepared by Jean Maugin de Richebourg, 4 vols. (Paris: Cailleau, 1741-1754). The most extensive collection of alchemical poetry in English is that of Elias Ashmole, *Theatrum chemicum Britannicum* (London, 1652), reprinted with an introduction by Allen G. Debus (New York: Johnson Reprint, 1967).

[11]The complexity of the Morienos texts has been most recently discussed by Lee Stavenhagen in "The Original Text of the Latin *Morienus*," *Ambix*, **17** (1970), 1-12, and in *A Testament of Alchemy: Being the Revelations of Morienus to Khālid ibn Yazīd*, ed. and trans. Lee Stavenhagen (Hanover, N.H.: Brandeis University Press by the University Press of New England, 1974).

both accepted and denied the possibility of transmutation. But although Albertus himself believed in the truth of transmutation, he remained skeptical of the "transmuted" metals he had actually seen, none of which had been able to withstand the heat of the assayer's fire. With Albertus we also have early evidence of the use of the sulfur-mercury theory in the West. In his *De mineralibus*[12] he referred to the ancient concept of the exhalations, but he went on to discuss a "new" theory that attributed the origin of metals to sulfur and mercury.

Some of the most interesting medieval alchemical treatises date from the late thirteenth and early fourteenth centuries. Among them were the *Summa perfectionis* and other works ascribed to the eighth-century scholar Jābir ibn Hayyān (Latinized as Geber). A comparison reveals a possible knowledge of some of the Arabic texts, but the work of the Latin Geber is essentially new. Thus, the sulfur-mercury theory forms the basis for an understanding of the metals: the alchemist must arrange these two substances in perfect proportions for the consummation of the Great Work. But the mystical, quantitative relationship existing between sulfur and mercury that was present in the Arabic of Jābir is no longer evident in the Latin. Furthermore, in the Latin Geber's detailed description of laboratory processes and equipment, there is reflected an important change in distillation techniques that seems to have originated among twelfth- and thirteenth-century chemists. The introduction of condensation made possible the collection of low boiling fractions for the first time, and as a result we find in the literature of the mid-twelfth century the first clear Western reference to alcohol.[13] The Latin Geber confirmed this change in equipment and procedure: he described condensation apparatus in detail, and he was also the first to give a method for the preparation of a mineral acid—our nitric acid. These substances plus mixtures of other mineral acids placed powerful new reagents in the hands of alchemists who were to use them regularly after this period.

---

[12]Conveniently available in English as the *Book of Minerals*, trans. Dorothy Wyckoff (Oxford: Clarendon Press, 1967).

[13]A discussion of the early literature on alcohol in the West—plus the much earlier Chinese discovery—will be found in Lu Gwei-Djen, J. and D. Needham, "Ardent Water."

Another Latin text that was to be frequently reprinted in the Renaissance was the *Pretiosa margarita novella* of Petrus Bonus of Ferrara (c. 1330).[14] This work reflects the influence of Scholasticism in its tripartite structure: arguments in favor of transmutation follow the initial refutations, and these in turn are followed by positive answers to the objections. Peter asserted that alchemy differed from the other sciences in its method of demonstration:

> ... nothing short of seeing a thing will help you know it. If you wish to know that pepper is hot and that vinegar is cooling, that colocynth and absinthe are bitter, that honey is sweet, and that aconite is poison; that the magnet attracts steel, that arsenic whitens brass, and that tutia turns it of an orange color, you will, in every one of those cases, have to verify the assertion by experience.[15]

Peter's insistence on the importance of observation and experience had little influence on his deep belief in transmutation. Indeed, he stated that the true process might easily be learned in an hour. But while being honest enough to admit that he did not know how to produce gold himself, he remained convinced that alchemy was "the key of all good things, the Art of Arts, the science of sciences."[16]

No less sweeping were the claims of the pseudo-Raymond Lull. His alchemical *Testament* was also frequently referred to and made a part of collected editions of alchemical classics down to the eighteenth century. This text began with the statement that though logicians might be ever so wise, they would never know anything of nature without personal observations and experience. Only through knowledge obtained in this fashion will the intellect stand

> ... uncloathed of superfluities and errours, which do ordinarily remove it from the Truth, by reason of presumptions, and prejudiced or fore-judged things, believed in the conclusions. For from hence it is, that our *Philosophers* or followers, have directed themselves to enter through any kind of Science, into all experience, by Art, according to

---

[14] An English abridgement by Arthur Edward Waite is available as *The New Pearl of Great Price* (London: James Elliott, 1894; reprint London: Vincent Stuart, 1963).

[15] *Ibid.*, pp. 86-87.

[16] *Ibid.*, p. 138.

the course of Nature in its Univocal or single Principles. For *Alchymie* alone, is the Glass of true understanding; and shews how to touch, and see the truths of those things in the clear Light. Neither doth it bring Logical arguments: because they are too remote and far off from the clear Light.[17]

In this way the passage is reproduced in the *De lithiasi* (1644) of Jean Baptiste van Helmont. The attack on the logic of the Schools, the emphasis on its replacement with observation and experience as a guide to truth, and the insistence that alchemy is the proper key to all nature gave this fourteenth-century text a current relevance to the iatrochemists of the seventeenth century.

The alchemy of the late fourteenth and the fifteenth centuries indicates an increasing interest in allegorical and mystical themes. Thomas Norton's *Ordinall of Alchimy* (1477) is little concerned with clear-cut descriptions of chemical processes or laboratory equipment.[18] Rather, there is a lengthy poetical account of the difficult nature of the work, the need of virtue for its successful conclusion, and veiled descriptions of the true process. These and similar texts were accompanied by a widespread reaction against alchemy. The unsavory characterization of the alchemist in medieval literature knows no better example than Chaucer's "Canon's Yeoman's Tale" (c. 1390), while on an official level there were the decrees and statutes of Pope John XXII (1317) and Henry IV of England (1404) directed against those who attempted to multiply gold.

Medieval Medical Chemistry: The Analysis of Spa Waters

Closely associated with the medieval interest in transmutation was a parallel trend that led to the employment of chemical techniques in medicine. One significant area of interest was the analysis of spa waters.[19] The importance of tests for the purity of

---

[17]J.B. van Helmont, "De lithiasi" (Ch. 3, Sect. 1), *Opuscula medica inaudita* (Frankfurt: Joh. Just. Erythropili, 1682), p. 12; *Oriatrike or Physick Refined*, trans. J(ohn) C(handler) (London: Lodowick Loyd, 1662), pp. 839-840.

[18]In Ashmole, ed., *Theatrum chemicum Britannicum*, pp. 1-106.

[19]See Allen G. Debus, "Solution Analyses Prior to Robert Boyle," *Chymia*, **8** (1962), 41-61, upon which the present account is based.

water had been recognized in antiquity by both Vitruvius and Pliny, who referred to the examination of residues after evaporation, the use of oak galls as a color indicator in the testing of alum and verdigris, and the comparison of equal volumes of water by weight. Yet, although a few such early references were readily available, the medieval interest in spa waters seems to have derived from studies of local springs by Italian physicians. Surely the most significant developments were carried out by those physicians associated with the University of Padua who sought to determine the health-giving ingredients in the mineral water springs to which they sent their patients.

The earliest of the Italian texts appears to be a late-twelfth-century work by Peter of Eboli,[20] which was followed by a series of similar treatises. Thirteenth-century texts for the most part simply suggested evaporating a water sample and examining the residue by the senses. Giacomo de Dondi (1289-1359) went on to recommend that the residue be tested on a red-hot coal to release any hidden odors that might aid in the identification,[21] and by the beginning of the fifteenth century distillation had replaced evaporation as a means of obtaining the solids. Bartolomeo Montagnana (d. 1460) analyzed the waters of seven Italian spas with the aid of distillation, and his work was frequently reprinted in the sixteenth century.[22]

Far more influential for the development of analytic methods was the work of Montagnana's contemporary, Michael Savonarola (c. 1390-1462). His treatise on balneology, the *De balneis et termis naturalibus omnibus ytalie*,[23] was probably the most advanced text yet

[20] Pietro da Eboli, *I Bagni di Pozzuoli* (1195) (Naples, 1887).

[21] Giacomo de Dondi, *Tractatus de causa salsedinis aquarum, et modo* in *De balneis omnia quae extant apud Graecos, Latinos conficiendi salis ex eis et Arabas* (Venice: Apud Juntas, 1553), fol. 109. On de Dondi see Max Neuburger, *Geschichte der Medizin* (Stuttgart: Ferdinand Enke, 1911), Vol. II, p. 486.

[22] Bartolomeo Montagnana, *Tractatus tres de balneis patavinis*, included in his *Consilia CCCV*, ed. J. de Vitalibus (Venice: G.L. per Bonetū Locatellū, 1564), fols. 350-353. Montagnana wrote (fol. 350): "Cum n. aquam huius loci diligenti distillatione distillaverim maximā faeculentiae partem sulfuris inveni: parum vero cineris: velut si calx diu in aqua coquetur." On Montagnana's life see Neuburger, *Geschichte der Medizin*, pp. 507-508.

[23] Johannes Michael Savonarola, *De balneis et termis naturalibus omnibus ytalie* (Ferrara: 10 Nov. 1485). This work is also included in his *Practica canonica de febribus* (Venice, 1552), fols. 1-26. On Savonarola see Neuburger, *Geschichte der*

written on solution analysis, and its influence can be traced for over a century. Savonarola admitted that it had been customary for many years to investigate mineral waters by "alembication," and he suggested that the residue from this process be dried in the sun. This would coagulate the salts so that they might be seen more easily,[24] and the sunlight would also cause any sulfur to shine with greater brilliance, making it easier to identify. Further tests were then to be made by casting the residue on the fire and noting any odor or change of color in the solid or the flame (he knew that sulfur burned with a green flame).[25] He referred to Aristotle's tests of color and taste but warned that for positive identification this was not enough; as an example he took the case of sulfur and stated that only if

> ... this water has a sulfureous odor and a citrine color, only if its residue burns when thrown on the fire, and in so doing emits a sulfureous odor—and on its surface there swims an oil which when placed on the fire immediately bursts into flame—and its taste is that of sulfur—and near it is the mineral sulfur out of which skilled workers refine sulfur. Only then when all of these things are found is there no doubt that the water ought to be called sulfureous.[26]

Savonarola distinguished between niter and salt by the crackling noise the latter emitted when strongly heated. He used similar methods to identify alum, gypsum, and iron.

The Italian treatises of de Dondi, Montagnana, and Savonarola were frequently reprinted and their analytical procedures repeatedly copied by other authors. By the early sixteenth century this work is reflected in the German medical literature on spa waters. Johann Widman (1513) and Johann Dryander (1535) described specific baths, while Paracelsus advocated the use of oak

---

Medizin, pp. 506-507, and also Lynn Thorndike, *A History of Magic and Experimental Science*, 8 vols. (New York: Columbia University Press, 1923-1958), Vol. IV, pp. 211-212.

[24]Savonarola, *De balneis*, Ch. 7: "Apud veteres nostros mos fuit usquam ad hanc servatus etatē q ĩnvestigatio balnearum nälium per alēbicationē fieri debeat."

[25]The analytical procedure is described on the final leaf, but the specific reference to the characteristic flame color of sulfur is on fol. b5.

[26]*Ibid.*

galls as a test in his treatise on natural baths (c. 1520).[27] These analytic procedures are also seen in the mining and the metallurgical works of the period. Both Biringuccio (1540) and Georgius Agricola (1556) described methods of isolating the residue and testing for its contents.[28]

It is clear that this medieval tradition became an integral part of Renaissance medicine. Perhaps the most influential text of this genre was the work on medicated waters written in 1564 by the Paduan anatomist Gabriel Fallopius (1523-1563).[29] Although he relied heavily on the earlier texts of Savonarola and Montagnana, Fallopius went far beyond these authors in his analytical procedure. Like them he advised reliance first on identification by the senses, but most of his tests were to be made on the residue after distillation. The investigator was told to dry the solids in the sun on a smooth table so that any bright crystals might be seen. The size of the grains aid in the identification, for alum appears as small crystals, while niter and salt are larger. Sulfur, he wrote, may be seen by its color, but other metals, minerals, and earths cannot be determined by this method. Therefore, for further identification some of the residue was to be placed on a red-hot iron. Gypsum, lime, and marble do not burn, but they may be differentiated by the fact that gypsum quickly becomes much whiter during the heating, while the others only slowly become whiter. On the other hand, brimstone melts and gives off its peculiar odor.

[27] Johannes Widman dicti Mechinger, *Tractatus de balneis thermarum ferinarum (vulgo Uuildbaden) perutilis belneari volentibus ibidem* (Tubingen, 1513), fols. A-Aii; Johann Dryander, *Vom Eymser Bade* (Koblenz, 1538); Paracelsus, *Eilff Tractat oder Bücher vom Ursprung und ursacher* in *Opera Bücher und Schrifften*, Vol. I, p. 521. Additional material on the German sources is found in Gernot Rath, "Die Anfänge der Mineralquellenanalyse," *Medizinischen Monatsschrift*, 3 (1949), 539-541 and "Die Mineralquellenanalyse im 17. Jahrhundert," *Sudhoffs Archiv für Geschichte der Medizin und der Naturwissenschaften*, 41 (1957), 1-9.

[28] Vannoccio Biringuccio, *The Pirotechnia*, trans. C.S. Smith and M.T. Gnudi (New York: American Institute of Mining and Metallurgical Engineers, 1942), p. 15; Georgius Agricola, *De natura eorum quae effluunt ex terra libri IV* (1545) in *Schriften zur Geologie und Mineralogie* [from the *Ausgewählte Werke*], ed. Hans Prescher (Berlin: VEB Deutsche Verlag der Wissenschaften, 1956), Vol. I, pp. 228-231, 240-241, 246. Agricola discussed identification by taste in the *De re metallica*. See the translation by H.C. and L.H. Hoover (New York: Dover, 1950), pp. 34, 235.

[29] Gabriel Fallopius Mutinensis, *De medicatis aquis atque de fossilibus* (Venice, 1564). In the 1569 edition the analytical procedure is on fols. 30-37.

Both salt and niter sparkle in this test, but salt gives off a crackling sound which niter does not. Finally, if the sediment melts and becomes as white as milk, it is alum. This, for Fallopius, was the limit of usefulness for the hot iron.

Fallopius then turned to another method and suggested that vitriol or alum might be determined by the use of a decoction of oak galls. Metals were more of a problem, but they were to be identified by acidifying the residue with aqua fortis (nitric acid) and then distilling off the acid so that the solids could be examined. Here he suggested that iron and copper might be determined, copper in particular by the green substance it leaves.[30] He urged the testing of this method by comparing residues from iron filings treated in this manner with those obtained from natural solutions of iron salts.

The analytical work of Fallopius had a far-reaching influence, not so much in its original form as in a lengthy abstract which was included by Conrad Gesner in the second part of his *Euonymus* (1569).[31] Like the work of Fallopius, this volume was published posthumously. The abstract included all of the analytical method and was credited fully to the Paduan anatomist. Almost unbelievably popular, Gesner's *Euonymus* was soon translated into French by Jean Liébaut (1573), English by Thomas Hill and George Baker (1576), and German by Nüscheler (1583).[32] However, by the third quarter of the sixteenth century the work of Paracelsus had become a subject of intense interest and debate among physicians, and the further history of the chemical analysis of spa waters is found primarily in the works of those authors who quickly adopted Paracelsian methods as part of their chemical philosophy.

---

[30]It is interesting that Fallopius cautioned the analyst that the green color of the residue is not enough to ascertain the presence of copper, and he also belittled the importance of taste, since he felt that there are so many tastes that no one man can know them all.

[31]Conrad Gesner, *Euonymus . . . de remediis secretis, liber secundus* (Zurich, 1569), fols. 24-27. In his abstract Gesner fully attributed this analytical method to Fallopius. On Gesner's work see Bernhard Milt, "Conrad Gesner als Balneologe," *Gesnerus*, **2** (1945), 1-16.

[32]These translations are succinctly discussed in Ferguson, *Bibliotheca chemica*, Vol. I, pp. 315-316.

The Chemically Prepared Medicines

No less important than the development of chemical analysis by the physicians of medieval Europe was their interest in alchemy as a source for new medicines. This is seen as early as the mid-thirteenth century in the work of Roger Bacon (c. 1214-c. 1294).[33] Bacon considered alchemy to be a major division of the sciences, and he was well read in the already copious Latin literature of the field. Bacon's general discussion of speculative and operative alchemy separated the chemical study of nature as a whole from more practical laboratory procedures.[34] He defended those authors who maintained their secrets through anagrams and allegories, and he expressed the hope that the benefits of transmutation might be employed to alleviate the poverty of mankind.

More significant at this point is the fact that Bacon connected the search for the prolongation of life with alchemy.[35] He referred to Holy Scripture to prove that at one time men lived nearly one thousand years, a time span that had been shortened because of the corruption of the human race. Bacon was able to cite only a few examples of extreme longevity in recent times, the most noteworthy being the alchemical author Artephius, who claimed to have reached the age of 1025. Bacon accepted this uncritically, since it supported his thesis that alchemy and medicine were fundamentally united. He went on to suggest that a true remedy against human corruption might lie in the complete governance of health from youth, and here he emphasized a sane and moderate diet. He argued that even if it was impossible to repair the total

[33] Bacon is discussed by most authors who write on alchemy. Especially helpful is E. J. Holmyard's treatment in *Alchemy* (Harmondsworth: Penguin, 1957), pp. 108-119. For Bacon's chapter on alchemy in the *Opus tertium*, which forms the basis of most of these discussions, see Fr. Rogeri Bacon, *Opera quaedam hactenus inedita*, Vol. 1, ed. J.S. Brewer (London: Longman, Green, Longman, and Roberts, 1859), pp. 39-43. Also of great interest for the clarification of Bacon's views on alchemy is the *Epistola fratris Rogerii Baconis de secretis operibus artis et naturae, et de nullitate magiae*. Although the authenticity of this work has been questioned, it is generally agreed that it is "Baconian" in tone. This text is also in the Brewer volume, pp. 523-551.

[34] Bacon, *Opus tertium* in *Opera*, pp. 39-43.

[35] Bacon's views on the prolongation of life can be quickly surveyed by reading his chapter "De retardatione accidentium senectutis, et de prolongatione vitae humanae" in the *Epistola*, in *Opera*, pp. 538-542.

loss that had occurred in the six thousand years of human existence, it should still be possible to lengthen human life in the course of several hundred years. But the call for abstinence was only one approach; alchemy was a basic tool as well. In the *Opus tertium* (1267) Bacon spoke of those physicians who teach how medicines should be distilled, sublimed, resolved, and submitted to other alchemical processes, and he commented that although "almost no one knows how to make metals, . . . even fewer men know how to perform these works which are of value for the prolongation of life."[36]

Similar themes occur in the work of Bacon's younger contemporary Arnald of Villanova (1235-1311), a man whose life has been compared with that of Paracelsus.[37] A restless man, he was constantly on the move—and not infrequently in trouble with the authorities. And if he praised the academics on occasion, he also acknowledged his deep debt to the common people, from whom he believed much real knowledge was to be obtained. Accordingly, one finds that Arnald sought a balanced approach to knowledge, utilizing both reason and experience.[38] He complained bitterly about those who rejected either *in toto*. The proper method was to study nature to uncover its miraculous workings—and then to ascend to more occult matters.

Arnald did not doubt that God was the source of medicine and its effects.[39] Nor did he question the belief that the true physician was himself touched with divinity. While the action of both disease and cure was to be explained through the macrocosmic influences of the stars, Arnald also believed that the physician could interfere with this normal pattern to prevent or cure disease. Significant is the fact that he wrote within the framework of the Galenic system of medicine.[40] He wrote that the welfare of the soul was dependent upon the proper mixture of the humors, and he placed special emphasis upon the blood and the spirits which ensured health. As

---

[36]Bacon, *Opus tertium* in *Opera*, p. 41.

[37]See the treatment of Arnald as a predecessor of Paracelsus in Pagel, *Paracelsus*, pp. 248-258. Here the extensive secondary literature on Arnald is cited in detail. Of special interest is the recent paper by Michael McVaugh, "Arnald of Villanova and Bradwardine's Law," *Isis*, **58** (1967), 56-64.

[38]Pagel, *Paracelsus*, p. 250.

[39]*Ibid.*, p. 253

[40]*Ibid.*, pp. 257-258.

a Galenist, Arnald called for an alteration of the humoral balance when illness occurred. But here alchemy was to play a special role.⁴¹ He claimed to possess a chemical elixir that had virtues far in excess of any other medicine known, one which cured all disease and sickness. Arnald was one of the earliest authors to refer frequently to alcohol for its medicinal value. In the preparation of alcohol and also the new mineral acids distillation was essential. Surely from this date medieval medical chemistry was best characterized by its dependence on distillation processes.

With John of Rupescissa (mid-fourteenth century) the main line of medieval iatrochemistry becomes more sharply defined.⁴² Once again we meet an unorthodox man who found himself frequently in trouble with the authorities. He, too, emphasized a macrocosmic-microcosmic universe and sought to gain knowledge of the virtues given to man by God in nature. Indeed, here was the real purpose of alchemy: to benefit mankind through medicine. In this quest John of Rupescissa sought "quintessences," or pure virtues of natural substances, which would preserve the body. His works abound with medicinal preparations derived from metals and minerals, and he emphasized distillation processes which seemingly separated these quintessences from the gross matter of the substances collected in nature.

It was this medieval tradition of medical chemistry that bore fruit in the Renaissance "distillation books." Michael Puff von Schrick's book on distilled waters (1478 and later) is probably the earliest printed book of this type, but without question the most popular was the *Distillierbuch* (1500) of Hieronymus Brunschwig (c. 1440-c. 1512). This and his much larger *Liber de arte distillandi* (1512) are the two most influential distillation texts of the sixteenth century.⁴³ At least fourteen editions in three languages are

---

⁴¹*Ibid.*, p. 263. See esp. the *Rosarius philosophorum* in Arnald's *Opera* (Lugduni: Scipionem de gabiano, 1532), fols. 296^r-300^v. The English influence of Arnald may be noted in the early translation, *Here is a newe boke, called the defence of age and recovery of youth*, trans. J. Drummond (London: R. Wyer, 1540).

⁴²Pagel is important for his discussion of John as a predecessor of Paracelsus (*Paracelsus*, pp. 263-266) while R. P. Multhauf discusses the significance of this author's chemistry in John of Rupescissa and the Origin of Medical Chemistry," *Isis*, **45** (1954), 359-367.

⁴³The English translation (c. 1530) is conveniently available in the *Book of Distillation* by Hieronymus Brunschwig, with a new introduction by Harold J.

known to 1536, while other printings exist as late as 1610. These books are practical in nature, with the main emphasis on obtaining the desired waters and oils. Brunschwig offered the reader a useful description with illustrations of the laboratory equipment (furnaces, lutes, alembics, and cucurbits) requisite for the isolation of the desired products. In addition he described the plants that were to be distilled and the medical application of their waters. As a chemist, Brunschwig was convinced that the distilled waters were far more potent than the crude herbs so frequently employed by contemporary pharmacists. Yet at the same time his almost exclusive interest in the medicinal properties of plants shows this branch of medical chemistry to be part of a much older herbal tradition.

Numerous works on distillation by other authors were printed in the sixteenth century, but there is little doubt that Brunschwig's texts remained the most popular throughout most of this period. It is necessary here only to refer to one other, Conrad Gesner's *Thesaurus Euonymi Philiatri, de remediis secretis* (part 1, 1552; part 2, 1569). The posthumous second part has already been referred to for its important paraphrase of Fallopius on the analysis of spa waters, but the two parts as a whole are no less significant for their summary of the medieval distillation tradition. Gesner wrote that "waters and oyles secreate by the singuler industrie and wit of Chymists, are of most great vertues," but he explained that some physicians rightly held them in contempt because they had been incorrectly prepared in the past.[44] Attesting to the overwhelming influence of Brunschwig, Gesner rejected the views of those who ascribed the introduction of this art to him. Among his authorities he listed Dioscorides, Geber, Arnald of Villanova, Raymond Lull, and John of Rupescissa—thus correctly recognizing the earlier medieval tradition.[45] In contrast to Brunschwig, Gesner was interested in the medicinal virtues of substances other than plants,

---

Abrahams (New York/London: Johnson Reprint, 1971). The various editions and the relationship of this edition to the German ones are discussed in the introduction. The large *Liber de arte distillandi de compositis* (1512) has been recently reprinted in its entirety by the Zentralantiquariat der Deutschen Demokratischen Republik (Leipzig, 1972).

[44]Conrad Gesner, *The Treasure of Euonymus: Conteyninge the wonderfull hid secretes of nature*, trans. P. Morwyng (London: John Daie, 1559), pp. 293, 412.

[45]*Ibid.*, p. 421.

and he refused to limit himself to distillation processes alone; one section of his work is devoted to metallic preparations and some of these were isolated in the form of precipitates.[46]

This sixteenth-century interest in chemicals of medicinal value was to become more closely identified with the herbal tradition; in fact, both Brunschwig's and Gesner's texts can be characterized as chemically modified herbals. But even Petrus Andreas Matthioli's (1501-1577) massive edition of the six books of materia medica by Dioscorides with its commentary included an appended description of distillation equipment and procedures. Matthioli considered this information essential since distillation was now a recognized method of preparation—one that had not been known in antiquity.[47] And so important was this section that later editions were to replace the short appendix with a much longer chemical text on the distillation of vegetable substances.[48] In England John Gerarde's noted *Herball* (1597) included a prefatory letter from Stephen Bredwell, who proposed the establishment of a chemical lectureship at the London College of Physicians. Converted to, but still somewhat skeptical of the chemists of his day, Bredwell wrote:

> I say in like manner, the art of Chimistrie is in it selfe the most noble instrument of naturall knowledges; but through the ignoraunce and impietie, partly of those that most audaciously professe it without skill and partly of them that impudently condemne that they knowe not, it is of all others most basely despised and scornfully reiected.[49]

[46]Here we have evidence of the isolation of the products of chemical reactions prior to the time of the late-sixteenth-century Paracelsians. For Gesner's place in the history of distillation chemistry see Robert P. Multhauf, "The Significance of Distillation in Renaissance Medical Chemistry," *Bulletin of the History of Medicine*, 30 (1956), 329-346 (341).

[47]Petrus Andreas Matthioli, "De ratione distillandi aquas ex omnibus plantis: Et quomodo genuini odores in ipsis aquis conservari possint" in *Opera quae extant omnia* . . . (Frankfurt: N. Bassaei, 1598), pp. 1029-1033 (1030).

[48]See the entirely new (with references to Charas, Glaser, Lemery—and even John Mayow's aerial niter) *Traité de chymie en abregé, contenant quelques operations les plus faciles & les plus necessaires à la pratique de medecine*, "Par un Docteur en Medecine" in *Les commentaires de M.P. André Matthiole, medecin sienois, sur les six livres de la matiere medecinale de Pedacius Dioscoride, Anazarbéen*, trans. M. Antoine du Pinet (Lyon: Jean Baptiste de Ville, 1680), pp. 599-636.

[49]John Gerarde, *The Herball or Generall Historie of Plantes* (London, 1597), introductory nonpaginated leaf titled "To the well affected Reader and peruser of this booke."

If chemical methods were quickly adopted by herbalists in the sixteenth century, they were also seen as a necessary addition to the traditional materia medica by surgeons. This will become more evident to the reader in subsequent chapters, but here reference need only be made to the surgical writings of Ambroise Paré (c. 1517-1590). In the first French edition of his *Oeuvres* (1575) Paré pointedly added a section on distillation procedures and the medical importance of distilled products to his book of materia medica. These medicines

> ... are such as consist of a certain fift essence separated from their earthie impurities by Distillation, in which there is a singular, and almost divine efficacie in the cure of diseases. So that of so great an abundance of the medicines there is scarce anie which at this daie Chymists do not distil, or otherwise make them more strong and effectual then they were before.[50]

In a list of chemically prepared substances Paré referred to various salts, metals, and minerals,[51] but his detailed preparations were limited solely to substances of vegetable origin. Paré's source of chemical information was, once again, Conrad Gesner. The identification is certain, since in his introduction to this section Paré referred with praise to Jean Liébaut, the translator of the French edition (1573) of the *Thesaurus Euonymi*.[52]

It is important to note that in the medieval alchemical and iatrochemical tradition there was no urgent sense of a righteous crusade, no feeling of imminent conflict with the medical establishment or with the proponents of the ancient philosophers. Arnald of Villanova, and others, may well have complained about the conservatism of physicians, but his own views were fundamentally Galenic and he could still speak with praise of the academics.

---

[50] Ambroise Parey, *The Workes of that famous Chirurgion* ..., trans. Thomas Johnson (London: Richard Cotes and W. Dugard, 1649), p. 735 (from the 27th Book, "Of Distillations"). Ambroise Paré, *Oeuvres complètes d'Ambroise Paré* ..., notes and introduction by J.F. Malgaigne, 3 vols. (Paris: J.B. Baillière, 1840-1841). Here the book "Des Distillations" is in Vol. III, pp. 614-633, with the quotation on p. 614. Although reflecting the current interest in chemistry, Paré was no Paracelsian. In a list of 2,168 citations given by Paré, Malgaigne has found only two to Paracelsus (pp. xx-xxi).

[51] *Oeuvres*, ed. Malgaigne, Vol. III, pp. 635-636.

[52] *Ibid.*, p. 614, n. 1.

Indeed, alchemy had come to the West together with the other texts of antiquity; it had been cultivated along with classical philosophy and medicine in the Near East, and it was not to be immediately divorced from this earlier union. Even the Lullian claim that alchemy formed the basis of natural philosophy caused no problem. Nor is there any indication that chemistry was viewed as a rival discipline by physicians. Paduan physicians were the first to apply chemical methods to the analysis of spa waters, while the commentators on traditional herbals and the authors of the more recent distillation books in no way considered that they were attacking the revered authors of antiquity. Paré argued for the elements and humors of Aristotle and Galen, and neither he nor Gesner felt that they had to reject the work of the ancients simply because they themselves favored chemically prepared medicines. In effect, for many physicians of the Renaissance there was no need to argue over the application of chemistry to medicine. The conflict that developed in the late sixteenth century occurred only after the works of Paracelsus had become widely known.

Renaissance Factors: The Educational Problem

Although the alchemical tradition had not encountered frontal opposition from the educational or the medical establishment, there were surely new factors that were to contribute to an increasing dissatisfaction in the fourteenth and the fifteenth centuries. The growing educational problem was a matter of particular concern. The roots of this can be sought less in scientific history per se than in the more general field of the history of learning.

Eugenio Garin has identified two major themes in Renaissance scholarship. The first was centered on medical and philosophical problems; this was essentially a scholastic self-criticism within the Aristotelian-Galenic tradition. The second he has characterized as predominantly "rhetorical and philological" or basically "humanist" in nature.[53] In the period after 1550 the chemical

[53]Eugenio Garin, *L'educazione in Europa (1400-1600) problemi e programmi* (Bari: Editori Laterza, 1957), p. 31. This section on education is based upon my discussion in *Science and Education in the Seventeenth Century. The Webster-Ward Debate* (London: Macdonald/New York: American Elsevier, 1970), pp. 5-14.

philosophers were to reflect both of these themes in their attack on the sciences and the entrenched educational system.

Scholastic self-criticism was nothing new. We know that all of the universities of the sixteenth century were not as reactionary as they have been described by modern historians or their contemporary critics. Of particular importance has been the research of John Herman Randall, who has linked the fourteenth-century critical reevaluation of Aristotelian philosophy at Oxford and Paris with the northern Italian universities in the fifteenth and sixteenth centuries.[54] This he has emphasized as a living self-critical Aristotelianism rather than a shift of interest from Aristotle to Plato.[55] Paul Kristeller has gone even farther to argue that as far as Italy was concerned, "Aristotelian scholasticism . . . is fundamentally a phenomenon of the Renaissance period whose roots can be traced to the very latest phase of the Middle Ages."[56] Indeed,

> . . . if we pass from the humanist scholars to the professional philosophers and scientists, it appears that the most advanced work of Aristotle's logic, the *Posterior Analytics,* received greater attention in the sixteenth century than before, and that at the same time an increased study of Aristotle's biological writings accompanied the contemporary progress in botany, zoology, and natural history.[57]

One may thus seek this Aristotelian tradition in the work of major scientific figures of the seventeenth century, men such as Harvey, Fabricius, and Galileo. There may be no better example than Harvey, who Walter Pagel has identified as an Aristotelian, a man who turned to the method of the *Posterior Analytics* and the *Physics—* works which called for a dual role of experiment and reason.[58]

The progressive self-criticism of Padua and other Italian uni-

---

[54] John Herman Randall, Jr., "The Development of Scientific Method in the School of Padua," *Journal of the History of Ideas,* 1 (1940), 177-206.

[55] John Herman Randall, Jr., "Scientific Method in the School of Padua" in *Roots of Scientific Thought. A Cultural Perspective,* ed. Philip P. Wiener and Aaron Noland (New York: Basic Books, 1957), pp. 139-146 (145).

[56] Paul Oskar Kristeller, *Renaissance Thought. The Classic, Scholastic and Humanist Strains* (New York: Harper & Row, Harper Torchbooks, 1961), p. 40.

[57] *Ibid.,* p. 40. See also Charles B. Schmitt, "Towards a Reassessment of Renaissance Aristotelianism," *History of Science,* 11 (1973), 159-193.

[58] The case of Harvey is described in detail by Walter Pagel in *William Harvey's Biological Ideas. Selected Aspects and Historical Background* (Basel/New York: S. Karger, 1967), pp. 28-47. On method the reader is also referred to the pertinent

versities in the sixteenth century may be contrasted with the very conservative approach of such universities as Oxford, Cambridge, and Paris.[59] There scholars felt that the main function of the university was to preserve the learning of the past rather than seek new knowledge through fresh observations, an attitude in keeping with the spirit of the literary Renaissance. At these universities the infallibility of Galen was still maintained, and the inflexibility of medical education was to contribute to heated conflicts in which the chemical philosophers demanded fundamental educational changes. In these schools it may have been acceptable, even commendable, for humanists to criticize the vulgar annotations and emendations foisted on the works of the ancients by the Arabs and the Scholastic philosophers, for the original and pure texts were like an impregnable fortress of truth not to be altered in any way.

Garin's second theme, Renaissance humanism, was to redirect the entire course of the educational process. The moral character of man became important rather than the study of philosophy, theology, and the sciences, which remained of prime interest to the university scholar. The desire to mold a youth's character placed a new interest on elementary education, and for many humanists the problems of the universities receded into a dim background. At the humanist school of Vittorino da Feltre (1378-1446) in Mantua the student was urged to excel at games and military exercises to strengthen his body. In the classroom he studied rhetoric, music, geography, and, above all, history, from which he could find examples of moral principle and political action.[60] For some—if not all—this meant that the study of natural philosophy was of little

---

sections in W. P. D. Wightman, *Science and the Renaissance. An Introduction to the Study of the Emergence of the Sciences in the Sixteenth Century*, 2 vols. (Edinburgh/London: Oliver and Boyd, 1962). See also Wightman's views as expounded in his paper "Myth and Method in Seventeenth-Century Biological Thought," *Journal of the History of Biology*, 2 (1969), 321-336.

[59]See, e.g., Phyllis Allen, "Scientific Studies in the English Universities of the Seventeenth Century," *Journal of the History of Ideas*, 10 (1949), 219-253; William T. Costello, S.J., *The Scholastic Curriculum at Early Seventeenth-Century Cambridge* (Cambridge, Mass.: Harvard University Press, 1958), p. 9; Mark H. Curtis, *Oxford and Cambridge in Transition 1558-1642* (Oxford: Clarendon Press, 1959), p. 227.

[60]William Harrison Woodward, *Vittorino da Feltre, and Other Humanist Educators. Essays and Versions* (Cambridge: Cambridge University Press, 1905); William

concern. Thus, Matteo Palmieri (1406-1476) considered the investigation of natural philosophy far less important than that of moral philosophy. While it is laudable, he wrote, to study the secrets of nature, "we hold them still of the minutest interest in the supreme task of solving the problem how to live."[61] Nor was Erasmus (1466-1536) far removed from Palmieri on this point. For him the young student would learn all he would ever need to know about nature through the diligent study of sound classical authors.[62]

This was an educational setting far removed from that of the Italian universities. The humanist may have expressed a love of nature, but he found the logic and philosophical debates of the universities of little value for moral growth.[63] This widespread dismissal of the traditional study of natural philosophy was also extended to mathematics. Erasmus felt that this subject was of relatively little importance for the educated man, and this thought was echoed by his friend Juan Luis Vives (1492-1540), the noted Spanish humanist and philosopher. As a professor at Louvain Vives admitted that medical students might benefit from observations and that educated men should not neglect natural philosophy. Nevertheless, they need not go beyond Aristotle in their search for scientific truths. As for mathematics, this subject was not of great value since it tended to "withdraw the mind from the practical concerns of life, and render it less fit to fuse concrete and mundane reali-

---

Harrison Woodward, *Studies in Education during the Age of the Renaissance 1400-1600* (Cambridge: Cambridge University Press, 1906), pp. 10-23.

[61]Woodward, *Studies in Education*, p. 76.

[62]*Ibid.*, p. 123. See also William Harrison Woodward, *Desiderius Erasmus Concerning the Aim and Method of Education* (Cambridge: Cambridge University Press, 1904), and Peter Krivatsky, "Erasmus' Medical Milieu," *Bulletin of the History of Medicine*, **47** (1973), 113-154.

[63]Both themes may be seen early in the Renaissance in the writings of Petrarch (1304-1374), who attacked the learning of the Scholastics and turned instead for inspiration to classical antiquity and nature. Although he was willing to grant that the original Aristotle surpassed the works of his more recent followers, he still turned with far more enthusiasm to Plato, the "Prince of Philosophers." For him the knowledge of God was the main goal of philosophy, and he found little difficulty in branding the studies of the Scholastics as basically irreligious. With a special interest in the nature of man and his character, Petrarch and his followers turned to Cicero, Virgil, Seneca, and Homer rather than the philosophical texts in vogue at the universities.

ties."⁶⁴ Nor need the student concern himself with research into causes and principles. Such abstract contemplation would only lead to metaphysical subtleties—of interest to those at the universities, but hardly preparation for the practical affairs of life or for wholesome intellectual development.

One cannot avoid being impressed by a growing sense of disillusionment in many of the authors of the sixteenth century. No longer did the aims of elementary and university education coincide. The humanist educators sought the moral development of youth and had scarcely concerned themselves with the reform of higher education. The latter—although subject to internal criticism and the search for better texts—remained largely tied to traditional methods of study and time-honored authorities. Most humanists found such practices of little interest, and they frequently complained of the logic, philosophy, and mathematics taught there which did little or nothing to prepare most men for the realities of life. Few exhibit the discontent of this period better than Peter Ramus (1515-1572). It was with a sense of despair that he recalled his own academic training:

> After having devoted three years and six months to scholastic philosophy, according to the rules of our university: after having read, discussed, and meditated on the various treatises of the *Organon* (for of all the books of Aristotle those especially which treated of dialectic were read and re-read during the course of three years); even after, I say, having put in all that time, reckoning up the years completely occupied by the study of the scholastic arts, I sought to learn to what end I could, as a consequence, apply the knowledge I had acquired with so much toil and fatigue. I soon perceived that all this dialectic had not rendered me more learned in history and the knowledge of antiquity, nor more skilful in eloquence, nor a better poet, nor wiser in anything. Ah, what a stupefaction, what a grief! How did I accuse my deficiencies! How did I deplore the misfortune of my destiny, the barrenness of a mind that after so much labour could not gather or even perceive the fruits of that wisdom which was alleged to be found so abundantly in the dialectic of Aristotle!⁶⁵

⁶⁴Woodward, *Studies in Education*, p. 203. Note also Juan Luis Vives, *On Education. A Translation of the De Tradendis Disciplinis* (1531), introduction by Foster Watson (Cambridge: Cambridge University Press, 1913).

⁶⁵Frank Pierrepont Graves, *Peter Ramus and the Educational Reformation of the Sixteenth Century* (New York: Macmillan, 1912), pp. 23-24. Important for the study of

These words were to be closely paralleled by the chemical philosopher Jean Baptiste van Helmont a century later.[66] However, Ramus in his own criticism turned first to the very basis of the scholastic system by proposing a new system of logic. And, as he proceeded, he rejected the natural science and physics of Aristotle while calling for new observations which were to be tested and arranged methodically.[67]

The Hermetic Revival and the Study of Nature

The concern for higher learning expressed by Ramus was soon widespread. It is especially important to find the rejection of ancient philosophy accompanied by the recognition of a need for a new approach to nature and medicine: observation rather than books was widely stressed. Thus, Rabelais (c. 1495-1553) argued that the student should concentrate on learning the facts of nature,

> ... so that there be no sea, river or lake of which thou knowest not the fish; so that all the birds of the air, and all the plants and fruits of the forest, all the flowers of the soil, all the metals hid in the bowels of the earth, all the gems of the East and the South, none shall be foreign to thee. Most carefully pursue the writings of physicians, Greek, Arab, Latin, despising not even the Talmudists and Cabalists; and by frequent searching gain perfect knowledge of the microcosm, man.[68]

Here we may note the similarity between Rabelais and his younger Paracelsian contemporary Peter Severinus, who urged students of nature to discard their books and seek the truth of nature directly

---

Ramus is W.J. Ong's *Ramus: Method and the Decay of Dialogue* (Cambridge, Mass.: Harvard University Press, 1958) and R. Hooykaas, *Humanisme, science et reforme: P. de la Ramée, 1515-72* (Leiden: E.J. Brill, 1958). The latter work compares Ramus and Paracelsus.

[66]J. B. van Helmont, "Confessio Authoris" (Ch. 2, Sects. 1-5), *Ortus medicinae. Id est, initia physicae inaudita. Progressus medicinae novus, in morborum ultionem, ad vitam longam* (Amsterdam: Ludovicus Elzevier, 1648; reprint Brussels: Culture et Civilisation, 1966), p. 16; the English translation of this passage by John Chandler (from the *Oriatrike*, p. 12) is quoted at length below in Ch. 5.

[67]Graves, *Ramus*, p. 169.

[68]Rabelais, *Les oeuvres*, ed. Ch. Marty-Laveaux, 6 vols. (Paris: Lemerre, 1870-1903), Vol. I, p. 256.

through observation.[69] However, the ultimate source for both authors is the Hermetic text the *Asclepius*. Here the true philosopher was told to confirm his belief in the Creator through a study of his Creation. The process was indeed seen as a form of divine worship:

> ... the student of philosophy undefiled, which is dependent on devotion to God, and on that alone, ought to direct his attention to the other sciences only so far as he may thereby learn to see and marvel how the returns of the heavenly bodies to their former places, their halts in pre-ordained positions, and the variations of their movements, are true to the reckonings of number; only so far as, learning the measurements of the earth, the depth of the sea, (the ... of air), the force of fire, and the properties, magnitudes, workings, and natures of all material things, he may be led to revere, adore, and praise God's skill and wisdom.[70]

It was this Christianized search for God's truth in nature that was to appeal to Renaissance nature philosophers. To be sure, this theme was already present in the earlier alchemical texts, but the work of Marsilio Ficino (1433-1499) and the Platonic Academy in Florence gave a new impetus to the study of the mystical texts of late antiquity.[71] Ficino did not concern himself with the Hermetic and neo-Platonic fragments that might be isolated from medieval texts; rather, he saw his task to be the preparation of new translations and commentaries. The discovery of a nearly complete copy

---

[69] Petrus Severinus, *Idea medicinae philosophicae* (3rd ed., The Hague: Adrian Clacq, 1660; 1st ed., 1571), p. 39. A translation of this passage by the present author is given below in Ch. 2.

[70] Hermes Trismegistus, *Asclepius I* in *Hermetica*, ed. and trans. Walter Scott (Vol. IV with A.S. Ferguson), 4 vols. (Oxford: Clarendon Press, 1924-1936), Vol. I, p. 311.

[71] See P.O. Kristeller, *The Philosophy of Marsilio Ficino*, trans. Virginia Conant (New York: Columbia University Press, 1943); R. Klibansky, *The Continuity of the Platonic Tradition during the Middle Ages: Outlines of a Corpus Platonicum Medii Aevi* (London: Warburg Institute, 1939); D.P. Walker, *Spiritual and Demonic Magic from Ficino to Campanella* (London: Warburg Institute, 1958); Frances A. Yates, *Giordano Bruno and the Hermetic Tradition* (Chicago: University of Chicago Press, 1964); Walter Pagel, *Das Medizinische Weltbild des Paracelsus: seine Zusammenhänge Neuplatonismus und Gnosis* (Wiesbaden: Franz Steiner, 1962); Charles B. Schmitt, *Gianfrancesco Pico della Mirandola (1469-1533) and His Critique of Aristotle* (The Hague: Martinus Nijhoff, 1967).

of the *Corpus Hermeticum* in 1460 offered a unique opportunity. So important did the recovery of this manuscript seem to Ficino's patron, Cosimo d'Medici, that he requested an immediate translation—before the already planned translations of Plato's *Republic* and the *Symposium*.[72] The completed work (1463) was to prove highly influential in the Renaissance revival of natural magic, astrology, and alchemy.

Ficino fully concurred in the widespread belief that Hermes Trismegistus was an author of great antiquity. In the Augustinian tradition he pictured Hermes as a contemporary of Moses and the founder of a line of sages that could be traced directly to Plato.[73] Here then might be sought a Gentile tradition comparable to the wisdom of the Hebrew prophets, a body of work that should properly be studied with the same diligence as the Old Testament. The Hermetic texts do indeed reflect important themes found in the Holy Scriptures. The *Pymander* opened with a dream sequence, and the substance of the work dealt with the Creation epic. Here, as in an alchemical text, the question of the elements was introduced and from there the author proceeded to the creation of other substances, of minerals, vegetables, and animals.[74] Both the Creation and the Fall of man were discussed, and for Ficino the similarity with Genesis seemed so striking that he could not resist referring to Hermes as the Egyptian Moses. It is of the greatest importance to note that the "Hermetic Adam" had within him a divine creative power—which was never entirely lost.[75]

Ficino's views were further developed in the *Platonica theologia* (1482) and the *De triplici vita* (1489), works in which he combined neo-Platonic, Gnostic, Cabalistic, and Hermetic elements with more traditional (Aristotelian) views of form and matter.[76] Throughout these works one meets with the doctrine of the two worlds—the concept that man is in all respects an accurate microcosm of the great world about him. Ficino believed that these two worlds were united through influences which descend from the higher world to man by means of the stars. Yet, he was no ad-

[72]Yates, *Bruno*, pp. 12-13.
[73]*Ibid*., pp. 14-15.
[74]Hermes, *Poimandres* in *Hermetica*, ed. Scott, Vol. I, pp. 115-133.
[75]Yates, *Bruno*, pp. 26, 27-28.
[76]Walker, *Spiritual and Demonic Magic*, pp. 3-59; Pagel, *Paracelsus*, pp. 221-226.

vocate of traditional astrology; rather, he argued that the divine nature of man permits him to overcome these astral emanations. All action in nature is based upon sympathy and antipathy, and man has the ability to seek out the hidden virtues of things which he may then use to overcome evil emanations or strengthen beneficial ones.

As a physician himself, Ficino had a special interest in medicine. In the Holy Scriptures we are told to honor the physician because God has created him to serve man's needs. Thus, as the physician heals man, he acts simultaneously as a priest in honoring God. Yet, this is true—and natural—magic. The magician is told to seek out all Godly knowledge so that he might better understand his Creator. Thus, in the Hermetic corpus it is clearly indicated that man is able to learn all things, but that to accomplish this he must strive to make himself the equal of his Creator:

> If then you do not make yourself equal to God, you cannot apprehend God; for like is known by like . . . . Think that for you too nothing is impossible; deem that you too are immortal, and that you are able to grasp all things in your thought, to know every craft and every science, find your home in the haunts of every living creature; make yourself higher than all heights, and lower than all depths; bring together in yourself all opposites of quality, heat and cold, dryness and fluidity . . . grasp in your thought all this at once, all times and places, all substances and qualities and magnitudes together; then you can apprehend God.[77]

Natural magic was interpreted as a legitimate investigation of nature, one that was closely allied with religion. And although the importance of magical texts had been noted by Roger Bacon in the thirteenth century,[78] the subject became again one of intense interest in the late fifteenth and sixteenth centuries. In the most

---

[77] Hermes, *A Discourse of Mind to Hermes* in *Hermetica*, ed. Scott, Vol. I, p. 221.

[78] Bacon did condemn the books of the magicians as containing dangerous knowledge in many places, but he also insisted that they must be studied "because many books are considered magical which are not such, but contain the worthiness of wisdom; which therefore are esteemed, and those which are not, will not give us the experience of any knowledge. For if anyone comes into the work of nature or of art in any of these he takes it; if not, he leaves it as suspected . . ." (*Epistola* in *Opera*, p. 526).

detailed Elizabethan survey of witchcraft (1584), Reginald Scot (c. 1538-1599) asserted that "naturall magicke is nothing else, but the worke of nature."[79] Henry Cornelius Agrippa (c. 1486-1535), the author of the frequently cited *De occulta philosophia* (1533) and the *De incertitudine et vanitate scientiarum* (1530), offered the following definition:

> Magic is a faculty of wonderful virtue, full of most high mysteries, containing the most profound contemplation of most secret things, together with the nature, power, quality, substance and virtues thereof, as also the knowledge of whole Nature.... This is the most perfect and chief Science, that sacred and sublimer kind of Philosophy, and lastly the most absolute perfection of all most excellent Philosophy.[80]

Yet, this sublime philosophy was nothing else but natural philosophy, as John Baptista Porta (c. 1535-1615) had noted in his *Magia naturalis* (1558), a work that went through many editions for over a century.

> ... I think that Magick is nothing else but the survey of the whole course of nature. For whilst we consider the Heavens, the Stars, the Elements, how they are moved, and how they are changed, by this means we find out the hidden secrecies of living creatures, of plants, of metals, and of their generation and corruption; so that this whole Science seems merely to depend upon the view of Nature....[81]

Again, we see here the close connection of the macrocosmic and the microcosmic worlds. Natural magic, with roots deep in the newly recovered Hermetic writings, as well as in the mystical texts of late antiquity which achieved great popularity in the Renaissance, was also closely related with Christianity. It was thought to be an essential branch of knowledge; indeed, it was the legitimate and proper way for man to learn of the universe about him.

[79]Reginald Scot, *The Discoverie of Witchcraft* (1584), introduction by the Rev. Montague Summers (1930; New York; Dover, 1972), p. 165.

[80]Henry Cornelius Agrippa, *Three Books of Occult Philosophy or Magic. Book One—Natural Magic* ..., (1533), ed. Willis F. Whitehead (1897; London: Aquarian Press, 1971), pp. 34-35. On Agrippa see Charles G. Nauert, Jr., *Agrippa and the Crisis of Renaissance Thought* (Urbana: University of Illinois Press, 1965).

[81]John Baptista Porta, *Natural Magick* (London: Thomas Young and Samuel Speed, 1658), p. 2.

Magic, Mathematics, and Nature

The Renaissance interest in natural magic raised an unavoidable theological conflict. Such magicians rarely denied the existence of evil or demonic magic, but they argued that all forces in the universe must be natural in their operation, since they all derive from the omnipotent Creator. It was only the goal of the magician that might be questioned, and they went on to affirm that their task was not to prepare spells, charms, and incantations but to explore the wonders of God's created universe so that they might better comprehend their Creator.

This revival in magic also led to a new interest in mathematics as a key to nature. But while it fostered an essentially modern application of mathematics to the interpretation of natural phenomena, it also pointed to a more lofty Platonic approach, one that was related to mystical theology and provided great appeal for Hermetic philosophers and neo-Platonists. Thus Pico della Mirandola insisted that it was only this true natural magic that interested him, but, he added, essential to this magic was the corresponding true mathematics which was the formal part of physics.[82] Here Cabalistic number mysticism, a popular study of Renaissance scholars, played as strong a role as the more formal Pythagorean tradition.[83] We may be certain, said Pico, that a knowledge of numbers provides a key to the proper understanding

---

[82]Pico della Mirandola, *Opera quae extant omnia*, 2 vols. (Basel: Sebastianum Henricpetri, 1601), Vol. II, p. 114. The present section is based upon Allen G. Debus, "Mathematics and Nature in the Chemical Texts of the Renaissance," *Ambix*, **15** (1968), 1-28, 211 (3-21).

[83]Fifteenth- and sixteenth-century Christian scholars found a new stimulus for mystical numerical and geometrical studies in their fascination with the Hebrew Cabala and the Lullian system. Included in the Cabalistic tradition were Reuchlin, Agrippa of Nettesheym, and Georgius Venetus ("Zorzi") besides Pico. Because these studies were quickly applied to alchemical thought, they are of importance for an understanding of Paracelsian mathematics. Thus J.A. Pantheus discussed the art and theory of metallic transmutation in a Cabalistic system which assigned numbers to Latin, Greek, and Hebrew letters. In a discussion of the four elements he gave each letter a number according to its place in the alphabet, added the total, and multiplied by eight. See J. A. Pantheus, *Ars & theoria transmutationis metallicae cum voarachadumia, proportionibus, numeris, & iconibus rei accomodis illustrata* (1518) in *Theatrum chemicum*, Vol. II, pp 459-549, esp. p. 472. The Renaissance symbolic and Cabalistic thought is well treated in Desirée

of everything that may be learned by man.⁸⁴ Indeed, arising from mathematical books is a contemplative philosophy requisite for our further studies in theology, and since mathematics includes in its sphere the whole universe, earlier philosophers were correct in placing astronomy and astrology within its fold.⁸⁵

An equally interesting example is in the work of Nicholas Cusanus (1401-1464). For him both philosophical truth and Holy Writ led to the same conclusion: that a knowledge of mundane and divine truth may be had only through a knowledge of numbers. He quoted Wisdom 11:17 (Douay version 11:21) to show that God created "all things in number, weight and measure."⁸⁶ The Creation itself could then be interpreted as a mathematical process, since arithmetic, geometry, music, and astronomy were the "same sciences God employed when He made the world."⁸⁷ Such a

---

Hirst, *Hidden Riches* (London: Eyre & Spottiswoode, 1963), and of special significance is F. Secret, *L'astrologie et les kabbalistes chrétiens à la Renaissance* (La Tour St.-Jacques,1956).

Raymond Lull claimed that encyclopedic knowledge would be at one's fingertips by means of tables, movable circles, and geometrical patterns. Although Lull himself had little patience with alchemy, his disciples wrote numerous alchemical tracts which are filled with Lullian geometrical figures to aid the scholar in his understanding of divine secrets. The authorship of these tracts was—in typical alchemical fashion—assigned to Lull. An example of this is the widely known *Testamentum theoria & practica*, already referred to through van Helmont's lengthy quotation (see above, n. 66; the entire work is printed in the *Theatrum chemicum*, Vol. IV, pp. 1-170), where typical Lullian figures appear throughout. Giordano Bruno, an enthusiastic Lullist himself, found a strong Lullist strain in the work of Paracelsus (Pagel, *Paracelsus*, p. 245). Indispensable for this subject is Frances A. Yates, "The Art of Ramon Lull: An Approach to it through Lull's Theory of the Elements," *Journal of the Warburg Institute*, **17** (1954), 115-173.

⁸⁴Pico, as quoted by John Dee in his "Mathematicall Preface" to *The Elements of Geometrie of the most auncient Philosopher Euclide of Megara*, trans. H. Billingsley (London: John Daye, 1570), sig. *i^v.

⁸⁵Pico, *Opera*, Vol. II, pp. 285, 488.

⁸⁶Nicolas Cusanus, *The Idiot in Four Books. The first and second of Wisdome. The third of the Minde. The fourth of statick Experiments, or Experiments of the Ballance* (London: William Leake, 1650), p. 172. Nicolas Cusanus, *Of Learned Ignorance*, trans. Fr. Germain Heron (London: Routledge and Kegan Paul, 1954), p. 119.

⁸⁷Cusanus, *Of Learned Ignorance*, pp. 118-119. "With arithmetic He adjusted the World into unity, with geometry He gave it a balanced design upon which depends its stability and its power of controlled movement; with music He allotted its parts that there should be no more earth in the earth than water in the water,

process could surely be studied in a mathematical way, and thus we could truly learn of our Creator, since we would be using the method of Creation as a guide. Furthermore, if we properly study the creatures and objects of our world, we will discover an underlying unity. We do not know why, but "we know for a fact that all things stand in some sort of relation to one another, that, in virtue of this inter-relation, all the individuals constitute one universe, and that in the one Absolute the multiplicity of beings is unity itself."[88]

Observation and experience were thus sanctioned as an approach to truth and readily used by Cusanus in his book of static experiments which he included in *The Idiot*. Here in addition to Wisdom 11:17 he cited Proverbs 16:11 ("The weight and the Ballance are the judgements of that Lord, who hath created") and Proverbs 8:28 ("Who weighed the fountaines of waters, and the greatnesse of the Earth, in a Ballance, as the wise man saith").[89] Although such quantification could hardly compare with the search for the inner meaning of a divine mathematics, these scriptural references alone gave a divine sanction for such studies. Cusanus suggested that "although nothing in this world can reach precision, yet wee finde by experience, the judgement of the Ballance, one of the truest things amongst us. . . ."[90] With this justification he proceeded to propose the analysis of spring waters, blood, and urine by weight comparisons. The natural philosopher should also weigh different samples of air and similarly test the statements of the alchemists. In the most famous example of all, he suggested that the investigator plant seeds in a weighed earthen pot, and after they have grown, "hee woulde finde the earth but

---

than air in the air or than fire in the fire, so that no element could be wholly transmuted into another, whence it comes that the physical system cannot sink into chaos."

[88]*Ibid.*, p. 25. For a comparison of man with nature, Cusanus turned to Plato: "The earth, as Plato says, is like some vast animal whose bones are stones, whose veins are rivers and whose hairs are the trees; and the animals that feed among the hairs of the earth are as the vermin to be found in the hair of beasts" (p. 119).

[89]Cusanus, *The Idiot*, p. 172. The neo-Platonic and Paracelsian background to Harvey's quantification (including Cusanus and van Helmont) has been discussed by Pagel, *Harvey's Biological Ideas*, pp. 73-82. Here Pagel shows that for Harvey also, mathematics was basically a tool for observation.

[90]Cusanus, *The Idiot*, p. 171.

very little diminished, when he came to weigh it againe: by which he might gather, that all the aforesaid herbs had their weight from the water."[91]

For Cusanus such experiments in no way conflicted with his grander scheme: his search for the divine plan of the universe. Here the search should begin in the perfect world of ideas, where mathematics was a valid image of reality, a perfect example of abstract truth. Cusanus found support in Boethius, who had stated that a knowledge of things divine was impossible without some knowledge of mathematics.[92] But above all he looked to the Pythagoreans as the ultimate source for those who believed in mathematics and number as the true key to nature—a belief that Aristotle had discussed and rejected in the *Metaphysics*.[93]

"Cosmic" mathematics proved to be a persistent theme in Renaissance texts and one which was referred to by authors of widely differing views. For some mathematics was equated with astrology or astronomy while for others the word really meant the use of astrology in medicine.[94] Agrippa argued that natural magic was fundamentally a mathematical subject. A magician, he wrote, could do nothing without mathematics, since all things were made and continue to be ruled over by number, weight, measure, harmony, motion, and light. When such mathematical studies are mastered and put into practice with the mechanical arts the result will be the production of marvels and wonders.[95] In addition, Agrippa added, man himself might fruitfully be subjected to a numerical analysis. He is the most nearly perfect product of God's Creation, and in him will be found the same proportions and harmonies that exist in the macrocosm.[96] For Agrippa, these truths could properly be related to Cabalistic studies.

There are few authors, however, whose work is more il-

[91]*Ibid.*, pp. 188-189.

[92]Cusanus, *Of Learned Ignorance*, p. 26.

[93]Aristotle, *Metaphysics* 989b, trans. Richard Hope (New York: Columbia University Press, 1952), p. 25.

[94]On the history of this subject see Karl Sudhoff, *Iatromathematiker vor nehmlich im 15. und 16. Jahrhundert* (Breslau: J.U. Kern, 1902). During this period the term "iatromathematics" was interchangeable with "astrological medicine."

[95]Henricus Cornelius Agrippa, *De occulta philosophia. Libri tres* ([Cologne], 1533), pp. 99-101.

[96]*Ibid.*, p. 160.

PLATE I Although the chemical philosophers rejected the use of mathematical abstraction in the study of nature, they were heir to a long alchemical tradition of quantification by means of weight determinations. The earliest illustration of an enclosed analytical balance occurs in the *Theatrum chemicum Britannicum* . . ., collected and annotated by Elias Ashmole (First Part, London: J. Grismond for Nathaniel Brooke, 1652), p. 51.

lustrative of Renaissance mathematics than John Dee. Clearly an adherent of the natural magic tradition, Dee has a special significance in any study of the chemical philosophy because he was an author well known to alchemists.[97] Here the discussion will be confined to two works, the "Mathematicall Preface" to the first English translation of Euclid (1570) and the *Monas hieroglyphica* (1564). In the first of these, Dee proceeded to divide and subdivide the many disciplines he felt belonged within the realm of mathematics. He has frequently been praised for his description of *Archemastrie* which, he wrote, "is named of some *Scientia Experimentalis*. The *Experimental Science*."[98] Obviously aware of recent publications of significance, he noted for instance that Benedetti had exposed the error in the general belief that heavier bodies fall faster than lighter ones through his knowledge of the mathematical science of *Statike*.[99] Dee has been lauded also for his practical

[97]There has been considerable recent interest in the work of John Dee, to a large extent because of the key position Frances Yates has given him in her discussion of the development of English science. Here see particularly her *Theatre of the World* (London: Routledge & Kegan Paul, 1969), pp. 1-41 and *The Rosicrucian Enlightenment* (London/Boston: Routledge & Kegan Paul, 1972), pp. 31-40 and *passim*. The Yates position has been more fully stated in Peter J. French, *John Dee. The World of an Elizabethan Magus* (London: Routledge & Kegan Paul, 1972). The views of the present author—upon which the following pages are based—will be found in "Mathematics and Nature," pp. 10-12, and in his introduction to the reprint of the *Mathematicall Praeface* (New York: Science History Publications, 1975). A balanced statement of John Dee's approach to mathematics appears in Nicholas Clulee's doctoral dissertation, "The Glas of Creation: Renaissance Mathematicism and Natural Philosophy in the Work of John Dee" (University of Chicago, 1972). All students of Dee in recent years have a deep debt to I. R. F. Calder's University of London dissertation "John Dee Studied as an English Neo-Platonist" (1952).

[98]Dee, "Mathematicall Preface," sig. Aiii^r. Archemastrie "teacheth to bryng to actuall experience sensible, all worthy conclusions by all the Artes Mathematicall purposed, & by true Naturall Philosophie concluded: & both added to them a farder scope, in the termes of the same Artes, & also by hys propre Method, and in peculiar termes, procedeth, with helpe of the foresayd Artes, to the performance of complet Experiêces, which of no particular Art are hable (Formally) to be challenged: . . . And bycause it procedeth by *Experiences*: and searcheth forth the causes of Conclusions, by *Experiences*: and also putteth the Conclusions them selves, in *Experience*, it is named of some *Scientia Experimentalis*. The *Experimentall Science*."

[99]*Ibid.*, sig. ci^r. "By these verities, great Errors may be reformed, in Opinion of

writings on astronomy, navigation, and mathematics, which give an early indication of the significance and interrelation of technology and science.

If, however, we read his works in greater detail, we find a repetition of the now familiar Platonic arguments. Mathematics is a divine science because the Creation was essentially a mathematical process. The Creator's "*Numbryng* ... was his Creatyng of all thinges. And his continuall *Numbryng*, of all thinges, is the Conseruation of them in being." We find therefore that "the constant law of nūbers ... is planted in thyngs Naturall and Supernaturall: and is prescribed to all Creatures, inviolably to be kept."[100] The mathematical sciences then rightly become a means to learn of our Creator: arithmetic is, indeed, a subject little less useful than theology,[101] while the primary use of geometry, as Plato teaches us, is

> ... to conceiue, discourse, and conclude of things, Intellectuall, Spirituall, aeternall, and such as concerne our Blisse euerlasting: which otherwise (without Speciall priuiledge of Illumination, or Reuelation frõ heauen) No mortall mans wyt (naturally) is hable to reach unto, or to Compasse.[102]

Christian scholars should study numbers so that

> ... we may both winde and draw our selues into the inward and deepe search and vew, of all creatures distinct vertues, natures, properties, and *Formes*: And also, farder, arise, clime, ascend, and mount up (with Speculative winges) in spirit, to behold in the Glas of Creation, the *Forme* of *Formes*, the *Exemplar Number* of all things

---

the Naturall Motion of thinges, Light and Heauy. Which errors, are in Naturall Philosophie (almost) of all mē allowed: to much trusting to Authority: and false Suppositions. As Of any two bodyes, the heauyer, to moue downward faster than the lighter. This error, is not first by me, Noted, but by one *Iohn Baptist de Benedictus* [Benedetti]. The chief of his propositions, is this: which seemeth a Paradox.

If there be two bodyes of one forme, and of one kynde, aequall in quantitie or unaequall, they will moue by aequall space, in aequall tyme: So that both theyr mouynges be in ayre, or both in water: or in any one Middle."

[100] *Ibid.*, sig. *i^v.
[101] *Ibid.*, sig. ai^v.
[102] *Ibid.*, sig. aiii^r.

Numerable: both visible and invisible: mortall and immortall, Corporall and Spirituall.[103]

Surely with this admitted, "No man . . . can doute, but towards the atteyning of knowledge incomparable, and Heauenly Wisdome: Mathematicall Speculations, both of Numbers and Magnitudes: are means, aydes, and guides: ready, certain, and necessary."[104]

Dee was a sincere alchemist and astrologer, and with this broader understanding of the primary purpose of mathematics his views in the "Mathematicall Preface" can be seen to fit in with his attitude toward other sciences. His description of astronomy related this science to its major goal: while astronomy teaches us the distances, magnitudes, and motions of the stars and planets, more important, we learn thereby of our Creator so that we may better glorify him. As a science it could well be used "for Consideratiõ of Sacred Prophesies, accomplished in due time, foretold: as for high Mysticall Solemnites holding."[105]

We see the familiar macrocosm-microcosm relationship develop in other mathematical sciences discussed by Dee. Astrology was such a science, for it treats quite simply of the "operations and effectes, of the naturall beames, of light, and secret influence: of the Sterres and Planets: in euery element and elementall body."[106] It was thus concerned with the actual connection of the celestial and terrestrial worlds. Anthropography in turn "is the description of the Number, Measure, Waight, figure, Situation, and colour of euery diuerse thing, conteyned in the perfect body of MAN,"[107] a microcosm which deserved a mathematical description no less than the greater world. Natural magic, called *Thaumaturgike* by Dee, was devoted to the construction of "straunge workes, of the sense to be percieued, and of men greatly to be wondred at." Dee's views here are not dissimilar to those of Roger Bacon, Agrippa, and Porta. Natural magic was truly a mathematical science pursued by men who "seketh . . . in the Creatures Properties, and wonderfull vertues, to finde juste cause,

[103]*Ibid.*, sig. *i^{r,v}.
[104]*Ibid.*, sig. aiii^r.
[105]*Ibid.*, sig. bii^r, bii^v.
[106]*Ibid.*, sig. biii^r.
[107]*Ibid.*, sig. ciiii^r.

to glorifie the Aeternall, and Almightie Creator...."[108] It was wicked and blasphemous for the ignorant to call these pious men conjurers (a term he had been called himself).

Of special interest is Dee's *Monas hieroglyphica* (1564). Here Dee sought a new approach to alchemy (*astronomia inferior*) through mathematics. Although the result remains largely unintelligible to Dee's growing band of twentieth-century commentators,[109] it is clear that the work was widely known and quoted by alchemists in the century after it was first printed. Among those who found special power in Dee's symbolism was Robert Fludd, who advanced the hieroglyphic monad against Kepler in their conflict over the correct use of mathematics in the study of natural phenomena.[110]

John Dee's hieroglyphic monad from the *Monas hieroglyphica* (Antwerp: William Sylvius, 1564), fol. 25^r.

Arguing that "Pythagorean" figures have an underlying clarity and strength "almost mathematical," Dee proceeded to construct step by step through a series of "theorems" a figure based on

[108] *Ibid.*, sig. Ai^r, Aii^r.
[109] C.H. Josten, "A Translation of John Dee's 'Monas Hieroglyphica' (Antwerp, 1564), with an Introduction and Annotations," *Ambix*, **12** (1964), 84-220. Even Josten (p. 111) admits to "being at a loss" to provide any elucidation of Dee's deliberately veiled mysteries. The first edition of the *Monas hieroglyphica* appeared at Antwerp in 1564, a second was printed at Frankfurt in 1591, and the text was later to be included in Zetzner's *Theatrum chemicum*. Number mysticism in early alchemical texts has been discussed by H. E. Stapleton, "The Antiquity of Alchemy," *Ambix*, **5** (1953), 1-43.
[110] See the discussion of this debate below in Ch. 4.

points, straight lines, and circles. The finished product closely resembled the alchemical figure for mercury, and Dee openly acknowledged that this was truly "the rebuilder and restorer of all astronomy."[111] In the process of construction the scholar had himself produced the sun, the moon, and the earth;[112] he had symbolically retraced the first stages of Creation. From here the mysteries of the elements were developed. Throughout, Dee digressed to offer numerical and Cabalistic analyses, and the latter part of the slim volume appears to be a veiled representation of the alchemical process itself.[113] Proper understanding of the monad promised to reveal the true mysteries of the physical world, since the symbol was alleged to include the hidden secrets of both celestial and terrestrial astronomy.[114] However, Dee's object was also closely allied to that of Renaissance natural magic in a broader sense. The mysteries of the monad would enable the artificer to produce wonderful effects in engineering, music, optics, the manipulation of weights, and hydraulics. It would also unite the different parts of the Cabala and make easily available all art and medicine.[115]

The *Monas hieroglyphica* indicates how Renaissance Hermeticism and natural magic might be applied to alchemy. Dee's emphasis was clearly spiritual and closely in tune with the mystical mathematics favored by those who sought a key to Creation in the Pythagorean tradition. Here the lofty truths of magic were avowed while more conventional mathematical proofs, chemical laboratory techniques, and practical medical applications held relatively little interest.

Paracelsus: The Man

If John Dee exemplifies one way in which Hermetic, mystical mathematical, and alchemical strains were blended in Renaissance natural philosophy, Paracelsus may be chosen as a very

[111] Dee, *Monas hieroglyphica*, trans. Josten, pp. 119, 121, 123.
[112] *Ibid.*, p. 125.
[113] See Josten's introduction to the translation, pp. 84-111.
[114] *Ibid.*, p. 175.
[115] *Ibid.*, pp. 131, 137.

different example. To be sure, both men were well versed in the alchemical literature and both knew the more recent Renaissance Hermetic, neo-Platonic, and magical literature. In addition, both claimed that their own interpretation of alchemy or chemistry would lead to a new understanding of nature. Paracelsus, however, linked his plea to an unyielding rejection of the ancients and to a demand that the new science be firmly united with medicine. The proper key to both was to be chemistry. His work and that of his followers triggered a deep-rooted conflict in the scholarly world of the sixteenth and seventeenth centuries, a conflict that was to make permanent the union of chemistry and medicine. The work of John Dee, in contrast, remained known primarily to alchemists and was largely forgotten by the end of the seventeenth century.

The life of Paracelsus—more properly, Philippus Aureolus Theophrastus Bombastus von Hohenheim (1493-1541)—reveals a man of mercurial disposition who was never permanently settled.[116] In a sense his life is reflected in his works. These touched on many subjects and they seemed difficult and contradictory to his contemporaries, even to those who were deeply influenced by him. Above all, these works reveal a man who was steeped in all aspects of Renaissance philosophical thought. For us it is particularly significant that he was impressed at an early age with the importance of both chemistry and medicine.

Paracelsus was born in the small town of Einsiedeln, not far from Zurich. Here his father served officially as a physician, and whenever he could find time, he also busied himself with alchemy. This joint interest of his father in chemistry and medicine was to be shared by Paracelsus throughout his life. As luck would have it, both subjects were also fostered outside his home while he was still very young. It has been traditionally accepted that several bishops helped to train Paracelsus, among them the famed alchemist Johannes Trithemius. At the age of nine, the boy moved to the town of Villach in Austria with his father. Here he served as an apprentice in the Fugger mines and undoubtedly became familiar with the mining and metallurgical practices that were about to be

---

[116]The present account is a shortened and a slightly altered version of the brief life originally presented in *The English Paracelsians*, pp. 14-18, which in turn was based on a number of different sources, primarily on the discussion presented by Pagel in his *Paracelsus*, pp. 5-35.

described for the first time in print.[117] In addition, he was able to observe the characteristic illnesses of the workers who sought help from his father. This experience was later to bear fruit in his speculations on the growth of metals and in his book on occupational diseases, the first book ever written on the diseases common to miners.

At the age of fourteen he left home as a wandering journeyman scholar and apparently visited a number of continental universities. Although he may have taken an M.D. at Ferrara, there is no solid proof that he did so.[118] He surely was employed in the following years as a surgeon in a mercenary army, a position which would hardly have been considered desirable or prestigious by a Doctor of Medicine, but it did make it possible for Paracelsus to visit many areas of Europe. Attempts have been made to trace these travels in the period after 1510.[119] He himself refers to Holland, Scandinavia, Prussia, Tartary, and the countries under Venetian influence; and he states that he was on the last boat out of Rhodes before that island fell to the Turks. Still, these references have not been verified, and the period is difficult to reconstruct.

It is only in the last fifteen years of his life that the records become clearer. In the third decade of the century we see him in Central Europe, occasionally making an attempt to settle, but always moving on after a short time. A difficult and blunt-spoken man, he would brook no opposition from authorities in any sphere. The results were uniformly disastrous. Perhaps this is best illustrated in his stay at Basel in 1527. Called there from Strasbourg

[117] An excellent discussion of the early-sixteenth-century metallurgical and mining literature will be found in Agricola, *De re metallica*, trans. H.C. and L.H. Hoover, App. B, pp. 609-615.

[118] Pagel, *Paracelsus*, p. 10. Evidence in support of Paracelsus' statement that he had been made a *Doctor utriusque medicinae* at Ferrara is discussed by W. Pagel and P. Rattansi in "Vesalius and Paracelsus," *Medical History*, **8** (1964), 309-328 (317). Here the acceptance of Paracelsus' affirmation to this affect by a magistrate at Basel and the address to Paracelsus as *beider arznei doctori* in the introductory letter to the *Grosse wundarznei* by Dr. Wolfgang Thalhauser, municipal physician at Augsburg, are cited as being of special importance.

[119] Pagel, *Paracelsus*, pp. 13-14. For detailed accounts of the Paracelsian travels see Basilio de Telepnef, *Paracelsus: A Genius Amidst a Troubled World* (St. Gall: Zollikofer, 1945) and Otto Zekert, *Die grosse Wanderung des Paracelsus. De Peregrinatione Paracelsi Magna. Von Einsiedeln nach Salzburg* (Ingelheim am Rhein: C.H. Boehringer Sohn, 1965). These accounts have to be used with caution.

by the famed humanist publisher Frobenius, his task was to treat a leg ailment that the local physicians had been unable to improve. Never did it seem more likely that Paracelsus might prosper. In the home of Frobenius he was placed in contact with one of the most enlightened groups of scholars in Europe, including the humanists Oecolampadius and Erasmus. He became the personal physician to the latter while the former was influential in having him appointed city physician and professor of medicine. Although it was a city appointment, it carried with it the right to lecture at the famed university.

It was at the University of Basel that he immediately ran into trouble. Announcing that he would ignore the ancient medical authors, Paracelsus proceeded to lecture on his own medical experience—and this he did in common Swiss-German dialect rather than the customary Latin. To further show his contempt for medical authority he cast the revered medical *Canon* of Avicenna on the St. John's Day bonfire. These actions hardly inspired the love of the regular medical faculty, nor did they result in a following among the students—often a more conservative group than their teachers. He was attacked as "Cacophrastus" in a poem supposedly written by Galen and sent from Hell. True, "Galen" admitted,

> ... I know not thy mad alchemical vapourings,
> I know not what *Ares* may be, nor what *Yliadus*,
> Know not thy tinctures, thy liquors divine of *Taphneus*
> Nor *Archeus*, thy spirit preserver of everything living in all things.

But these were not worth knowing, and this "Galen" judged Paracelsus as scarcely worthy of tending his swine.[120] Not long after this his protection vanished with the sudden death of Frobenius (October 1527). The final blow came with his involvement in a lawsuit with a church dignitary who refused to pay his fee. By now his position was untenable, and he fled Basel in haste, even leaving behind his manuscripts.

The final years reveal a man busily writing, but ever moving. His work on syphilis was composed at Nuremberg. Here he attacked the most commonly accepted remedies—liquid mercury

---

[120] The entire poem is quoted in A.M. Stoddart, *The Life of Paracelsus: Theophrastus von Hohenheim 1493-1541* (London: William Rider, 1915), p. 133.

PLATE II Paracelsus at the age of forty-seven (by August Hirschvogel). From Dr. Fritz Jaeger et al., *Theophrastus Paracelsus: 1493-1541* (Salzburg: M. Mora, 1941), p. 41.

and guaiac wood—and suggested that a chemically altered form of mercury be substituted with careful attention paid to dosage. Laudable though these views were, they constituted a tactical error, for the influential Fugger interests held the guaiac monopoly. Although Paracelsus saw through the press several short tracts on this subject, he was prohibited from publishing his *Eight Books on the French Disease* by a decree based on the opinion of the Leipzig Medical Faculty.

At Beratzhausen he was to write the *Paragranum*, in which he described the bases of medicine as philosophy, astronomy, alchemy, and ethics. Other works were written rapidly in other cities, and the sum of the scientific, medical, and philosophical texts comes to an imposing fourteen volumes in the Sudhoff edition. (Other theological manuscripts are still in the process of being published.) Paracelsus' end finally came at the early age of forty-eight (September 1541) in Salzburg, where he had been called at the request of the suffragan bishop Ernest of Wittelsbach.

The Paracelsian System

The Paracelsian chemical philosophy as developed in the decades after his death will be the subject of the following chapter. In this section we will do no more than point to some fundamental themes in his work that were generally adopted by his disciples. His rejection of authority, his concept of a unified nature, and his emphasis on chemistry as a key to nature and to medicine all form essential ingredients for the conflict that was to occur.

The attitude of Paracelsus toward the ancients was uncompromising. He argued that Aristotle, Galen, and later Avicenna had been ignorant of true philosophy and their followers cared little for the truth of nature or the lessons that might be learned through observation. A specific point was chemistry. While Gesner simply added chemical knowledge to the ancient medical corpus, Paracelsus pointed to the ancients' ignorance of chemistry as a reason for discarding their work.[121] His attitude may be judged in the contemptuous address he penned to the physicians of his day:

[121] See Pagel's discussion of Paracelsus' censure of Aristotle and Avicenna in *Paracelsus*, pp. 58-59.

> I am Theophrastus, and greater than those to whom you liken me; I am Theophrastus, and in addition I am *monarcha medicorum* and I can prove to you what you cannot prove. I will let Luther defend his cause and I will defend my cause, and I will defeat those of my colleagues who turn against me; and this I shall do with the help of the *arcana* .... It was not the constellations that made me a physician; God made me.... I need not don a coat of mail or a buckler against you, for you are not learned or experienced enough to refute even one word of mine. I wish I could protect my bald head against the flies as effectively as I can defend my monarchy ... I will not defend my monarchy with empty talk, but with *arcana*. And I do not take my medicines from the apothecaries; their shops are but foul sculleries, from which comes nothing but foul broths. As for you, you defend your kingdom with belly-crawling and flattery. How long do you think this will last? ... Let me tell you this: every little hair on my neck knows more than you and all your scribes, and my shoe-buckles are more learned than your Galen and Avicenna, and my beard has more experience than all your high colleges.[122]

Outbursts of this sort were to result in our adjective "bombastic," but it should also be kept in mind that such attacks were common for the period—and here the sincerity of Paracelsus cannot be questioned. It is evident from the quotation that he was aware of the reference to the divine nature of the physician in Ecclesiasticus. Furthermore, he was fully convinced that it was an abomination to base one's study of medicine upon heathen authors condemned by the Church.

But what was to be substituted? The Bible surely, but also one must base his understanding on the essential unity of nature as seen in the macrocosm-microcosm analogy. As Pagel has commented,

> The distinguishing feature of Paracelsus' own philosophy is the consequential view of cosmology, theology, natural philosophy and medicine in the light of analogies and correspondences between

---

[122] This famous passage from the *Paragranum* has been translated often. Rather than offer a new version, I have reproduced the translation of Norbert Guterman in Paracelsus, *Selected Writings*, ed. Jacobi, pp. 79-80. For the original text see Paracelsus, *Sämtliche Werke*, ed. Karl Sudhoff and Wilhelm Matthiessen, 15 vols. (Munich/Berlin: R. Oldenbourg [Vols. VI-IX: O.W. Barth], 1922-1933), Vol. VIII, pp. 63-65.

macrocosm and microcosm. Speculation about such analogies had seriously engaged the human mind since pre-Socratic and Platonic times and throughout the Middle Ages. Paracelsus was the first to apply such speculation to the knowledge of Nature systematically.[123]

The doctrine of the microcosm was outlined succinctly by Paracelsus in his *Volumen medicinae paramirum* (c. 1520), where the natural philosopher was told to learn everything about the firmament, the earth, and the things it contains.

> All this you should know exists in man and realize that the firmament is within man, the firmament with its great movements of bodily planets and stars which result in exaltations, conjunctions, oppositions and the like, as you call these phenomena as you understand them. Everything which astronomical theory has searched deeply and gravely by aspects, astronomical tables and so forth,—this self-same knowledge should be a lesson and teaching to you concerning the bodily firmament. For, none among you who is devoid of astronomical knowledge may be filled with medical knowledge. . . . You are aware that the earth exists solely for the purpose of bearing fruit and for the sake of man. With the same logic the body also exists solely for the same reason. Thus from within the body grows all the food which is to be used by the members that belong to the body. These grow like the fruit of the earth.[124]

The interrelation of the two worlds was made possible by astral emanations, and in this connection a special emphasis was placed on the atmosphere, recognized as necessary both for fire and life by Paracelsus.[125] The seventeenth-century search for life-giving emanations was to result in the concept of an aerial saltpeter, directly traced back to this Paracelsian concept. On the other hand the astral emanations were offered by Paracelsus as the means by which the Creator had impressed earthly things with "signatures," which man might discover to his own benefit. Here in effect was a plea for a new investigation of nature. For Paracelsus the universe was a living unit with occult or magical forces everywhere at

---

[123] Pagel, *Paracelsus*, p. 50.

[124] Paracelsus, *Volumen medicinae paramirum*, trans. and Preface by Kurt F. Leidecker, *Supplement to Bulletin of the History of Medicine*, No. 11 (Baltimore: Johns Hopkins Press, 1949), p. 36.

[125] See below, Ch. 2.

work.[126] Man himself was part of the chain of life, and his task was not to read the books of the ancients but to seek out the hidden gifts implanted in the earth by the Lord. "A Physitian ought not to rest only in that bare knowledge which their Schools teach, but to learn of old Women, Egyptians, and such-like persons; for they have greater experience in such things than all the Academians."[127] And yet, this knowledge was not available to everyone; it was acquired only through Divine Grace, either by some direct mystical experience or through personal observation in nature. As Pagel had summarized this vision:

> . . . by means of unprejudiced experiment inspired by divine revelation, the adept may attain his end. Thus, knowledge is a divine favour, science and research divine service, the connecting link with divinity. Grace from above meets human aspiration from below. Natural research is the search for God.[128]

For Paracelsus the very basis of his philosophy rested upon the pillars of prayer, faith, and imagination.[129] And if the existence of black magic was never denied, it is clear that a deep gulf separated it from natural magic, which was nothing else but natural philosophy, the proper way for man to learn of the universe about him.[130]

The Paracelsian rejection of the ancients carried with it a rejec-

---

[126] Pagel, *Paracelsus*, p. 223. Alexandre Koyré, *Mystiques, spirituels, alchimistes: Schwenckfeld, Seb. Franck, Weigel, Paracelse* (Paris: A. Colin, 1955), p. 50 (n. 2). Giorgio de Santillana, *The Age of Adventure: The Renaissance Philosophers* (New York: Mentor, 1956), p. 14. The Pagel citation is of special interest for the relationship of magic to medicine.

[127] Paracelsus, *Of the Supreme Mysteries of Nature*, trans. R. Turner (London, 1655), p. 88; *Sämtliche Werke*, Vol. XIV, p. 541.

[128] Walter Pagel, "Religious Motives in the Medical Biology of the XVIIth Century," *Bulletin of the Institute of the History of Medicine*, 3 (1935), 98 f.

[129] Paracelsus, *Of the Supreme Mysteries of Nature*, pp. 29-31. This work includes "Of the Secrets of Alchymy: Discovered, in the Nature of the Planets," pp. 1-28; "Of Occult Philosophy," pp. 29-90; and "Of the Mysteries of the Signes of the Zodiack," pp. 91-158. Although possibly written by Gerhard Dorn, the English translation was ascribed to Paracelsus, and the *De occulta philosophia* was included in the Huser *Opera*. On the authenticity of this work see Paracelsus, *Sämtliche Werke*, Vol. XIV, p. 516.

[130] Paracelsus, *Of the Supreme Mysteries of Nature*, pp. 39, 81-82; *Sämtliche Werke*, Vol. XIV, pp. 516, 538.

tion of their logic. For Paracelsus logical methods were of no use in learning: they were confined to the explanation of given statements and thus added nothing new. This attitude was to affect deeply his approach to the most "logically oriented" branch of knowledge, mathematics. In common with many of his contemporaries he relied on Wisdom 11:17 when he insisted that "disease stands on weight, number and measure,"[131] but for the most part, mathematics was of value to him when interpreted as a form of natural magic. The Paracelsian "sidereal mathematician" was a student of the universe,"[132] only a lesser scholar would confine himself to the study of numbers and geometrical theorems. But if traditional mathematical reflection aided little in the search for the secrets of nature, just what would reveal the hidden signatures? The answer was found in chemistry. Primary observations in nature would lead directly to further observations in the chemical laboratory or to analogies based upon them. In the *Paragranum* Paracelsus stated that one of the four fundamental pillars of medicine was alchemy. Here he clearly indicated that through chemical means man might understand the elements, and beyond them, the cosmos as a whole.[133]

This theme was most fully developed in the *Philosophia ad Athenienses* (first published in 1564), where the Creation was discussed in alchemical terms. Here all things were said to have come from the primal matter or *mysterium magnum*, the great generating substance from which all other "mysteries" proceed.[134] But God

[131] Pagel, *Paracelsus*, pp. 281-282.
[132] Allen G. Debus, "Mathematics and Nature," pp. 13-14. This topic is returned to in greater depth in Ch. 2.
[133] Paracelsus, *Sämtliche Werke*, Vol. VIII, pp. 55 f.
[134] Paracelsus, *Philosophia ad Athenienses* in *Opera* (1616), Vol. II, p. 2. See also Kurt Goldammer, "Die Paracelsische Kosmologie und Materietheorie in ihrer wissenschaftsgeschichtlichen Stellung und Eigenart," *Medizin historisches Journal*, 6 (1971), 5-35; "Bemerkungen zur Struktur des Kosmos und der Materie bei Paracelsus" in *Medizingeschichte in unserer Zeit. Festgabe für Edith Heischkel und Walter Artelt zum 65. Geburtstag*, ed. Hans-Heinz Eulner et al. (Stuttgart: Ferdinand Enke, 1971), pp. 121-144. And of special significance here are Walter Pagel's studies, "The Prime Matter of Paracelsus," *Ambix*, 9 (1961), 117-135; "The Eightness of Adam and Related 'Gnostic' Ideas in the Paracelsian Corpus" (with Marianne Winder), *Ambix*, 16 (1969), 119-139; "The Higher Elements and Prime Matter in Renaissance Naturalism and the Paracelsian Corpus," *Ambix*, 21 (1974), 93-127.

was likened at the same time to a sculptor who carves a block of stone, or to a chemist who separates one thing from another:

> The principle . . . of all generation was Separation. . . . If vinegar be mixed with warm milk, there begins a separation of the heterogeneous matters in many ways. The truphat of the minerals brings each metal to its own nature. So it was in the Mystery. Like macerated tincture of Silver, so the Great Mystery, by penetrating, reduced every single thing to its own special essence. With wonderful skill it divided and separated everything, so that each substance was assigned to its due form.[135]

The Creation itself was pictured as a cyclical process. First the elements were formed, then the firmament was separated from the fire. Further separations resulted in spirits and dreams (from the air); water plants, salts, and marine animals (from water); wood, stone, land plants, and animals (from earth). Other substances were then separated from those already created and the process continued until the original primal matter was once more obtained. This concept of Creation as an essentially chemical process of separation was to be of special interest to the followers of Paracelsus.

The emphasis of Paracelsus on the Creation made it imperative for him to consider the elements. Indeed, a never-ending publication of polemical literature on the elements was to characterize the Paracelsian debates for more than a century after his death—and as a result the subject is one to which we shall frequently return. The seemingly contradictory texts of Paracelsus himself provided fuel for the fires of debate. He had both accepted and used the four Aristotelian elements, even though he identified fire with heaven and thus gave it a separate position outside the ken of the traditional elements. He further altered the definition of the elements by rejecting the traditional conjunction of qualities: rather than insisting that an element is a combination of paired qualities (such as "moist" and "hot"), Paracelsus asserted that there was only one quality—the singleness making the word *element* more applicable. Again, he questioned the commonly held view that the elements were material substances. Instead he argued that the air and water we see were only crude approximations of the true

---

[135] Paracelsus, *Philosophia ad Athenienses* in *Opera* (1616), Vol. II, p. 3.

spiritual elements. He thought of the elements as the "mothers" of objects in the sense that the earth was the mother of men and of that which grows on the earth. Once again we are back to the concept of the microcosm, since it is thus that man could be said to contain the impression of all earthly objects, which derive from the same "mother." In an analogous fashion water was seen as the matrix for metals, stones, and gems, as well as being the main substance of plants.[136]

While Paracelsus accepted—with significant alterations—the traditional elements, he also introduced a second elementary system, the three principles: salt, sulfur, and mercury. The *tria prima* were to account for all things. To this extent the system differed from the sulfur-mercury theory still being employed as an explanation for the metals, but like it, the Paracelsian principles were sophic in nature. Although they might never be isolated, it was possible to identify their properties in other substances. Thus Paracelsus was able to demonstrate their existence by burning a twig: the vaporous fumes denoted mercury, the flame was sulfur, and the final ashes were salt.[137] But Paracelsus also insisted that the principles were different from one substance to another[138]—thus negating their potential value as a basis for analytical theory.

Even more problematical was the discussion of elementary substances on the two levels of body and soul. Whereas in one text Paracelsus might speak of the four elements on the highest level, as imperceptible elements or matrices, in another he might discuss them as the perceptible substances we see.[139] The same confusion was found in his use of the principles. Although it seems possible that he meant salt, sulfur, and mercury to act as *vulcani* within the

[136] Pagel, *Paracelsus*, pp. 93 ff., 96-98.

[137] Paracelsus, *Die 9 Bücher de Natura Rerum* in *Sämtliche Werke*, Vol. XI, p. 348. ". . . dan alles was im feur reucht und verrecht ist mercurius, was brennet und verbrennet ist sulphur und alles was aschern ist, das ist auch ein sal." A translation of this work is in *The Hermetic and Alchemical Writings of Paracelsus*, trans. Waite, Vol. I, p. 150.

[138] Paracelsus, *De mineralibus* in *Sämtliche Werke*, Vol. III, pp. 42-43. In addition to this chemical meaning, the three principles were often understood on a more lofty plain. In this case a comparison was frequently drawn between them and the Holy Trinity.

[139] This is discussed at greater length in *The English Paracelsians*, pp. 28-29. In the *Archidoxis* Paracelsus discusses the elements on both levels of "body" and

elements, this was not clear to all of his followers. For them the relationship of the two elemental systems was difficult to understand; indeed, it was even possible to cite contradictory passages from within the Paracelsian corpus.

If Paracelsus still frequently utilized the elements as a means of explanation, he bluntly rejected the analogous system of the four humors. Galenic medicine normally attributed disease to an imbalance of blood, phlegm, yellow and black bile. When disease occurred, the duty of the physician was to restore the proper balance—and thus health.[140] In contrast, the emphasis of Paracelsus was on local malfunctions which, he argued, were due to external causes.[141] He sought to relate specific diseases to specific agents. The body itself was described in chemical terms with individual organs endowed with their own *archei*—or life forces acting as internal alchemists—which separated the useful from the nonuseful substances supplied. The body's chief alchemist resided in the stomach, and it was here that the first separation of food took place:

> A person eating meat, wherein both poison and nourishment are contained, deems everything good while he eats. For, the poison lies hidden among the good and there is nothing good among the poison. When thus the food, that is to say the meat, reaches the stomach, the alchemist is ready and eliminates that which is not conducive to the well-being of the body. This the alchemist conveys to a special place, and the good where it belongs. This is as the Creator ordained it. In

---

"soul," differentiating between "predestined" and "material" elements. Here the principles are not mentioned. In the *Philosophia de generationibus et fructibus quatuor elementorum* he stated the three principles "form everything that lies in the four elements" (Paracelsus, *Astronomia magna* in *Sämtliche Werke*, Vol. XIII, pp. 12-13). On this problem see also Hooykaas, "Elementenlehre des Paracelsus," pp. 175-177.

[140]See Owsei Temkin, *Galenism. Rise and Decline of a Medical Philosophy* (Ithaca/London: Cornell University Press, 1973), pp. 17-18.

[141]A useful summary of Paracelsus' medical thought is found in Lester S. King, *The Growth of Medical Thought* (Chicago: University of Chicago Press, 1963), pp. 86-138. More detail will be found in the masterful discussion in Pagel, *Paracelsus*, pp. 126-203, supplemented with his "Van Helmont's Concept of Disease—To Be Or Not To Be? The Influence of Paracelsus," *Bulletin of the History of Medicine*, **46** (1972), 419-454. Here Pagel has reassessed the positions of van Helmont and Paracelsus in relation to ancient humoral pathology.

this manner the body is taken care of so that no harm will befall it from the poison which it takes in by eating, the poison being eliminated from the body by the alchemist without man's cooperation. Of such a nature are thus virtue and power of the alchemist in man.[142]

In short, disease was no longer considered an imbalance of fluids; it was local in nature and directly related to bodily malfunctions which were essentially chemical in nature.

Medicine was to be indebted in other ways to this useful science of chemistry through works attributed to Paracelsus. "Water-casting," the diagnosis of urine samples, was replaced by a distillation process,[143] while Paracelsus followed an older tradition in his search to identify the chemical ingredients of spa waters.[144] Most important, he was convinced of the superior efficacy of chemically prepared medicines. Such medical chemistry, not transmutation, should be the major practical goal of the alchemist in his laboratory. Although the remedies of Paracelsus reflected the distillation products of the earlier medieval tradition of John of Rupescissa and Hieronymus Brunschwig,[145] he also emphasized the value of mercury, antimony, and iron salts—which became the subject of bitter debate in ensuing years.[146] Authority

---

[142] Paracelsus, *Paramirum*, trans. Leidecker, p. 29.

[143] Pagel, *Paracelsus*, pp. 189-194. The early history of the distillation of urine is discussed at greater length by Pagel, *Das medizinische Weltbild des Paracelsus*, pp. 19-20. A work by Gerhard Dorn on this subject was incorporated into the Huser edition of Paracelsus.

[144] Debus, "Solution Analyses Prior to Robert Boyle," p. 45.

[145] Sudhoff considered the *Archidoxis* to be the chief work of Paracelsus dealing with medicinal chemistry and the "fundamental text of the new specific therapeutics on a chemical basis." See "The Literary Remains of Paracelsus," *Essays in the History of Medicine* (New York: Medical File Press, 1916), p. 275. In addition to the works of Multhauf already cited, see his "Medical Chemistry and the Paracelsians," *Bulletin of the History of Medicine*, **30** (1956), 329-346.

[146] Wolfgang Schneider, "A Bibliographical Review of the History of Pharmaceutical Chemistry (with particular reference to German Literature)," *American Journal of Pharmaceutical Education*, **23** (1959), 161-172; "Die deutschen Pharmakopöen des 16. Jahrhunderts und Paracelsus," *Pharmazeutische Zeitung*, **106** (1961), 3-15; "Der Wandel des Arzneischatzes im 17. Jahrhundert und Paracelsus," *Sudhoffs Archiv*, **45** (1961), 201-215. Schneider's most recent thoughts on the chemical remedies of Paracelsus are in his *Geschichte der pharmazeutische Chemie* (Weinheim: Verlag Chemie, 1972), pp. 48-49 and *passim*.

for the medicinal use of poisons was sought in the acceptance of the folk tradition of cure by similitude—another break with the Galenic tradition that affirmed cure by contraries.[147] Although Paracelsus argued that attention must be given to proper dosage with these more potent remedies,[148] the warning was too often ignored by his followers—a fact that his enemies were to duly note.

Conclusion

The chemical philosophy of the sixteenth century is seen best in the work of the first-generation disciples of Paracelsus rather than in the work of their master, for they were the ones who were forced to collect, interpret, and systematize the Paracelsian corpus. Our own description of this universal chemical system in the following chapter will be drawn primarily from these late-sixteenth- and early-seventeenth-century texts.

While in this chapter the work of Paracelsus himself has not been discussed in detail, perhaps enough has been said to indicate that his work was at the same time conservative and radical. He and his followers were accused by some of being innovators of the worst sort—men who would destroy the accumulated knowledge of the past. There were others, however, who sought harmony between the new breed of chemists and ancient authority; they went to great lengths to show that many concepts employed by Paracelsus had been held by the ancient philosophers and physicians.

The conservative nature of Paracelsian natural philosophy and medicine will be noted frequently in the following chapters. Although Paracelsus forcefully argued against the ancients, it is possible to show how he frequently did little more than echo the views of Aristotle and Galen.[149] In addition, his "occult philosophy" clearly reflected Renaissance natural magic and the recently revived neo-Platonic and Hermetic texts of late antiquity.

[147]Paracelsus, *Paragranum* in the *Sämtliche Werke*, Vol. VIII, p. 107.
[148]*Ibid.*, Vol. VII, pp. 300-301; Pagel, *Paracelsus*, pp. 275-276; Henry M. Pachter, *Paracelsus, Magic Into Science* (New York: Schuman, 1951), p. 128.
[149]Lester S. King and Walter Pagel have been particularly concerned with establishing the Aristotelian basis of the thought of Renaissance medical figures.

Still, it is with his chemical heritage that we are most concerned, and here it is quite evident that Paracelsus was well grounded in both the chemical medicine of the Middle Ages and traditional cosmic alchemy. From these sources he became convinced that chemistry was the key to nature on all levels. Specifically, the physician-magus might employ his knowledge of the macrocosm to cure the ills of the microcosm.

Although it is thus not difficult to show that Paracelsus was deeply dependent on the work of earlier authors, in contrast to the medieval alchemists and alchemical physicians, his work aroused bitter debate. There was good reason for this. Paracelsus had sharply denounced the ancients and their followers, and in so doing he had condemned the educational system of his day and ridiculed the medical establishment. He wanted both to be replaced with a "Christian" approach to nature, one that relied upon the Holy Scriptures and on fresh observations. In the course of his attack Paracelsus launched a body blow at logic, the basis of ancient philosophy, and the foundation of the educational system. He was no less gentle on Galenic medicine in his outright rejection of the humoral system. In contrast, chemistry seemed to offer a method that was unchained to the logic and mathematics of the past but was dependent for its truths upon new observations in laboratory and field. It was a subject all the more to be cultivated since alchemy was undefiled by the classical authors studied in the universities. Before proceeding to discuss the inevitable conflict such ideas produced, it will be necessary to give a more detailed account of chemical philosophy as it was developed by Paracelsians of the late sixteenth century.

2
THE CHEMICAL PHILOSOPHY

> ... *without this Chymicall Phylosophy all Physick is but liveless.*
> Oswald Crollius (1609)*

AT THE DEATH of Paracelsus in 1541 there was little to indicate that his work would become the focal point for debate among scholars for more than a century. True, he had been a controversial figure during his lifetime, but relatively few of his voluminous writings had been published while he was alive. The flood of Paracelsian texts began to appear from the presses only later. In 1550 Cyriacus Jacobus spoke of the wondrous cures performed by this man, while in 1553 the important *Labyrinthus medicorum errantium* appeared from the press of Bernhard Vischer in Nuremberg. In the following years a host of other manuscripts were sought out and published, often with extensive notes and commentaries. As the number of Paracelsian titles grew, collected editions were prepared and offered to the public. Perna's edition of 1575 filled two volumes and contained twenty-six treatises, and before the end of the century the definitive ten-volume quarto edition of the *Opera* appeared under the editorship of Johannes Huser, a physician at Glogau (1589-1591). Huser's text was reprinted twice in folio (1603, 1605 and 1616, 1618), and it remained influential long enough to warrant a Latin translation, which was prepared under the direction of Fridericus Bitiskius as late as 1658.[1]

---

*Oswald Crollius, *Discovering the Great and Deep Mysteries of Nature* in *Philosophy Reformed and Improved*, trans. H. Pinnell (London: M.S. for Lodowick Lloyd, 1657), p. 94. In the original Latin text the complete sentence reads as follows: "Digna provincia in quâ omnes medici senescant, & plena exercitationis, sine siquidem Philosophia Chymica omnis Medicina mortua est: Absque Alchymiae cognitione neque Theorica neque Practica Medicina esse potest." *Basilica chymica* (Frankfurt: Godfrid Tampach, n.d. [1623]), p. 49.

[1] The basic guide to the editions of Paracelsus is Karl Sudhoff's *Bibliographia Paracelsica* (Berlin: Verlag Georg Reimer, 1894; reprinted Graz: Akademische Druck- u. Verlagsanstalt, 1958). The major editions are described by Walter Pagel in his *Paracelsus. An Introduction to Philosophical Medicine in the Era of the Renaissance* (Basel: S. Karger, 1958), pp. 31-35. Johannes Huser's edition of the *Bücher und Schrifften* of Paracelsus was printed at Basel by Conrad Waldkirch in ten quarto volumes in 1589-1591. Lazarus Zetzner reprinted these works in three

Along with the demand for books there was an ever-increasing need for Paracelsian physicians.[2] In 1585 the English theorist R. Bostocke spoke of the "great number of learned Philosophers and Phisitions, as well as weare *Galenists*, as others, which at this daie

---

folio volumes at Strasbourg in 1603 and 1605 and again in 1616 and 1618. The Latin *Opera omnia* was the result of the work of Fridericus Bitiskius and was printed by De Tournes at Geneva in three volumes in 1658. In 1678 Richard Russell announced that he had already completed the translation of two of the three volumes of the works of Paracelsus. These were never printed. (*The Works of Geber*, trans. Richard Russell, London: William Cooper, 1686, sig. A2^V. From "The Translator to the Reader" dated May 3, 1678.) I have referred to the Huser and Bitiskius editions as well as that of Sudhoff: Paracelsus, *Sämtliche Werke*, ed. Karl Sudhoff and Wilhelm Matthiessen, 15 vols. (Munich/Berlin: R. Oldenbourg [Vols. VI-IX: O. W. Barth], 1922-1933).

[2] A useful guide to Central European Paracelsians is Karl Sudhoff's "Ein Beitrag zur Bibliographie der Paracelsisten im 16. Jahrhundert," *Centralblatt für Bibliothekswesen*, **10**(1893), 316-326, 386-407. Another useful guide to elusive titles is E. Morwitz, *Geschichte der Medicin* (Wiesbaden: Dr. Martin Sändig oHG, n.d.), Vol. I, pp. 159-178. A number of regional studies of Paracelsians have appeared in recent years, among them the following: Allen G. Debus, *The English Paracelsians* (London: Oldbourne Press, 1965; New York: Franklin Watts, 1966); P.H. Kocher, "Paracelsan Medicine in England (ca. 1570-1600)," *Journal of the History of Medicine*, **2** (1947), 451-480; Wlodzimierz Hubicki, "Chemie und Alchemie des 16. Jahrhunderts in Polen," *Annalen Universitatis Mariae Curie-Sklodowska Polonia-Lublin*, **10** (1955), 61-100; W. Hubicki, "Chemistry and Alchemy in Sixteenth-Century Cracow," *Endeavour*, **17** (1958), 204-207; José Guillermo Merck Luengo, *La quimiatría en España* (Madrid: Instituto Arnaldo de Vilanova, 1959); José María López Piñero, "Juan de Cabriada y las primeras etapas de la iatroquímica y de la medicina moderna en España," *Cuadernos de Historia de Medicina de España*, **2** (1962), 129-154; Allen G. Debus, "The Paracelsian Compromise in Elizabethan Medicine," *Ambix*, **8** (1960), 71-97; Allen G. Debus, "Paracelsian Doctrine in English Medicine," in *Chemistry in the Service of Medicine*, ed. F.N.L. Poynter (London: Pitman, 1963), pp. 5-26; N. A. Figurovskij, "Die Chemie in Russland im Zeitalter der Iatrochemie," *Nova Acta Leopoldina*, **27** (1963), 351-366. Gerhard Eis has discussed a number of Central European Paracelsians in his *Vor und nach Paracelsus. Untersuchungen über Hohenheims Traditionsverbundenheit und Nachrichten über seine Anhänger* (Medizin in Geschichte und Kultur, Band 8) Stuttgart: Gustav Fischer Verlag, 1965). Here see his "Zur Paracelsusnachfolge im Sudetenraum," which appeared first in the *Südostforschungen*, **6** (1941), 440-462. Leo Norpoth has discussed "Kölner Paracelsismus in der 2. Hälfte des 16. Jahrhunderts," in the *Jahrbuch des Kölnischen Geschichtsvereins*, **27** (1953), 133-146, and "Paracelsismus und Antiparacelsismus in Köln in der 2. Hälfte des 16. Jahrhunderts" in *Medicinae et artibus. Festschrift für Professor Dr. phil. Dr. med. Wilhelm Katner zu seinem 65. Geburtstag, Düsseldorfer Arbeiten zur Geschichte der Medizin*, Beiheft 1 (Düsseldorf: Michael Triltsch Verlag, 1968), pp. 91-102, while Scandinavian Paracelsian thought is the subject of Sten Lindroth's *Paracelsismen i*

doe embrace, follow, and practise, the doctrine, methods and wayes of curyng of this Chimicall Phisicke."[3] His list of authorities included the names of well-known Paracelsians such as Adam of Bodenstein (1528-1577), Gerhard Dorn (fl. c. 1566-1584), Petrus Severinus (Sørenson) (1542-1602), Leonhard Thurneisser (1530-1596), Michael Toxites (1515-1581), and Joseph Duchesne (Quercetanus) (c. 1544-1609), as well as the more famous Theodore Zwinger (1533-1588) and Johannes Guintherius von Andernach (1505-1574), the teacher of Vesalius who had applied himself to the study of chemistry late in life. A contemporary of Bostocke's, the London surgeon George Baker (1540-1600), made haste to inform his readers where they might find reputable apothecaries in the capital who would prepare the new chemical remedies.[4]

On the continent a similar list was composed by George Bernard Penotus (1520?-1620?), who singled out the works of Dorn, Severinus, Thurneisser, and Duchesne as being of special merit,[5] while Olaus Borrichius (1626-1690) arranged contemporary chemical authors by country in his *De Ortu et progressu chemiae* (1668).[6]

Sverige till 1600—talets mitt (Uppsala: Akademish Avhandling, 1943). A number of important essays on sixteenth- and seventeenth-century Paracelsians appear in *Science, Medicine and Society in the Renaissance. Essays in Honor of Walter Pagel*, ed. Allen G. Debus (New York: Neale Watson Press; London: Heinemann, 1972).

[3]R. B. Esquire [R. Bostocke], *The difference betwene the aunicient Phisicke, first taught by the godly forefathers, consisting in vnitie peace and concord: and the latter Phisicke proceeding from Idolaters, Ethnikes, and Heathen: as Gallen, and such other consisting in dualitie, discorde, and contrarietie* . . . (London: Robert Walley, 1585), sigs. Ii^v-Iiii^r.

[4]Conrad Gesner, *The New Jewell of Health*, corr. and pub. by George Baker (London: H. Denham, 1576), sig. iv. From the prefatory "George Baker to the Reader."

[5]Mr. Bernard G. Londrada A. Portu Aquitanus (Penotus), "An Apologeticall Preface" to the *One Hundred and Fourteen Experiments and Cures of the Famous Physitian Theophrastus Paracelsus*, in Leonard Phioravant (Fioravanti), *Three Exact Pieces of Leonard Phioravant Knight and Doctor in Physick* . . . *Whereunto is Annexed Paracelsus his One hundred and fourteen Experiments: With certain Excellent Works of B.G. à Portu Aquitans, Also Isaac Hollandus his Secrets concerning his Vegetall and Animall Work With Quercetanus his Spagyrick Antidotary for Gunshot* (London: G. Dawson for William Nealand, 1652), sig. Ccc 4^r-Ddd 4^r. This collected edition is a reprint of John Hester's sixteenth-century translations which are discussed in Debus, *English Paracelsians*, pp. 66-69.

[6]Olaus Borrichius, *De ortu et progressu chemiae dissertatio*, in J. J. Manget, *Bibliotheca chemica curiosa*, 2 vols. (Geneva: Chouet, C. De Tournes, Cramer, Perachon, Ritter & S. De Tournes, 1702), Vol. I, pp. 1-37. Here sixteenth- and seventeenth-century authors are discussed on pp. 35 and 36.

In the *Basilica chymica* (1609)—surely one of the most influential of all of the Paracelsian treatises on chemical medicine—Oswald Crollius (c. 1560-1609) wrote that works of the Dane Severinus, the Frenchman Duchesne, the Italian Zefiriele Tommaso Bovio (1521-1609), and the Englishman Thomas Moffet (1553-1604) should stand next to those of Paracelsus himself as the foundation of this new science.[7]

There is little doubt that for these authors it was of major importance to point out that there *was* an alternative to the Galenism of the schools, that texts and commentaries for this new system of medicine were available in print, and that there were physicians who practiced the new medicine. For Crollius the future seemed to promise great rewards, and he predicted that the true philosophical (or chemical) medicine would be "restored to its true lustre & ancient splendor" with heavenly assistance and the combined efforts of chemists and those Galenists who were willing to learn.[8]

The Paracelsian Universe

What did these men propose? Above all, they sought to overturn the traditional Aristotelian-Galenic bias of the entrenched educational system. In that intensely religious age, Aristotle and Galen could be severely attacked: Aristotle as a heathen whose work had rightly been condemned repeatedly in church councils; Galen, who had uncritically accepted his work; and their combined system which had subsequently become the basis of medical training throughout Europe.[9]

The Paracelsians hoped to replace all of this with a philosophy

---

[7] Crollius, *Mysteries of Nature*, pp. 74-75; *Basilica chymica*, pp. 39-40.

[8] *Mysteries*, p. 226; *Basilica*, p. 110.

[9] This theme is a common one in the literature of the period, but one might turn especially to Robert Fludd and to Bostocke, who wrote that "The heathnish Phisicke of *Galen* doth depende uppon that heathnish Philosophie of *Aristotle*, (for where the Philosopher endeth, there beginneth the Phisition) therefore is that Phisicke as false and injurious to thine honor and glory, as is the Philosophie." Bostocke, *Auncient Phisicke*, sig. Av^V (from "The Authors obtestation to almightie God"). For Bostocke, Galen is "that heathen and professed enemy of Christ" (sig. H$^{V\text{-}r}$).

PLATE III   The true chemical philosopher worships his God through prayer and the study of His written word as well as through the study of His Creation, nature, in the chemical laboratory. From Heinrich Khunrath, *Amphitheatrum Sapientiae aeternae, solius verae, Christiano-Kabalisticum, divino-magicum, nec non physico-chymicum, tertriunum, catholicon . . .* (Hanover: Gulielmus Antonius, 1609), folding plate at end.

strongly influenced by the newly translated Christian neo-Platonic and Hermetic texts. It was to be a philosophy which could account for all natural phenomena. Subjects rejected by Aristotle were to be opened for discussion. For this reason it is with Hermetic and chemical philosophers of the late sixteenth and the early seventeenth centuries that we first see a new growth of interest in the subject of atomism.[10]

Rejecting the knowledge of Aristotle and Galen, they sought instead the pristine knowledge granted to Adam, which they believed could be discovered in the wisdom of the Old Testament and in the knowledge of Hermes Trismegistus and his disciples. This wisdom, partially preserved in the views of the pre-Socratics and Plato, they contrasted with the works of Aristotle and Galen—Aristotle, who had traitorously attacked Plato, his own teacher, and Galen, the persecutor of Christians. The Paracelsian historian believed that in the intervening centuries philosophical truths had been preserved by a dedicated few (for the most part Byzantine and Islamic alchemists), who passed on from master to pupil the true ancient wisdom. Thus, it was possible for Bostocke in his history to compare the work of Paracelsus in medicine with that of Copernicus in astronomy and with Luther, Melanchthon, Zwingli, and Calvin in theology. For him all these scholars had devoted their energies primarily to the restoration of the true knowledge of an earlier age.[11]

The natural philosophy of the Paracelsians was founded on the two-book concept of nature. The true physician should turn first to the book of divine revelation—the Holy Scriptures—and then to the book of divine Creation—nature.[12] As the English Paracelsist Thomas Tymme (d. 1620) wrote in 1605,

---

[10] See the discussion of the revival of Democritean atomism by J. E. McGuire and P. M. Rattansi in "Newton and the 'Pipes of Pan,'" *Notes and Records of the Royal Society of London*, **21** (1966), 108-143 (130). Here the authors discuss the identification of the Phoenician Moschus with Moses among iatrochemists and other authors. The atomistic views of Robert Fludd and other chemists will be discussed later in this volume.

[11] On Bostocke's work as an example of a Paracelsian history see Allen G. Debus, "An Elizabethan History of Medical Chemistry," *Annals of Science*, **18** (1962, published 1964), 1-29.

[12] This approach is evident in the *Labyrinthus medicorum errantium* of Paracelsus. See the *Opera omnia*, Bitiskius ed., Vol. I, pp. 264-288 (275).

> The Almighty Creatour of the Heauens and the Earth, (Christian Reader), hath set before our eyes two most principall Bookes: the one of Nature, the other of his written Word . . . The wisedome of Natures booke men commonly call Naturall Philosophie, which serueth to allure to the contemplation of that great and incomprehensible God, that wee might glorifie him in the greatnesse of his worke. For the ruled motions of the Orbes . . . the connexion, agreement, force, vertue, and beauty of the Elements . . . are so many sundry natures and creatures in the world, are so many interpreters to teach us, that God is the efficient cause of them, and that he is manifested in them, and by them, as their finall cause to whom they also tend.[13]

On the one hand, the Paracelsians applied themselves to a form of Biblical exegesis that was strongly tinged with their Hermetic background; on the other their religious orientation led them to the call for a new philosophy of nature based on fresh observations and "experiments."[14] An excellent example of this is found in the work of the important early systematizer of the Paracelsian corpus, Peter Severinus, the physician to the king of Denmark, who re-echoed an earlier Hermetic theme when he admonished his readers to

> . . . sell your lands, your houses, your clothes and your jewelry; burn up your books. On the other hand, buy yourselves stout shoes, travel to the mountains, search the valleys, the deserts, the shores of the sea, and the deepest depressions of the earth; note with care the distinctions between animals, the differences of plants, the various kinds of minerals, the properties and mode of origin of everything that exists. Be not ashamed to study diligently the astronomy and terrestrial philosophy of the peasantry. Lastly, purchase coal, build furnaces, watch and operate with the fire without wearying. In this way and no other, you will arrive at a knowledge of things and their properties.[15]

---

[13] Thomas Tymme, *A Dialogue Philosophicall* (London, 1612), sig. A3 (from the dedication to Sir Edward Coke, Lord Chief Justice of the Court of Common Pleas).

[14] The Latin *experientia* was commonly translated as "experiment" in works of the period. Accordingly, the word is used here, but with the caveat that its meaning might vary considerably from one text to another. Examples of Paracelsian thought on "experiments" will be offered in later chapters.

[15] Petrus Severinus, *Idea medicinae philosophicae* (3rd ed., Hagae Comitis: Adrian Clacq, 1660), p. 39. The first edition appeared in 1571.

EMBLEMA XLII. *De secretis Naturæ.*
In Chymicis versanti Natura, Ratio, Experientia & lectio,
 sint Dux, scipio, perspicilia & lampas.

PLATE IV The chemical philosopher following in the footsteps of nature. From Michael Maier, *Atalanta fugiens, hoc est, Emblemata nova de secretis chymica* (Oppenheim, J. T. De Bry, 1618).

Paracelsian "Mathematics"

It was in the *Astronomia magna* that Paracelsus discussed the relationship of mathematics to natural philosophy[16] in greatest detail. Here we read that "Mathematica est *Magica, Nigromantica, Astrologia, Signata, Artium incertarum, medicinae adeptae, Philosophia adeptae.*"[17] If mathematics is magic, then true magic in turn may be equated with nature. The godly magus may concentrate in himself celestial virtues which are the hidden powers of nature, and he may then use these powers to work wonders and learn of his Creator.[18]

But we must distinguish, Paracelsus continued, between elementary mathematics and the mathematics of the adept, for only the latter can be called magic. Elementary mathematics is concerned with visible and tangible objects such as the growth of a tree, the motion of the sun, and the phases of the moon. The mathematics of the adept goes beyond this to study visible but nonpalpable objects such as the image of the sun in a river or the moon in a well. Also called sidereal mathematics, this subject examines generation and creation as well as essences and archidoxies which the practitioner may then associate with their proper geometrical figures according to their height, profundity, amplitude, and angles. The sidereal mathematician is the master of a subject which would hardly be given that name in the twentieth century. He can show that paper is nothing but wood, discover the real substance of the sun, and learn the differences of type and species. But such studies are possible only for the "Adepta Mathematica, Philosophia & Medicina."[19]

When late-sixteenth-century Paracelsians mentioned mathematics other than that of weight measurements it was generally this sidereal mathematics. Bostocke in 1585 wrote that "the true

---

[16]This topic is treated in more detail in Allen G. Debus, "Mathematics and Nature in the Chemical Texts of the Renaissance," *Ambix*, **15** (1968), 1-28, 211.

[17]Paracelsus, *Opera omnia*, Bitiskius ed., Vol. II, p. 549.

[18]*Ibid.*, pp. 558 and 555: "Nihil autem est alius, quàm potentia, virtutes coelestes in Medium inducendi, & in illo operationis perficiendi. Medium hoc est centrum. Centrum est homo. Sic ergo per hominem vis coelestis in hominem induci potest, ita, ut in homine tali virtus, constellationi illi congrua, inueniatur."

[19]*Ibid.*, pp. 577-578.

and auncient phisicke" may be sought in nature "and is collected out of Mathematicall and supernaturall precepts." This was a sacred art properly called *chymia,*

> Which sheweth foorth the compositions of all maner bodies, and their dissolutions, their natures & properties by labour by the fire, following Nature diligently. So that Philosophie naturall and supernaturall, the Mathematicals *Chimia* and *Medicina* be so combined together, that one of them can not be without the other.[20]

Mathematics was truly a mystical science for Bostocke, and as the goals of the sidereal mathematics of Paracelsus would indicate, it could be easily equated with the mystical chemistry of the philosophical alchemists.

Among the early Paracelsians few authors were more concerned with numerical studies than Gerhard Dorn. In his description of the Creation, and in his work on the Physics of Trithemius, he continually referred to the power of mystical numerical analysis,[21] while his *Monarchia triadis, in vnitate soli Deo sacra* (1577) is a complex tract which attempted to show and relate the triune nature of divinity with the universe as a whole.[22] Beyond this, Dorn found a correspondence between divine unity, the Paracelsian principles, and the medicine of the Paracelsians. In contrast to this unity, the evil of a binary system was related to the devil and found its expression in the corrupt medicine of the Galenists. Although elements of the thought of Cusanus are present,[23] the progression of geometrical figures which illustrates this tract is strongly reminiscent of John Dee's *Monas hieroglyphica*. Dorn's mathemat-

---

[20] Bostocke, *Auncient Phisicke,* sig. Bi^r and Bi^v.

[21] The *Physica Genesis* and the *Physica Trithemii* form parts of the *Liber de natura luce physica, ex Genesi desumpta* (1583), which is reprinted in the first volume of Lazarus Zetzner's *Theatrum chemicum,* 4 vols. (Strasbourg: Zetzner, 1613; 1st ed., 1602), Vol. I, pp. 352-532. Specifically note his interest in arithmetical progression (applied to the elements) and preoccupation with the sphere and other geometrical figures in the *Physica Genesis,* pp. 323-325, 362-363.

[22] Gerhard Dorn, *Monarchia triadis, in vnitate, soli Deo sacra* in Paracelsus, *Aurora thesaurusque philosophorum* (Basel: Palma Guarini, 1577), pp. 65-127.

[23] Pagel, *Paracelsus,* p. 135.

ical-geometrical theme is one which is recurrent among the Paracelsian authors.[24]

While it is true that some authors wrote at length on the mystical mathematical harmony of the universe, it was becoming increasingly customary in the late sixteenth century for the Paracelsians to react with distaste against the logical, "geometrical" method of argument employed by the Aristotelians and the Galenists. We find Peter Severinus making this a major point in his rejection of Galenic medicine in his *Idea medicinae philosophicae* (1571). Describing the decline of medical studies in antiquity, Severinus pictured Galen as a compiler faced with the task of putting order into the work of his predecessors. Seeking a unifying principle with laws and demonstrations, he at length came upon the writings of the geometricians. Seduced by their exquisite demonstrations, he attempted to make medicine a part of geometry with its own principles, axioms, and mathematical explanations. So influential had been his work that only in recent years had the fallacy of his approach been demonstrated. When Fernel tried Galen's methods and laws of cure he found that they failed in the treatment of those new diseases which were ravaging the continent. We now know, Severinus continued, that the only true medicine is that to be found in the writings of Paracelsus, and far from being mathematically inspired, this science finds its roots in the observations of the chemists.[25]

The limited interest in the cosmic "mathematics" of Paracelsus and the deep distrust of a logico-mathematical approach to nature was coupled with an increased interest in the use of the balance in weight determinations. This was part of alchemical tradition, and it could also be supported through scriptural reference. Thus we find a real interest on the part of sixteenth-century iatrochemists in the analysis by weight of mineral waters and urine,[26] while in the next century van Helmont was to perform his willow tree experiment based upon the earlier suggestion made

---

[24] The incompatibility of the *medicine ternarii* (or unity) with the *medicine binarii* was a major theme of Bostocke's Paracelsian apology of 1585.

[25] Severinus, *Idea*, pp. 2, 3, 21.

[26] See Allen G. Debus, "Solution Analyses Prior to Robert Boyle," *Chymia*, **8** (1962), 41-61. Pagel, *Paracelsus*, pp. 190-94, 281-282.

by Cusanus in *The Idiot*.[27] This mood is reflected in the work of the well-known iatrochemist Daniel Sennert (1572-1637), who paid little attention to the study of mathematics in his comparison of Aristotelian, Galenic, and Paracelsian thought (1619), and in his *Epitome naturalis scientiae* (1618) discussed the subject primarily as a practical art to be used whenever quantity is considered.[28] Surely this made sense to Paracelsian physicians, chemists, and pharmacists—men who weighed and measured regularly in the course of their work.

Chemistry and the New Science

While the Paracelsians rejected the logico-mathematical method of the schools, they turned to chemistry with the conviction that this science could be the basis for a new understanding of nature. It is the emphasis on chemistry and medicine that distinctly separates them from other Renaissance natural philosophers and Hermeticists. For Paracelsus alchemy offered an "adequate explanation of all the four elements,"[29] and this meant literally that alchemy and chemistry could be used as a key to the cosmos either through laboratory techniques or through analogy. In the *Philosophia ad Athenienses* (published first in 1564) Paracelsus pictured the Creation itself as a chemical unfolding of nature. His followers agreed and amplified this theme. In England Thomas Tymme insisted that "*Halchymie* should have concurrence and antiquitie with *Theologie*," since Moses

> ... tels us that the *Spirit of God moued upon the water*: which was an indigested Chaos or masse created before by God, with confused Earth

---

[27]See Nicolas Cusanus, *The Idiot in Four Books. The first and second of Wisdome. The third of the Minde. The fourth of statick Experiments, Or Experiments of the Ballance* (London: William Leake, 1650), p. 172.

[28]Daniel Sennert, *De chymicorum cum Aristotelicis et Galenicis consensu ac dissensu* (3rd ed., Paris: Apud Societatem, 1633); *Epitome naturalis scientiae* (Paris: Apud Societatem, 1633), pp. 6-7.

[29]Paracelsus, *Philosophia ad Athenienses* in the *Opera omnia*, Bitiskius ed., Vol. II, pp. 239-252. Two English translations have appeared: the first, by H. Pinnell, is *Three Books of Philosophy Written to the Athenians* in *Philosophy Reformed and Improved*; the second is in *The Hermetic and Alchemical Writings of . . . Paracelsus the Great*, ed. Arthur Waite, 2 vols. (London: James Elliott, 1894), Vol. II, pp. 249-282.

in mixture: yet, by his Halchymicall Extraction, Seperation, Sublimation, and Coniunction, so ordered and conioyned againe, as they are manifestly seene a part and sundered: in Earth, Fyer included, (which is a third Element) and Ayre, [and] (a fourth) in Water, howbeit inuisibly.[30]

For Tymme the Creation was a chemical process that accounted for the production of the three principles as well as the four elements. In like fashion the Last Judgment would be a chemical process. For this reason it was a wise man who based his study of nature on this science and who would consequently learn that through separation salt, sulfur, and mercury will always be obtained. In this fact we may "discerne the holy and most glorious Trinitie in the Vnitie of one *Hupostasis* Diuine. For the inuisible things of God (saith the Apostle) this is, his eternal power and Godhead, are seene by the creation of the world, being considered in his workes." The study of chemistry then "is not of that kind which tendeth to vanity and deceit, but rather to profit and to edification, inducing first the knowledge of God, & secondly the way to find out true medicine in his creatures."[31]

This theme was developed at much greater length by Gerhard Dorn in his *Liber de naturae luce physica, ex Genesi desumpta* (1583), which was to reach a large audience through its later inclusion in Lazarus Zetzner's collected edition of alchemical classics, the *Theatrum chemicum* (first edition in 4 volumes, 1602). In this work Dorn offered a lengthy commentary on the first six days of Creation. Here his principal guides were clearly the Hermetic corpus and the *Philosophia ad Athenienses*. Commenting on the division of the waters (Second Day), Dorn insisted that this could be nothing other than a chemical process.[32] It was, in effect, the familiar separation of the pure from the impure on a divine level. While the process might be illustrated through any simple distillation, the

---

[30] From the dedication to Sir Charles Blunt in Joseph Duchesne (Quercetanus), *The Practise of Chymicall, and Hermeticall Physicke, for the preseruation of health*, trans. Thomas Tymme (London: Thomas Creede, 1605), sig. A3^r.
[31] *Ibid.*, sig. A4^r.
[32] Gerhard Dorn, *Liber de naturae luce physica, ex Genesi desumpta* in Zetzner, ed., *Theatrum chemicum*, Vol. I (1613 ed.), p. 373.

chemical Creation had resulted in products that were no less than light and heaven itself on the one hand and earth and the elements on the other. Dorn carried this chemical theme further in his discussion of the physics of Hermes Trismegistus, where the Creation was again discussed.[33] Finally, in the closing book of the *Liber de naturae luce physica*, Dorn stated that the philosophical key utilized by the chemist could on its highest level be considered as nothing other than a heavenly operation.[34]

Chemical Theory and the Elements

This chemical interpretation of Genesis helped to focus attention on the problem of the elements as the first fruit of the Creation. The Aristotelian elements (earth, water, air, and fire) served as the very basis of the traditional cosmological system. They were used by the alchemists to explain the composition of matter, by the physicians (through the humors) to interpret disease, and by the natural philosophers as the basis for all motion. The introduction of a new elemental system—the Paracelsian *tria prima* (salt, sulfur, and mercury)—thus called into question the whole framework of ancient medicine and natural philosophy.

Although the theory of the *tria prima* was a modification of earlier theories, it has a special significance in the rise of modern science. Clearly part of an attack on Scholastic philosophy, the introduction of the new principles also led to considerable confusion. Here much of the blame can be placed upon Paracelsus himself.

In his *Philosophia ad Athenienses* the four Aristotelian elements play a fundamental role: they are the first product of the divine alchemical separation; all more complex forms of matter are dependent on them. Renaissance Aristotelians and Galenists generally spoke of earth, water, air, and fire as material substances, each of which was associated with two of the four qualities hot, cold, moist, and dry. It was a system of great antiquity which had been extended to cosmological and physiological as well as chemical problems. Paracelsus in his attack on ancient thought

[33]*Ibid.*, pp. 392-420 (417-418).
[34]*Ibid.*, p. 461.

had no desire to dispose of these traditional elements, yet for him they were generally considered to be spiritual in nature and only crude approximations of the objects we call by these names.[35]

Paracelsus' introduction of the *tria prima* was basically an extension of the old sulfur-mercury theory of the metals—but now broadened to provide an explanation for all nature. Not only were the sophic mercury, sulfur, and salt spiritual rather than material, they also differed qualitatively in all substances: there were "as many sulfurs, salts and mercuries as there are objects."[36] Further confusion was added by Paracelsus' failure to make clear the relationship of the new triad to the traditional elements. Perhaps the three substances had been offered originally as internal *vulcani* (innate powers) of the elements, but contradictions permitted any number of interpretations. Varying between spiritual and material meanings, the elements and principles were often pictured as almost indefinable aspects of a primal stuff that was the basis for the more complex things of this world. Little more agreement existed. Some scholars who pored over his darker passages felt that Paracelsus had meant that the elements were formed of the principles; others insisted that he really meant the opposite.[37]

The dispute over the elements posed a potential threat to traditional cosmology, but little could be definitely settled without sound analytical evidence. Later iatrochemists, finding little convincing proof for either system, felt free to utilize the four elements and the three principles as they saw fit.[38] Although a proponent of the three principles himself, the English Paracelsian Thomas Moffett complained of the growing number of elemental systems when he wrote (1584) that "Some wish that there should be but

[35] Pagel, *Paracelsus*, pp. 93 ff.

[36] Paracelsus, *Sämtliche Werke*, Vol. III, pp. 42-43, from the *De mineralibus*. It should be noted also that an analogy was often drawn between the Holy Trinity and the three principles.

[37] On Paracelsus' meaning of the principles see Pagel, *Paracelsus*, pp. 100-104.

[38] For the variant element-principle systems presented in the seventeenth century the reader is referred to Hélène Metzger, *Les doctrines chimiques en France du début du XVIIe à la fin du XVIIIe siècle*, Pt. 1 (Paris: P.U.F., 1923). R. Hooykaas has pointed out that Joseph Duchesne was the first to develop the concept of the "five principles" which was so prevalent among chemists and alchemists in the seventeenth century ("Die Elementenlehre der Iatrochemiker," *Janus*, **41** (1937), 26-28).

one element, while others think they are many, and some even think they are infinite, innumerable and immovable; these assert there are two, those three, some others say four, while others demand eight."[39] Nearly a century later, Johann Joachim Becher could still deplore the lack of agreement among chemists regarding the elements in his own widely read treatise on this subject, the *Oedipus chemicus* (1664).[40] With contention among chemists themselves, it is little wonder that the subject became a major one for those attacking the chemical philosophers.

Nevertheless, from the texts of this period we can see that the chemical philosophers were turning in increasing numbers to the three principles as a means of explanation. While they, no less than the Aristotelians, were willing to appeal to broad generalizations, logic, and time-honored authority, they often demanded—and were accustomed to—more concrete facts as practicing physicians and chemists. The three principles were assigned qualities that made it possible to identify their presence through normal chemical operations. With few exceptions this meant the application of heat either through decomposition (simple combustion) or through distillation.[41] Sulfur was considered to be the cause of combustibility, structure, and substance. Solidity and often color were due to salt, and the vaporous quality was always assigned to mercury. Here at least was a working concept. Some might speculate about the existence of the various systems, but the working chemist actually saw vaporous, inflammable, and ashy portions every time he performed an organic distillation. The difference between the philosopher of the Schools and the new chemical philosopher seemed clear to Bostocke, who gloried that the latter should

[39]Thomas Moffett, *De jure et praestantia chemicorum medicamentorum*, in the *Theatrum chemicum* (1613 ed.), Vol. I, pp. 99-100.

[40]J. J. Becher, *Oedipus chymicus, obscuriorum terminorum & principiorum mysteria aperiens & resolvens*, in Manget, *Bibliotheca chemica curiosa*, Vol. I, pp. 306-336.

[41]The history of the concept is discussed at greater length in Allen G. Debus, "Fire Analysis and the Elements in the Sixteenth and the Seventeenth Centuries," *Annals of Science*, **23** (1967), 127-147. A seventeenth-century example of the use of the properties of the principles as a guide to chemical investigations is found in Allen G. Debus, "Sir Thomas Browne and the Study of Colour Indicators," *Ambix*, **10** (1962), 29-36.

... knowe all things by visible and palpable experience, so that the true proofe and tryal shal appeare to his eyes & touched with his hands. So shall he have ☿ three *Principia*, ech of them separated frō the other, in such sort, ÿ he may see them, & touch them in their efficacie and strength, then shal he have eyes, wherewith the phisition ought to looke and reade with al. Then shall he have that he may taste and not before. For thē shall he know, not by his owne braines, nor by reading, or by reporte, or hearesay of others, but by experience, by dissolution of Nature, and by examyning and search of the causes, beginnings and foundations of the properties and vertues of thinges, which he shall finde out not to be attributed to colde or heate, but to the properties of the three substanties of each thing and his *Arcanum*. . . . Experience is the maysteries [mistress] of thinges.[42]

Indeed, the Paracelsians referred to both the four elements and the three principles, but because of a real interest in actual chemical operations, they spoke most often of the *tria prima*, whose existence seemed to be confirmed by their work. In an inflammatory address against the Galenists, Bernard G. Penotus ranted in a style reminiscent of Paracelsus:

I say, every *Paracelsian*, which doth but onely carry coals unto the work, can shew you by eye three principles of *Theophrastus* Physick. Have you tasted the most sharp Salt, or the most sweet Oil, or the Balm, that most delicate liquor? All those being hidden in every thing that is created, you have not once perceived. The metalline spirits, in whom Physick doth consist, by no means can be found out, neither what force they have, or fellowship with mans nature, but onely by fire. . . .[43]

In the literature of the period the importance of fire as a means of analysis is everywhere stressed. Chemical processes in the laboratory were almost exclusively carried out with the aid of heat, and alchemical tradition affirmed that fire caused a separation, not generation. The concept of separation is one that can be traced back to Aristotle,[44] and summing up mid-sixteenth-century thought on this point, Conrad Gesner stated that

[42] Bostocke, *Auncient Phisicke*, sig. Dv^r and Dv^v.
[43] Penotus, "Apologeticall Preface," sig. Ddd 1^r.
[44] Discussing Empedocles on the elements, Aristotle went on to say that the principles cannot be the same: "If however they are different, one difficulty is whether they too are to be regarded as imperishable or as perishable. For if they

... in destillaciō we seke ỹ separation of the elemēts either for one or mo of them ... separation truely can not be don withoute heate. For heat uniteth and gathereth together suche thinges as be of one kinde and nature, and they that do differ and disagree it separateth....[45]

It was because of this conviction that the chemists could argue that their distilled medicines were superior to the traditional Galenic herbs: obviously the pure medicinal quintessence had been separated from the gross vegetable matter.

The belief that analysis by fire would duplicate the separation obtained by a slow natural putrefaction was widespread. Jean Beguin wrote:

> Le Chymiste doit proceder en tous ses examens, theories, & operations par ces trois principles autrement ses cognoissances & artifices seroient sons fondement, & hors de ses principes ... la corruption qui est la resolution naturelle des choses, s'arrestoit à ces principes, & ne les pouvoit plus resoudre en d'autres.[46]

And, in his theoretical introduction to the *Basilica chymica*, Oswald Crollius insisted that visible experiments by the spagyrists provided indisputable evidence that natural bodies could not be divided into more than three substances, "mercury or liquor, sulfur or oil, and salt."[47] Both Beguin and Crollius pointed to the example of the burning twig: here the smoke indicated the presence

---

are perishable, it is clearly necessary that they too must be derived from something else, *since everything passes upon dissolution into that from which it is derived.* Hence it follows that there are other principles prior to the first principles; but this is impossible, whether the series stops or proceeds to infinity." *The Metaphysics*, Bks. I-IX, trans. Hugh Tredennick (London: Heinemann; New York: Putnam's Sons, 1933), pp. 131-133 (italics added). On the other hand, elsewhere Aristotle stated that heat and cold "are observed to determine, combine and change things both of the same and of different kind...." *Meteorologica*, trans. H.D.P. Lee (London: Heinemann; Cambridge, Mass.: Harvard University Press, 1952), p. 291, Bk. IV, Ch. 1.

[45][Conrad Gesner], *The Treasure of Euonymus* ..., trans. Peter Morwyng (London: John Daie, 1559), p. 67.

[46]Jean Beguin, *Les elemens de chymie* (Lyon: Pierre & Claude Rigaud, 1666), p. 32.

[47]Crollius, *Basilica chymica* (1623), p. 18.

The Chemical Philosophy

of mercury, the flame, sulfur, and the ash, salt.[48] This was little comfort to the Aristotelians, who had traditionally utilized the same example to show the four elements: earth (ash), water (sap), air (smoke), and fire (visible in burning).[49]

Perhaps even more influential was the judgment of the Paracelsian scholar Peter Severinus. For him there were three main classes of bodies: solids, inflammable oils, and ordinary liquors. These could be generally classified as salt, sulfur, and mercury provided that the reader did not take these words to mean the substances commonly designated by them. In his plea for fresh observations and the collection of samples to be subjected to analysis by fire, Severinus concluded that all such chemical dissections or "anatomies" would result in visible evidence of the three principles.[50]

Some turned to the principles because they were attracted by

[48] *Ibid.* Beguin, *Les elemens*, p. 42. In both cases the source here is clearly Paracelsus, who stated in *Die 9 Bücher de Natura rerum* (see the *Sämtliche Werke*, Vol. XI, p. 348): "dan alles was im feur reucht und verreucht ist mercurius, was brennet und verbrennet ist sulphur und alles was aschern ist, das ist auch ein sal." Paracelsus, *Nine Books of the Nature of Things* in Michael Sendivogius, *A New Light of Alchymy* . . . , trans. J[ohn] F[rench] (London: A. Clark for Tho. Williams, 1674), p. 219.

[49] In 1661 Robert Boyle used this example as that proposed by the Aristotelians in their defense of the four elements. *The Sceptical Chymist* . . . (London: J. Cadwell for J. Crooke, 1661; reprinted London: Dawson, 1965), pp. 21-22.

[50] Severinus, *Idea*, pp. 36, 40. It is interesting to note the direct influence of Severinus on Francis Bacon. Willing to attack both Galen and Paracelsus and their followers, he added, "Yet I count them not all alike; forasmuch as there is a useful sort of them, who not very solicitous about theories, do by a kind of mechanic subtlety lay hold of the extensions of things; such is Severinus." *Of the Interpretation of Nature,* in Francis Bacon, *Works,* ed. Basil Montagu, 3 vols. (Philadelphia, 1842), Vol. II, pp. 546-547. The influence is seen further in Bacon's paraphrase of Severinus in *On the Advancement of Learning*: "Another defect I note, wherein I shall need some alchemist to help me, who call upon men to sell their books and to build furnaces; quitting and forsaking Minerva and the Muses as barren virgins, and relying upon Vulcan. But certain it is that unto the deep, fruitful, and operative study of many sciences, specially natural philosophy and physic, books be not only the instrumentals. . . ." *The Philosophical Works of Francis Bacon . . . Reprinted from the Texts and Translations, with the Notes and Prefaces of Ellis and Spedding,* ed. John M. Robertson (London: Routledge and Sons; New York: Dutton, 1905), p. 77.

the trinitarian analogy of body, soul, and spirit (in the same way that others had upheld the four elements through their analogy with the four evangelists);[51] others turned to them in search of an alternative to the humors. For some chemical theorists they represented philosophical substances which could never be isolated in reality; for others they were precisely what should be sought in the laboratory. The Polish philosopher-chemist Michael Sendivogius (1556-1636) advised his readers in 1604 to avoid spending an undue amount of time investigating the elements, because only God could create from them:

> ... neither canst thou out of them produce any thing but these three Principles, seeing Nature her self can produce nothing else out of them. If therefore thou canst out of the Elements produce nothing but these three Principles, wherefore then is that vain labour of thine to seek after, or to endeavour to make that which Nature hath already made to thy hands? Is it not better to go three mile then four? Let it suffice thee then to have three Principles, out of which Nature doth produce all things in the Earth, and upon the Earth; which three we find to be entirely in every thing.[52]

There is little doubt that the practical pharmacist and others who worked in the laboratory sought evidence of these principles in their distillation products, whatever their ultimate argument might have been—theological, medical, chemical, or pharmaceutical.

Chemistry and the Geocosm

The concept of a chemical universe went beyond the chemical interpretation of the Creation and the problems of element theory.

[51] On "equating" the four elements with the four evangelists see *The Last Will and Testament of Basil Valentine, Monke of the Order of St. Bennet . . . To which is added Two Treatises The First declaring his Manual Operations. The Second shewing things Natural and Supernatural. Never before Published in English*, trans. J(ohn) W(ebster) (London: S.G. and B.G. for Edward Brewster, 1671), p. 123.

[52] Sendivogius, *New Light of Alchymy*, p. 116-117. Although it is generally assumed that the *Novum lumen chymicum* should be ascribed to the Scottish alchemist Alexander Seton, this has recently been disputed: W. Hubicki, "Michael Sendivogius's Theory, its Origin and Significance in the History of Chemistry," *Actes du Dixième Congrès International d'Histoire des Sciences*, 2 vols. (Paris: Hermann, 1964), Vol. II, pp. 829-833. See also W. Hubicki, "Alexander von Suchten," *Sudhoffs Archiv für Geschichte der Medizin und der Naturwissenschaften*, **44** (1960), 54-63.

PLATE V Michael Sendivogius (1556-1636/1646). Engraving by J. C. de Reinsperger (1763) from an earlier original.

Since the divine Creation was best understood as a chemical process, then nature must continue to operate in chemical terms. Chemistry was the true key to nature—all created nature. There was a concurrent belief in the macrocosm-microcosm relationship: the universe was conceived to be a vast system with all its parts interconnected; in particular, man was seen as a small copy of the great world, and therefore it was valid to draw analogies between him and the earth or the universe as a whole.

In this system life and motion derived only from the Creator. It was the spirit of the Lord reaching through the air that made this total life system possible.[53] In the *Liber azoth* (1591, possibly spurious) Paracelsus wrote about a valid analogy seen in life and combustion:

> Since fire cannot burn without the presence of air, one may say of the element of fire that it is of itself nothing other than a body to the soul or perhaps a house in which the soul of man lives. Therefore fire is the true man about which our whole philosophy is concerned. And if I have now said that no fire can burn without air, I will go on to say that this should be understood correctly—that one should understand under this burning always life. So if I should say by way of an example that something cannot burn, then I also mean that it cannot live.[54]

From the Paracelsian corpus we may conclude that air is necessary both for combustion and respiration, and further that these two superficially different processes are essentially the same. Beyond this, the vital property in air is really a starry emanation or a heavenly fire which was in a limited fashion referred to by

---

[53] See Allen G. Debus, "Motion in the Chemical Texts of the Renaissance," *Isis*, 64 (1973), 4-17, and "The Paracelsian Aerial Niter," *Isis*, 55 (1964), 43-61.

[54] Paracelsus, *Liber azoth* in the *Opera Bücher und Schrifften*, Huser ed., Vol. II, p. 520. "Dieweil kein Fewr brennen kan / es habe dann Lufft: also ist es nun zu reden von dem Elemento Ignis, welches nichts anders an ihm selbs ist / dann ein Leib der Seelen / oder ein Hauss / darinnen die Seel dess Menschen wohnet: Und also ist dasselbige Fewr der rechte Mensch / von dem da geredt werden soll in der gantzen Philosophia. Darumb wie gemeldet / dass kein Fewr brennen kan / es habe dann Lufft: Also will ich / das man mich in diesem Ort sonderlich wol verstehe / dass ich solch Brennen allwegen für dass Leben will verstanden haben. Als wann ich sage: Es kan nicht brennen / ist gleich so viel / als sagte ich / es kan nicht leben."

Paracelsus as an aerial niter or saltpeter.[55] His followers were to develop this concept much further.

The Paracelsians seem to have agreed that the font of life was the heavenly spirit, which was found in the sun. Surely no life could survive without the rays of sunlight. For Jean d'Espagnet (1608) it was not unlikely that "the soul of the World was in the Sun, and the Sun in the Centre of the whole."[56] But, in what fashion did these rays benefit the earth? Again, chemical analogies were advanced as a means of explanation. There were numerous chemical circulations in nature and these moved substances and purified them. D'Espagnet, for instance, described the formation of rain in the following fashion:

> The lower Region of the Air is like unto the neck or higher part of an Alembick, for through it the Vapours climbing up, and being brought to the top, receive their condensation from Cold, and being resolved into water, fall down by reason of their own weight. So Nature through continued distillations by sublimation of the Water, by cohobation, or by often drawing off the liquors being often poured on, the body doth rectifie and abound it. In these operations of Nature, the Earth is the Vessel receiving. Therefore the Region of the Air that is nearer to us, being bounded by the Region of Clouds, as by a vaulted chamber, is of a greater thickness and impurity than those Regions above.[57]

For Sendivogius there was a central sun in the earth corresponding to the celestial sun. The four elements distilled into the center of the earth a radical moisture which the central sun then sublimed to the surface. There the beams of the two suns joined "to produce flowers and all things."

> Therefore when there is Rain made, it receives from the Air that power of life, and joyns it with the Salt-nitre of the Earth . . . and by how much the more abundantly the Beams of the Sun beat upon it, the greater quantity of Saltnitre is made, and by consequence the greater plenty of Corn grows, and is increased, and this is done daily.[58]

Again this theme is one that derives from Paracelsus, who dis-

---

[55] See Debus, "Paracelsian Aerial Niter," pp. 46-48.
[56] Jean d'Espagnet, *Enchyridion physicae restitutae; or, the Summary of Physicks Recovered* (London, 1651), p. 19. This work was published first in Paris in 1608.
[57] *Ibid.*, p. 52.
[58] Sendivogius, *New Light of Alchymy*, pp. 44-45.

cussed the relationship of water, salt, and dung to growth in his *Von den Natürlichen Wassern*.[59] It is likely from this source that the French Paracelsist Bernard Palissy developed his concept that fertilizers derive their importance from water-soluble salts.[60] In England, as early as 1594, Hugh Plat quoted Palissy at length in order to promote the use of artificial and natural fertilizers for the benefit of the economy of the country.[61]

Although internal distillations within the earth might be utilized to explain the growth of vegetable matter, they could explain other natural phenomena as well. For Sendivogius such distillations were the cause of mountain streams. These were produced by the Lord God, pictured as the eternal Distiller. Thus, he argued that springs result not from starry emanations or virtues, but rather from a divine earthly process:

> ... there are in the bowels of the Earth places in which Nature distills and separates a sulphureous Mine, where, by the Central Fire it is kindled. The Waters running through these burning places, according to the neerness or remoteness are more or less hot, and so breaks forth into the superficies of the Earth, and retains the tast of Sulphur, as all broth doth of the Flesh that is boiled in it. After the same manner it is, when Water passing through places where are Minerals, as Copper, Allum, doth acquire the savour of them. Such therefore is the Distiller, the Maker of all things, in whose hands is this Distillatory, according to the example of which all Distillations have been invented by Philosophers; which thing the most High God himself out of pity, without doubt, hath inspired into the Sons of Men: and he can, when it is his holy Will, either extinguish the Central Fire, or break the Vessel, and then there will be an end of all.[62]

---

[59]Paracelsus, *Opera*, Huser ed., Vol. II, p. 150; in Latin translation see the *Opera omnia* (1658), Vol. II, pp. 359-360.

[60]Discussed by Allen G. Debus in "Palissy, Plat and English Agricultural Chemistry in the 16th and 17th Centuries," *Archives Internationales d'Histoire des Sciences*, **21** (1968), 67-88 (74-76). See also below, Ch. 6. A survey of the literature relating to chemical theory and agriculture is contained in the bibliography of G.E. Fussell's *Crop Nutrition: Science and Practice before Liebig* (Lawrence, Kansas: Coronado Press, 1971).

[61]Debus, "Palissy," pp. 69-82. Hugh Plat's "Paracelsian agriculture" is discussed in his *The Jewell House of Art and Nature. Conteining diuers rare and profitable Inuentions, together with sundry new Experiments in the Art of Husbandry, Distillation and Moulding* (London: Peter Short, 1594).

[62]Sendivogius, *New Light of Alchymy*, pp. 94-95. On seventeenth-century

It was perhaps inevitable that these men should turn to the old belief in a great central fire as an explanation for earthly phenomena.[63] The effects of volcanic action seemed to provide visible proof of such a fire, while the high temperatures of many mineral water spas and deep mines gave further evidence. Yet, how might such a fire be kindled and maintained?[64] Some said that this fire originated through the celestial bodies, with rays of heat from the stars and the planets focusing on the earth and generating fire by friction with one another. These celestial virtues then met with otherwise inert matter to give rise to the metals. Thus, Sendivogius spoke of the influence of celestial rays and of the internal distillations and sublimations that produced the metals within the earth,[65] while Johann Rudolph Glauber (1603/1604-1668/1670) wrote of the sidereal rays that strike the earth and "are contracted into a streight room, and (driven back) from thence are sublimed and distilled throughout the whole Orb, from which all kind of Metals and Minerals (by the help of the Earth and Water corporifying them) are produced."[66] For Glauber the internal fire differed greatly from the superficial fires that brought forth volcanoes, but all would not agree. The question of fuel was an important one. Agricola could not accept a sidereal explanation, and he insisted on more natural causes: bitumen and sulfur deposits beneath the surface were ignited and gave the only possible natural explanation of the observed phenomena. As Adams has summarized this argument, these deposits existing beneath volcanic centers "were ignited by intensely heated vapors which, in their

---

geology and the central fire see Robert Lenoble, *La géologie au milieu du XVIIe siècle* (Paris: University of Paris, 1954). J. J. Becher discusses Fromondus, Cabeus, and other authors who upheld the distillation theory of the origin of mountain streams in his *Physica subterranea* (1st ed., 1669; 3rd ed., Leipzig: Weidmann, 1738), p. 35.

[63] See Francis M. Cornford's discussion of the concept of the central fire among the Pythagoreans in his *Plato's Cosmology* (New York: Liberal Arts Press, 1957), pp. 126-130.

[64] See Frank Dawson Adams' discussion in *The Birth and Development of the Geological Sciences* (New York: Dover, 1954), pp. 279-286.

[65] Sendivogius, *New Light of Alchymy*, pp. 40-47.

[66] John Rudolph Glauber, *The Works . . . containing, Great Variety of Choice Secrets in Medicine and Alchymy . . .*, trans. Christopher Packe (London: Thomas Milbourn, 1689), p. 116.

PLATE VI Volcanoes depicted as being the result of the central fire in the earth. From Athanasius Kircher, *Mundus subterraneus* (Amsterdam: Janssonio-Waesbergiana, 1678), pp. 104-105.

turn, derived their heat from friction which was set up within the gaseous mass itself or by its contact with the walls of narrow spaces through which it was forced when in rapid movement within the earth."[67]

The significance of heat as a source of natural phenomena, combined with the chemists' search for laboratory examples, is seen to good advantage in Gabriel Plattes' explanation of the formation of the earth (1639).[68] Rejecting the common belief that rocks and minerals were growths on the land like warts and tumors on man, he argued instead that they were features formed from the vapors of bituminous and sulfureous substances. The veins of metals were engendered subsequently in the cracks and crannies of the mountains; hills and valleys resulted from the gradual action of the sea. In support of these views he suggested the following experiment:

> Let there bee had a great retort of Glasse, and let the same be halfe filled with Brimstone, Sea-coale, and as many bituminous and sulphurious subterraneall substances as can bee gotten: then fill the necke thereof halfe full with the most free earth from stones that can be found, but thrust it not in too hard, then let it be luted, and set in an open Furnace to distill with a temperate Fire, which may onely kindle the said substances, and if you worke exquisitely, you shall finde the said earth petrified, and turned into a Stone: you shall also finde cracks and chinkes in it, filled with the most tenacious, clammy, and viscous parts of the said vapours, which ascended from the subterraneall combustible substances.[69]

The interest in the origin of metals expressed by Plattes was shared by most Paracelsians, who borrowed heavily from their alchemical and metallurgical heritage. Indeed, their most valuable—and at the same time most controversial—contribution to medicine was their new emphasis on metallic compounds as medicinals. Again scriptural authority could be brought to bear, since it was possible to cite Ecclesiasticus 38:4, "The most High hath created medicines out of the earth: and a wise man will not abhor them" (Douay version).

---

[67]Adams, *Birth and Development*, p. 281.

[68]This is discussed in greater detail in Allen G. Debus, "Gabriel Plattes and his Chemical Theory of the Formation of the Earth's Crust," *Ambix*, **9** (1961), 162-165.

[69]Gabriel Plattes, *A Discovery of Subterraneall Treasure* (London, 1639), p. 6.

Few discussed the origin of metals at greater length than "Basil Valentine," an author whose works date from the late sixteenth and early seventeenth centuries, even though they were assigned a date nearly two centuries earlier.[70] The Basilian corpus may be best understood in a Paracelsian context. "Basil" was a thoroughgoing vitalist who assumed that few would disagree with his version of the macrocosm-microcosm universe. He wrote primarily to "promote the health of thy Neighbor Christian,"[71] and it is clear that his interest in his subject was motivated by religious as well as medical reasons. There is no doubt that the author was a practicing chemist with an extensive knowledge of medicine and contemporary metallurgical techniques.

All of these factors contribute to make Basil Valentine's views of considerable interest. For him heavenly and earthly phenomena were intimately connected:

> For you are to understand, that Heaven worketh upon the Earth, and the Earth keepeth correspondency with Heaven: for the Earth hath likewise seven Planets in it, which are brought forth and wrought upon by the seven Heavenly Planets, only by a spiritual impression and infusion; and in this manner all the Minerals are wrought by the Stars. This now is done spiritually beyond our apprehension, and therefore to be accounted not Natural, in the manner of two, that are enamoured. . . . when Heaven beareth love to the Earth, and the Earth hath love, inclination, and affection for Man, as the great World for the little one, because the little World is taken out of the great, and when the Earth through the desire of an invisible imagination doth attract such Love of the Heavens, then is there a conjunction made of the Superiour with the Inferiour, like unto a Husband and his Wife, which are accounted one body: And after such a conjunction the

---

[70] The arguments against the fifteenth-century date of the texts ascribed to Basil Valentine were marshalled by Hermann Kopp in his various publications, but see especially his *Beiträge zur Geschichte der Chemie*, 3 pts. (Braunschweig, 1869-1875). A discussion of the literature accompanies the present author's article, "Basil Valentine," in the *Dictionary of Scientific Biography*, Vol. XIII (New York: Scribner's, 1976). The authorship of "Basil Valentine's" texts is now most commonly attributed to Johann Thölde, a councillor and salt boiler of Frankenhausen in Thuringia, whose *Haligraphia* (1603 and 1612) closely resembles the *Letzes Testament* (1626), one of the principal works ascribed to Basil Valentine.

[71] Basil Valentine, "Preface" to *The Last Will and Testament* (1671), sig. A8^r. See also the "Darinnen angezeiget werden die Berg-Wercke . . ." in *Chymische Schrifften* (3rd ed., 2 vols., Hamburg: Gottfried Leibezeits, 1700), Vol. II, p. 10.

Earth becometh impregnated by such infusion of the superiour Heaven, and beginneth to bear a birth according to the infusion, which birth is ripened, after its conception, by the Elements, and is digested to a perfect maturity.[72]

... And as the Seed of a Man doth fall into the Womb, and toucheth the *Menstruum*, which is its earth; but the Seed, which goeth out of the Man into the Woman, is wrought in both by the Stars and Elements, that it may be united and nourished by the Earth to a generation: So you are likewise to understand, that the soul of Metals, which is conceived by an unperceiveable, invisible, incomprehensible, abstruse, and supernatural Celestial composition, as out of Water and Air, which are formed out of the Chaos, and then further digested and ripened by that heavenly Elemental light and fire of the Sun, whereby the Stars do move the Powers, when its heat in the inward parts of the Earth, as in the Womb is perceived:[73]

Thus, as the human fetus grows in its mother, so metals grow in the earth-mother after impregnation by the heavens. For Basil all metals have their appropriate seed. This derives from a divine celestial influence mixed with astral properties, which "two beget an Earthly substance, as a third thing, which is the beginning of our seed."[74] The growing metallic ore must have proper nutriment, which it finds in moist subterranean liquors and other growths (salts, vitriols, and *flores*) which abound in mines.[75] And, again in a fashion similar to animal life, metals must inhale the life-giving spirit which originates in the solar rays. Although this inspiration is invisible, it may readily be detected by the correct use of a rod (similar to the more common dowsing rod).[76] Metallic exhalation, in contrast, is often noted by miners at night. With normal growth, ores will progress to perfection (gold). That this natural transmutation is underway is verified by the presence of mixed ores in

---

[72]Basil Valentine, "Of Things Natural and Supernatural" in the *Last Will*, pp. 469-470; "Von den natürlichen und übernatürlichen Dingen" in *Chymische Schrifften*, Vol. I, pp. 216-217.

[73]*Ibid.* (English), p. 512; (German), p. 264.

[74]Basil Valentine, "Of the Great Stone of the Philosophers" in the *Last Will*, p. 221; "Von dem grossen Stein der uhralten Weisen" in *Chymische Schrifften*, Vol. I, p. 11.

[75]Basil Valentine, *Last Will* (Pt. I), pp. 7-8. "Berg-Wercke," in *Chymische Schrifften*, Vol. II, pp. 22-23.

[76]*Ibid.* (English), pp. 41-43, 56; (German), Vol. II, pp. 64-67, 91.

most mines.⁷⁷ However, metallic ores are also subject to infirmities no less than human beings. An indication of such inorganic illness is the malignant breathing (vapors) occasionally present in mines which endangers the life of miners. These diseases are normally associated with the presence of salt.⁷⁸

The Microcosm and Medical Theory

The Paracelsian chemical philosophy was rightly considered a call for a new observational approach to all nature, but from the beginning it carried a special appeal for physicians. Paracelsus insisted that God rather than the constellations created him a physician, and later van Helmont was to repeat this and add that because of its divine origin medicine stood above the other sciences.⁷⁹ Here they both reflected the priest-physician concept which was a fundamental part of Renaissance neo-Platonism, and it is likely that their ultimate source was Ecclesiasticus 38:1, "Honor the physician for the need thou hast of him: for the most high hath created him."

For Paracelsus the role of the physician was properly compared to that of the true natural magician. Basing his views on the interrelation of the macrocosm and the microcosm, he explained that the godly magus may concentrate in himself celestial virtues which are the hidden powers of nature. He may then use these powers to work wonders and to learn of his Creation.⁸⁰ The magus transfers the powers of a celestial field into a small stone; the physician extracts the hidden virtues of herbs and prepares powerful remedies which he then uses to cure the sick and the infirm. While a saint operates directly by means of God's grace, the magus and the physician operate by means of God's Creation—nature.⁸¹

The microcosm concept was indeed fundamental for the

---

⁷⁷*Ibid.* (English), pp. 43-45, 31; (German), Vol. II, pp. 67-70, 51.

⁷⁸*Ibid.* (English), p. 69; (German), Vol. II, p. 115.

⁷⁹From the *Paragranum*, in *Sämtliche Werke*, Vol. VIII, pp. 63-65. This passage is quoted at length above in Ch. 1. J.B. van Helmont, *Oriatrike or Physick Refined*, trans. John Chandler (London: Lodowick Loyd, 1662), p. 4.

⁸⁰*Philosophia sagax* in the *Opera omnia*, Bitiskius ed., Vol. II, p. 555.

⁸¹See Walter Pagel's description of *magia naturalis* in his *Paracelsus*, pp. 62-65. See also D. P. Walker, *Spiritual and Demonic Magic from Ficino to Campanella* (London: Warburg Institute, 1958).

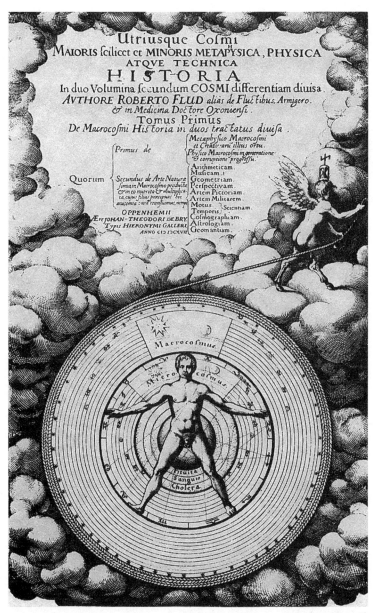

PLATE VII The macrocosm and the microcosm as portrayed on the title page of Robert Fludd's *Utriusque cosmi maioris scilicet minoris metaphysica, physica atque technica historia* (Oppenheim: J. T. De Bry, 1617).

orthodox Paracelsian: man is a small replica of the great universe and within him are represented all parts of that universe.[82] We have seen how Basil Valentine confidently described the origin of metals in terms of the analogy between microcosm and macrocosm. In his most influential work, the *Triumph Wagen des Antimonii* (1604), he again wrote with conviction about the correspondence of the two worlds, this time revealed in an alchemical dream sequence:

> Being removed from all worldly care by the fervour of prayer and heavenly thoughts, I determined to yield up my soul to those spiritual inspirations without which it is impossible to have a right knowledge of created things. I proposed to furnish myself with wings wherewith I might ascend to the stars and inspect the heavens, as Icarus had done before me, if we may believe the old writers.
>
> But, when I approached the sun too closely, my feathers were consumed by his burning heat, and I fell headlong into the sea; then in answer to my prayer, God sent an angel to help me, who bade the waves be still, and caused a great mountain to arise in the midst of the water. I ascended it to see whether there really existed that correspondence between things below and things above of which so much has been said, and whether the stars possess the power of producing things resembling them on earth. As a result of my investigation I found that what the ancient Doctors have delivered to their disciples was God's own truth. Therefore I rendered profound thanks to the Lord of Heaven and earth for His wonderful works.[83]

Although this passage could be dismissed as the poetical digression of an author more profitably remembered for his relation of more factual material, there is no doubt that the macrocosm-microcosm universe enjoyed widespread popularity throughout most of the seventeenth century. Even a chemist as prominent as Johann Rudolf Glauber was able to accept this as a basic article of faith in the second half of the century.[84]

[82]Paracelsus' doctrine of the microcosm is discussed in the *Philosophia sagax* (*Opera omnia*, Bitiskius ed., Vol. II, p. 601) and is described by Pagel in *Paracelsus*, pp. 65-68. In his attack on Paracelsus in 1572, Thomas Erastus placed special emphasis on this doctrine (*Paracelsus*, pp. 323-324).

[83]Basil Valentine, *The Triumphal Chariot of Antimony with the Commentary of Theodore Kerckringius*, trans. Arthur Edward Waite (London: James Elliott, 1893), pp. 183-184; *Triumph Wagen des Antimonii*, in the *Chymische Schrifften*, Vol. I, pp. 439-440.

[84]Glauber, *Works* (Packe translation), Pt. 2, pp. 30-31.

Those chemists who believed in the two-world doctrine placed a new emphasis on the ancient doctrine of signatures. It was easy for them to believe that God had "signed" all things on earth for the benefit of His Creation—man—and that the true magus would be performing a service to his Creator by discovering these divine secrets. However, they disagreed intensely with those earlier authors who sought to understand and to identify these signs only through external appearances. Rather, they argued, internal essences were the true signatures, and these could be revealed in the chemist's laboratory. The literature of the period contains frequent exhortations to seek natural truths with the aid of the fire. Thus, Crollius in 1609 stated that the internal signature, force, or occult virtue

> ... (which as Natures Gift is insited, and infused by the most high God, into the Plant or Anima, from the Signature and mutual Analogick Sympathy and harmonious concordance of Plants, with the Members of the Human Body,) is by the prudent Physitian only inquired into: and thence by the industrious help of *Vulcan*, or *Anatomick Knife*, is drawn out and applied to its proper use, not drousily passed over in noxious Silence, as is by Vulgar *Herbarists* too frequently done.[85]

In the search for these signatures, evidence for the resuscitation of plants from their ashes (palingenesis) assumed a special significance.[86] Like Paracelsus, Joseph Duchesne rejected external signs in favor of those internal essences which could be uncovered through the application of fire.[87] His account of the resuscitation of

[85]Oswald Crollius and John Hartman, *Bazilica chymica, & Praxis chymiatricae or Royal and Practical Chymistry in Three Treatises*, trans. Richard Russell (London: John Starkey and Thomas Passinger, 1669, 1670), sig. A2^V from the preface to *A Treatise of Oswaldus Crollius of Signatures of Internal Things; or, a True and Lively Anatomy of the Greater and Lesser World* (1669). The Latin text is in Crollius, *Tractatus de Signaturis internis rerum, seu de vera et viva anatomia majoris & minoris mundi*, pp. 1-2, found with a separate title page and pagination in his *Basilica chymica* (1623).

[86]On this subject see Jacques Marx, "Alchimie et palingénésie," *Isis*, **62** (1971), 274-289, and Allen G. Debus, "A Further Note on Palingenesis: The Account of Ebenezer Sibly in the *Illustration of Astrology* (1972)," *Isis*, **64** (1973), 226-230.

[87]Joseph Duchesne (Quercetanus), "De signaturis rerum internis seu specificis, ab Hermeticis philosophis multa cura singulariéque industria comparatis, atque introductis," pp. 104-153 (104-106) in *Liber de priscorum philsophorum verae medicinae materia, praeparationis modo, atque in curandis morbis, praestantia* (n.p. Thomas Schürer and Barthol. Voight, 1613; 1st ed., 1603).

PLATE VIII An eighteenth-century chemist exhibiting the phenomenon of palingenesis. Ebenezer Sibly, *A New and Complete Illustration of Astrology* . . ., 4 parts (London: W. Nicoll for M. and E. Sibly, 1784, 1792), Pt. 4, Plt. 39, opposite p. 1115.

plants from their ashes at a demonstration at Cracow proved to be especially significant, for it was widely quoted in the literature of the seventeenth and early eighteenth centuries.[88] Although admitting that he could not repeat this phenomenon himself, Duchesne did seek observational evidence that such essences would not be destroyed by heat. Accordingly, after burning nettles he dissolved their ashes in caustic, which was subsequently cooled until the onset of freezing. Noting the jagged spikes of ice, Duchesne argued that through their influence, the ashes had recreated their original form. Although the account was questioned by some of Duchesne's contemporaries,[89] the example does indicate a search for evidence to support the chemist's interpretation of the doctrine of signatures.

For the Paracelsians it was at all times considered a fruitful field of inquiry to seek out correspondences between the greater and lesser worlds, and the theory of sympathy and antipathy was employed to explain universal interaction. In contrast to Aristotelians, who insisted on action through contact, the Paracel-

---

[88]E.g., James Gaffarel, *Vnheard-of Curiosities: Concerning the Talismanical Sculpture of the Persians; The Horoscope of the Patriarkes; And the Reading of the Stars,* trans. Edmund Chilmead (London: Humphrey Moseley, 1650), pp. 135-138. Duchesne's account of this observation will be found in his *Ad veritatem Hermeticae medicinae ex Hippocratis veterumque decretis ac therapeusi; nec non viuae rerum anatomiae exegesi, ipsiusque naturae luce stabiliendam, adversus cuiusdam Anonymi phantasmata responsio* (Frankfurt: Wolffgang Richter and Conrad Nebenius, 1605; 1st ed., Paris, 1604), pp. 233-235.

[89]The English Helmontian Noah Biggs argued that all ice begins to crystallize in this fashion and that if the seminal substance really did exert this influence, the resultant form would resemble, not the leaves of the plant, but either its root or fruit. See Allen G. Debus, "Paracelsian Medicine: Noah Biggs and the Problem of Medical Reform," in *Medicine in Seventeenth-Century England,* ed. Allen G. Debus (Berkeley: University of California Press, 1974), pp. 33-48. Other aspects of Biggs' work are discussed below, Chs. 6 and 7. It appears that Biggs borrowed his criticism from van Helmont, who faulted Duchesne precisely on this point. J. B. van Helmont, "Pharmacopolium ac dispensatorium modernum," in *Ortus medicinae* (Amsterdam: Elzevier, 1648; reprint Brussels: Culture et Civilisation, 1966), pp. 459-460. See also van Helmont's letter to Mersenne, Sept. 26, 1630 (*Correspondance du P. Marin Mersenne. Religieux Minime,* publiée par Mme. Paul Tannery, éditée et annotée par Cornelis de Waard, [Paris: P.U.F. Centre National de la Recherche Scientifique, 1932—], Vol. II [1945], p. 209), where the phenomenon is discussed as part of van Helmont's critique of the work of Gaffarel.

sians thought of natural phenomena largely in terms of action at a distance. Here too, there was an important break with the past, and it is easy to understand why Paracelsian-Hermeticists should have been among the first to defend the experimental research of William Gilbert (1544-1603) on the magnet. For the Paracelsians *De magnete* offered numerous experiments with the loadstone—which could best be explained in terms of their cosmology, in which sympathetic action played a fundamental part. In medicine this meant specifically that the weapon-salve might be accepted as a valid cure. Here it was argued that a salve including blood from the injured person would prove effective when applied to the weapon that caused the wound. This cure was described by many authors and repeatedly ascribed to magnetic or sympathetic influences. The complex seventeenth-century controversy on this subject may be seen as a test case involving the broader implications of action at a distance.[90]

The medical theory of the Paracelsians was influenced by religious principles, by the macrocosm-microcosm theory, and also by current concepts of the elements and the principles. For the Paracelsian the traditional Galenic explanation of disease as an internal imbalance of the four humors was no longer adequate. Some Paracelsians, most notably Duchesne, proposed a reformed system of three humors which were to be closely connected with the three principles.[91] Others, criticizing methods of diagnosis such as "water-casting," argued that if the analysis of urine had any value at all, it must be based upon chemical distillation rather than simple visual observation.

Even more indicative of change was a new concept of disease which sought the cause less in internal imbalances of fluids than in external factors which entered the body and became localized in

[90] The subject is discussed in greater detail in Allen G. Debus, "Robert Fludd and the Use of Gilbert's *De Magnete* in the Weapon-Salve Controversy," *Journal of the History of Medicine and Allied Sciences*, **19** (1964), 389-417, and below, Ch. 4. Robert Fludd discusses the relationship of magnetic to astral force in his *Philosophia Moysaica* (Gouda: Petrus Rammazenius, 1638), fol. 116^v. The most comprehensive seventeenth-century collection of the weapon-salve texts is in the *Theatrum sympatheticum auctum* (Nuremburg: J. A. Endterum & Wolfgangi Juniores Haeredes, 1662), where it is interesting to note discussions of the similarities and differences between magnetic and electrical attraction (pp. 302-303).

[91] Duchesne, *Practise of Chymicall . . . Physicke*, sig. L1^r.

PLATE IX Examples of sympathetic action in nature. From *Theatrum sympatheticum auctum, exhibens varios authores. De pulvere sympathetico quidem: Digbaeum, Straussium, Papinium, et Mohyum. De unguento verò armario: Goclenium, Robertum, Helmontium, Robertum Fluddum,* .... (Nuremberg: Endter, 1662), p. 125.

specific organs. Here again was the analogy between the macrocosm and the microcosm: in the same fashion that metallic "seeds" in the earth resulted in the growth of the veins of metals, so too did "seeds" of disease grow within the body while they combatted the local life force, or *archeus*, of a specific organ.[92] Under normal conditions of health these *archei* acted much as alchemists separating a pure substance from its impurities (as in a distillation). The *archeus* of the stomach served as an excellent example: here it seemed obvious that the life force separated the useless from the nutritional portions of the food, the former being discarded while the latter was directed to those parts of the body where it was needed. However, should the *archei* be impaired, either through disease or some other cause, they could no longer act in this fashion, and the result would be a build-up of poisonous waste within the body. A classic example was the "tartaric" disease of Paracelsus—malfunctions manifested in the growth of stony precipitates, including the calculus, the tartar of the teeth, and the calcifying material in the lungs in a case of tuberculosis.[93]

The relationship of the macrocosm to man also had chemical implications. Duchesne reflected both the Aristotelian position on catarrh and the persistent Paracelsian search for chemical analogies when he spoke of respiratory diseases (characterized by the formation of phlegm at the nose and a high bodily temperature) in terms of the same distillation analogy used by other iatrochemists in explaining the origin of mountain streams.[94] This was a long-lived theme. In the second half of the century Thomas Willis was still willing to use the distillation analogy when describing the separation of the animal spirits in the brain.[95]

Special significance was attached to the air. Paracelsus, as has already been mentioned, recognized that air is essential for the maintenance of both fire and life. He also postulated the existence of an aerial sulfur and an aerial niter which could react in the

[92] The Paracelsian *archeus* is discussed by Pagel in his *Paracelsus*, pp. 104-110.
[93] *Ibid.*, pp. 153-161.
[94] Joseph Duchesne, *Traicté de la matiere, preparation et excellente vertu de la medecine balsamique des anciens philosophes* (Paris: C. Morel, 1626: French trans. of the *De priscorum philosophorum*, 1603, noted above, n. 87), p. 183.
[95] Thomas Willis, *Of Fermentation* in *Practice of Physick*, trans. S. Pordage (London, 1684), pp. 12-13 (separately paginated). The *De Fermentatione* appeared first in 1659. This is quoted at length in the section on Willis below, Ch. 7.

atmosphere—as they did analogously in solid form in gunpowder—in the form of an explosion.[96] This was a common explanation for thunder and lightning. Peter Severinus was willing to ascribe the flux and reflux of the sea to nitroso-sulfureous spirits,[97] and Henricus de Rochas, in his search for an alternative to the central fire in the earth, suggested that the heat of mineral water springs derived from the reaction of sulfur and a nitrous salt in the earth.[98] Within the human body the gunpowder analogy was utilized by Paracelsus and his followers to explain diseases characterized by hot and burning qualities.[99] And, again, as late as John Mayow's *Tractatus quinque* (1674), the reaction of aerial nitrous particles with aerial sulfureous particles was discussed as a cause of illness and directly compared with thunder and lightning in the atmosphere.[100] Similarly, as Paracelsus had insisted on the need for a sal nitric food for muscular action, Mayow based his tract on muscular motion on his views of the nitro-aerial particles.[101]

By the early years of the seventeenth century the aerial niter

[96]Paracelsus, *Grossen Wundarznei* in the *Chirurgische Bücher und Schrifften*, ed. J. Huser (Strasbourg: Zetzner, 1618), p. 52. For other references see, in the *Opera*, Huser ed., Vol. II, *Philosophiae de generationibus & fructibus elementorum*, p. 32; *Dess Buchs Meteorum*, pp. 82-83, where he speaks of a "windischer Salpeter"; pp. 89-92, for an "etherisch Salniter", and p. 89: "In diesen Stralischen sternen ist der *Sulphur* in höchsten Grade / der Salpeter in der höchsten Feiste / der *Mercurius* in der höchsten Contrarietet: Dann der Sulphur ist das *Corpus* / *Nitrum* ist der *Spiritus*, und *Mercurius* ist *Contrarium* in sie beide / also / scherpffer dann der *Nitrum*, und gewaltiger dann *Sulphur*. Von den dreien wird der Donner / Strall / Wetter / das das aller grössest ist des gantzen Firmaments."

[97]Severinus, *Idea* (1660), p. 91.

[98]As described by John French in his *Art of Distillation* (1650) (4th ed., London: E. Cotes for T. Williams, 1674), pp. 94-95.

[99]Paracelsus, *Bertheonae* in the *Chirurgische Bücher*, p. 354. For Paracelsus some fistulas and ulcers were due exclusively to sal niter. See the *Grossen Wundarznei* in *Chirurgische Bücher*, p. 85. Severinus insisted that burning fevers resulted from "nitroso-sulphureous" astral emanations. (*Idea*, pp. 91, 99, 133, 164.)

[100]John Mayow, *Tractatus quinque*, translated as the *Medico-Physical Works* by A.C.B. and L.D. (Edinburgh: Alembic Club, 1907), pp. 256-257, 259, 279, 147-152.

[101]Paracelsus, *Liber azoth* in *Opera*, Huser ed., Vol. II, p. 525. "Und das dritte wesen *Salis Nitri* genannt / das seind die *Musculi* an allen orten / die essen auch Salnitrische speise / *Particulariter* und nit *Universaliter*." Mayow, *Tractatus*, pp. 229-302.

had become associated with a life force requisite for man and on occasion identified with the *spiritus mundi*.[102] Already hinted at in the Paracelsian *Liber azoth* (1591),[103] the presence of the life force in saltpeter was specifically discussed by Duchesne in 1603.[104] And for Robert Fludd (whose work will be discussed in more detail later) this vital spirit of the air was also identified with saltpeter, which, after having been separated from gross air in the respiration process, was formed into arterial blood. With the survival of this concept it is little wonder that we find later seventeenth-century Helmontians rejecting the common practice of bloodletting.[105] The rejection of bloodletting also reflected the Paracelsian-Helmontian opposition to traditional humoral pathology.

The Chemical Analysis of Spa Waters

If the Paracelsian chemical philosophy of nature provided a conceptual framework for the iatrochemist, it also provided a basis for his practical work. Diagnosis by the inspection of urine was replaced by chemical distillation and (with van Helmont) by a

[102]Debus, "The Paracelsian Aerial Niter," pp. 43-61; see especially the work of Nicasius Le Febure, Johann Rudolph Glauber, and William Simpson described on pp. 58-59.

[103]In the possibly spurious *Liber azoth* Paracelsus refers to two lives: the Iliastric (prior to the Fall of Man) and the Cagastric (post-Fall). The body needs air as a cooling agent for the Cagastric soul, even though a remnant of the Iliastric life and soul remains. At one point in the *Liber azoth* the Cagastric life is equated with sal niter (*Opera*, Huser ed., Vol. II, p. 539), but elsewhere Paracelsus separates the two terms and insists that while the Cagastric soul depends or feeds on the soul of the greater world, the sal nitric life feeds on a sal nitric food. However, even in this case it is clearly stated that "It is particularly important to speak of the air . . . for in the air is the power of all life," and he includes here "the other life which we call S.S. nitri" (p. 525).

[104]Duchesne, *Practise of Chymicall . . . Physicke*, sigs. $D2^V$, $O4^r$, P1-P4. Duchesne argues that there is a volatile sulfur in saltpeter, and this is the *nectar vitae* for animals. See also the *De priscorum philosophorum* (S. Gervasii, 1603), p. 20. This passage was alluded to again by Duchesne in his *Ad veritatem Hermeticae medicinae* (Paris, 1604), p. 209. (These texts are discussed in Debus, "Paracelsian Aerial Niter," pp. 52-54.)

[105]The examples of Noah Biggs (1651) and George Starkey (1657) are discussed by Debus in "Paracelsian Doctrine in English Medicine" (see n. 2 above), p. 17. On this subject see also Peter Niebyl, "Galen, van Helmont and Bloodletting" in *Science, Medicine and Society*, pp. 13-23.

specific gravity test which was to become fundamental for modern urinalysis.[106] Paracelsians also turned to the healing powers of mineral water spas and sought to determine the ingredients of these springs by greatly expanding the known analytical tests for aqueous solutions.[107] The long medieval tradition in this field had resulted in the development not only of isolated tests but of real analytical procedures.[108] Perhaps the best example of this non-Paracelsian chemical tradition is the *De medicatis aquis* (1564) of Gabriel Fallopius (1523-1563), which includes both physical and chemical tests as part of a detailed method of analysis.[109]

The Paracelsians readily adopted this chemical tradition.[110] By 1571 Leonard Thurneisser was using quantitative methods, solubility tests, crystallographic evidence, and some flame tests in an attempt to determine the active substances in spa waters.[111] The

---

[106] Pagel, *Paracelsus*, pp. 189-200. See also Debus, *English Paracelsians*, pp. 157-158.

[107] Gernot Rath, "Die Anfänge der Mineral Quellenanalyse," *Medizinischen Monatsschrift*, 3 (1949), 539-541; "Die Mineralquellenanalyse im 17. Jahrhundert," *Sudhoffs Archiv für Geschichte der Medizin und der Naturwissenschaften*, 41 (1957), 1-9; Debus, "Solution Analyses" (n. 27 above), pp. 41-61.

[108] This medieval background is discussed above in Ch. 1.

[109] Gabriel Fallopius Mutinensis, *De medicatis aquis atque de fossilibus* (Venice, 1569; 1st ed., 1564), described also in Ch. 1 above. The analytical procedure is found on fols. 30-37. Conrad Gesner reproduced this material in his *Euonymus . . . Liber secundus* (Zurich, 1569), fols. 24-27, where he fully attributed the procedure to Fallopius. Gesner's work was in turn translated into French by Jean Liébaut (1573), into English by Thomas Hill and George Baker (1576), and into German by Nüscheler (1583). See also Bernhard Milt, "Conrad Gesner als Balneologe," *Gesnerus*, 2 (1945), 1-16.

[110] The pertinent literature is described by Debus, "Solution Analyses," p. 45.

[111] Leonhart Thurneisser zum Thurn, *Pison* (Frankfurt an der Oder, 1572), pp. 31-38, 43-46. In his actual analysis Thurneisser took a measured volume of water, weighed it (using as his unit of weight the Nuremberg mark), and filtered it into a clean receptacle. The water was then distilled and the distillate measured and saved while a series of tests were run on the residue. First the powder was weighed and recrystallized with the aid of a small stick put into the solution. After three days, if there was nothing on the wood, the sample was presumed to be salt water without any other mineral present, but if a white solid appeared, it was examined on the fire or a glowing coal. If it was saltpeter it burned and sputtered on the fire; if not, it was left on the glowing coal for an hour. If during that time, it became red, it was either lead or vitriol; with the addition of warm water, vitriol would dissolve and lead or red lead would not. Other tests might be tried

The Chemical Philosophy

work of Libavius at the end of the century contains many of the same methods outlined by Thurneisser: he, too, used measured samples and insisted on taking weights of the water, the residue, and other quantities as he proceeded.[112] He believed that waters containing minerals must be heavier than pure or "spiritous" waters, and he used a procedure first described by Andrea Baccius in 1571 as a semiquantitative test for the presence of dissolved salts.[113] In this test Baccius soaked a cloth in the spa water, then dried and reweighed it to determine any weight increase. Libavius isolated the contents of the water through distillation, then followed this by recrystallization, suggesting that alum might thus be separated from atrament and salt from saltpeter. The account of Libavius is more practical than that of Thurneisser, and he did not depend on the distillate to obtain further evidence of dissolved minerals. Also important was his use of oak galls, which he observed to react in the presence of copper as well as iron vitriol to

---

on the original residue. If its color was blue, Thurneisser felt that there must be either gold or silver present, and like the metallurgists, he knew that silver dissolved in nitric acid while gold did not. A white or greenish substance might be alum, which could be tested on the fire, since it reacts like the puffing or blowing of baking bread. He also observed that the weight of alum heated on the fire decreased by about half (p. 46), which today's analyst would accept as a fairly close estimate for many forms of alum. Like many of his predecessors, he recognized sulfur by its peculiar odor and taste and especially by its characteristic flame color. He also observed a yellow flame for saltpeter. Besides these tests, Thurneisser relied on a new analytical tool, the crystalline forms of his recrystallized salts. He mentioned that vitriol particles normally formed in "squares" or "triangles," while saltpeter appeared as long spikes (p. 46). He considered this such a certain means of identification that even when several salts were in the same solution it was possible to tell one from another. *Pison* is an important text because of Thurneisser's interest in crystalline forms, flame tests, and his careful attention to weight measurements. However, Gernot Rath points out that Thurneisser marred his work by insisting that metals and minerals could be found in the distillate; he also thinks Thurneisser's reported analyses might have been more the result of his imagination than of his chemical methods (Rath, "Die Anfänge," p. 540). Thurneisser's chemical investigation of herbs is found in his *Historia und Beschreibung Elementischer und Natürlicher wirckungen aller Erdgewechssen* (Berlin: M. Hentzske, 1578).

[112]Andreas Libavius, "De iudicio aquarum mineralium et horum quae cum illis inveniuntur," in the *Alchemia* (Frankfurt, 1597), pp. 275-392; see pp. 309-311.

[113]*Ibid.*, pp. 310-311. Andrea Baccius, *De thermis* (2nd ed., Venice, 1588), p. 76.

darken the solution.¹¹⁴ One can trace a growing interest in analytical color tests in the work of Paracelsians throughout the early years of the seventeenth century.¹¹⁵ By 1631 the English chemist Edward Jorden, who had been in touch with Libavius as early as 1597, was advocating the red-blue color change of "scarlet cloth" as a regular test for those liquids which are now classified as acids and bases.¹¹⁶

The work of the iatrochemical analysts of the late sixteenth and early seventeenth centuries provided the basic information for Robert Boyle's analytic research later in the century. In addition, these chemical physicians sought to apply their new-found knowledge: for instance, they believed that once they had identified the ingredients of the natural spas, it should be possible to prepare equally effective artificial baths. In his *Art of Distillation* (1650) John French (1616-1657) gave directions for making several types of waters, including "artificial Tunbridge, and Epsome Water."¹¹⁷

The New Medicines

Above all, the Paracelsians were noted for their use of chemically prepared medicines. They saw that their new and violent age had spawned ravaging diseases unknown to the

---

¹¹⁴Hermann Kopp, *Geschichte der Chemie*, 4 vols. (Braunschweig, 1843, 1844, 1845, 1847), Vol. II, p. 56. The use of oak galls is especially important because it was the first color indicator in common use. In one of his earliest works, the *Eilff Tractat oder Bücher vom Ursprung und ursacher* (c. 1520) Paracelsus described the oak-gall test as a color indicator for vitriol (*Opera*, Huser ed., Vol. I, p. 521). I have seen no reference in Thurneisser's work, however, to the gall test. Until the mid-nineteenth century this test was one of the leading indicators for iron. See William Henry, *The Elements of Experimental Chemistry*, 2 vols. (1st American ed. from the 8th London ed., Philadelphia, 1819), Vol. II, p. 310.

¹¹⁵See Debus, "Solution Analyses," pp. 50-51 for reference to the seventeenth-century accounts of Henry de Heer, Jean Beguin, and Pierre Gassendi.

¹¹⁶Edward Jorden, *A Discourse of Naturall Bathes, and Minerall Waters* (London: Thomas Harper, 1632), p. 124. The analytical work of Jorden is discussed by Debus in "Solution Analyses," pp. 53-56, and in his "Edward Jorden and the Fermentation of the Metals: An Iatrochemical Study of Terrestrial Phenomena," in *Toward a History of Geology*, ed. Cecil J. Schneer (Cambridge, Mass.: M.I.T. Press, 1969), pp. 100-121.

¹¹⁷French, *Art of Distillation*, pp. 182-184.

VIVORVM

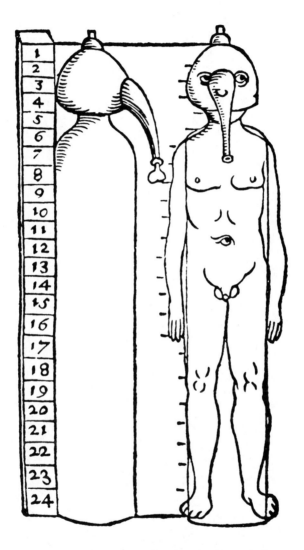

PLATE X   A comparison of the human body and the proper distillation equipment for urine: an example of Paracelsian "chemical anatomy," taken from a work by Gerhard Dorn incorporated into the Huser edition of the works of Paracelsus. From Walter Pagel, *Paracelsus: An Introduction to Philosophical Medicine in the Era of the Renaissance* (Basel/New York: S. Karger, 1958), p. 193.

ancients—they were particularly appalled by venereal diseases—
and they countered with new and more potent medicines. In his
preface to Phillip Hermann's *Treatise teaching howe to cure the French-
Pockes . . . Drawen out of that learned Doctor and Prince of Phisitions,
Theophrastus Paracelsus* (1590), the English Paracelsist John Hester
argued that it was hopeless to apply the methods of Hippocrates
and Galen to diseases they never knew of.

> Now that the diseases of the French Pocks was neyther knowne to
> them, nor to theyr successors for many yeeres . . . is a matter so far out
> of question, that it refuseth all shew of disputation, and therefore as
> this latter age of ours sustaineth the scourge thereof, a iust whyp of
> our lycentiousness, so let it (if ther be any to be had) carry the credite
> of the cure, as some rewarde of some mens industries.[118]

An ever-increasing number of pharmacists and surgeons called for
new, more potent medicines, for ingredients and formulas other
than those described centuries earlier by Dioscorides and Mesue.
Here again analytical evidence seemed to point the way to the
proper type of medicine—to chemicals rather than herbs. It was
known, for instance, that mineral waters contained "niter, alum,
vitriol, sulphur, pitch, antimony and lead," wrote Duchesne, so
therefore such chemicals would doubtless prove helpful to the sick
and the infirm in the form of medicine.[119]

At the same time most Paracelsians were aware of their
heritage from medieval alchemy; they knew and cited Islamic and
Latin authorities who had urged the use of chemicals in the practice of medicine. This medieval tradition of separating pure "virtues" from inactive residues continued to flourish into the sixteenth century, and, in addition to influencing Paracelsus, it was
strongly evident in the writings of his contemporaries Hieronymus
Brunschwig (c. 1450-1533), Philip Ulstadius (fl. first half of the
sixteenth century), and Conrad Gesner (1516-1565), who all advocated the use of distilled products of all sorts in medical practice.[120] Evidence of the use of antimony pills, solid metals, and even

---

[118]See the preface to Philip Herman's *An Excellent Treatise,* trans. John Hester
(London, 1590), p. 31.
[119]Duchesne, *Practise of Chymicall . . . Physicke,* sig. Q3^V.
[120]Robert Multhauf, "The Significance of Distillation in Renaissance Medical
Chemistry," *Bulletin of the History of Medicine,* **30** (1956), 329-346.

elemental mercury abounds in the literature of the period. But as Bostocke pointed out (1585), the true Paracelsian could be distinguished by his careful attention to dosage and his use of the chemical art to extract only the valuable essence of dangerous minerals.[121] It was Paracelsus himself who described the process of preparing potassium arsenate from the fusion of arsenic with saltpeter,[122] and the appearance of calomel (mercurous chloride) in the first London *Pharmacopoeia* (1618) derived from a description in the *Basilica chymica* of Crollius (1609).[123]

Recent research points to a significant change in the preparation of chemical medicines under the followers of Paracelsus. Although Schneider has shown that certain inorganic compounds of key importance (especially antimony, mercury, and iron salts) derived directly from authentic Paracelsian texts,[124] Paracelsus' influential *Archidoxis* (first published in 1570) was still dominated by medieval distillation techniques.[125] The search for pure "virtues" in distillates was not entirely useless as long as the operator confined himself to organic substances, but little more than the original solvent would be recovered from any inorganic reactions which might be distilled. Interest in residues rather than distillates

[121]Bostocke, *Auncient Phisicke*, sig. Ev^V.

[122]Pagel, *Paracelsus*, p. 145, with reference to *Von der natürlichen Dingen*. See also J. R. Partington, *A History of Chemistry*, Vol. II (London: Macmillan, 1961), p. 146.

[123]*Pharmacopoeia Londinensis of 1618 Reproduced in Facsimile with a Historical Introduction by George Urdang* (Madison: University of Wisconsin Press, 1944), p. 72.

[124]Wolfgang Schneider, "A Bibliographical Review of the History of Pharmaceutical Chemistry (with particular reference to German Literature)," *American Journal of Pharmaceutical Education*, **23** (1959), 161-172; "Die deutschen Pharmakopöen des 16. Jahrhunderts und Paracelsus," *Pharmazeutische Zeitung*, **106** (1961), 3-16; "Probleme und neuere Ansichten in der Alchemiegeschichte," *Chemiker-Zeitung—Chem. Apparatur*, **85** (1961), 643-652; "Der Wandel des Arzneischatzes im 17. Jahrhundert und Paracelsus," *Sudhoffs Archiv für Geschichte der Medizin und der Naturwissenschaften*, **45** (1961), 201-215 (see 208-214), *Geschichte der pharmazeutischen Chemie* (Weinheim: Verlag Chemie, 1972).

[125]Sudhoff considered the *Archidoxis* to be the chief work of Paracelsus dealing with medicinal chemistry and the "fundamental text of the new specific therapeutics on a chemical basis": "The Literary Remains of Paracelsus" in *Essays in the History of Medicine* (New York: Medical File Press, 1926), p. 275. For the views of Multhauf see the "Significance of Distillation," pp. 337-344; "Medical Chemistry and the Paracelsians," *Bulletin of the History of Medicine*, **28** (1954), 101-126 (105).

became more evident in the late sixteenth and early seventeenth centuries. In the work of Libavius and Crollius many significant inorganic preparations were described, and a number of these in turn found their way into the official pharmacopoeias of the seventeenth century.[126] The earlier chemical preparations used by physicians had caused no problems with the medical establishment, but the same was not to be true for the Paracelsians. Their bitter attack on the medicine of the Galenists was a threat to older physicians, as was their insistence that the new chemical remedies were not only to be added to the materia medica, they were to supersede the traditional methods of cure. The battle was intensified by a frontal attack on Galenic theory of medicinal action, that is, on the claim that "contraries cure."[127] It had been an article of faith that a medicine with an excess of the "hot" quality would be required to restore to balance a system that was predominantly "cold." Germanic folk tradition suggested an opposed theory—that like cures like, that the poison in proper dosage would also be the cure.[128] With his distrust of the ancients and his sympathy for folk medicine, Paracelsus accepted this as valid, and the homeopathic principle became a hallmark of his followers. Although the Paracelsian position was to result in the introduction of important new inorganic medicines, it also set the stage for an inevitable conflict with the medical establishment.

The *Basilica chymica* of Oswald Crollius (1609)

If one were to choose a single author as representative of the late-sixteenth-century Paracelsian view of nature and medicine, it might well be Oswald Crollius, whose *Basilica chymica* has already been frequently cited because of its extensive influence. Both in its original form and in a revision made by J. and G. E. Hartmann

[126]Robert P. Multhauf, *The Origins of Chemistry* (London: Oldbourne, 1966), pp. 222-232.

[127]Discussed by Paracelsus in the *Paragranum*, in *Sämtliche Werke*, Vol. VIII, p. 107. See also Pagel, *Paracelsus*, pp. 146-148.

[128]A brief survey of the use of the "like-like" principle in antiquity will be found in Jack Lindsay, *The Origins of Alchemy in Graeco-Roman Egypt* (London: Frederick Muller, 1970), pp. 4-7.

and J. Michaelis the *Basilica chymica* went through some eighteen editions between 1609 and 1658 and appeared in French, English, and German translations.[129]

In contrast to many Paracelsian authors, Crollius had a sound university training.[130] Born near Marburg about 1560, he entered the University of Marburg in 1576, later attended the universities at Heidelberg, Strasbourg, and Geneva, and received his doctorate in medicine in 1582. Originally in service to a number of wealthy families of the nobility, Crollius eventually turned to the life of the medical nomad like Paracelsus and many of his followers. For a decade (1593-1602) he traveled about Eastern Europe, but after his successful cure of Prince Christian I of Anhalt-Bernburg, who named him *Archiat*, Crollius settled in Prague. He was well known in the court circles of the empire and was frequently consulted by the Emperor, Rudolf II.

The *Basilica chymica* is the only work published by Crollius. It appeared at the end of his life, and it contains many of the themes that one might expect to find discussed by a late-sixteenth-century chemical physician. The book is divided into three major sections: a lengthy introductory preface describing the Paracelsian system of nature and medicine; the *Basilica chymica* proper, a practical guide to chemical medicines describing their methods of preparation; and a treatise on signatures.

Although Crollius spoke often of the advancement of medical knowledge through the united efforts of all sects, it is clear from the outset that his sympathies were decidedly pro-Paracelsian. Indeed, he remarked that the soundest part of medicine was that based upon the work of the spagyrists, "those who know how to discern between true and false, who are well grown and exercised in *Chymicall* employments . . . [and who] have bathed themselves in

---

[129]For a discussion of the various editions see Partington, *History of Chemistry*, Vol. II, pp. 174-177 and J. Ferguson, *Bibliotheca chemica*, 2 vols. (Glasgow, 1906 and London: Academic and Bibliographical Publications, 1954), Vol. I, pp. 185-187.

[130]In addition to Partington, see the following on the life and work of Crollius: L. Vannier, "L'oeuvre de O. Crollius (1580-1609)," *Bulletin de la Société Française d'Histoire de la Médecine*, **31** (1937), 91-108 and G. Schröder, "Oswald Crollius," *Pharmaceutical Industry*, **21** (1959), 405-408; "Studien zur Geschichte der Chemiatrie," *Pharmazeutische Zeitung*, **111**, No. 35 (1966), 1246 ff.; and his article on Crollius in the *Dictionary of Scientific Biography*, Vol. III (1971), pp. 471-472.

the springs of true *Philosophy*." Attacking the "common rabble" of alchemists as well as the "spurious and adulterous Theophrasteans," Crollius went on to describe "a true and noble Alchymistry" which is the basis for true philosophy.[131]

It was clear to him that the physician (described as "Nature's Minister")[132] must be able correctly to interpret nature as a whole, since this philosophy was one that seeks an understanding of all Creation. Fundamentally, the quest of the physician was as much religious as scientific. He was a man of God, always maintaining before him the two-book theory of knowledge:

> And next after the great book of Grace, wherein the eternall health of our soules consisteth, more diligently to pry into that other also of Nature, treating of those Secrets which respect our bodily health, not passing by without taking notice of those choise treasures of Nature, wherein the most High hath laid up medicines for our greatest and worst diseases.[133]

It is the destiny of man to manage the earth. With this in mind the physician

> ... should through Nature search out the wonderfull works of God by consideration of Temporall and Caelestiall observations, thereby to make known the invisible works of God, celebrating the infinite Wisdome, Power, and perpetuall Goodnesse of the Creator in admiration of his marvellous works, wonders and mysteries which he hath revealed.[134]

Man may learn of his Creator through the study of nature, because true medicine is a gift of God, founded not upon the arguments of scholars but on "Nature, upon which God hath written with his own sacred finger in sublunary things, but especially in perfect Mettalls ... [This] Medicine ... is contained in the vitall Sulphur ... and is founded in the Balsam of Vegetables, Minerals and Animalls." The Creator has stamped his Creation with hid-

---

[131]In citing Crollius, the English translation, *Mysteries of Nature* (1657), will be given first, followed by the Latin text of *Basilica chymica* (1623): 1657, pp. 7, 10-12; 1623, pp. 6, 7-8.

[132]1657, p. 126; 1623, p. 65.

[133]1657, p. 18; 1623, p. 11.

[134]1657, p. 97; 1623, pp. 50-51.

den signs, which can be revealed to man because of the eternal harmony between the macrocosm and the microcosm:

> The Foundation of this Physick is according to the agreement of the lesser World Man with the greater and eternall world, as we are sufficiently instructed by Astronomy and Philosophy, which explain those two globes, the superiour and inferiour ... [indeed] Heaven and Earth are Man's Parents, out of which Man last of all was created; He that knowes the parents, and can Anotomize them, hath attained the true knowledge of their child Man, the most perfect creature in all his properties; because all things of the whole Universe meet in him as in the Centre, and the Anotomy of him in his Nature is the Anotomy of the whole world.[135]

It was precisely his belief in the two-world doctrine that caused Crollius to rule out the study of the anatomy of dead bodies as a basis for medical training; the physician instead was encouraged to study "The Essentiated and Elemented Anotomy of the World and man," from which he will uncover both the disease and the cure.

> The Members or parts of the great world are the Remedies of the members and parts of man by an agreement between the externall and internall Anotomy.... And though the hidden virtue of Hearbs, or the Stars of that Physitian Heaven may be known to us, yet the chiefest thing that the Physitian is also to consider is to know the Concordance of Nature, viz. how he may make the Astrum of the Physick or of the magicall Heaven agree with the internall Astrum and Olimpos of Man....[136]

The physician was warned not to search for the cause of diseases in the four "fictitious" humors, but rather in their "seed," or *astrum*; for each disease, like other living things, has its own specific seed origin:

> ... by reason of the Analogy or proportion that is between the great and the little world; he that knoweth the diseases of the great world, cannot be ignorant of the distempers of man; As many kinds of Mineralls as are in the world, so many there be in Man; So many kinds of diseases are there, as there be sorts, bodies and seeds of things

[135] 1657, p. 23-25; 1623, pp. 13-14.
[136] 1657, p. 43; 1623, p. 24.

that grow; No man knows the number of diseases but he can tell the number of all things that grow.[137]

The true physician-anatomist, being basically a chemist, knows that he must perform an alchemical separation if he is to find the desired internal *astrum*. It is only thus that the true principles of nature and man can be discovered—principles which are, without question, the familiar Paracelsian three: "no body compos'd by Nature can by any dissolving skill be parted into more or lesse then *Three*, viz. Into *Mercury* or liquor, *Sulphur* or *Oyle*, and *Salt*; every created thing is generated and preserved in these three."[138]

The proper method of dissolving or anatomizing bodies is clearly the use of fire, as it always has been for chemists.

> As all things are proved by Fire, so also the Tryall of the knowledge of Physick is to be made by Fire: Physick and Chymistry cannot be separated. For Chymistry . . . doth make manifest not onely the true Simples . . . Secrets, Mysteries . . . respecting health, but also in imitation of the Archaeon Ventricle or Naturall In-bred Chymist, it teacheth to segregate every mystery into its own reservacle, and to free the medicines from those scurvy raggs wherein they were wrapt up by a due separation from the impurities. . . .[139]

The theory of cure occupies many pages in the preface to the *Basilica*. Crollius denied that contraries cure; it was obvious that such a procedure could only agitate internal dissension. Satisfactory remedies would be found only when physicians accepted the existence of a complete harmony between the two worlds. The true physician knew that when the fruits of the elements in man are sick, "they must be restored by fruits like themselves of the Macrocosm," for "Nature doth strengthen and help its own Nature. . . ." The method of applying macrocosmic principles to the microcosm was less clear, since they were recognized as being different in the two worlds. Nevertheless, the salt, sulfur, and mercury of the microcosm "analogically" agreed with the corresponding principles in the macrocosm. Thus, although differences were recognized, true similarities were also postulated, and the latter were to be discovered through the use of the fire. The physician

[137] 1657, p. 44; 1623, p. 24.
[138] 1657, p. 32; 1623, p. 18.
[139] 1657, pp. 93-94; 1623, pp. 48-49.

was thus told once again to seek out the internal *astrum* of a medicine through an alchemical separation of the pure from the impure.[140]

Crollius was convinced that no real medicine could be possible without an understanding of the true philosophy, a philosophy which encompassed a knowledge of the correspondence of the macrocosm and the microcosm, but this knowledge would benefit man in other ways also: because he "is a Quintessence of the greater world, it follows that Man may not onely imitate Heaven, but rule it also at his beck, and reigne over it at his pleasure."[141] Chemistry—only true chemistry—would make this possible. And although he was willing to concede that "what our industry hath found out should not shut the door against that of the Ancients as if all the strength of Nature were hatched in us only,"[142] it is clear that Crollius believed that the greatest published truths of this philosophy were found in the writings of Paracelsus, "that Monarch of Mysteryes."[143]

The *Basilica chymica* is a work of special interest not only because of its informative theoretical preface but also because of its two other sections, the treatise of signatures and the chemical practice. In the section on signatures, which has already been mentioned, Crollius attacked those herbalists who concerned themselves only with the exterior appearance of a subject, taking this to reflect the signature impressed by the Creator. Crollius argued that the much desired internal virtue could only be discovered through proper chemical anatomies in which fire strips away the impure outer shell.[144] While the external sign might aid in indicating the inner virtue, this kernel can be obtained only through true natural magic, the art of the chemist.

The subject was fundamental for physicians, Crollius continued, because all things have their hidden virtues—from the lowly murex which harbors the royal purple of kings to that weak reed, wheat, which is the staff of life. The Creator furnished every place with all the medicines it needs; it was the duty of the physi-

---

[140] 1657, pp. 116-121; 1623, pp. 60, 62-63.
[141] 1657, pp. 74, 71; 1623, pp. 39-40, 38.
[142] 1657, p. 148; 1623, p. 76.
[143] 1657, p. 132; 1623, p. 68.
[144] *Treatise . . . of Signatures* (1669), sigs. A2-A3; *De signaturis* (1623), pp. 1-3.

The Chemical Philosophy

cian, the chemist, the natural magician, to seek them for the benefit of man and the glory of God.[145]

In presenting a lengthy list of substances known to have a sympathy for each other, Crollius gives little evidence of having applied his own rules. Rather, his text seems based primarily on other, earlier discussions of signatures and sympathetic magic. Much derives from current folk medicine; we are told, for instance, that "a living *Crab* tyed with a *Ligature* upon the *Cancer* until it die, the *Cancer* is bettered thereby; if ulcerated, it is afterwards Cured with any *Opodeldoch*: Thus one *Cancer* both kills and Cures another."[146] Again, in a section on the "Correspondencies of Signatures of the Greater and Lesser World," Crollius writes that

> The same which causeth Stormes and Tempests in the greater World, the same also causeth the *Epilepsie* in the lesser World; and as a Tempest changes and weakens the Animal Sense and Intellect, apparent in the unusual crowing of *Cocks*, also by the unaccustomed singing and noise of other Birds and Beasts, and by the more vehement stingings of *Flies* and other Insects, so in like manner it is in *Epilepticks*.[147]

The main section of the *Basilica chymica*—the chemical pharmacopoeia accompanying the preface and the tract on signatures—has attracted considerable attention in recent years. Partington has pointed to the important descriptions of the preparation of potassium sulfate, succinic acid, "luna cornea," fulminating gold, calcium acetate, and numerous mercury compounds.[148] Multhauf has noted that the chemistry of antimony is described in considerably more detail by Crollius than by Basil Valentine in the contemporaneous *Triumph Wagen des Antimonii* (1604).[149] Even more important, Multhauf shows that the emphasis on distillation techniques characteristic of medieval alchemists and still dominant in the work of Paracelsus has disappeared with Crollius. Instead, Crollius "in his treatment of gold, mercury, and antimony (sulphide), very nearly exhausts the possibilities of these [mineral] acids. That is to say, he seems to have

---

[145]1669, sigs, A4^V, A4^r; *De signaturis*, pp. 7, 5.
[146]1669, p. 18; *De signaturis*, p. 43.
[147]1669, p. 24; *De signaturis*, pp. 52-53.
[148]Partington, *History of Chemistry*, Vol. II, p. 183.
[149]Multhauf, "Medical Chemistry" p. 119.

prepared most of the simple compounds obtainable by direct reaction."[150] The mineral acids, earlier used as remedies themselves, have become in the *Basilica* agents for the production of chemical change.[151] In short, Oswald Crollius is one of the first chemical writers to emphasize the real chemical products of a reaction—rather than the distillate or the quintessence—as the desired substances to be collected.

Although it may be argued that there is little obvious connection between the theoretical views expressed in Crollius' "Admonitory Preface" and the very real chemical knowledge displayed in the appended pharmaceutical preparations,[152] it is doubtful that Crollius would have admitted this. Clearly the divine signatures had to be discovered and their internal virtues determined. However, once this was done, it was the task of the practical chemist to learn to prepare the substances in the most effective manner so that their virtues might best be applied for the benefit of man. In this way the chemical philosopher could relate his detailed chemical procedures to the lofty goals of natural philosophy expressed by Paracelsus.

Conclusion

Although Crollius thought that his work would not offend the "holy and Learned Physician" and that "subtle Galenists" would embrace the chemical truths,[153] he was surely well aware of the deep divisions that separated the Paracelsians from the Galenists. His hope for a new medical unity might have been doomed from the outset, but his work remains of considerable interest because it expresses many of the major themes germane to late-sixteenth-century Paracelsism.

Any general assessment of the work of the early Paracelsians would have to begin by noting the physician's task as a form of religious quest: the secrets to be discovered would benefit man not only in the form of new medicines, they would also reveal the virtues which were deposited for man's use in all things at the time of

[150]*Ibid.*, p. 117.
[151]*Ibid.*, p. 116.
[152]As it has by Multhauf, *ibid.*, p. 108.
[153]Crollius, 1657, p. 20; 1623, p. 12.

the Creation. It is little wonder that the Paracelsists turned to the account of Creation in Genesis and interpreted this as a divine chemical separation. Their special interest in the Creation led naturally to the problem of the elements and the principles, since of necessity these were among the first fruits of the Creation.

Their unquestioned belief in the harmony of the universe was expressed in the macrocosm-microcosm analogy: anything that existed in the great world had its analogue in man. Thus, the astronomy of the heavens was reflected in a terrestrial astronomy which postulated cosmic interaction and the existence of similar processes in the two worlds.[154] And because all things in the earth were found in man, the new chemically uncovered knowledge of the inner virtues of natural substances (the divine signatures) could be applied to their counterparts in man. With this in mind, it is easy to understand why the Paracelsians infused a new interest into the study of metals and all other earthly processes. Although these men were physicians, they were convinced in their thoroughly vitalistic universe that "metallic illnesses" might be examined advantageously because analogous diseases had to exist in man.

For the Paracelsist the medical implications of this new philosophy were of overriding importance. Although the medieval medical alchemists had augmented the traditional materia medica with numerous new applications of chemistry to medicine, they did not seek a major confrontation with the medical authorities. This was hardly the case with the Paracelsists. They accused the Galenists of maintaining the status quo at the universities and of refusing to lecture on any of the results of the new Paracelsian investigations.[155] They demanded that the old texts be scrapped—or at least completely reexamined in the light of their own views. Although recent research has revealed their work was strongly influenced by the ancients, they would hardly have agreed; they

[154]"Terrestrial astronomy" was a term frequently used by chemists, particularly after its use by Tycho Brahe. See Ch. 34, "De Tychonis astronomia terrestri" in Samuel Reyher, *Dissertatio de nummis quibusdam ex chymico metallo factis* (Kiel: Joachim Reumann, 1692), pp. 123-125.

[155]It is interesting to note that the first professor of chemistry in Europe, Johann Hartmann (1568-1631), who became professor of mathematics in 1592 and of *chymiatria* in 1609 at Marburg, prepared notes to accompany the *Basilica chymica*, and his *Praxis chymiatrica* (1633) was frequently printed with Crollius' work.

would instead have pointed to the differences which separated them from those who relied on Galen, Averroës, and Avicenna. These Paracelsists found no value in the humoral pathology, they insisted on discarding the time-honored belief that contraries cure, and they soundly damned the whole of Scholastic method as being too closely dependent upon logic and mathematical reasoning. They clearly sought a new understanding of nature in which mathematics would play a lesser rather than a greater role.

In practice these Paracelsians sought to turn to their advantage whatever they could find in the earlier literature. They pored over the treatises of the medical alchemists, the surgeons, and the herbalists on chemically prepared medicines, and they kept much of what they found. Many of them, like Crollius, added new metallic and mineral compounds to the materia medica. Similarly they adopted the earlier literature on the analysis of spa waters, while greatly improving upon the methods described. They took to heart the call by Paracelsus to learn from nature and from common folk rather than from books, and their texts indicate that their views were indeed influenced deeply by traditional folk medicine as well as by metallurgical information that could have come only from a thorough acquaintance with miners and mining techniques.

Had the views of the Paracelsists been offered simply as additions to the work of the ancients—as had those of their chemical predecessors—it is unlikely that the confrontation that developed in the late sixteenth and seventeenth centuries would have occurred at all. However, their work was interpreted at the time as they meant it to be interpreted—as a call for the overturn of traditional medicine and natural philosophy and for its replacement by a chemical philosophy with strong religious overtones. This was a direct challenge to the powerful medical establishment and to those who were entrenched at the universities. It would not go unanswered.

3
THE PARACELSIAN DEBATES

> The innumerable dissentions amongst the learned concerning the Arabicke and Chymicke remedies at this day infinitely, with opposite and contradictorie writings, and invectives, burthen the whole world.
>
> John Cotta (1612)*

HISTORIANS OF SCIENCE have noted the lack of debate over the Copernican hypothesis during the latter half of the sixteenth century. No one would question the fact that the heliocentric system presented a new set of problems centering on motion which was directly connected with the emergence of the new science of the seventeenth century. Yet, these same historians have largely ignored the often heated Paracelsian debates of the same period. It is, of course, difficult to defend the mystical cosmology of the Paracelsians from our post-Newtonian vantage point. Still, these debates are of importance if we are properly to reconstruct the period of the scientific revolution. For many scholars in the sixteenth century the work of Copernicus could be safely passed over

---

*John Cotta of Northampton, Doctor in Phisicke, *A Short Discoverie of the Unobserved Dangers of severall sorts of ignorant and unconsiderate Practisers of Physicke in England* (London: [R. Field] for W. Jones and R. Boyle, 1612), pp. 82-83. The author continues:

Some learned Phisitions and writers extoll and magnifie them as of incomparable use and divine efficacie. Some with execration accuse and curse them as damned and hellish poysons. Some because they find not these remedies in the common & vulgar readings of the Ancients (the famous and learned Grecians) with feare and horror endure their very mention, farre therein unlike and differing from that ingenuous spirit of the thrise worthy and renowned Pergamene Claudius Galen who ... did ... with humble and daigning desire search & entertaine from any sort of people, yea from the most unlearned Empiricke himselfe, any their particular remedies or medicines, which after by his purer and more eminent iudgement, and clearer light of understanding, refining, he reduced to more proper worth, and thereby gave admired president of their wondered ods in his learned prescription and accomodation. Some contrarily contemning the learning and knowledge of the Grecian, and with horrid superstition, deifying an absolute sufficiencie in Chymicke remedies, reiect the care or respect of discreet and prudent dispensation. A third and more commendable sort differeth from both these,

in silence, because he seemed to offer little more than a technical modification of Ptolemaic astronomy. In contrast, the work of the Paracelsians appeared to present a new philosophy of nature opposed to the Aristotelian-Galenic heritage taught at the universities. The work of Copernicus was of interest to the astronomer, but the work of Paracelsus was of interest to all concerned with the future of science and medicine.

Disputes over the new medicine arose only after the published Paracelsian literature began to accumulate in quantity. Here the crucial period is the decade 1565-1575. Although Sudhoff lists only twenty-three works by Paracelsus published during his lifetime (1493-1541) and another forty-two up to 1565, he found an additional one hundred and three titles for the years 1565 through 1575. Appearing at the same time were many other texts which equal in interest those penned by the master. This was the decade that witnessed the publication of the first works whose goal was the synthesis of the often verbose and seemingly contradictory writings of Paracelsus. It was also the period during which the representatives of the medical establishment began to react with alarm to what seemed a dangerous medical heresy. And, finally, one finds in this decade a sincere attempt on the part of some to find common ground for all—Aristotelians, Galenists, Paracelsians alike.

Synthesis and Reaction: The Work of Severinus (1571) and Erastus (1572-1574)

Early examples of Paracelsian syntheses appear in the works of the French author Jacques Gohory (Leo Suavius), who prepared a compendium of Paracelsian philosophy and medicine in 1567, and in the *Idea medicinae philosophicae* of Peter Severinus. The work of Suavius will be discussed later, but the importance of Severinus warrants a more detailed examination at this point.[1]

---

and leaving in the one his learned morositie and disdainfull impatience or different hearing, and in the other his ignorant and perverse Hermeticall monopoly, with impartiall and ingenuous desire free from sectarie affectation, doth from both draw whatsoever may in either seeme good or profitable unto health or physicke use: from the Grecian deriving the sound & ancient truth, & from both Greeke, Chymicke, or Arabian, borrowing with thankfull diligence any helpfull good to needfull use.

[1]The following discussion is based upon the biography of Severinus prepared

Born in Jutland in 1542, Severinus (Sørenson) was educated at the University of Copenhagen, where he was appointed Professor Paedagogicus after taking his degree of master of arts. Still unsettled, he left his post and with financial support from his university he proceeded to travel throughout Germany, France, and Italy attending various universities. After taking an M.D. degree in France, he returned to Italy and completed his *Idea medicinae philosophicae* (1571), which he shrewdly dedicated to the Danish King, Frederick II. On his return to Denmark he became a physician to the court, and at the time of his death (1602) he had just been offered the Chair of Professor of Medicine at Copenhagen.

Although his short *Epistola scripta Theophrasto Paracelso* (1572) was well known,[2] the fame of Severinus was based primarily on the *Idea medicinae philosophicae*, published first in 1571, again in 1616, and finally in 1660 with a long commentary by the Scottish iatrochemist William Davisson. The *Idea* is a defense of Paracelsian doctrines in opposition to the traditional medicine. It begins with an attack on Galen and his logical, "geometrical" approach. This system, Severinus charged, totally failed to control the new diseases of his day and had to be replaced by the "nonmathematical," observational, and chemically based medicine of Paracelsus.[3] Severinus fully accepted Paracelsus' endorsement of the macrocosm-microcosm universe, and he wrote that man has

---

by the present author for the *Dictionary of Scientific Biography* (New York: Scribner's, 1975), Vol. XII. Biographical material on Severinus is found in Eyvind Bastholm's "Petrus Severinus (1542-1602). En dansk paracelsist," *Särtryck ur Sydvenska Medicinhistoriska Sällskapets Arsskrift* (1970), 53-72 and "Petrus Severinus (1542-1602). A Danish Paracelsist," *Proceedings of the XXI International Congress of the History of Medicine* (Siena, 22-28 September, 1968), pp. 1080-1085.

[2] Due less to its original edition than to its inclusion in the prefatory material of the Latin *Opera omnia* of Paracelsus, 3 vols. (Geneva, 1658), Vol. I, fols. ¶¶¶3ᵛ to ¶¶¶¶2ʳ.

[3] Petrus Severinus, *Idea medicinae philosophicae. Continens fundamenta totius doctrinae Paracelsicae Hippocraticae & Galenicae* (3rd ed., Hagae-Comitis, 1660), pp. 2-3, 21. An early-seventeenth-century English translation of this text exists in the British Museum as "A Mappe of Medicyne or Philosophicall Pathe Conteininge The Grounds of all ye doctrine of Paracelsus. Hippocrates & Galen" (Sloane Collection, No. 11). The translation is of interest for its short summaries of the various chapters and is being prepared for publication by the present author and Bruce Burgess. See also the discussion by Allen G. Debus in "Mathematics and Nature in the Chemical Texts of the Renaissance," *Ambix*, **15** (1968), 1-28 (19-20).

within him rivers, seas, mountains, and valleys analogous to the greater world.⁴ He accepted the doctrine of signatures and firmly condemned the humoral pathology of the ancients. Again, like Paracelsus, he broke with traditional medicine in affirming that the harmony of nature requires that like must cure like.⁵

Much of the *Idea medicinae philosophicae* is devoted to the problem of the elements and the principles. The influence of traditional alchemy is reflected in the acceptance of both material and insensible elements, while the inconsistencies of Paracelsus regarding the relationship of the Aristotelian elements and the Paracelsian principles is mirrored also. For practical purposes, Severinus spoke of main classes of substances—liquors, oils, and solids—commonly understood to be mercury, sulfur, and salt. All things could be divided into these three: "Haec vera est corporum Analysis. In hac Anatomia exercitatum oportet esse Medicum."⁶ Severinus returned again and again to an aerial, vital, and generative principle which is brought into the body and distributed throughout the arteries. This principle, primarily identified with a vital sulfur, is responsible for thunder, rain, dew, and other aerial phenomena, while as a nitrososulfureous spirit it also accounts for the flux and reflux of the sea in the geocosm and fevers in man.⁷

Severinus viewed the universe as a vitalist and believed that naturalists should seek out the vital principles in all substances. The elements contained certain forces, or *astra*, which in connection with the chemical principles formed *semina*. These *semina* are in all parts of a given body; in man, however, they are perfected in the generative organs. The seed is properly called "astral" in nature because it has the magisterial power of life which cannot be destroyed even through the processes of putrefaction and dissolution.⁸

As Pagel has recently shown, Severinus was the most eloquent exponent of epigenesis in the period between Aristotle and Harvey.

⁴Although Severinus' views on the relation of the greater to the lesser worlds appear throughout the *Idea*, the reader is directed particularly to the third chapter, "Universalis totius medicinae adumbratio," pp. 19-21.
⁵*Ibid.*, pp. 186-188.
⁶*Ibid.*, pp. 36-39, 40.
⁷*Ibid.*, p. 95, 77, 66-68, 91, 99.
⁸Walter Pagel, *William Harvey's Biological Ideas. Selected Aspects and Historical Background* (Basel/New York: S. Karger, 1967), pp. 239-247 (241-244).

He believed that the *semen* could give rise to a complex organism, not by virtue of the matter present, but through its internal endowment and the intrinsic "knowledge" within it. While his views here are significant, they were not based on embryological observations. In his role as the defender of Paracelsus, Severinus placed strong emphasis on the supremacy of the heart because of its relationship to the vital spirit, but even more important was the blood, since this is the vehicle through which the essential life force reaches all parts of the body. One can say that although Severinus adopted the hard line of the Paracelsians against the ancients, his views on the primacy of the heart and the blood—as well as his espousal of epigenesis—show an Aristotelian influence. Here his views also mark him as a significant precursor of Harvey.[9]

For a century after its publication the *Idea medicinae philosophicae* was a highly influential book. For those who could not, or would not, wade through the many and varied texts of Paracelsus then flooding the book stalls, Severinus seemed to offer a much saner key to the whole. Widely quoted by those who favored the new medicine, his book was no less well known to those who opposed Paracelsus. Among the early continental opponents were Bernhard Dessenius, who wrote a *Defensio medicinae veteris et rationalis adversus Georgium Phaedronem et sectam Paracelsi* (Cologne, 1573), and Thomas Erastus (Liebler) (1524-1583). It is the work of Erastus that is of most interest, for if the *Idea medicinae philosophicae* and the *Basilica chymica* were to become major Paracelsian apologies for nearly a century, the *Disputationes de medicina nova Paracelsi* (4 parts, 1572-1574) was a fundamental text for the opponents of Paracelsus during the same period.[10]

Born at Baden, Switzerland, in 1523, Erastus was trained first in theology and philosophy at Basel and then in medicine at Bologna, where he received his M.D. in 1552. Six years later he became Professor of Medicine at Heidelberg, where, as a strong

[9] *Ibid.*, pp. 245-247.

[10] On Erastus see particularly Walter Pagel, *Paracelsus. An Introduction to Philosophical Medicine in the Era of the Renaissance* (Basel/New York: S. Karger, 1958), pp. 311-332. My earlier account in *The English Paracelsians* (London: Oldbourne, 1965), pp. 37-39, upon which the present discussion is based, has been strongly influenced by Pagel's research. A useful summary of Erastus' views is in Pagel's biography in the *Dictionary of Scientific Biography*, Vol. IV (1971), pp. 386-388.

anti-Calvinist, he advocated state supremacy in church affairs. So involved did he become in theological matters that by 1580 he moved back to Basel as Professor of Theology and Moral Philosophy. Erastus wrote at length against astrology and natural magic, and although he condemned the curative use of human blood and parts of corpses as amulets to prevent epilepsy, his was a firm voice in support of witchcraft.

In the *Disputationes* Erastus granted some credit to Paracelsus for his chemical skill and for having pointed out some errors in Galen.[11] Nevertheless, he pictured Paracelsus primarily as an ignorant man driven by ambition and vanity, a *magus* informed by the devil and evil spirits,[12] an untrustworthy charlatan who continually contradicted himself, a man who knew no logic and consequently wrote in a totally disorganized and incomprehensible manner.[13]

Erastus was wholly opposed to the philosophic system of Paracelsus. He objected on theological grounds to the neo-Platonic unification of the corporeal and the spiritual with their continuous transition and conversion. For him this was the very basis of natural magic. Nor could Erastus accept the Paracelsian position that the divine Creation could be likened to a chemical separation[14] or that the universe was best interpreted in terms of the macrocosm and the microcosm. Indeed, if man's body contained the virtues and materials of all parts of the world, then why was it

---

[11]Thomas Erastus, *Disputationum de nova Philippi Paracelsi medicina. Pars altera: In qua philosophiae Paracelsicae principia & elementa explorantur* ([Basel], 1572), p. 3. "Arbitror etiam Chemiam ipsum non leviter solùm & perfunctoriè attigirse, sed prorsus studiosè coluisse. Si quid fuit in eo laudabile, unum hoc fermè videtur commendari posse, quòd pharmacorum quorundam praeparationes docuit." Pagel refers to Erastus' approval of Paracelsus for having pointed out an error made by Galen in a case of epilepsy (*Paracelsus*, pp. 328-329).

[12]Erastus, *Disputationum . . . Pars altera*, p. 2, where Paracelsus is accused of being a magician and of having called the Devil his friend.

[13]Thomas Erastus, *Disputationum de medicina nova Philippi Paracelsi. Pars prima: in qua, quae de remediis superstitiosis et magicis curationibus ille prodidit, praecipue examinantur* (Basel, s.a. [1572]), p. 16. At this point Erastus spoke of the "monstrosae Paracelsi cõtradictiones."

[14]*Ibid.*, p. 3 In referring to the account of Creation in the *Philosophia ad Athenienses* and its doctrine that the Creation was a separation, Erastus objected that "Nihil tam praeposterè, tam inconditè, tam impie, tam sacrilegè scriptum excogitari potest, quam sunt quae in hunc librum ex spurcissima mente expuit."

not possible for him to fly, to lay eggs, to live in the sea, to be able to do all the things that the other creatures could do?[15]

In medicine Erastus objected to the views of Paracelsus on disease. The latter had rejected the traditional view that disease was a disturbance of the humoral balance and had emphasized instead the importance of outside agents which entered the body and like parasites took possession of an organ and gradually consumed it. Not so, countered Erastus; Paracelsus has confused the disease with its cause and has disregarded the functions of the organs, which decide the character of the disease. Humoral medicine was a crowning glory of the Galenists, and to assume that diseases are separate entities which enter man from the outside was unthinkable.[16]

Hardly less objectionable was the role of innovator assumed by Paracelsus in advocating chemically prepared medicines. His theory of cure had led him to prescribe all kinds of minerals and metallic substances—here Erastus pointed especially to mercury compounds—which were nothing less than lethal poisons. How was it possible that so many were being drawn to this medical heresy by the reports of the wonderful cures performed by Paracelsus, when the fact was that these "cures" were at best temporary ones. In his search of the archives at Basel he was gratified to find that all those treated there with Paracelsian methods had died within a year, even if they had shown an initial improvement.[17]

Erastus placed special emphasis on the Paracelsian principles—a truly monstrous innovation that deserved to be refuted at length. He felt that since sulfur, salt, and mercury are corporeal objects, they could not be considered principles. And even if they were incorporeal, as Paracelsus wished, they could not be considered the origin of the four elements because corporeal matter cannot be created from that which is incorporeal.[18] Paracelsus had stated that all things were composed of what they may be dissolved into, and he had affirmed his ability to convert everything into his three principles. But, Erastus replied, this could not be true—and

[15] See Pagel, *Paracelsus*, pp. 323-324.
[16] *Ibid.*, pp. 324-326.
[17] *Ibid.*, p. 327.
[18] Erastus, *Disputationum . . . Pars altera*, pp. 37-39.

here he turned to the evidence offered by analysis and thus to the whole problem of the validity of analysis by fire. He argued that substances need not consist of the matter from which they are generated; for example, worms are not necessarily composed of the decaying body from which they have developed.[19] Above all, Erastus asserted that heat should not be considered the universal agent of separation which the alchemists considered it to be. Rather, heat changes bodies into substances that are not the constituents of the original bodies. He argued further that the degree of decomposition of a body varies in direct proportion to the degree of heat applied,[20] and he complained that the actual substances salt, sulfur, and mercury are never found as the products of a decomposition performed by heat. These arguments, which were to be incorporated into *The Sceptical Chymist*, were marred by Erastus' insistence that the chemical art reduces bodies not to the three principles but to the traditional four elements.[21]

Erastus went on to ask how Paracelsus could have made these errors, and he found the answer in the Paracelsian conception of an element. Convinced that an element could be noncorporeal, Paracelsus had been led to believe that it was similar in nature to the spirit or soul of a substance. This concept Erastus vehemently denied.[22] In his concern with element theory Erastus may well have been reacting in part to the recently published *Archidoxis* of Paracelsus. Here the elements were discussed on two levels—body (four elements as matter) and soul (quintessence). Elsewhere in the Paracelsian *Opera* the elements are treated in an immaterial cosmic sense in comparison to the material principles.[23] The publication of the immensely popular *Archidoxis* (Cracow, 1569, followed by four editions the following year and many thereafter) ensured a continued emphasis on the chemistry of Paracelsus and the difficult problem of the elements and the principles.

---

[19]*Ibid.*, pp. 42-43.
[20]*Ibid.*, p. 72.
[21]*Ibid.*, p. 73. "Ultimè in elementa resolvi cernimus omnia: in salem, sulphur, & mercurium, solvi nullus homo verax constanter defendet."
[22]See *ibid.*, pp. 44-48 for a general discussion of this problem.
[23]See above, Ch. 1, n. 139 and also *The English Paracelsians*, pp. 26-29.

The Search for Common Ground: Albertus Wimpenaeus (1569) and Guinter von Andernach (1571)

Of the editions of the *Archidoxis* published in 1570, two are those prepared by Johannes Albertus Wimpenaeus (von Wimpfen), a doctor of medicine and philosophy at Munich.[24] While these are of interest for their own sake, of even more importance is the short work Wimpenaeus published the previous year, the *De concordia Hippocraticorum et Paracelsistarum*, which was dedicated to Prince William of the Rhenish Palatinate and Duke of Bavaria. This book is one of the earliest, perhaps the earliest, of those that attempted to restore some harmony to a medical profession that was rapidly losing its earlier sense of unity. The answer of Wimpenaeus was to incorporate the work of Paracelsus and his followers into that of the ancients.

Complaining that he had been unjustly accused by his opponents of departing from the old and true medicine of the ancients to espouse the work of Paracelsus, Wimpenaeus replied that such charges had been made through utter ignorance.[25] Anyone who wished to understand Paracelsus must properly be a philosopher, a mathematician, an alchemist, a Cabalist, as well as a physician. His adversaries, however, had not read Paracelsus and they had no desire to do so. Yet even if they had so exerted themselves, they would not have understood the texts, since they knew nothing of alchemy, mathematics, or the Cabalistic art.[26] These are men who have complained bitterly of the metallic medicines of the chemists, yet if they had truly read the texts that they supposedly revered—Arnald of Villanova, Cardan, and Scaliger—they would have known that these authorities also appreciated the medical virtues of metals.[27] Wimpenaeus placed special emphasis on antimony—

---

[24] Nos. 119 and 129 in Karl Sudhoff's *Bibliographia Paracelsica*.

[25] Johannes Albertus Wimpenaeus, *De concordia Hippocraticorum et Paracelsistarum libri magni excursiones defensiuae, cum appendice, quid medico sit faciundum* (Munich: Adamus Berg, 1569), sig. B3^r and D4^v. I wish to express my thanks to the Bayerische Staatsbibliothek for furnishing a microfilm copy of this extremely rare volume.

[26] *Ibid.*, sigs. D5^r, D5^v.

[27] *Ibid.*, sig. E3^r.

already a subject of heated debate in 1569 because of its strong purgative action—and he similarly connected this metal with earlier or less controversial authors.[28]

To be sure, "I follow Paracelsus,"[29] wrote Wimpenaeus, but he added that he also followed Hippocrates, Aristotle, Vesalius, Cardan, and Gesner among others. Wisdom was to be found in all of the older authorities; it should be sought out and utilized whether it be Greek or Paracelsian in origin.[30] The physicians who are most dangerous are not those who read all authors but those who fear and reject everything that they do not know or understand.[31]

The learned should seek harmony rather than dissension; in truth, Wimpenaeus continued, it would not be too difficult to integrate the doctrines of the ancients with those of the Paracelsians. In regard to the elements there is no problem, since "that which the ancients understood by the four elements Paracelsus understood through the tria prima." Thus, sulfur corresponds to fire, salt to earth, mercury or liquor to water, and air participates in proportion with them all. The only real difference is that the Paracelsian view is more subtle and useful.[32]

The ancients taught that man is made of the four elements: is that radically different from the Paracelsian view that man is a small world containing all things to be found in the macrocosm? The ancients spoke of four humors which nourish the body and become—when in improper balance—the basis of disease. Paracelsus stated that there are three principles; and if he investigated them through artificial examinations (by fire), is this really far removed from the examinations by sight proposed by the doctors of our medical schools?[33] And if cures for gout, dropsy, epilepsy, and paralysis had been ascribed to Paracelsus, in what way did this conflict with the ancients? With these afflictions their methods had accomplished little or nothing at all. In short, for Wimpenaeus the solution to the problem was simple: the medical

[28]*Ibid.*, sigs. F2 ff.
[29]*Ibid.*, sig. A6^V.
[30]*Ibid.*, sigs. C2^r, C3^r, A4^V.
[31]*Ibid.*, sigs. F5^r ff.
[32]*Ibid.*, sigs. G1^r, G1^V ff.
[33]*Ibid.*, sig. G3^r.

PLATE XI Johannes Albertus Wimpenaeus at the age of thirty. From Paracelsus, *Archidoxa ex Theophrastiu*, ed. J. A. Wimpenaeus (Munich: Adam Berg, 1570), sig. *iii^r.

profession needed both the work of the ancients for its long-standing authority and the work of the Paracelsians for its certainty of truth.[34]

Although it was an early work of conciliation, the *De concordia* did not pass without notice. In the closing decades of the century the book was frequently cited by Paracelsian apologists, and it was sufficiently in demand to require a second edition in 1615. Indeed, it was to this work that Crollius referred in 1608 when he wished to show that the most difficult diseases were cured by the Paracelsians rather than the Galenists. Quoting Wimpenaeus, he wrote that the reasons for this were threefold: (1) diseases were now more perfectly known, (2) there were more effective medicines, and (3) the harmony of the macrocosm and the microcosm was well understood through the efforts of the Paracelsians.[35]

Although the work of Wimpenaeus is important, that of Joannes Guinter of Andernach (c. 1505-1574) is even more so. Born of a poor family, Guinter proceeded to Paris, where he took his M.D. degree in 1532 and became one of the two professors of medicine two years later. Among his duties was the teaching of the winter course in anatomy, which he offered along thoroughly traditional lines. It was sometime during the period 1533-1536 that he taught anatomy to the young Andreas Vesalius—who ever after had a low opinion of his teacher on this subject.[36] Guinter, however, stands as one of the major medical humanists of the first half of the sixteenth

---

[34]*Ibid.*, sig. G3^V.

[35]Oswald Crollius, *Discovering the Great and Deep Mysteries of Nature* in *Philosophy Reformed and Improved*, trans. H. Pinnell (London: M.S. for Lodowick Lloyd, 1657), pp. 142-147; *Basilica chymica* (Frankfurt: Godfrid Tampach, n.d. [1623]), pp. 73-75.

[36]On the life of Guinter and his relationship to Vesalius see C. D. O'Malley, *Andreas Vesalius of Brussels* (Berkeley/Los Angeles: University of California Press, 1964), pp. 54-61, and also his biography of Guinter in the *Dictionary of Scientific Biography*, Vol. V (1972), pp. 585-586. O'Malley did not refer to the chemical and Paracelsian views of Guinter, although W. P. D. Wightman had called attention to them in his *Science and the Renaissance*, 2 vols. (Edinburgh/London: Oliver and Boyd; New York: Hafner, 1962), Vol. I, p. 256 n. 1. The following account of Guinter's *De medicina veteri et noua* . . . is based upon my earlier paper "Guintherius—Libavius—Sennert: The Chemical Compromise in Early Modern Medicine," *Science, Medicine and Society in the Renaissance. Essays to Honor Walter Pagel*, ed. Allen G. Debus, 2 vols. (New York: Science History Publications, 1972), Vol. I, pp. 151-165.

century. Not only did he translate the larger part of Galen, he prepared editions of Paul of Aegina, Caelius Aurelianus, Oribasius, and Alexander of Tralles. In addition he published books on the plague, medical spas, and obstetrics.

Given Guinter's catholic interests, it is not too surprising that he should have examined the work of the Paracelsians when that became the subject of debate in medical circles. After carefully reading this new literature, Guinter composed his *De medicina veteri et noua tum cognoscenda, tum faciunda commentarij duo*. Published in 1571, the same year as the Paracelsian synthesis of Severinus, this is a vast work touching all aspects of medicine in some 1,700 folio pages. Here again we find a genuine effort to conciliate the warring sects. There is little doubt that Guinter hoped to make the useful results of the Paracelsian chemists acceptable to those physicians who wanted to retain a more traditional foundation for their medicine. With this aim in mind he sought the best in both systems.

It is true, Guinter began, that the writings of Paracelsus are a strange mixture of valuable and false information. They contain much magic, which is rightly condemned. Furthermore, the Paracelsians are a difficult lot—arrogant men who publish incomplete and worthless writings which they ascribe to their master and sell at exorbitant prices,[37] at the same time retaining for their own profit any valuable and genuine manuscripts they might find. Yet, he continued, there is much of value in the new medicine. Surely the best of the chemical remedies have effected cures that verge on the miraculous. When men of good will seek the best in both schools as a basis for their medical practice they will see that the "old" and the "new" have much in common. That this has not been pointed out earlier may indeed be the fault of the Paracelsians, for Paracelsus did not know Greek and wrote only in German,[38] while his followers are no better. It is little wonder that the medical world is swamped with conflicting claims rather than sincere attempts at conciliation.

The three principles—so important to the Paracelsians—may

---

[37] J. Guintherius (Guinter) von Andernach, *De medicina veteri et noua tum cognoscenda, tum faciunda commentarij duo*, 2 vols. (Basel: Henricpetrina, 1571), Vol. II, pp. 11, 651.

[38] *Ibid.*, pp. 28, 680.

ΟὙΒΕΡΤΟΣ ΔΑΜΙΟΣ Ο‛ ΑΝΔΕΡΝΑ-
κῶς φιλιατρὸς εἰς τὴν εἰκόνα.

PLATE XII Joannes Guinter of Andernach (c. 1505-1574). Portrait from *De medicina veteri et noua tum cognoscenda, tum faciunda commentarij duo* (Basel: Ex off. Henricpetrina, 1571).

serve as an example, continued Guinter. Surely few students of antiquity would argue against equating the fire of the ancients with the new sulfur, and one may similarly pair salt and earth, liquor (water) and mercury, and air and spirit. The real difference lies not in a new set of substances but in the new emphasis placed by Paracelsus on the use of the senses. While both schools are speaking essentially of the same things, the Aristotelians base their proof on reason and the Paracelsians emphasize the authority of observation and experience.[39] Guinter was convinced of the importance of these principles, and he described at considerable length the means for demonstrating their existence through the distillation of ebony wood.[40]

We need not censure Paracelsus for his use of the macrocosm-microcosm analogy, he added. The Swiss reformer had taught that this was the foundation of all medicine and would lead to the true anatomy of all things—something far different from the false anatomy of the schools which was limited to the cutting of the human body. But here again, Guinter stated, the doctrine of the two worlds was known to the ancients; and although they used it seldom, he was certain that they used it more accurately than had Paracelsus himself.[41]

What of the seemingly deep division over the methods of cure? The Galenists argue that contraries cure and the Paracelsians object, but once again the apparent disagreement may be only superficial. Paracelsus speaks of seeking the inner nature of substances through his "anatomy," assuming that his treatment with fire will not change these essences.[42] In reality there is every reason to believe that his "similar" cures will be changed to "contrary" ones through the process of distillation, and here again the Galenists and the Paracelsians may well be speaking of the same thing while arguing heatedly.

While Guinter was well aware that the role assigned to chemistry was a crucial point of contention, he felt no need to debate the merits of a chemical philosophy but limited the question to chemically prepared medicines. He was willing to grant

[39] *Ibid.*, p. 31.
[40] *Ibid.*, p. 629.
[41] *Ibid.*, pp. 28-31.
[42] *Ibid.*, pp. 25-26.

that some chemical remedies appeared in the writings of Hippocrates and Galen, but it seemed to him that more recent authors—especially Paracelsus and his followers—placed a new emphasis on them.[43] For Guinter this seemed right and proper: he agreed with the Paracelsian argument that their new and violent age needed new medicines. The earliest men, he wrote, were strong and needed only simple and mild remedies for retaining their health. More debilitating diseases arose only later when the accumulated luxuries of centuries resulted in a permanent corruption of mankind. Then we see different medicines—resins and aromatic substances—introduced in the texts of the Islamic and Indian authors. It was the destiny of Paracelsus not only to restore to use chemicals known to earlier authors but also to enrich them with a treasury of new waters, liquors, salts, and oils[44]—medicines often more efficacious than the traditional ones.

In the case of chemically prepared medicines, no compromise with antiquity was sought; it was quite plain that "chemicorum medicamēta plus quam diuina."[45] The hundreds of pages of medical preparations in the *De medicina veteri et noua* are filled with recipes for chemically altered organic and inorganic substances. Compounds of iron, antimony, mercury, and other substances are described, and the true preparation of *crocus martis* is given with the understanding that this secret will be made available to poor patients at no charge.[46] A method for the preparation of *turpeth minerale*[47] is given, and it is noted that the force of mercury may be modified so that it could safely be taken as a true cure for the French disease.[48] Guinter was convinced that the preparation of *aurum potabile* was above all the greatest gift bestowed upon medicine by the chemists; it was of almost divine power, a cure for most diseases.[49]

For Guinter there was only one valid conclusion—and here he was in complete agreement with Wimpenaeus—both medicines,

[43]*Ibid.*, pp. 26, 621-622, 28.
[44]*Ibid.*, p. 28.
[45]*Ibid.*, p. 650. This quotation is particularly effective as a marginal notation.
[46]*Ibid.*, pp. 192-198.
[47]*Ibid.*, p. 673.
[48]*Ibid.*, p. 195.
[49]*Ibid.*, pp. 650-651.

veteri and *nova*, were needed. "The ancients on account of time-honored authority are to be given first place," but there was much of great value in the work of the more recent chemists.[50] Would that Galen had been more brief and more accurate; would that Theophrastus has been more open and candid![51] There are faults and virtues in the work of both factions, and physicians must choose the best from each.

French Paracelsism in the Late Sixteenth Century

Guinter's interest in the new medicine was typical of the widespread concern felt by physicians in the third quarter of the sixteenth century. And, although he had left the Paris Medical Faculty first for Metz (1538) and later for Strasbourg (c. 1540), his association with this major medical school properly draws our attention to the French reaction to both the new chemical medicines and the Paracelsian "philosophy of nature." As Henry Guerlac has noted, very few of the early French texts have been examined in the light of recent scholarship.[52] Nevertheless, it is important to discuss a few of these works because of the light they shed on the important Parisian medical conflict in the early seventeenth century.

The earliest translation of a work by Paracelsus into French noted by Sudhoff is *La grande, vraye et parfaicte chirurgie* (Anvers, 1567, reprinted 1568) by Pierre Hassard, a physician and surgeon of Armentières who had published earlier works on astrology, the effects of an earthquake, and the cure of venereal disease. In his preface (dated Brussels, July 10, 1566) Hassard pictured Paracelsus as a tireless scholar who had written hundreds of volumes on philosophy, medicine, astrology, theology, political justice, natural magic, and many other subjects. The goal of Paracelsus had been "experimenter de tous, en tous, & par tout,

---

[50]*Ibid.*, pp. 31-32. "Veterum sanè medicina propter auctoritatem primum obtinere locum debet: recentiorem, si quid primae attulerit, aut in ea correxerit, contemnenda non est, sed ambae simul conferri debent, & quicquid melius in utraque fuerit retineri."

[51]*Ibid.*, p. 32.

[52]Henry Guerlac, "Guy De La Brosse and the French Paracelsians," *Science, Medicine and Society in the Renaissance*, Vol. I, pp. 177-199.

pour trouver le vray fondement de tous artz & sciences, & principalemēt de Medecine & Chirurgie...."[53] But although his was the true natural magic and "la pure fontaine cabaline," Hassard found his work of special interest because of its medical value. He was to place no particular emphasis on chemistry. After preparing a second translation from Paracelsus three years later, the *De la peste* (Anvers, 1570),[54] Hassard disappears from view, and his edition of the *Grossen Wundartzney* became superseded by that made by Claude Dariot in 1588.

Far different from the practical information to be gleaned from Hassard's translations was the *Compendium* of Paracelsian philosophy and medicine written by Jacques Gohory (1520-1576) under the pseudonym Leo Suavius, in 1567 (reprinted 1568).[55] A diplomat and an advocate at the Parlement of Paris, Gohory was at the same time a literary figure with a decided interest in the occult arts. He wrote on music and he has a special place in the history of French literature for his translations of Machiavelli and *Amadis de Gaule*, the latter prepared because the translator felt the work was an alchemical allegory.[56]

Although Gohory displays an acquaintance with the traditional neo-Platonic and Hermetic texts, he was also well aware of the most recent literature relating to the work of

[53]Paracelsus, *La grande, vraye, et parfaicte chirurgie*, trans. M. Pierre Hassard d'Armentieres, medicin et chirurgien (Anvers: Guillaume Silvius, 1568), sig. A5^V. I used a microfilm of the copy in the Bibliothèque St. Géneviève, Paris, kindly furnished to me by the librarian. The first edition of this extremely rare volume is cited by Sudhoff (No. 83 in the *Bibliographia Paracelsica*) and was also printed at Anvers by the same publisher.

[54]No. 114 in Sudhoff's *Bibliographia Paracelsica*.

[55]Nos. 89 and 99 in the *Bibliographia Paracelsica*.

[56]A good survey of the literature is found in Owen Hannaway's biography of Gohory in the *Dictionary of Scientific Biography*, Vol. V, pp. 447-448. The basic study is E. T. Hamy, "Un précurseur de Guy de la Brosse. Jacques Gohory et le Lycium Philosophal de Saint-Marceau-les-Paris (1571-1576)," *Nouvelles Archives du Museum d'Historie Naturelle*, 4eme, Ser. I (1899), 1-26. Gohory's relationship to Paracelsism, neo-Platonism, and Hermeticism is discussed in D. P. Walker, *Spiritual and Demonic Magic from Ficino to Campanella* (London: The Warburg Institute, 1958), pp. 96-106 and in Walter Pagel, *Das medizinische Weltbild des Paracelsus seine Zusammenhänge mit Neuplatonismus und Gnosis*, Kosmosophie, Band I (Wiesbaden: Franz Steiner, 1962), pp. 128-129. Gohory's place in the history of sixteenth-century chemistry is discussed by J. R. Partington, *A History of Chemistry* (London: Macmillan, 1961), Vol. II, pp. 162-163.

Paracelsus. He cited the preface to Hassard's translation of Paracelsus and the work of Adam of Bodenstein. He referred to the work of Thomas Erastus and strongly criticized the Paracelsian editions prepared by Gerhard Dorn. In 1571 Gohory founded a small philosophical society, the Lycium Philosophal San Marcellin, at his home in the Faubourg Saint-Marcel, where the members prepared Paracelsian medicines and carried out alchemical experiments. Gohory was in touch with the most important medical figures of the French capital, and he mentioned having discussed Paracelsian medicine with Jean Fernel, Ambroise Paré, Jean Chastellan, and Leonardo Botal.

It is clear in the *Compendium* that Gohory was strongly influenced by the *Philosophia ad Athenienses*. He discussed in detail the generation of the four elements from the *mysterium magnum,* then proceeded to describe the Paracelsian account of the spirits of the universe associated with the elements—the Nymphs, Tritons, Lorinds, Durdales, Melosynes, and others.[57] These, he suggested, were borrowed by Paracelsus from his master, Johannes Trithemius. As for the Paracelsian principles, Gohory attributed them to the earlier alchemists, particularly Ioannes Valentianus.[58]

The *Compendium* is clearly a Paracelsian work, yet it lacks a satisfying depth. He wrote with approval of the macrocosm-microcosm analogy[59] and insisted that all diseases are curable, recommending such remedies as antimony and potable gold. He rejected the humors as a basis for medicine,[60] but he did this without discussing his reasons in any great detail. Magic was clearly of great importance: the Spirit of Life, the Cabala, talismans, seals, characters, and amulets all play a part in his description.[61]

The *Compendium* was published along with the *De vita longa* of Paracelsus and Gohory's commentary on the text. From this it is

---

[57]Leo Suavius (Jacques Gohory), *Theophrasti Paracelsi philosophiae et medicinae utriusque universae, compendium, ex optimi quibusque eius libris: Cum scholijs in libros IIII eiusdem De vita longa, plenos mysteriorum, parabolorum, aenigmatum* (Basel, 1568), pp. 22-24.

[58]Partington, *History*, p. 162.

[59]Gohory, *Compendium*, p. 32.

[60]*Ibid.*, p. 33.

[61]As an example, see the chapter "De vera magia carminibus & characteribus adversus Wieri aliorumque calumnias," *ibid.*, pp. 275 ff.

apparent that although Gohory was impressed by the wondrous cures of Paracelsus and his special remedies for the gout, leprosy, and epilepsy, he was not really concerned with the practical problems of the physician or the chemist. Nor was his interest in alchemy quite the same as that of someone like Crollius who was to discuss the chemical philosophy as the basis of medicine. Rather, Gohory treated Paracelsus as an Hermetic magus, and he related *De vita longa* to the earlier discussions of the prolongation of life by Roger Bacon and Marsilio Ficino.[62] To this extent the work of Gohory does not fall in the mainstream of Paracelsian thought; Walker has perhaps correctly categorized this author as one of the "Demonic magicians" in the Hermetic tradition rather than a chemist or chemical philosopher.[63]

Although Gerhard Dorn replied to Gohory in an appendix to the Basel edition of the *Compendium*, in France there was no strong reaction to the new medicine from the medical establishment until the following decade. In 1566 both the Medical Faculty and the Parlement of Paris forbade the internal use of antimony after an exchange between Loys d l'Aumay and Jacques Grevin (1564).[64] Antimony and mercury compounds, both highly poisonous, serve as irritants in small doses and were widely employed as purgatives by Paracelsians whose theory of cure by similitude sanctioned these substances.

A strong critique of the views of Paracelsus on chemical remedies and the origin of metals (1575) written by Jacques Aubert (d. 1586)[65] proved to be the occasion for the first publication by Joseph Duchesne (Quercetanus) (c. 1544-1609), who was to become the key figure in the Paracelsian-Galenist debate at

---

[62]Roger Bacon is a major source and is referred to throughout Gohory's work. The comparison of Paracelsus and Ficino will be found on pp. 199 ff.

[63]Walker, *Spiritual and Demonic Magic*, pp. 96-106.

[64]Jacques Grévin, *Discours . . . sur les vertus et facultez de l'antimoine, contre de qu'en a escrit maistre Loys de Launay* (Paris: A. Wechel, 1566); *La second Discours . . . sur les vertus et facultez de l'antimoine . . . pour la confirmation de l'advis des médecins de Paris et pour servir d'apologie contre ce qu'a escrit M. Loïs de Launoy . . .* (Paris: J. Du Puys, s.d.). These works, the 1566 Paris decree, and the rivalry between Paris and Montpellier are discussed by A. G. Chevalier in his "The 'Antimony-War'—A Dispute Between Montpellier and Paris," *Ciba Symposia*, 2 (1940), 418-423.

[65]Iacobus Aubertus Vindonis, *De metallorum ortu & causis contra chemistas brevis & dilucida explicatio* (Lyon, 1575).

Paris in the first decade of the following century.⁶⁶ Like most chemical philosophers a Calvinist, Duchesne was forced to live many years away from his homeland. After receiving his medical degree at Basel (1573), he settled first at Cassel, where the Grand Dukes had for some time patronized Paracelsian physicians, and then was received as a citizen of Geneva (1584), where he served the Council of the Two Hundred as a diplomat. Only after Henry of Navarre took Paris (1593) did he return to France, where he was appointed physician in ordinary to the King.

Duchesne's *Responsio* to Aubert (1575) was a short work in which he argued against the Aristotelian position presented by his opponent. There is no doubt that he considered chemistry the key to an understanding of nature, for chemistry

> ... openeth unto us so many works of the almightie God, it laieth open so many secretes of nature, and preparations of herbes, beastes and mineralls hetherto unknowen, and sheweth the uses almost of all things, which were hidden and laid up on the bosome of Nature, that they shew themselves unkinde toward man, that would have this art buried.

Nevertheless, he continued,

> As touching *Paracelsus*, I have not taken upon mee the defence of his divinitie, neither did I ever thinke to agree with him in all points, as though I were sworne to his doctrine: but ... he teacheth many things almost divinely, in Phisicke, which the thankfull posteritie can neither commend and praise sufficientlie.... ⁶⁷

Neither in 1575 nor a quarter of a century later would Duchesne

---

⁶⁶Joseph Duchesne (Quercetanus), *Ad Iacobi Auberti Vindonis de ortu et causis metallorum contra chymicos explicationem ... eiusdem de exquisita mineralium, animalium, & vegetibilium medicamentorum spagyrica praeparatione & usu, perspecua tractatio* (Lyon: Apud Ioannem Lertotium, 1575). A discussion of the literature on Duchesne will be found in the present author's biography of him in the *Dictionary of Scientific Biography*, Vol. IV, pp. 208-210. His very considerable influence on English chemistry and medicine is discussed in *The English Paracelsians*, pp. 87-101.

⁶⁷Joseph Quercetanus, *A Breefe Aunswere ... to the exposition of Iacobus Aubertus Vindonis, concerning the original, and causes of Mettalles. Set foorth against Chimistes. Another exquisite and plaine Treatise of the same Josephus, concerning the Spagericall preparations, and use of minerall, animall, and vegitable secretes, not heeretofore knowne of many.* By Iohn Hester, practitioner in the Spagericall Arte (London, 1591), fol. 1; Latin (1575), pp. 2-3.

consider himself a blind follower of Paracelsus. Rather—and on this point he was adamant—he argued that the true chemical physician should appreciate the work of Hippocrates and Galen while at the same time recognizing that much had been discovered since their time. Those who charged that Paracelsian medicines were limited to metallic and mineral substances were being both unfair and untrue, since all chemists regularly prescribe animal and mineral preparations. And as to the propriety of using metallic preparations, when properly prepared by chemists they are neither sharp nor violent in their action; rather, they are "sweet and familiar to our nature."[68]

Aubert had questioned the validity of the art of alchemy by arguing that it was customarily accepted that metals were engendered by the stars—a force not at the disposal of the alchemist. Not so, replied Duchesne, who countered that the metals have a nearer efficient cause "that is heate, by force whereof mettales congealed in the bowels of the earth are disposed, digested and made perfect."[69] And since only gold is perfect, this may be considered a normal and natural process.[70] Aubert's refutation of alchemy, he felt, could hardly stand on such an argument.

Duchesne's *Responsio* was published with his *De mineralium, animalium, et vegetabilium medicamentorum spagyrica praeparatione & usu*, a practical guide to chemical preparations in which he stated that medicines are prepared from antimony not only for external but also for internal usage. The following year Duchesne published his *Sclopetarius*, which emphasized chemical oils and balms of special value for the cure of gunshot wounds, a problem of increasing concern for Renaissance physicians and surgeons. A popular treatise for well over half a century, the English translation prepared by John Hester (1590) was prefaced by a "Sonet" praising the author:

> That Monk or Frier accurst may be, that bent his wit so wickedly,
> By salt petre and other stuffe, through shot of gunne most cruelly
> To murther men at unawares, by sodaine stroke of bullet shot
> So as no force can now prevaile, & strength of men is quite forgot,

[68]Duchesne, English, fols. 6^r, 6^v; Latin, pp. 20, 22.
[69]Duchesne, English, fol. 14^v; Latin, pp. 53-54.
[70]Duchesne, English, fol. 15^r; Latin, p. 57.

Hæc Quercetani, corpus quæ pinxit Imago est
Ingenio at melius pingitur ille suo.
Iunge animam membris, quæ docta pingitur arte
Scriptorum, et totus tum tibi pictus erit.

PLATE XIII   Joseph Duchesne (Quercetanus) (c. 1544-1609). Portrait from *Recueil des plus curieux et rares secrets* (Paris: J. Brunet, 1641).

For where these engines come in place, mens bodies there are rent and torne,
And many then loose leg or arme, & bodies maimed are left forlorne.
But monsieur du Chesne we wel may blesse, whose godly care and happy skill,
Hath found a mean to ease this griefe, & how the rage of shot to stil.
Yes he the meane hath well set down, to heale all Gunshot with good speede,
Which was a worthy deed of his, & comfort yields to such a need.[71]

In this work Duchesne answered those who had suggested that there were already satisfactory remedies to be found in the writings of the ancients, and he went on,

> ... to what purpose is it that they obiect unto us ♄ sulphurie metaline, & venomous stinckes (as they call them) by whose smell and drawne breath (for these are their contumelius words) they be almost strangled that come into the dennes of those *Cyclops*? But is it unknowne unto those slaunderers and sicophantes, that the olde Phisitions made verie many medicines of most filthie thinges, as of the filth of eares, sweate of the body, of womens mentrewes (and that which is horrible to be spoken) of the doong of man, and other beastes, spittle, urine, flyes, mise, the ashes of an Owles head, the heues of Goates and Asses, ♄ wormes of a rotten tree, and the scurfe of Mules, as may be gathered out of the writings of *Galen, Aetii, Aegineta, Diosc. Plinius, Serap.*: to passe the metalines which it is euident they did also use.[72]

In the cure of gunshot wounds he advocated careful diet and the evacuation of humoral excess through the proper application of medicines, which for the most part were chemically prepared. He warned against the use of hot oil in the case of amputations, advocating the constant use of running water instead. Here again chemical remedies were prescribed to help stay the blood.[73]

Duchesne's attack on Aubert was criticized by John Antony

---

[71]Joseph Duchesne, *The Sclopotarie of Iosephus Quercetanus, Phisition, or His booke containing the cure of wounds received by shot of Gunne or such like Engines of warre, whereunto is Added his Spagericke antidotary of medicines against the aforesayd wounds*, trans. John Hester, practitioner in the said spagiricall Arte (London, 1590), sig. A5.

[72]*Ibid.*, p. 74.

[73]*Ibid.*, p. 19, 31.

Fenot in 1575, and Aubert himself replied the following year.[74] In 1579 the Parisian physician Courtin published an attack on Paracelsus, the three principles, potable gold, and the whole of pyrotechny.[75] It is clear from the *Discours admirables* of Bernard Palissy (1580) that he had carefully read Paracelsus on the relationship of salt to life and growth; it formed the very basis of Palissy's essay on the use of marl in agricultural practice.[76] Indeed, by the end of this decade it appears that the French chemical physicians were deeply divided from the medical establishment. In the preface to his edition of the *Centum quindecim curationes experimētáque* of Paracelsus (Lyon, 1582), the translator, Bernard Gabriel Penot, or Penotus, complained bitterly of the treatment accorded to the chemists:

> For when as reports are spread of the strange cures of sundry grievous diseases, which are wrought by the benefit of tinctures, and vegetall, and minerall spirits, by the cunning and labour of those whom the common sort at this day call Chymists, or Alchymists; by and by on the contrary part they cry out that those colliar Phisitions can doe no good, but kill all men that put themselves into their hands with their venomous Medicines, so that they ought to be driven out of the Common-wealth, and that they are deceivers, and that their extractions, and preparations, their subtill, and thinne spirits will profit nothing, and that the spirit of *Vitrioll* is poyson, the essence of *Antimony* and *Mercury* is nothing, the extraction of *Sulphur* is nothing worth, neither the liquor of Gold, and to be breif, that all things are contrary to the nature of man, and more to be avoided, then the eyes of a Basilisk: and yet they, in the meantime, like cunning and crafty Theeves, privily, and with fair promises, pick out from the poor Chymists the secrets of Physick, and secretly learn those things that

[74] John Antony Fenot, *Alexipharmacum, sive antidotus Apologetica, ad virulentias Josephi cuiusdam Quercetani Armeniaci, euomitas in libellum Jacobi Auberti, de ortu & causis Metallorum contra Chymistas . . . In quo . . . omnia argumēta refelluntur, quibus Chymistae probare conantur, aurum argentumq; arte fieri posse . . .* (Basel, s.d. [1575]); Iacobus Aubertus Vindonis, *Duae apologeticae responsiones ad Iosephum Quercetanum* (Lyon, 1576).

[75] Germani Courtin, Medici Parisiensis, *Adversus Paracelsi de tribus principiis, auro potabile totáque pyrotechniā, portentosas opiniones, disputatio* (Paris: Ex Officina Petri L'Hillier, via Iacobaea sub signo Oliviae, 1579).

[76] On Palissy see below, Ch. 6, and Allen G. Debus, "Palissy, Plat and English Agricultural Chemistry in the 16th and 17th Centuries," *Archives Internationales d'Histoire des Sciences*, **21** (1968), 67-88 (74-76).

they forbid the common people as poysons, afterwards challenging them for their own practices . . . ."[77]

Here Penotus claimed that most dangerous diseases had been cured by properly prepared chemical medicines and that the chemists could not be praised enough: "We have brought into Physick, Essences, Oiles, Balmes, and Salts, all which the Alchymists schools have found out. And how great light is come into Physick onely by true distillation, it is known unto all men, and daily experience teacheth, how great commodity hath redounded thereby unto the sick."[78]

Penotus warned against the "counterfeit Paracelsian" who offered dangerous remedies never prescribed by the true Paracelsian. The problem he referred to had been intensified by a chemical "quack" who had been active in Paris only a few years earlier. Roch le Baillif, sieur de la Rivière, a native of Normandy, had published his summary of Paracelsian medicine, *Le demosterion*, in 1578.[79] Here the author had praised the still living Adam of Bodenstein, Gerhard Dorn, and Pierre Hassard while he wrote of the "certain" cures of the alchemists for leprosy, dropsy, paralysis, and gout.[80] Stating that "La science ["medecine," marginal note] est de la creation de Dieu, & partant certaine & veritable," le Baillif offered a number of scriptural references that seemed to support

---

[77] From the "Apologeticall Preface of Maister Bernard G. Londrada A Portu Aquitanus" to the *A Hundred and fourteene Experiments and Cures of the famous Physitian Phillipus Aureolus Theophrastus Paracelsus,* trans. John Hester (London: Vallentine Sims, 1596) as reprinted in Leonard Phioravant, *Three Exact Pieces of Leonard Phioravant Knight and Doctor in Physick, viz.: His Rationall Secrets & Chirurgery, Reviewed and Revived, Together with a Book of Excellent Experiments and Secrets, Collected out of the Practices of severall Expert Men in both Faculties, Whereunto is Annexed Paracelsus his One hundred and fourteen Experiments With certain Excellent works of B.G. a Portu Aquitans* . . . . trans. John Hester (London, 1652), sig. Ccc 4^V.

[78] *Ibid.*

[79] Roch le Baillif, Edelphe Medecin Spagiric, *Le demosterion . . . Auquel sont contenuz Trois cens Aphorismes Latins & François. Sommaire veritable de la Medecine Paracelsique, extraicte de luy en la plus part, par le dict Baillif. Le Sommaire duquel se trouvera a fueillet suyvant* (Rennes: Pierre le Bret, 1578). The present account of le Baillif's career is based primarily on Hugh Trevor-Roper's "The Sieur de la Rivière, Paracelsian Physician of Henry IV," *Science, Medicine and Society in the Renaissance,* Vol. II, pp. 227-250. See also Dietlinde Goltz, "Die Paracelsisten und die Sprache," *Sudhoffs Archiv,* **56**, 337-352.

[80] Le Baillif, *Demosterion,* from the "Au lecteur."

the divine nature of the art.⁸¹ He briefly discussed the three principles, and, as would be expected, denied the existence of the humors.⁸² In a series of three hundred aphorisms le Baillif offered a small compendium of Paracelsian dogma, but with very little in the way of explanation. *Le demosterion* concludes with a short work on chiromancy and a Paracelsian/chemical lexicon. While the work is of interest as an early French effort of this sort, one could not say that the text is profound.

The same year he published *Le demosterion* le Baillif left Brittany for Paris, where he was appointed *médecin ordinaire* to Henry III. Here he began work on his short *Traicte de l'homme et son essentielle anatomie*, which was published in 1580. In this text le Baillif ignored traditional anatomical studies in order to concentrate on the relationship of the macrocosm and the microcosm, particularly on the influence of the planets on man.⁸³

Le Baillif's stay in Paris was short and disastrous. He was at first attacked for coining false money; even more calamitous, his medical views and activities became a matter of concern to the members of the Parisian Medical Faculty—physicians who had the right to limit practitioners in the city to Paris graduates. Shortly after his arrival in the capital le Baillif was ordered to halt both his practice and his lecturing (1578). On his refusal to do either, he was forthwith summoned to appear before the Parlement. Here in an emotional three-day trial le Baillif and his counsel, the distinguished Étienne Pasquier, defended Paracelsian medicine. The Galenists, desiring to destroy an ever-increasing Paracelsian threat to their views, hoped to quash the new sect decisively. They found that they had pitted their strength against a man who "a toute question proposée, tousjour chantoit l'une de ses trois chansons ... à sçavoir, de ses trois principes, Sel, Soufre et Mercure; de la séparation du pur et de l'impur; et du microcosme."⁸⁴ The trial proved to be a victory for the medical establishment, and on June 2, 1579, le Baillif was ordered to leave the capital. Shortly

---

⁸¹*Ibid.*, p. 4.

⁸²*Ibid.*, pp. 14 ff., 18.

⁸³Roch le Baillif, *Premier traicte de l'homme, et son essentielle anatomie, avec les elemens, & ce qui est en eux: De ses maladies, medecine, & absoluts remedes ès tainctures d'or, corail, & antimoine: & Magistere des perles: & de leur extraction* (Paris: Abel l'Angelier Libraire, 1580), p. 26.

⁸⁴Trevor-Roper, "The Sieur de la Rivière," p. 236.

thereafter he returned to Brittany, where he established himself at Rennes and continued to write and to practice his unorthordox medicine.

The Galenist victory at Paris in 1579 did not stay the incipient interest in chemistry and Paracelsian medicine. In 1581 a graduate of Montpellier, Claude Dariot (1533-1594), published his *De praeparatione medicamentorum*.[85] Although Dariot paid tribute to the efforts of both Severinus and Guinter, he deplored the fact that there was still confusion in the medical profession about these remedies. Acknowledging his debt in particular to Guinter, Dariot nevertheless complained of the lack of explicit descriptions in the preparation of mineral substances in the *De medicina veteri et nova*.[86] Although it was unfortunate that, because of "envious opponents," Paracelsus had written in an obscure style, the medical profession would have gained much "if the physicians of his time, rather than belaboring him and chasing him from their company, would have received him and exhorted him to write his secrets more clearly."[87] Dariot obviously felt there was a great need for a simpler text on the Paracelsian remedies; this was his goal and the purpose of his book. And, writing of his preparation for this task, Dariot referred to his long work in the search of the secrets of Hermes, Arnald of Villanova, Geber, and Lullius, as well as those of Paracelsus and his followers.

Dariot refused to agree with those who argued that the cure by similitude advocated by the Paracelsists conflicted with the work of the ancients.[88] Paracelsus referred here neither to primary nor secondary qualities of the Aristotelians, but rather to substances

[85]Claude Dariot, *De praeparatione medicamentorum* (Lyon, 1582). Dariot matriculated at Montpellier in 1553 and, as in the case of so many other French Protestants, left his country after the St. Bartholomew's Day Massacre. He was admitted a citizen of Geneva in 1573. He served as the town physician of Beaune. Information on pharmaceutical practice in Renaissance Montpellier is contained in Louis Dulieu, *La pharmacie à Montpellier* (Avignon: Les Presses Universelles, 1973).

[86]I have used the French translation of the *De praeparatione medicamentorum* which forms the third part of *La grand chirurgie de Philippe Aoreole Paracelse . . . traduite en Francois de la version Latin de Iosquin d'Alhem . . . par M. Claude Dariot plus un discours de la goutte . . . Item III. Traittez de la preparation des medicaments* (3rd ed., Montbeliart: Iaques Foillet, 1608), separate pagination, pp. 4, 13.

[87]*Ibid.*, p. 12.

[88]*Ibid.*, pp. 14-15.

and virtues. At the same time he introduced a tripartite division of diseases based upon the three principles in such a way that when he stated that "les semblables sont gueris par leurs semblables," he was in reality speaking of salt, sulfur, and mercury.[89] Little wonder that the identification and cure of the disease must now be assigned to those familiar with chemistry. For Dariot it was definitely not a question of overturning one system for another, but one of adding a valuable tool and means of explanation to medicine. He was concerned especially with showing that there was no real conflict over the question of the elements. Paracelsus himself had taught how to separate earth, water, air, and fire from bodies, and then the Aristotelian matter, form, and privation might be compared with the Paracelsian principles. For Dariot the example of burning wood could be used not just to show one system or the other, but rather the two systems at once.[90]

A practical work that claimed to present chemical preparations in a more open form than ever before, Dariot's text was printed at least twice in French translation as the *Trois discours de la preparation des medicamens*. In 1588 Dariot prepared a short treatise on the gout which was also oriented toward Paracelsian medicine and which discussed the origin of the disease in terms of the functions of the *archei*.[91]

Also in 1588 Dariot finished his translation of Paracelsus' *La grand chirurgie*, a translation he made not from the original German, but from the Latin translation of Josquin d'Alhem printed first at Lyon in 1573. Once again he referred to the difficult nature of the alchemical texts, which he ascribed to their original hieroglyphic form. Over the years a few alchemists had maintained an understanding of these characters—a knowledge which they had passed on only to their disciples. It was, indeed, the glory of Paracelsus to have found and have collected these alchemical secrets.[92] The chemical bias of Dariot's translation is evident from the outset

[89]*Ibid.*, pp. 16-17.
[90]*Ibid.*, p. 21, 22.
[91]*Discours de la goutte. Auquel les causes d'icelle sont amplement declarees, avec sa guerison et precaution* (1588; ed. used, Montbeliart, 1608), with separate pagination.
[92]*La grand chirurgie* . . ., trans. Dariot (1588; ed. used, Montbeliart, 1608), with separate pagination: sig. ***5^r.

when he stresses that alchemy and medicine can never be separated.[93] This emphasis is strengthened by his lengthy annotations—comments on the results of his own medical practice and upon theoretical problems concerning the Paracelsian principles and the relation of the two worlds.[94]

The Paris Confrontation (1603)

Interest in Paracelsian medicine was clearly growing in the closing decades of the sixteenth century. There was no question where the medical establishment stood in relation to the new sect, but their victory over le Baillif in 1579 had done little to discourage the continued activities and publications of other Paracelsists. Gaston Duclo (b. c. 1530) published a number of alchemical works and prepared a lengthy refutation of Erastus (1590) which was later included in Zetzner's *Theatrum chemicum* (1602).[95] Blaise de Vigenère (1523-1596) was also well read in alchemical texts, and his *Discourse of Fire and Salt* (1608) refers to saltpeter in words that are reminiscent of the Paracelsian *Liber azoth*.[96] In addition to these authors, Penotus continued to publish his violent attacks on the Galenists, while Dariot's translation of Paracelsus was reprinted in 1593, 1603, and 1608.

The entry of Henry IV into Paris in 1593 proved to have unexpected medical overtones. French physicians who had an interest in the work of Paracelsus and in the application of chemistry to medicine were, almost without exception, Huguenots. These Protestants, many of whom had been in exile for years, now felt that they could return to France, even to the capital itself.

In 1594 Jean Ribit, sieur de la Rivière (c. 1571-1605), a

---

[93]*Ibid.*, p. 12.

[94]Excellent examples are Dariot's annotations on pp. 245-252.

[95]Partington, *A History of Chemistry*, Vol. II, p. 158. Duclo's (Claveus) *Apologia chrysopoeiae & argyropoeiae adversus Thomam Erastum* will be found in the *Theatrum chemicum* (Strasbourg: Zetzner, 1613), Vol. II, pp. 1-83.

[96]Allen G. Debus, "The Paracelsian Aerial Niter," *Isis*, **55** (1964), 43-61 (49). Although this work was first published at Paris in 1608, I had recourse only to the English translation of Edward Stephens, *A Discourse of Fire and Salt, Discovering Many Secret Mysteries, as well Philosophicall as Theologicall* (London, 1649).

Huguenot, was appointed first physician to the King.[97] Critical of the many unlearned Paracelsists he had met on his travels throughout Europe, Ribit de la Rivière nevertheless had a genuine interest in the new chemically prepared drugs which seemed to offer much for the future of medicine. Ribit's appointment was symptomatic of change in the new reign, and it is not surprising to find that his friend, Joseph Duchesne, was appointed physician in ordinary to the King. However, if Ribit published nothing, this was hardly the case with Duchesne. Long recognized as a major advocate of chemical medicine, Duchesne proceeded to publish a new, largely theoretical work in 1603, the *De priscorum philosophorum verae medicinae materia*. Here and in the *Ad veritatem hermeticae medicinae ex Hippocratis veterumque decretis ac therapeusi* (1604)[98] Duchesne defended the new remedies and the Hermetic approach to medicine. Discussing the various sects of medicine, he con-

[97]The differentiation of Jean Ribit, sieur de la Rivière, from Roch le Baillif, sieur de la Rivière, earlier thought to be identical, is the subject of Trevor-Roper's paper cited above, n. 79.

[98]In addition to the original editions of 1603 and 1604, the following Latin editions have been used: *Liber De Priscorum Philosophorum verae medicinae materia, praeparationis modo, atque in curandis morbis, praestantia. Déque simplicium, & rerum signaturis, tum externis, tum internis, seu specificis, à priscis & Hermeticis Philosophis multa cura, singularíque industria comparatis, atq introductis, duo tractatus. His accesserunt ejusdem Ios. Quercetani de dogmaticorum medicorum legitima, & restituta, medicamentorum praeparatione, libri duo.* (Leipzig: Thom. Schürer and Barthol. Voight, 1613); *Ad Veritatem Hermeticae medicinae ex Hippocratis veterumque decretis ac Therapeusi, nec non vivae rerum anatomiae exegesi, ipsiusque naturae luce stabiliendam, adversus cuiusdam Anonymi phantasmata Respondio* (Frankfurt: Ex Officina Typographica Wolffgangi Richteri, Impensâ Conradi Nebenii, 1605). A French edition of the first title referred to is the *Traicté de la matiere, preparation et excellente vertu de la Medecine balsamique des Anciens Philosophes. Auquel sont adioustez deux traictez, l'un des Signatures externes des choses, l'autre des internes & specifiques conformément à la doctrine & pratique des Hermetiques* (Paris: C. Morel, 1626). The English translation by Thomas Tymme (Timme) containing large sections of both works is *The Practise of Chymicall, and Hermeticall Physicke, for the preseruation of health* (London: Thomas Creede, 1605). The above editions will be referred to by language and date for identification in the following notes

General accounts of the Parisian debate in the first decade of the seventeenth century will be found in Chevalier, " 'The Antimony War' " (above, n. 64); Lynn Thorndike, *A History of Magic and Experimental Science* (New York: Columbia University Press, 1959), Vol. VI, pp. 247-253; Wightman, *Science and the Renaissance*, Vol. I, pp. 257-263.

trasted the dogmatics with the spagyrists. The former follow Galen, "and as if by a royal edict where the sentence is firm and without doubt, they pronounce that contraries cure."[99] The latter, however, are chemists who place their faith not in books but in reason and experience, rejecting external judgments and seeking instead the internal essence of bodies. This, Duchesne continued, is the real basis of spagyric medicine and the root of the disagreement between the Galenists and the chemists.[100]

Although Duchesne constantly strove to disassociate himself from doctrinaire Paracelsism, it is clear that he shared many views in common with other chemical philosophers. His cosmology, for instance, was founded on the Biblical story of Creation. God had first separated light from darkness and then waters from waters. This Duchesne interpreted as the separation of the three principles,[101] and throughout his works he emphasized the importance of salt, sulfur, and mercury. To them were assigned the active qualities while the passive ones were reserved for the traditional elements—less one since fire was conceived to be nothing other than heaven itself.[102] This rejection of fire as a true element was not uncommon for the period: both Paracelsus and Cardan had affirmed that fire was not a true element.[103] Yet, if Duchesne felt free to adopt a three-element system of nature, he nevertheless con-

[99] Duchesne (French, 1626), p. 4; (Latin, 1613), sig. A3^r.
[100] Duchesne (French, 1626), pp. 5-6; (Latin, 1613), sig. A3^v. The following discussion of Duchesne is based largely on Debus, *The English Paracelsians*, pp. 90-96.
[101] Duchesne (English, 1605), sig. H1^r; (Latin, 1605), p. 144.
[102] Duchesne (English, 1605), sig. G3^r; (Latin, 1605), p. 137.
[103] Myrtle Marguerite Cass has written that Cardan maintains that there are only three elements: earth, air, and water. "Fire is not an element, for it requires food, moves about, is of very fine substance, and nothing is produced from it." Girolamo Cardano, *The First Book of Jerome Cardan's De subtilitate*, Latin text, commentary, and translation by Myrtle Marguerite Cass (Williamsport, Pa., 1934), p. 15. See also Jerome Cardan, *De subtilitate Libri XXI* (Lugduni: Stephan Michael, 1580), pp. 45-46, where he states that "Sed certè sub coelo Lunae nullus est ignis . . . Natura enim semper extrema mediis iūgit . . . Nam inter duo extrema, non duo, sed tantum vnum solet assignari medium. Quod si statuatur, non quatuor, sed tria tantum erunt elementa," and his *De varietate libri XVII* (Basel, 1557), p. 21, Ch. 2, "Esse aũt tria, perspicuum est: terram solidissimam, aeram tenuissimum, aquam inter haec mediam." John Woodall referred to the *Meteorem*, cap. 1 as evidence of Paracelsus' support of three rather than four elements: *The Surgions Mate* (London, 1617), pp. 309-310.

tinued to make use of all four when it was convenient to do so. Showing the relationship of the two sets of basic matter, he paired salt with earth, sulfur with fire, and mercury with water and air.[104] While such comparisons might be made, the elements were pictured as a secondary form of matter from the three principles themselves.[105]

Although not spelled out with great clarity, Duchesne's system of elemental matter led to a three-principle/three-element theory based on the story of Creation, in which the three elements are essentially earth, heaven (fire), and water. By water could be understood the terrestrial waters and the "upper" waters (air). Thus, although there are only three true elements, they may be explained in terms of the traditional four. Duchesne considered it proper to discuss all of nature in terms of triads which might then be referred to the all-important *tria prima*. Quite simply, "There are three principall things mixed in euery Naturall bodie: to wit, *Salte, Sulphur,* and *Mercurie*. These are the beginnings of all Naturall things."[106] Still, "these three principles of Chymists are not the common Salt, Sulphur, and Mercurie: but some other thing of nature, most pure and simble, which neuerthelesse hath some conscience and agreement with cōmon Salt, Sulphur, and Mercurie."[107]

These principles could clearly be compared to body, soul, and spirit,[108] but Duchesne thought it helpful to be more specific as to their characteristics:

> *Mercurie* is a sharpe liquor, passable, and penetrable, and a most pure & *Aetheriall* substantiall body: a substance ayrie, most subtill, quickning, and ful of Spirit, the foode of life, and the Essence, or terme, the next instrument.
>
> *Sulphur* is that moyst, sweet, oyly, clammy, original, which giueth substance to it selfe; the nourishment of fire, or of natural heate, endued with the force of mollifying, and of gluing together.
>
> *Salt*, is that dry body, saltish, meerely earththy, representing the nature of *Salt*, endued with wonderfull vertues of dissolving, congeal-

---

[104]Duchesne (English, 1605), sig. T4^r; (Latin, 1613), pp. 116-117.
[105]Duchesne (English, 1605), sig. G2^v; (Latin, 1605), p. 135.
[106]Duchesne (English, 1605), sig. B3^r.
[107]Duchesne (English, 1605), sig. H3^v; (Latin, 1605), p. 150.
[108]Duchesne (English, 1605), sig. P4^v; (Latin, 1613), p. 28.

ing, clensing, emptying, and with other infinite faculties, which it exerciseth in the Indiuiduals, and separated in other bodyes, from their Indiuiduals.[109]

Here Duchesne introduced the Platonic "argument of the mean" to show why it was essential that all three principles should be present in every material substance.

> For as a man can neuer make good closing morter, of water and sand onely, without the mixture of lime, which bindeth the other two together like oile and glue: so Sulphur or the oily substance, is the mediator of Salt and Mercurie, and coupleth them both together: neither doth it onely couple them to death, but it also represse and contemperate the acrimonie of Salt, and the sharpnesse of Mercurie, which is found to bee very much therein.[110]

Since they are said to be mixed in all substances, Duchesne went on to show how they might be recognized by their properties. He suggested that enough is known of the properties of these principles to estimate roughly their proportions in the metals.[111] Even in the three elements we may discern the three principles:

> For out of the element of Water, the iuyces and metallick substances do daily break forth in sight: the vapours of whose moysture or iuyce more spiritous, do set forth *Mercury*: the more dry exhalations, *Sulphur*: and their coagulated or congealed matter, *Salt*.[112]

In air the activity and power of the winds betray a mercurial spirit, while the comets and lightning are an evidence of sulfur. Even salt may be identified in air through the thunderbolt or the "stone of lightening." Finally, in earth "the Mercurial spirits shewe themselves in the leaues and fruits; the Sulphurus, in the flowers, seedes, and kirnels: The salts, in the wood, barke and rootes."[113] These same principles may be found in all living creatures, for the properties of sulfur may be seen in grease, tallow, and marrow,

---

[109]Duchesne (English, 1605), sig. D1^V; (Latin, 1605), p. 131.

[110]Duchesne (English, 1605), sig. T4^V; (Latin, 1613), p. 108.

[111]Duchesne (English, 1605), sigs. H3-H4; (Latin, 1605), pp. 148-152. He also observed that since the three principles remain in a substance no matter how it is acted on or changed, in the same way copper or iron must remain in vitriol even though its outward form has been changed.

[112]Duchesne (English, 1605), sig. H2^V; (Latin, 1605), pp. 147-148.

[113]Duchesne (English, 1605), sig. I3^V; (Latin, 1605), p. 166.

which burn, while the salt is the bones and other hard parts, and mercury can be identified with the blood, the humors, and the vaporous substances.[114] Besides these observational tests for the three principles, Duchesne attributed to them various other qualities such as "tastes in Salt, most chiefely; odours in Sulphur: colours out of both, but most chiefely out of Mercurie: because Mercurie hath the volatile Salt of al things, ioyned unto it."[115]

Unmoved by the arguments of Erastus to the contrary, Duchesne argued that the three principles are made manifest by fire analysis which separates without corruption.[116] And although he was somewhat disturbed by the difficulty of altering gold and silver in the fire, Duchesne felt certain that these "noble and perfit" metals are also composed of the three principles; the equality and purity of their mixture makes it seem that "they are one substance, not three, or consisting of three."[117] In *The Sceptical Chymist* the case of gold was to be one of Boyle's main examples against the existence of the three principles, but for Duchesne there was no serious problem, since the noble metals could be explained as an extension of the like-like principle which had generally been limited by Paracelsians to medical explanations:

> ... in Golde, the sulphur which is fixed and incombustible, of a fiery nature, bringeth to passe that it standeth invincible against all force of fier, and looseth not the least waite thereof, because like wil never oppresse his like, but contrariwise do cherish and preserve one the other: whereby it commeth to passe that it ioyeth in the fier, and alwaies commeth out of the same, more pure and noble then it went in.[118]

The traditional four humors—blood, yellow bile, black bile, and phlegm—were summarily rejected by Duchesne. Rather, he suggested that there are only three humors—analogous to the three principles. Duchesne's humors were based upon the three fluids most closely associated with the Galenic vascular system: the first was chyle; the second, which he called blood, corre-

---

[114]Duchesne (English, 1605), sig. K2^v; (Latin, 1605), pp. 172-173.
[115]Duchesne (English, 1605), sigs. T3^v and T4^r (Latin, 1613), p. 106 (incorrectly numbered 116).
[116]Duchesne (English, 1605), sig. G3^r; (Latin, 1605), p. 136.
[117]Duchesne (English, 1605), sig. H4^r; (Latin, 1605), p. 151.
[118]Duchesne (English, 1605), sig. H4^r; (Latin, 1605), pp. 151-152.

sponded closely to what we would distinguish as the venous blood.

> The third of the humours, is that which after sundry reterations of the circulations, made by the much vital heate of the heart, doth very farre exceede in perfection of concoction: the other two, which may be called the elimentary or nourishing humour of life, and radical Sulphur: the which is dispearced by the arteries throughout the whole body, and is turned into the whole substance thereof, . . .[119]

Duchesne suggested that as there is a continual circulation of the elements in the macrocosm,[120] so also is there a circulation of the blood in man as well. However, what Duchesne was suggesting was not the circulation of the blood as we now understand it; it was more in the sense of the circulation of a liquid in a chemical distillation. His view of the blood flow remained essentially Galenic.[121]

Duchesne discussed the interrelation of the macrocosm and the microcosm with other examples that shed light on his medical theory. He described respiratory diseases in terms of an internal bodily distillation, a position which reflected both the Aristotelian

---

[119] Duchesne (English, 1605), sigs. Li^r-Li^v; (Latin, 1605), p. 179.

[120] Duchesne (English, 1605), sig. C4^r; (Latin, 1605), p. 128.

[121] The concept of the circulation of the blood as a chemical distillation has been discussed in some detail in regard to the work of Cesalpino by Walter Pagel in his "Harvey and the Purpose of the Circulation," *Isis*, **42** (1951), 25-26 and again in *William Harvey's Biological Ideas*, pp. 169-209. For further details on this problem and the relation of Duchesne and Fludd see the following chapter and also Allen G. Debus, "Robert Fludd and the Circulation of the Blood," *Journal of the History of Medicine and Allied Sciences*, **16** (1961), 374-393. In brief, with the exception of a reference which suggests that all of the blood circulates in the body as in a chemical pelican, Duchesne discusses in detail only the venous system. He states that the blood which is carried to the heart by the vena cava is there circulated and distilled over in a purer form to the brain where it is redistilled or circulated a second time. In the Tymme translation (sig. K4^v; Latin, pp. 177-178) Duchesne states that ". . . the same blood being carried into the heart, by the veyne called *Vena Caua*, which is as it were the Pellican of nature, or the vessel circulatory, is yet more subtilly concocted, and obtaineth the forces as it were of quintessence, or of a Sulphurus burning Aquavita, which is the original; which is the original of natural & unnatural heat. The same Aquavita being carried from hence by the arteries into the *Balneum Maria* of the Braine, is there exalted againe, in a wonderful maner by circulations; and is there changed into a spirit truly ethereal and heauenly, from whence the animal spirit proceedeth, the chiefe instrument of the soule . . . ."

explanation of catarrh and the persistent search for chemical analogies among the Paracelsians. But beyond this, he argued that these diseases were similar to the formation of clouds and rain; so if we investigate the microcosmic phenomena, we may also hope to learn of macrocosmic truths such as the source of winds, sleet, and snow.[122] Like most Paracelsians Duchesne upheld the principle of like curing like instead of the ancient method of seeking cures in contraries.[123] Also, although he was convinced that the purpose of chemistry was to prepare medicines, he was not above summarizing the directions laid down by Paracelsus in the *Aurora* for the transmutation of lead to antimony.[124]

Though Duchesne was thus not far removed ideologically from the Paracelsian chemists, he still wanted to indicate that his chemical and universal medicine was the true medicine of Galen and Hippocrates,[125] and—although he devoted a lifetime to the collection and publication of chemical recipes of medicinal value—he stated that he only wanted to enrich the older cures "with some chemical or spagyric ornaments" which had been found useful.[126] Indeed, for Duchesne, Paracelsus had been only one of many chemists and alchemists who had taught this true medicine in a line that stretched back to the greatest antiquity.[127]

This conciliatory note was not destined to satisfy Duchesne's more conservative colleagues at Paris. His correlation of the three principles with the Trinity, his emphasis on the importance of the microcosm analogy, and his overwhelming interest in chemistry as the basis of medicine offended many. The book was condemned and immediately answered by the elder Jean Riolan (c. 1538-1606), who incorrectly implied that this chemist had totally rejected the works of Galen and Hippocrates.

The remainder of the decade witnessed the publication of a large number of pamphlets, monographs, commentaries, and books that alternately defended and attacked Galenic, Paracelsian, and chemical medicine. The complex relationship of these works

---

[122] Duchesne, *Traicte de la matiere*, p. 183.
[123] Duchesne (Latin, 1613), pp. 14-15; see also (English, 1605), sig. N4^r.
[124] Duchesne (English, 1605), sigs. I1^v-I2^r; (Latin, 1605), pp. 157-158.
[125] Duchesne (French, 1626), pp. 104-105; (Latin, 1613), pp. 69-70.
[126] Duchesne (French, 1626), pp. 8-9; (Latin, 1613), sig. A4^v.
[127] Duchesne (French, 1626), pp. 22-27; (Latin, 1613), pp. 4-8.

has not been thoroughly examined, but at this point it will suffice to note that the *Apologia pro Hippocratis et Galeni medicina* (1603) of Riolan was answered by Duchesne's colleague Turquet de Mayerne, in an *Apologia in qua videre est inviolatis Hippocratis et Galenis legibus, remedia chymice preparate, tuto usurpare posse* (1603). The following year Duchesne penned his *Ad veritatem Hermeticae medicinae ex Hippocratis veterumque decretis ac therapeusi*. The year 1605 saw the publication of at least three works in support of the action of the Medical Faculty by Jean Riolan the younger, while Duchesne wrote an *Ad brevem Riolani excursum brevis incursio* and his *Ad veritatem* of the previous year was republished. Guillaume Bauchinet of Orléans wrote in defense of the chemists, and Israel Harvet contributed three texts supporting the chemical doctrine.

By 1605 the dispute was widely known beyond the borders of France. These texts were now being published in Latin in Germany, while in England Thomas Tymme translated major portions of Duchesne's books of 1603 and 1604 in his *Practise of Chymicall and Hermeticall Physicke for the Preservation of Health*. Andreas Libavius was so moved by the issues raised in the dispute that he felt it necessary to write a lengthy *Defensio alchemiae et refutatio objectionum ex censura Scholae Parisiensis* when he published an expanded version of his *Alchymia* (1597) in 1606. This was to result in an *Ad Libavimaniam . . . Responsio pro censura Scholae Parisiensis contra alchymiam lata* from the volatile elder Riolan. Riolan's death in 1606 did not deter Libavius from printing an *Alchymia triumphans* in 1607 which was a sentence-by-sentence reply to the French physician in a volume over nine hundred pages in length.[128]

While the above titles do not exhaust the number actually printed at the time, they should indicate the interest stirring in learned circles over the question of chemical medicine. It was also in the early years of the century that Jean Beguin arrived in Paris—possibly from Sedan. Here, with the influence of Ribit de la Rivière and Turquet de Mayerne, he was granted permission to set up a laboratory and give lectures on pharmaceutical preparations.[129] His *Tyrocinium chymicum* (1610), based on these lectures,

[128] Discussed by Wightman in *Science and the Renaissance*, Vol. I, pp. 259-263.

[129] The classic account of the work of Beguin and the various editions is that of T. S. Patterson, "Jean Beguin and his *Tyrocinium Chymicum*," *Annals of Science*, 2 (1937), 243-298. See also the discussion in Vol. III of Partington's *History of*

appears to have been derived ultimately from the *Alchymia* of Libavius,[130] but the direct connection with the Paracelsian corpus is also evident in the prefatory quotation from the *De tinctura physicorum*.[131] Primarily a practical work, it was surely one of the most successful publications of the century, with forty-one editions traced by Patterson in the period 1610-1690.

The works of Petrus Palmarius in support of the Parisian Medical Faculty were published in 1609, while Duchesne's most popular text on the preparation of chemical medicines, the *Pharmacopoea dogmaticorum*, appeared first in 1607 (twenty-five editions followed in the first half of the seventeenth century). And although Duchesne died in 1609, his work continued to attract avid readers for the rest of the century. Many of his individual texts were republished in the second and third decades of the century, a limited *Opera medica* went through at least two Latin editions (1602, 1614) and one German edition (1631), and an extensive three-volume collection, the *Quercetanus redivivus*, was prepared by Johann Schröder (three editions: 1648, 1667, and 1679).

And yet, if the number of titles—as well as their wording—give some indication of the intensity of this debate and help us to understand the subsequent development of chemical medicine in

---

Chemistry (1962), pp. 2-4, and Hélène Metzger, *Les doctrines chimiques en France du début du XVIIe à la fin du XVIIIe siècle*, première partie (Paris, 1923), pp. 35-44. A discussion of the literature as a whole will be found in P. M. Rattansi's biography in the *Dictionary of Scientific Biography*, Vol. I (1970), pp. 571-572.

[130]A. Kent and O. Hannaway, "Some New Considerations on Beguin and Libavius," *Annals of Science*, 16 (1960), 241-250.

[131]Although dropped from the French and Latin editions by 1615, Richard Russell's English translation, based on the earliest versions of the text, still included this practical admonition to the reader by Paracelsus: "First, you must Learn Digestions, Distillations, Sublimations, Reverberations, Extractions, Solutions, Coagulations, Fermentations, and Fixations; and you must also know what Instruments are required for use in this Work; as Glasses, Cucurbits, Circulatory Vessels, Vessels of *Hermes*, Earthen Vessels, Balneums, Wind Fornaces, Fornaces of Reverberation, and other such like: as also a Marble, Mortars, Coals, &c. So may you at length proceed in the Work of alchimy, and Medicine.

"But as long as you shall by Phantasie and Opinion adhere to feigned Books, you will be apt for, and Predestinated to none of these." [John Beguinus, *Tyrocinium Chymicum: or, Chymical Essays, Acquired from The Fountain of Nature, and Manual Experience*, trans. Richard Russell (London: Thomas Passenger, 1669), sig. A4^r-A4^v.]

France, it is particularly illuminating to pause for a moment to examine the work of Libavius.[132] A native of Halle in Saxony (c. 1540-1550), Libavius took his M.D. at Jena, where he was appointed Professor of History and Poetry (1586-1591). In later years he served as the town physician in Rothemburg. Attracted time and time again to current controversies, Libavius supported the Aristotelians in their conflict with the Ramists in 1591, and three years later he attacked the Paracelsians for their use of the weapon-salve. At all times an independent thinker, Libavius heatedly defended the ancients, but at the same time he saw much of value in the contemporary medical innovations.

Both in the *Alchymia* and in his historical review of the Parisian debate that forms the preface to the *Commentariorum alchymiae* (1606) we have the statement of a major chemist who was critical of most of his colleagues.[133] For Libavius alchemy was a divine art, but one which had been misled by obscure authors.[134] How hopeless, he wrote, is the task of those who seek our chemical meanings in the *Song of Songs* or the *Book of Alexander*.[135] Alchemy is nothing but the perfection of magisteries and the extraction of pure essences from mixed substances by separation.[136] With these views it is understandable that the bulk of the chemical work of Libavius

---

[132] On the life and work of Libavius in general see Partington, *History of Chemistry*, Vol. II, and Thorndike, *History of Magic*, Vol. VI, pp. 238-253.

[133] The present account of the views of Libavius in regard to Paracelsus and the chemists is based upon two earlier papers by the present author: "The Paracelsians and the Chemists: the Chemical Dilemma in Renaissance Medicine," *Clio Medica*, 7 (1972), 185-199, and "Guintherius—Libavius—Sennert."

[134] Andreas Libavius, *Alchymia, recognita, emendata, et aucta, tum dogmatibus & experimentis nonullis* ... (Frankfurt: Excudebat Joannes Saurius, impensis Petri Kopffii, 1606), sig. A2^r.

[135] *Ibid.*, sig. A6^v.

[136] *Ibid.*, sig. B5^r. "Alchemia est ars perficiendi magisteria, & essentias puras è mistis separato corpore, extrahendi." A difference between the two is noted in the contemporary definition by Martin Ruland the Elder: "MAGISTERIUM is a Chemical State which follows the process of extraction, and in which a matter is developed and exalted by the separation of its external impurities. In this manner are all the parts of natural and homogeneous concretion preserved. But they are so exalted that they almost attain the nobility of essences." [Martin Ruland, *A Lexicon of Alchemy*, trans. A.E. Waite (1st ed., 1893; London: John M. Watkins, 1964), p. 211. Martin Ruland, *Lexicon Alchemiae* (1st ed., 1612; Hildesheim: Georg Olms, 1964), p. 310.]

should fall into two categories: attacks on mysticism, and practical laboratory procedures. We need not be surprised to find in his work the first detailed description of an ideal chemical laboratory, and he wrote at length on the preparation of chemical medicines and the chemical analysis of mineral waters. But for him the transmutation of the base metals into gold seemed to be no less practical, and he did not hesitate to write on the philosopher's stone as well as the mercury and azoth of the alchemists.[137]

True, Libavius believed in metallic transmutation, but it was evident to him that the most interesting work in chemistry was related to medicine, and he accordingly directed his comments primarily to this subject. Reviewing the medical scene, he divided the profession into three medical sects: the Galenists, the Chemiatri, and the Paracelsians.[138] Surely all Galenists, he began, were not as bad as they had been pictured by their adversaries. They adhered to many important doctrines, and they did not follow the ancients to the letter; some at least were willing to judge the works of Galen with an open mind.[139]

The second medical sect, the Chemiatri, were clearly more interesting for Libavius. He divided this sect into two groups. The first were those physicians who advocated adding the benefits of pharmaceutical chemistry to the traditional materia medica. In this camp he listed authorities such as Avicenna, Mesue, Rhazes, Albertus Magnus, Arnald of Villanova, Raymond Lull, and Philip Ulstadius—men whom he commended for not desiring to introduce undue novelties into medicine and philosophy. However, there was a second branch of the Chemiatri—the Hermetic physicians—who searched for catholic principles and spoke at length of an all-encompassing science based on the macrocosm and the microcosm. They were accustomed to claim themselves the only true chemical physicians, but in reality they were but sophists, paying lip service to Hippocrates while distorting everything with their chemical explanations. No matter what they

---

[137] See Books 2, 3, and 4 of Part II of the *Commentariorum alchymiae* (Frankfurt: Saurius/Kopffii, 1606?). On the relationship of these substances see also the article on "Azoth" in Martin Ruland (English trans., pp. 66-67; Latin, pp. 96-97).
[138] *Ibid.*, sigs. Aa2^V-Aa5^V.
[139] *Ibid.*, sig. Aa2^V.

PLATE XIV   Andreas Libavius (c. 1540-1616). Photograph of an engraving by Fennitzer.

said, these authors were among the greatest enemies of the Galenists.[140]

The Paracelsians were the third major medical sect in Libavius' division. He condemned their writings, which he accused of having been founded on paradoxes, absurdities, and every possible consequence proceeding from delirium. These chemists, he wrote, rejected the commonly accepted philosophies outright, turning instead to the *Philosophia sagax* of Paracelsus, a work which was nothing but magic. Nor was this legitimate natural magic; it was what all knew to be damned—necromancy, fascination, commerce with spirits, and other monstrosities.[141] This they used as an explanation for terrestrial and celestial phenomena. But their very acceptance of the macrocosm-microcosm analogy invalidated their work, for no one with any wisdom could agree that celestial substances are found in man. Furthermore, there can be no mixture of the celestial and the terrestrial worlds.[142] It was true, Libavius conceded, that the three principles were described in the work of Paracelsus—substances indeed valuable for all chemists—but then they were clearly derived from the work of Aristotle.[143]

Libavius concluded that the Parisian Medical Faculty had rightly condemned the Paracelsians and the sophist chemists, yet their attack on Joseph Duchesne and Turquet de Mayerne had gone beyond bounds to become an attack on chemistry itself.[144] Clearly this was not warranted; the benefits of chemical medicine were too well known to be denied. As for himself, surely Libavius thought that he was one of the true Chemiatri who wished to apply the very real benefits of alchemy to the healing art without overturning the medical thought of antiquity.

The English Solution

If the Parisian debate was to become a matter of vital concern to Libavius and other Central European physicians, it was also to affect the medical scene in England. But, although the English

---

[140] *Ibid.*
[141] *Ibid.*, sigs. Aa2^V and Aa3^r.
[142] *Ibid.*, sig. Aa3^V.
[143] Libavius, *Alchymia, recognita*, p. 109.
[144] Libavius, *Commentariorum alchymiae*, sig. Aa4^r.

drama was played out against a background similar to the French one, a compromise solution was reached relatively early—and with comparatively little acrimony.[145]

To be sure, the London College of Physicians, founded in 1518, was originally a conservative group. Thomas Linacre (1460-1524), chief among the founders of the College, was a medical humanist who excelled in the collection and translation of Greek medical texts. The early members remained wedded, for the most part, to traditional authors; the prosecution of Dr. John Geynes for his suggestion that Galen was not infallible (1559) may be taken as typical of the period.[146] Indeed, the very purpose of the College was to maintain control over the profession, and from the beginning the members were given the right to examine all surgeons and physicians who practiced in London or within seven miles of the city.[147] In subsequent years this power was increased. In 1540 the physicians were permitted to include surgery in their practice, since this was considered to be a basic part of the study of medicine. At the same time the Company of Barber-Surgeons was founded, and it was reaffirmed that no surgeon could practice medicine in London without obtaining a College license. Also in 1540 the College was given control over the apothecaries in that the Censors of the College were empowered to enter their homes "to search, view, and see such Apothecary-wares, drugs and stuffs."[148] This privilege

---

[145] The reader is directed to *The English Paracelsians* for a more detailed treatment of the English scene to c. 1640.

[146] W. S. C. Copeman, *Doctors and Disease in Tudor Times* (London: Wm. Dawson and Sons, 1960), p. 36.

[147] In addition to the Statutes, I have relied on the following authorities: C. R. B. Barrett, *The History of the Society of Apothecaries of London* (London: Elliot Stock, 1905); Charles Goodall, *The Royal College of Physicians of London* (London: M. Flesher for W. Kettilby, 1684); O. M. Lloyd, "The Royal College of Physicians of London and Some City Livery Companies," *Journal of the History of Medicine and Allied Sciences*, **11** (1956), 412-421; Edward Kremers and George Urdang, *History of Pharmacy* (Philadelphia: Lippincott, 1951); and *Pharmacopoeia Londinensis of 1618 Reproduced in Facsimile with a Historical Introduction by George Urdang* (Madison: University of Wisconsin Press, 1944). There was only one exception to the licensing power the College exerted over physicians and surgeons in the London area in the sixteenth century: the same grant could be, and occasionally was, obtained from the Archbishop of Canterbury. This, however, was a practice that was becoming ever less common.

[148] *The Statutes at Large from the Thirty-second year of King Henry VIII to the Seventh*

was to be reaffirmed by Queen Mary at her accession in 1553.[149]

Yet, while the London area was under constant surveillance by the members of the College, a statute of 1543 permitted medical anarchy in the provinces. With a real shortage of trained physicians in rural areas, many local inhabitants turned to neighbors who were thought to have the ability to heal—or to travelling medicine men. Now "every person being the King's subject having knowledge and experience of the Nature of herbs, roots and waters . . . [might] . . . use and minister, . . . according to their cunning, experience and knowledge."[150] The result was clearly to permit all types of empirics to practice in England, since they were in effect controlled only by their own "experience and knowledge." Charles Goodall's history of the College (1684) shows that a significant proportion of the actions taken by the College were directed against medical quacks who had ventured into the London area. Many of these unlicensed practitioners adopted the title "Paracelsian," and it was to the credit of the members of the College that they sought to separate the learned from the unlearned chemists from the start.

Traditional applications of chemistry to medicine were well

Year of King Edward VI Inclusive, ed. Danby Pickering (Cambridge, 1763), 32 Hen. VIII, c. 40, p. 57.

[149]*The Statutes at Large from the First Year of Queen Mary to the Thirty-fifth year of Queen Elizabeth Inclusive*, ed. Danby Pickering (Cambridge, 1763), 1 Q.M. sessio second., c. 9, p. 16. George Urdang concluded that the Paracelsian reforms reached England by 1553 because this act gave the Censors of the College the right "to survey and examine the stocks of apothecaries, druggists, distillers and sellers of waters and oils, and preparers of chemical medicines" (Kremers and Urdang, *History of Pharmacy*, p. 138; also *Pharmacopoeia Londinensis . . . with a Historical Introduction by George Urdang*, p. 6). The mere reference to chemical medicines does not prove that they were Paracelsian in origin. Beyond this, the text of the document does not refer to chemical medicines at all. As quoted by Pickering in the *Statutes at Large*, the act of 1553 merely gave the Censors the right to "search and view of Poticarye Wares Drugges and Compositions" (see also Goodall, *Royal College of Physicians*, p. 33). Kremers and Urdang seem to have obtained their quotation from the 1618 Charter of the College which granted them the authority "to examine survey governe correct and punishe all and singular Physitians and practisers in the facultie of Physick Apothecaries Druggists Distillers and Sellers of waters or oyles Preparers of Chymical Medicynes . . .". (Goodall, p. 37).

[150]*The Statutes . . . from King Henry VIII to . . . King Edward VI*, 35 Hen. VIII, c. 8, pp. 143-144. See also Kremers and Urdang, *op. cit.*, p. 138.

known—and generally accepted—in England long before most physicians knew of Paracelsus. Hieronymus Brunschwig's book on distillation (1500) was translated into English by Lawrence Andrew in 1527, Arnald of Villanova's *Defense of Age and Recovery of Youth* (1540) stressed the medical virtues of gold and the "quintessence" (wine), and a work by Thomas Raynalde (1551) discussed a new chemically prepared oil.[151] But it was the translation of a book by the highly respected Conrad Gesner—*Treasure of Euonymus* (1559)—that effectively introduced to the English physician an armory of chemically prepared remedies.[152] Gesner's compendium included the recipes of many medieval medical alchemists, and he gave pointed emphasis to the use of metallic preparations.[153] Although this is so, it is worth repeating that Gesner was working within a tradition different from that of Paracelsus. There is no indication that he wished to overturn the works of the ancients. Rather, he noted elsewhere that Paracelsus had "condemned Galen, Hippocrates, and all the ancient doctors . . . I heard that he accomplished nothing worthwhile, indeed, rather he was an impostor."[154] But although Gesner had an obvious dislike for Paracelsus, he had to admit that "many were cured by him in desperate illnesses and that malignant ulcers were healed by him easily."[155]

The earliest references to Paracelsus by English authors are to be found in the works of Prostestants who fled abroad in the final years of Henry VIII and during the reign of Queen Mary—both

[151]Hieronymus Brunschwig, *The vertuose boke of distyllacyon* . . . (London: Laurens Andrewe, 1527). Arnaldus of Villanova. *Here is a newe boke, called defence of age and recovery of youth*, trans. J. Drummond (London: R. Wyer 1540). Thomas Raynalde, *A compendious declaration of the . . . vertues of a certain lateli inventid oile* . . . (Venice: J. Gryphius, 1551).

[152]Conrad Gesner, *The Treasure of Euonymus: Conteyninge the wonderfull hid secretes of nature*, trans. P. Morwyng (London: John Daie, 1559), pp. 293, 412, 421. Of interest is Gesner's discussion of distillation, pp. 1-6.

[153]Here we have evidence of the isolation of the products of chemical reactions prior to the time of the late-sixteenth-century Paracelsists. See the discussion of Gesner's chemical medicines above, Ch. 1. For Gesner's place in the history of distillation chemistry see Multhauf, "Significance of Distillation," p. 341.

[154]Conrad Gesner, *Bibliotheca universalis* (Tiguri, 1545), fol. 614.

[155]Thomas Erastus, *Disputationum de medicina nova Philippi Paracelsi: Pars prima* (Basel, 1572). From the short essay titled "Conradus Gesnerus Medicus Tigurinus de Theophrasto Paracelso" appended to this work.

periods of religious conservatism. Thus, we find William Turner referring to Paracelsus as an authority in his work on medicinal springs (1557),[156] and William Bullein doing the same in his text on surgery (1562).[157] The Cambridge physician John Jones gave an (unfilled) promise to refute Paracelsus in his *Galens Bookes of Elements* (1574), and the prominent surgeon George Baker similarly upheld Galenic theory while at the same time lauding the benefits of chemically prepared medicines.[158] The second part of the *Treasure of Euonymus* had appeared at Zurich in 1569, and so useful did it seem that Thomas Hill, an extraordinarily active translator, prepared an English version. However, by this time mortally ill, Hill bequeathed his manuscript to Baker. When the work was published in 1576 as *The Newe Jewell of Health*, it contained a preface by Baker in which he wrote that "the vertues of medicines by Chimicall distillation are made more vailable, better, and of more efficacie than those medicines which are in use and accustomed," yet, he continued, without a knowledge of Galen and Hippocrates the reader would be at a loss to properly apply the remedies in the book.[159]

As Thomas Hill had willed the Gesner translation to George Baker, he similarly left the translation of another book to the London apothecary John Hester. This text, Leonardo Fioravanti's *A Ioyfull Iewell*, was published in 1579 and it, too, was filled with chemical recipes. Hester proceeded to turn out numerous similar

[156]William Turner, *A booke of the natures and properties as well of the bathes in England as of other bathes in Germanye and Italye* (Collen: Arnold Birckmann, 1562), fol. iii. In the same year this was also issued with the second part of William Turner's *Herball*. In the 1568 edition of this work the preface is dated March 10, 1557, at Basel.

[157]William Bullein, *A Little Dialogue betwene Twoo Men, the one called Sorenes, and the other Chyrurgi* (London: Jhon Kyngston, 1562), fol. v. This work is the second part of *Bulleins Bulwarke of Defence against all Sicknes, Sornes and Woundes* . . . . Bullein referred to Paracelsus as "Theophrastus Peraselpus."

[158]John Jones, *Galens Bookes of Elementes* (London: William Jones, 1574). This is an appendix to *A briefe, excellent, and profitable discourse of the naturall beginning of all growing and living things; heate, generation . . . effects of the spirits, government, use and abuse of phisike, preservation*, . . . George Baker, ed. and trans., *The Composition or Making of the Moste Excellent and Pretious Oil called Oleum Magistrale* (London: John Alde, 1574); see the nonpaginated "To the Reader."

[159]Conrad Gesner, *The Newe Jewell of Health*, corr. and pub. by George Baker (London: H. Denham, 1576), sig. iii seq. (preface by George Baker).

works until his death (c. 1593).[160] Concentrating first on Fioravanti, he soon discovered other authors, including Isaac Hollandus, Philip Hermannus, Penotus, and some spurious works by Paracelsus. Most notably, Hester discovered Joseph Duchesne, whose *Aunswere . . . to . . . Aubertus* (1591) and *Sclopotarie . . .* (1590) appeared in English through his efforts.

It is clear that the earliest knowledge of Paracelsus in England was based upon his wonderful cures and the new chemical remedies. Few complaints were raised against these preparations, because they formed part of a much older tradition that was sanctioned by the names of Brunschwig and (to a much greater degree) Gesner. Furthermore, an ever-increasing number of medical men were finding the chemical compounds to be of great value in their practice. Among prominent surgeons there was again George Baker, who regularly advocated the use of chemically prepared oils and balms, while as early as 1589 John Banister cited Paracelsus, Duchesne, and Thurneisser no less than thirty-five times in his influential *Antidotarie*. William Clowes carefully distinguished between the "proud pratling Paracelsian" and "the good workes of the right Paracelsian" in his work on the *Morbus Gallicus* (1585).[161] Later, in his last publication (1602), Clowes confessed his inability to understand the philosophical thought of Paracelsus, but he reaffirmed his belief in the new medicines.

> I must confesse his Doctrine hath a more pregnant sence then my wit or reach is able to construe: onely this I can say by experience, that I haue practised certaine of his inuentions Chirugicall, the which I haue found to be singular good, & worthy of great commendations. How be it, much strife I know there is betwen the Galenistes and the Paracelsians, as was in times past betweene Aiax and Ulisses, for Achilles Armour. Notwithstanding, for my part I will heere set up my rest and contention, how impertinent and unseemely so euer it make shew: That is to say, if I find (eyther by reason or experience) any thing that may be to the good of the Patients, and better increase of my knowledge & skil in the Arte of Chirurgery, be it eyther in Galen or Paracelsus; yea, Turke, Iewe, or any other infidell: I will not refuse it, but be thankful to God for the same.[162]

[160]Debus, *English Paracelsians*, pp. 65-69.
[161]William Clowes, *A Briefe and necessarie Treatise Touching the Cure of the Disease called Morbus Gallicus* (London: T. Cadman, 1585), fol. 59.
[162]William Clowes, *A Right Frutefull and Approoved Treatise for the Artificiall Cure of*

The Paracelsian Debates

The chemical influence on surgeons continued into the new century. John Woodall's *The Surgions Mate* (1617), which contains an early and important passage urging all ships' surgeons to take on citrus fruits to combat scurvy, is also a Paracelsian treatise.[163] Both "Emplastrum stipticum Paracelsi" and "Laudanum opiat Paracelsi" (the latter cited from Crollius' *Basilica chymica*) are considered essential for the surgeons's chest.[164] However, beyond this, Woodall felt it important to discuss the elements and the principles: quoting Paracelsus in the *Meteorem*, he seemed to find proper authority for restricting the elements to three,[165] but he went on to attribute the current acceptance of this interpretation to the work of Duchesne. Later editions of *The Surgions Mate* (1639, 1655) found the alchemical section expanded, and Woodall offered in addition a short experimental manual giving directions for separating the three principles from samples of animal, vegetable, and mineral matter through distillation.[166]

English expositions of chemical theory were not at all so common as references to the new medicines, nor were the authors of these works medical men of prominence. R. Bostocke, whose *The Difference betwene the Aunciaent Phisicke . . . and the latter Phisicke* (1585) is the first Paracelsian apology in English, is known only by his name. Thomas Tymme, the translator of Duchesne's theoretical texts of 1603 and 1604 in *The Practice of Chymicall, and Hermeticall Physicke* (1605), was rector of St. Antholin's in London and later held the same position at Hasketon in Suffolk.

Tymme's translation of Duchesne—coupled with his own *Dialogue Philosophicall* (1612), which presents similar views—can be shown to have been influential, but then Duchesne's works have already been discussed. In some respects Bostocke remains the more interesting figure. His book, in addition to being an excellent Paracelsian *primer*, was also a conscious answer to the recent attack

---

that Malady called in Latin Struma (London: E. Allde, 1602), "Epistle to the Reader."

[163]Allen G. Debus, "John Woodall—Paracelsian Surgeon," *Ambix*, 10 (1962), 108-118.

[164]John Woodall, *The Surgions Mate* (London: E. Griffen for L. Lisle, 1617), pp. 40, 224-232.

[165]*Ibid.*, pp. 309-310.

[166]*Ibid.* (3rd ed., 1655), pp. 233-246.

of Thomas Erastus.[167] From the start Bostocke was bent upon comparing the work of the Paracelsians—whom he specifically called the chemical philosophers—with that of the "heathnish" Aristotelians and Galenists. The latter, he wrote, assume that nothing new has transpired for over a millennium, and they waste their time in disputations about the true meaning of contradictory passages from useless texts.[168] These Galenists, or "ethnikes," accept the preparations preserved in their ancient authorities, but in the search for the requisite ingredients for these recipes the physicians refuse to go beyond obvious outward signs of identification. How could such medicines be anything but raw and undigested?[169] Surely they must prove deadly to the recipients. The Galenists' constant emphasis on useless outer signs is in keeping with the traditional anatomy founded on the dissection of the human body; rather than this, physicians should be conversant with another form of anatomy, one that seeks a deeper understanding of man and nature.[170] It is because of such things that it is impossible for the Galenists to have any real understanding of the true differences between diseases. They attribute them to humors, complexions, and qualities—fantasies which are nothing but "dead accidents" unrelated to any existing malady.[171]

How different all this is, Bostocke continued, from the work of the chemical philosopher, who lays the foundation of his philosophy in God's Book, who knows that the principal cause in the world is the Creator.[172] In contrast to the Galenist, he seeks out divine unity in the Trinity—in the Paracelsian principles of salt, sulfur, and mercury, which form a basis for the study of nature and

[167] Although the present account of the work of Bostocke is drawn primarily from my paper in *Clio Medica* (see above, n. 133), a number of additional points will be found in *The English Paracelsians*, pp. 57-65, and in my paper "An Elizabethan History of Medical Chemistry," *Annals of Science*, **18** (1962, published 1964), 1-29.

[168] R.B., esq. (R. Bostocke), *The Difference betwene the aunciente Phisicke . . . and the latter Phisicke* (London: R. Walley, 1585). Here the eighth chapter (sigs. Dii^v-Fi^r) compares and contrasts the Galenic and the chemical medicines. On this see especially sigs. Div^v, $Eiii^v$, and $Eviii^r$.

[169] Ibid., sigs. Dvi, $Evii^v$.

[170] Ibid., sigs. Ev, $Cvii^r$.

[171] Ibid. sigs. $Dviii^v$-Ei^v.

[172] Ibid., sig. $Diii^r$.

medicine alike.[173] The existence of these three substances has been established through valid observations—and it is on this point that the Paracelsians differ most from their adversaries.[174]

It is through experience that the physician must find the secrets of the macrocosm which will benefit his patients. But these secrets can be found only through chemical means, that is, through the use of the fire. This tool can shear away the gross and deceiving outward signs of substances and reveal the previously hidden arcana of nature. By separating the pure from the impure in the laboratory the chemist imitates natural processes such as the growth of a seed or the action of the stomach on food, and in like fashion the fire makes it possible to examine the true anatomy of everything needed for our knowledge of the macrocosm and the microcosm.[175]

Any undue reliance on the traditional humors is to be avoided. Diseases are surely not the result of humoral imbalance; rather, they arise from seeds, which grow within the body as the earth brings forth fruit from seed. For a true understanding the physician must seek diseases in the macrocosm and then reduce them to their principles by chemical means.[176] Once the diseases of the great world have been identified, the physician can be certain that he will find them in man, because of the valid analogy of the macrocosm and the microcosm. Bostocke gave no real indication how this process was to be carried out, but he stated confidently that the need for harmony in the two worlds requires that the chemical anatomies of the disease and the remedy must be matched—that is, that like cures like.[177]

The chemical philosopher knows that the three principles correspond to three major categories of disease.[178] And while those

---

[173] *Ibid.*, sig. Diiii^r.
[174] *Ibid.*, sig. Fi^r.
[175] *Ibid.*, sigs. Dvi, Dvii, Dviii^v.
[176] *Ibid.*, sig. Ei^v.
[177] *Ibid.*, sigs. Dviii^v-Ei^r, Cvi^r. Bostocke explicitly states that the physician must not simply try possible remedies on the ill without some valid reason to do so. The Paracelsians had been strongly criticized for this by others. Rather (sig. Eviii^v), he insists that they first learn the nature of the disease before trying a substance on man. Walter Pagel (*Paracelsus*, pp. 130-147) discusses the relation of macrocosmic to microcosmic medicine.
[178] Bostocke, *Difference*, sigs. Ev^v-Evii^r, Liv^r.

diseases caused by supernatural influences are to be cured only through divine intervention, the other two, terrestrial and celestial, can be cured through a knowledge of vegetable and mineral preparations respectively. This true philosopher knows further that all medicines must be perfected by the fire; and if, in the preparation of medicines, the initial result of the fire is to accentuate a corrosive nature, the true chemist knows well that by continued chemical treatment he will be able to remove the undesired toxic qualities.[179] For Bostocke it was quite clear that medicine was a divine science and that a knowledge of disease was the only way to prepare a "perfect philosopher." But this same philosopher must of necessity be a chemical philosopher, "for Chymia and Medicina may not be separated asunder, no more thē can preparation or separation from knowledge or science."[180]

The College of Physicians and the *Pharmacopoeia* (1618)

In any comparison with the Parisian Medical Faculty one cannot help but be struck by the different attitude toward the chemical medicine taken by the members of the London College of Physicians. By the turn of the century they had already gone on record as favoring at least some of the new medicines, and there is little doubt that their membership included those who were urging their colleagues to adopt the new chemical therapy.

As early as 1585 the members of the College began to discuss the possibility of issuing an official pharmacopoeia.[181] An outline of this proposed volume, dating from 1589, still exists along with a list of those who were to be placed in charge of completing the various sections. Christopher Johnson (d. 1597), Thomas Langton (president of the College from 1604 until his death in 1606), and

---

[179]*Ibid.* sigs. Ev^v, Eiiii^r.

[180]*Ibid.*, sig. Ei^r.

[181]On the background to the first London *Pharmacopoeia* see George Urdang, "How Chemicals Entered the Official Pharmacopoeias," *Archives Internationales d'Histoire des Sciences,* **7** (1954), 303-314; "The Mystery About the First English (London) Pharmacopoeia (1618)," *Bulletin of the History of Medicine,* **12** (1942), 304-313; and his Introduction to the *Pharmacopoeia Londinensis of 1618.*

The Paracelsian Debates

Thomas Moffett (1553-1604) were to prepare the section on chemical medicines.

Although this pharmacopoeia was never printed, the prominent inclusion of chemical medicines is striking for so early a date. It is also significant that Moffett was one of the most noted defenders of the new remedies.[182] A former student of John Caius and Thomas Lorkin at Caius College, Moffett had gone abroad to study under Plater and Zwinger at Basel, where he obtained his M.D. in 1578. The following four years were devoted to travel in Italy and Germany—a time when he seems to have adopted many elements of the new medicine. Back in England (1582) he incorporated his M.D. at Cambridge and then journeyed to Denmark, where he met Peter Severinus and Tycho Brahe. He was elected both a fellow and censor of the London College of Physicians in 1588 and moved freely in the highest court circles.

Moffett's interests were broad. He assembled a work on insects from a manuscript begun by Wotton, Gesner, and Pennius dating from the 1550s (published in 1634); he wrote a digest of the Hippocratic corpus (1588) and a general guide to *Health's Improvement* (1655). Of more concern is the *De iure et praestantia chemicorum medicamentorum* (1584), which was later to be included in the three editions of Zetzner's *Theatrum chemicum*. This tract, dedicated to Severinus, is a strong defense of the new medicines which had been used by the ancient and Islamic authors as well as by "Gesner of blessed memory."

Moffett was well aware of the theoretical concerns of interest to Paracelsians. He believed that there was a clear connection between man and the greater world,[183] and he argued for the existence of the Paracelsian principles through observational evidence. Although many argued over the actual number of the elements, there was no doubt in his mind: "Henceforth let us say that the body of men consists of sulfur, mercury, and salt alone, not because we know this as perfectly as Adam, but because the resolution of all kinds of natural as well as artificial bodies shows it to be so."[184]

[182]See Debus, *English Paracelsians*, pp. 71-76.

[183]Thomas Moffett, *De iure et praestantia chemicorum medicamentorum* in *Theatrum chemicum*, ed. L. Zetzner, 6 vols. (Strasbourg: Zetzner, 1659-1661), Vol. I, pp. 65, 76.

[184]*Ibid.*, p. 100; for a similar passage see p. 95.

Moffett thought that the chemists differed from the ancients only in some points and in fact utilized those parts of ancient medicine that were of value.[185] And if in the end they turned to the work of Paracelsus, it was surely because he was a learned man rather than the magician and the imposter he had been accused of being. Much of his work and that of his disciples was of great merit, chiefly those "golden preparations of metals and minerals." Indeed,

> ... if you wish to stand your ground on the judgment of the ancients or of the more recent writers, it would have to be conceded that the mineral and metallic remedies not only should merit their place with the doctors, but that they should even be preferred.[186]

The case of Moffett points out that there was little effort by the London physicians to destroy learned men who favored the new medicines. If the censors singled out many who sold chemicals, it was primarily because of their constant search for unlicensed practitioners. Thus, the young Arthur Dee had exhibited on his door a list of medicines which he guaranteed as certain cures for many diseases; lacking a license, he should not have been surprised to have been called to account for his offense.[187]

More prominent was the case of Francis Anthony, a Cambridge M.D. and self-styled Paracelsian who was well known throughout Europe for his *aurum potabile*. Although he had battled intermittently with the College since 1598, the matter came to a head in 1610 with the publication of Anthony's *Medicinae chymiae*. Here he cited Paracelsus, Penotus, and Duchesne as he argued

---

[185] These points are discussed in the letter to "Philalethes" (4 cal. Feb., 1583; *ibid.*, pp. 89-93). Part of the letter is translated in *The English Paracelsians*, pp. 75-76.

[186] Moffett, *De iure* in *Theatricum chemicum*, Vol. I, p. 83.

[187] Goodall, *Royal College of Physicians*, p. 364. Others prosecuted were surgeons such as William Foster and William Turner and even members of the clergy who dabbled in medicine. One of these was Henoch Clapham, a nonconformist preacher who published a tract during the plague of 1603 stating that a Christian who dies of the pestilence does so through a lack of faith; this naturally offended the ecclesiastical authorities. After beginning to sell home-made remedies he was prosecuted both by the London College of Physicians and the Church. He was eventually sent to the Clink prison for three years. On Clapham see Goodall, p. 364, and the *DNB* article by Ronald Bayne, Vol. IV (London, 1949-1950), pp. 371-372.

that gold could be placed in solution by proper treatment. This was clearly the most noble of medicines since it was the most noble of metals.[188] Anthony was answered immediately by Matthew Gwinne, a fellow of the College who was to be elected censor four times prior to his death in 1627. Naming Anthony's medicine an *aurum putabile*, he cited Paracelsus by way of Libavius to show that gold cannot be dissolved except through the use of a highly corrosive acid which would render it unfit for medicinal purposes.[189] Angered though he was, Gwinne clearly wished to distinguish the "chymici" from the "Chemiatrae" or the "Galenochymici," among whom he classed himself.[190]

The Anthony debate continued until his death in 1623. He was to be attacked by a fellow chemist, Thomas Rawlin, in 1610, and Anthony himself replied to his detractors in an *Apology* published simultaneously in Latin and English in 1616.[191] As late as 1623 John Cotta felt compelled to publish an *Ant-Anthony*.[192] The case is perhaps most significant in showing that contemporary English physicians made a sincere effort to distinguish between learned chemists and alchemical charlatans. And if the physicians deplored the frequent adoption of the name "Paracelsian" by quacks, there was no attempt on their part to condemn his legacy by mounting a wholesale attack on all Paracelsians or on the chemical philosophy. In short, the members of the London College

---

[188] Francis Anthony, *Medicinae chymiae, et veri potabilis auri assertio* (Cambridge: C. Legge, 1610), pp. 25-31. On Anthony's life see the *DNB*, Vol. I, pp. 519-520. An article titled "The Contribution of Francis Anthony to Medicine" by George H. Evans, *Annals of Medical History*, 3rd Ser., **2** (1940), 171-173, is an unfortunate eulogy. See also Goodall, *Royal College of Physicians,* pp. 349-351.

[189] Matthew Gwinne, *In assertorem chymicae, sed verae medicinae desertorem, Fra. Anthonium* (London: R. Field, 1611), p. 109. For Gwinne's life see the article by Norman Moore in the *DNB*, Vol. VIII, pp. 842-843.

[190] Gwinne, *In assertorem chymicae*, p. 22. "Non dico equidem chemiatros, Galenochymicos; è quibus esse pervelim, nec quin sin dubito; nec enim illi ista pollicenter, pollicendo pelliciunt in fraudem rerum imperitos."

[191] Thomas Rawlin, *Admonitio pseudo-chymicis seu alphabetarium philosophicum* (London: E. Allde, c. 1610); Francis Anthony, *Apologia veritatis illucescentis pro auro potabile* (London: J. Legatt, 1616); Francis Anthony, *The Apologie, or, defence of a verity heretofore published concerning a medicine called Aurum Potabile* (London: J. Legatt, 1616).

[192] John Cotta, *Cotta contra Antonium. or an Ant-Antony . . .* (Oxford: J. Lichfield and J. Short for H. Cripps, 1623).

of Physicians were well aware of the benefits chemistry offered to medicine, and they were quite prepared to seek out and use whatever might be of value.

The medical world of London was clearly more tolerant than Paris of the new medicines in the early years of the new century. And it was just at this time that Theodore Turquet de Mayerne (1573-1655), a leading figure in the Parisian conflict, arrived in England. Another French figure of Huguenot stock, Mayerne's father had fled to Geneva with his family after the St. Bartholomew's Day Massacre (1572), and there Theodore was born. He attended Heidelberg and then Montpellier, which in contrast to Paris was becoming oriented toward iatrochemistry. At Montpellier he took his M.D. degree in 1597, then moved to Paris, where he became both a friend and disciple to the *premier médecin*, Ribit de la Rivière.[193] The latter obtained the post of *médecin ordinaire* for Mayerne, who proceeded to initiate a series of lectures on iatrochemistry for surgeons and apothecaries (1599). His views were well known by 1603, and it is not surprising to find him already in the forefront of those defending Duchesne against the attack of Riolan. As in the case of Duchesne, the Medical Faculty acted swiftly: Mayerne's colleagues were forbidden to consult with him, and the King was urged to rescind his public offices. It is to Henry IV's credit that he paid little attention to this directive.

In 1606 Mayerne attended an Englishman who had fallen ill in Paris. Following his recovery, the patient invited Mayerne to return with him to London, where he soon found himself honored by being appointed physician to the Queen. This was a brief stay, however, and Mayerne was soon settled again in Paris, where he remained until the assassination of his patron, Henry IV, in 1610. Then, summoned to London by James I, he was appointed chief physician to the King, a post he retained until his death in 1655.

Mayerne was elected a member of the London College of Physicians in 1616, two years after Henry Atkins had reactivated the earlier plan to publish an official pharmacopoeia. Quite naturally this project was one of considerable interest to the

---

[193]The best biography of Mayerne available is that of Thomas Gibson, "A Sketch of the Career of Theodore Turquet de Mayerne," *Annals of Medical History*, N.S. **5** (1933), 315-326. For his relationship to Ribit de la Rivière see Trevor-Roper, "The Sieur de la Rivière."

PLATE XV Sir Theodore Turquet de Mayerne (1573-1655).

French physician, who became deeply involved in the writing of the volume. At this time the plan to publish a pharmacopoeia was closely connected with the desire of the fellows of the College to supervise more closely the practices of the apothecaries; it was, in effect, part of a larger plan to extend the powers of the College. An earlier charter for the apothecaries and the grocers (1606) had already proved unsatisfactory,[194] and in 1617 Atkins and Mayerne persuaded James I to grant the pharmacists a separate charter giving them the right to dispense but not to prescribe medicines. The official *Pharmacopoeia*, at this time being readied for publication, was intended to serve as a guide not only for doctors in their prescriptions but also for apothecaries in the compounding of their medicines.

The first edition of the *Pharmacopoeia* was issued in May 1618. Urdang noted that it is likely that conflicts between the older and younger members of the College were responsible for its recall and the publication of a considerably altered second issue later in the same year. The latter represented a "more pretentious pharmacopoeial combination of formulary and textbook, with the purpose of giving general information, also a survey of the entire materia medica, *simplicia*, and *composita*."[195] The author of the unchanged preface (most likely Mayerne) wrote:

> ... we venerate the age-old learning of the ancients and for this reason we have placed their remedies at the beginning, but, on the other hand, we neither reject nor spurn the new subsidiary medicines of the more recent chemists and we have conceded to them a place and corner in the rear so that they might be as a servant to the dogmatic medicine, and thus they might act as auxiliaries.[196]

This edition was to remain the official one with few variations for many years. Even the second London *Pharmacopoeia* (1650) maintained the selections on chemical remedies with few alterations.[197]

---

[194] Barrett, *History of the Society of Apothecaries*, p. xvi.
[195] *Pharmacopoeia Londinensis of 1618 . . . Introduction by Urdang*, p. 78.
[196] *Pharmacopoeia Londinensis* (2nd ed., London: J. Marriott, 1618), "Candido Lectori."
[197] A comparison of the chemical sections of the *Pharmacopoeia Londinensis* in the editions of 1618 (2nd), 1627, 1638, and 1650 is offered in *The English Paracelsians*, p. 153. A similar discussion of the seventeenth-century German pharmacopoeias is presented by Wolfgang Schneider in "Der Wandel des Arzneischatzes im 17. Jahrhundert und Paracelsus," *Sudhoffs Archiv*, **45** (1961), 203 ff.

The chemical divisions, the *Sales, metallica, mineralia,* the *Olea chymica,* and the *Praeparationes chymicae magis usuales,* consisted of a series of preparations which had been borrowed from both ancient and modern authors and which remained fairly constant throughout this period. Of the eighty-five preparations in the first category, Urdang has shown that most of the minerals and metals had been borrowed from Dioscorides.[198] Among the twenty "chemical oils" there were only a few that involved real chemical reactions rather than simple distillations of crude substances. And even among the seventeen "more usual chemical preparations," many of the recipes were for mixtures which involved no chemical reactions of any sort. A number of the substances listed were well known to earlier alchemists or to contemporary iatrochemists. The *saccharum saturni* probably derived from Libavius, and a number of compounds of mercury and antimony were Paracelsian in origin. Mayerne was indebted to the *Basilica chymica* for his directions for preparing *tartarus vitriolatus* (K_2SO_4) and *mercurius dulcis* (calomel),[199] the latter being one of the most recent remedies in the work. Few sources are cited in the chemical sections of the volume, and the name of Paracelsus appears only once—for his famed *stipticum,* or wound plaster.[200]

The *Pharmacopoeia Londinensis* was not the first to approve of the new remedies: the *Pharmacopoeia Augustana* of 1613 antedated the London collection in the prescription of chemical remedies for internal use. Still, the two editions of the London publication in 1618 retain their significance as the earliest national pharmacopoeias, at the same time being clearly representative of the tolerant attitude of the English medical profession for the prior forty-year period. In effect, the English solution to the Paracelsian-Galenist debate was in the spirit of the work of earlier conciliators such as Guinter and Libavius. Here no attempt was made to wrestle with the Paracelsian philosophy of nature; this was a subject of widespread disagreement, and individuals might accept or reject it as they saw fit. For the fellows of the College the crux of the matter was confined

---

[198] *Pharmacopoeia Londinensis of 1618* . . . Introduction by Urdang, p. 51.
[199] *Ibid.*, p. 72.
[200] According to his manuscript notes, Mayerne borrowed this from Hester's translation of the "Spagerical preparations" of Duchesne. *Pharmacopoeia Londinensis* (2nd ed.; British Museum copy with Mayerne's notes), p. 180.

to the remedies themselves. The chemical medicines which seemed to be of value were accepted—regardless of their origin.

The *Agreement and Disagreement* of Daniel Sennert (1619)

The *De chymicorum cum Aristotelicis et Galenicis consensu ac dissensu* of Daniel Sennert will serve as a final example of the effort made during the late sixteenth and early seventeenth centuries to integrate the work of the chemical physicians into the mainstream of medical thought. An early publication of this widely read author, it was a book that went through numerous Latin editions in the seventeenth century and also appeared in English in a truncated form prepared by Nicholas Culpeper and Abdiah Cole.[201]

From 1602 professor of medicine at Wittenberg—where he had received his M.A. (1597) and M.D. (1601)—Sennert (1572-1637) was one of the most influential authorities in his field. His huge *Institutionum medicinae* (1611) in abbreviated form became one of the most frequently reprinted medical texts of the century, and his collected *Opera omnia* (1641) comprised four folio volumes in its third edition (Lyon, 1654-1656). It is evident that Sennert had read widely in the current literature. He is best known for having been one of the first chemists to adopt an atomic view of matter.[202] His

---

[201] Daniel Sennert, *De chymicorum cum Aristotelicis et Galenicis consensu ac dissensu liber 1., controversias plurimas tam philosophis quam medicis cognitu utiles continens* (Wittenberg: Apud Zachariam Schurerum, 1619). The Latin edition cited below as (Latin, 1633) is the third, printed at Paris (Apud Societatem). The English translations are from Daniel Sennert, Nich. Culpeper and Abdiah Cole, *Chymistry Made Easie and Useful. Or, The Agreement and Disagreement Of the Chymists and Galenists* (London: Peter Cole, 1662). The English translation is drastically cut in length; however, the text given is for the most part a direct translation rather than a paraphrase. A discussion of Sennert's work by Pagel appears in his *Paracelsus*, pp. 333-343. The reader is referred also to Thorndike, *History of Magic*, Vol. VII (1964), pp. 203-217, and Partington, *History of Chemistry*, Vol. II, pp. 271-276. The pertinent discussion of Sennert is based primarily upon Debus, "Guintherius—Libavius—Sennert," pp. 157-161. See T. Gregory, "Studi sull'atomismo del seicento," *Giornale Critico della Filosofia Italiana*, **43**, (1964), 38-65; **45** (1966), 43-63; **46** (1967), 528-541. Here a major section is devoted to Sennert (**45**, 51-63).

[202] The importance of the atomic views of Sennert was noted by Kurd Lasswitz, *Geschichte der Atomistik vom Mittelalter bis Newton*, 2 vols. (Hamburg/Leipzig:

"atoms," however, were more than purely mechanical in nature; associated with them were formative forces in a sense similar to the *archei* of Paracelsus. And yet, although clearly influenced by the medical chemistry of the Paracelsians, Sennert was far from ready to discard many basic aspects of ancient medicine. Indeed, as Pagel has noted, Sennert felt that the most serious defect in the work of Paracelsus was his denial of the humors.[203]

In his comparison of the chemists with the Aristotelians and the Galenists, Sennert was clearly influenced by the earlier publications of Libavius, even though he was not in complete agreement with his predecessor and was interested in many areas of medicine not discussed by Libavius. The two authors surely agreed on the practical nature of chemistry. Sennert wrote that "it is an art to resolve Natural compound bodies into their principles

---

Leopold Voss, 1890), Vol. I, pp. 436-454. Referring to the *De chymicorum cum Aristotelicis et Galenicis consensu ac dissensu,* he stated (p. 441) that "in diesen Ausführungen ist die Korpuskulartheorie so bewusst ausgesprochen, dass wir vom Jahre 1619 ab die Erneuerung der physikalischen Atomistik datieren müssen." Thorndike, *History of Magic,* Vol. VII (1964), pp. 206-207, noted that in the same text Sennert had given credit to Scaliger for this concept. In her "The Establishment of the Mechanical Philosophy," *Osiris,* **10** (1952), 412-541 (428), Marie Boas mentions Sennert briefly. More recently Hans Kangro has discussed Sennert's views in relation to Jungius in his *Joachim Jungius' Experimente und Gedanken zur Begründung der Chemie als Wissenschaft* (Wiesbaden: Franz Steiner: 1968), which was the subject of a lengthy essay review by Walter Pagel: "Chemistry at the Cross-Roads: The Ideas of Joachim Jungius," *Ambix,* **16** (1969), 100-108. Useful also are Hans Kangro, "Erklärungswert und Schwierigkeiten der Atomhypothese und ihrer Anwendung auf chemische Probleme in der ersten Hälfte des 17. Jahrhunderts," *Technikgeschichte,* **35** (1968), 14-36, and Ivor Leclerc, "Atomism, Substance, and the Concept of Body in Seventeenth Century Thought," *Filosofia della Scienza,* **27**, 1-16. Partington cites a volume seen neither by him nor by me: Rembert Ramsauer's *Die Atomistik des Daniel Sennert* (Braunschweig: Viewig, 1935). Nevertheless, with all of this very real—and merited—interest in sixteenth- and early-seventeenth-century atomistic and corpuscularian thought, we still have had far too few studies connecting these authors with the Hermetic and Paracelsian background. Here the notice taken by J.E. McGuire and P.M. Rattansi, in "Newton and the 'Pipes of Pan,' " *Notes and Records of the Royal Society of London,* **21** (1966), 108-143 (130), of the late-sixteenth-century identification of the Phoenician Moschus-Mochus with Moses—and the subsequent placing of the atomic hypothesis within the familiar *prisca* tradition—is of considerable importance and merits further research.

[203]Pagel, *Paracelsus,* p. 341.

Curando, dubitem, an fuerit Podalirius ægris,
Hippocratesne docens. Egit utrumq; simul.

PLATE XVI Daniel Sennert (1572-1637) at the age of fifty-five. (from an engraving by M. Merian).

of which they are made, and to make them pure and strong for Medicines, or to perfect or change Metals."[204] Sennert agreed also with Libavius that the truth of metallic transmutation could not be challenged. Reliable observations confirmed it, as did the analogous example of the transmutation of food in our bodies.[205]

However, the relationship of chemistry to medicine was of special importance to Sennert. There are some, he noted, who argue that there is no difference between the medicine of the ancients and that of the chemists, but this is not true, for even if the ancients knew chemical methods they did not know them as well as the moderns.[206] The physician should not forget that

> ... Chymistry is not a peculiar Art, but belongs to Physick, and is the perfection of it, for it is the part only of the Physitian to use and apply Chymical medicines for cure, and [he] may be called then a Chymical Physitian, and the Medicines Chymical, which are the perfection of Physick.[207]

Those who would go beyond this practical goal are not to be heeded. Surely there are some who

> ... enlarge Chymistry, and dispute principles and labour to bring in new operations into all the parts of Philosophy and Physick, but it is not for Chymists, as such, to dispute of principles, but for Physitians and Philosophers. And Chymistry doth but only bring some observations and experiments by working from which the Physitian and Philosopher makes conclusions.[208]

Therefore, valuable though chemistry is, it should not be considered the foundation of medicine. Furthermore, physicians must be doubly cautious of the works of the Paracelsians: while there is no truth in the claim that chemistry originated with Paracelsus,[209] it is true that there is dangerous magic in his work, as Libavius has rightly pointed out.[210] Indeed, there is a fundamental difference between the ancient authorities and the Paracelsians. The concepts of Hippocrates, Aristotle, and Galen are based upon the twin

[204]Sennert (Latin, 1633), p. 9; (English, 1662), p. 5.
[205]1633, pp. 10-12; 1662, pp. 5-6.
[206]1633, p. 7; 1662, pp. 3-4.
[207]1633, p. 5; 1662, p. 3.
[208]1633, p. 7; 1662, p. 4.
[209]1633, p. 31; 1662, p. 15.
[210]1633, pp. 216 ff; 1662, pp. 77-81.

pillars of reason and experience.²¹¹ In opposition to them, the Paracelsist Crollius insists that the true physician cannot proceed without the light of Nature and Grace. One can only conclude that this

> ... false Chymistry hath its peculiar Religion, for because they think they have reformed or perfected all Philosophy & Physick, they stay not there, but proceed to Divinity, and mix prophane and holy things together: and so, they bring in any absurdity, and dispute wonderfully of the Kingdom of the Blessed, the Angels Miracles, Faith and the Resurrection.²¹²

But, Sennert added, since the Fall of man we have had to learn not from revelation, but from the things of nature. Religion should form no part of our natural philosophy.

Much of Sennert's treatise was aimed at specific doctrines adopted by the Paracelsians. It is true, he wrote, that both the Galenists and the Paracelsians accepted the doctrine of the macrocosm and the microcosm, but the concept "is extended too large by the Chymists, because they make not an Analogie, but an identity, or the same thing."²¹³ How can one really argue that the existence of something in one world necessitates its existence in the other? Indeed, some chemists have stated that there can be no humors because they have not been found in the macrocosm, but such proof by analogy cannot be accepted seriously.²¹⁴

Like Guinter, Sennert took pains to point out doctrines which differed little between the Paracelsians and the ancients. He believed that the Paracelsians were not far removed from the Aristotelians on the question of the Creation; if only the chemists would modify their doctrine of the abyss, the two views would correspond.²¹⁵ Similarly they might well be advised to alter their concept of life. It is true that metals and minerals grow in the earth, but the Paracelsians' blanket vitalism is overly simplistic. The inherent force of self-multiplication in inorganic matter is not the same as the life-soul type of growth.²¹⁶ The chemists also make

---

[211] 1633, pp. 55-56; 1662, p. 23.
[212] 1633, p. 57; 1662, p. 24.
[213] 1633, p. 63; 1662, p. 26.
[214] 1633, pp. 247-248; 1662, p. 96.
[215] 1633, p. 197; 1662, pp. 68-69.
[216] 1633, pp. 113-114; 1662, pp. 43-44.

much of their disagreement with the Galenists over the true anatomy of bodies, arguing against dissection of corpses and praising instead their own dissection of substances into the three principles by chemical means. For Sennert the latter was not far removed from concepts developed by Galen in his discussion of the faculties of nature.[217]

Sennert clearly recognized the importance of the three principles for the Paracelsians. He considered them to be a first mixture of the Aristotelian elements—substances which had in fact been described by the ancients but not given the same names.[218] It was important that their existence could be proved by observation. The truth that bodies are composed of those substances they can be resolved into had been stated by Aristotle and Galen no less forcefully than by Sennert's Paracelsian contemporaries.[219] Developing this theme, Sennert specifically took up the question "whether the chemical principles are produced by an artificial resolution from the mixed."[220]

For him, fire could produce nothing but fire, and therefore its role could only be one of dissolving a compound body.[221] Like others before him, Sennert chose to illustrate this with the example of burning wood. The smoke and flame—alleged to be air and fire—could be explained differently. The flame derived from the fire, but the smoke was actually composed of an innumerable number of small particles. In the case of the sublimation of mercury or sal ammoniac the fumes have been shown to be the vapors of these substances—not air.[222] The smoke of burning wood should be interpreted similarly. And, as for the ashes remaining after combustion, they are neither salt nor earth, but a mixed solid composed of a variety of substances.[223]

Violent natural resolution (combustion) is therefore difficult to judge and should be avoided. Chemical resolution (analysis by dis-

[217] 1633, pp. 251-252; 1662, pp. 97-98.
[218] 1633, p. 132; 1662, p. 53.
[219] 1633, p. 147; 1662, p. 54. Also see Allen G. Debus, "Fire Analysis and the Elements in the Sixteenth and the Seventeenth Centuries," *Annals of Science*, **23** (1967), 127-147, esp. pp. 134-135 on Sennert.
[220] Sennert (Latin, 1633), p. 156.
[221] *Ibid.*, pp. 153 ff.
[222] *Ibid.*, p. 160.
[223] *Ibid.*, p. 161.

tillation) is far more satisfactory, since the operator, if he is careful, can collect all fractions and thereby better estimate the ingredients. This is important, for while most vapors look the same, actually they vary greatly—witness the difference between the spirit of wine and the condensed spirit of distilled rose petals. Similarly oils and residuous matter can be examined properly only after a distillation analysis.[224]

Sennert noted that there were some who held that the three principles are not separated but formed through the action of fire, writers such as Schegkius and Riolanus, who believed that fire acts on earth to form salt. But surely we are aware of the prior existence of the principles: we may observe salt in excrement, note the inflammable property of tree resin, be aware of odor everywhere. These are properties that belong only to the principles. And what might be said of metallurgical assays by fire: are we to assume that the pure metals are made by the analytical process? Surely only one conclusion is possible. Chemical analyses made with the aid of heat result only in separation, not generation. The natural philosopher may safely assume that the "chemical principles are separated from mixed substances. Therefore, they are present in them."[225] Sennert's discussion of the principles was such a strong defense that Robert Boyle was later to designate this account as their most learned justification.[226]

Though he defended the three principles, Sennert argued forcefully against much of the Paracelsian pathology. He would accept neither the dictum that disease was hereditary, ultimately deriving from the Fall of man,[227] nor the concept of the *semina*. Paracelsus, he wrote, argued that a disease has its own life cycle, growing within the body as a plant does on earth; for a proper cure the whole disease organism must be removed, not just the outward signs.[228] But what does this mean in relation to traditional medicine? In plain terms the Galenic physician is being told by the Paracelsian that he must pay attention to more than the humors,

---

[224]*Ibid.*, p. 162.

[225]*Ibid.*, p. 163.

[226]Robert Boyle, *The Sceptical Chymist* (London: J. Cadwell for J. Crooke, 1661; Dawson reprint, London, 1965), p. 268.

[227]Sennert (Latin, 1633), p. 257; (English, 1662), p. 103.

[228]1633, pp. 255-256; 1662, p. 101.

for they are only the outward signs of the disease, not its essence. Sennert did not hesitate to take sides on this question. The Paracelsians, he noted,

> ... reject the humors erroneously, because it is manifest to sense that divers humors are evacuated in divers diseases by vomit and stool by Nature or Art; when they cannot deny this *Severine* thinks that excrementitious humors are to be called by other names; and leaving the phantastick names of Vitelline, eruginous, salt and crude flegm, he flies to the kinds of *Salts, nitrous, aluminous, vitriolate, and to the properties of salts in Plants, as Cookowpints, Nettles, Celandine, & etc.*[229]

Still, there is no reason why the humors might not be designated by the principles when they are in excess. Indeed, the chemist might render the physician a great service by helping him to properly identify these substances.[230]

And yet, if the chemists might prove helpful by using their art to identify the humors, they surely should not be followed when they expound their views on the origin of disease,[231] for they contradict each other as much as Paracelsus contradicted himself. Here the physician is well advised to follow the pathology of Galen. But although Sennert agreed with Libavius in his adherence to the humors, he followed Paracelsus more closely in regard to the specific category of tartaric disease. In his discussion of the formation of stones in the body, Sennert rejected the influence of the qualities heat and cold. Rather, he argued that such stones could only originate through chemical causes resulting from an internal formation of salt.[232] Nor could Sennert accept Libavius' rejection of the doctrine of signatures.[233] Libavius had argued that it was only by chance that objects in nature answer to their external forms, but Sennert could not concur and supported the Paracelsians again on this point. In a similar fashion Sennert agreed that the aspects of the stars must be observed when medicinal herbs were to be collected, and he supported the defenders of magnetic cures.[234]

And what of the new chemical medicines? Sennert contrasted

[229] 1633, pp. 262-263; 1662, p. 104.
[230] 1633, p. 263; 1662, pp. 105-106.
[231] 1633, pp. 168 ff.; 1662, pp. 107-116.
[232] 1633 pp. 274-277; 1662, pp. 109-111.
[233] 1633, p. 333, 1662, p. 134.
[234] 1633, p. 348; 1662, pp. 138, 139.

the Galenists, who sought out the affected parts of the body before prescribing, with the Paracelsians, who argued that the proper medicine—found through a knowledge of the interrelation of the macrocosm and the microcosm—once administered would itself seek out the seat of the disease and destroy it.[235] But although chemicals were acceptable as specifics, it did not follow that only those medicines prepared by chemical means should be approved. On occasion when medicine must be directed throughout the body, then substances prescribed must be whole and unchanged. However, when the medicinal force is to be directed to one specific point, the substance is quite properly separated by the chemist with his fire.[236] Furthermore, the chemist does not lie when he claims that he has the ability to detoxicate poisons. For these reasons

> We reject not wholly Physick made of Minerals and Metals, being perswaded by the Spaw-waters that cure desperate diseases. The ancient Physitians gave them inwardly as well as outwardly, as Steel, Sulphur vive, scales of Brass, burnt Brass, and the like. Therefore we may better use them Chymically prepared.[237]

In general, he concluded, chemical medicines are more noble, but because of the great differences between diseases they are not always to be preferred to the Galenicals.[238]

Conclusion

The half-century 1570-1620 proved to be a difficult period for the medical profession. Some authors such as Severinus (1571) and Crollius (1609) sought to organize the work of Paracelsus in a form more easily digested by their more conservative colleagues. For them the Paracelsian system was a chemical philosophy that properly accounted for nature as a whole as well as medicine. Others such as Erastus (1572-1574) reacted with alarm both to the

---

[235]1633, pp. 305-306; 1662, p. 122.
[236]1633, pp. 371-372; p. 144.
[237]1633, p. 364; 1662, p. 143. A similar argument had been used by Joseph Duchesne. See his *Practise of Chymicall, and Hermeticall Physicke*, sig. Q3^V.
[238]Sennert (Latin, 1633), p. 373; (English, 1662), p. 145.

violence of the Paracelsian chemical remedies and to a system that called for the rejection of fundamental tenets of traditional medicine. For Erastus the natural magic, the macrocosm-microcosm analogy, and the denial of the humors by Paracelsus were all anathema. He, and a high percentage of those who shared his beliefs, focused on the Paracelsian principles as they sought to attack the basis of this new medical heresy.

Yet, alongside the advocating and the attacking forces there was also developing a spirit of conciliation. The Bavarian Paracelsist Albertus Wimpenaeus argued that the new medicine really did not conflict with that of the ancients and that truth should be sought wherever it might be found, both in the works of the ancients and that of Paracelsus (1569). Far more influential was the discussion of the old and the new medicine penned by Guinter of Andernach in 1571. Like Wimpenaeus he argued that the best of both systems was to be sought by the prudent physician, and in his own case this meant the bulk of ancient theory plus the wondrous new chemical medicines. Wimpenaeus, Guinter, and a large number of later authors made a distinct effort to show that the views of Paracelsus did not really conflict with those of the ancients and that apparent differences vanished when subjected to careful scrutiny.

The fortunes of the new medicine were clearly related to the prevailing climate of opinion in any given locality. In the case of the Parisian Medical Faculty the problems that developed could well have been predicted in advance. Here the conservative medical establishment called Roch le Baillif to account for his chemical sins as early as 1578, and a rapid survey of the early French Paracelsian literature quickly uncovers an underlying feeling of bitterness felt by the chemists. With this background it is no surprise to find that when Joseph Duchesne defended the chemical medicine in 1603 he initiated a new period of sharp debate. The deep interest in the medical issues at stake helps to account for the exceptionally large number of Paracelsian, chemical, and alchemical works printed in France between 1610 and 1640.

And yet, for all of the sharp language written, the Parisian conflict again displays the spirit of compromise that existed within the ranks of the chemists themselves. The review of the medical war at Paris by Andreas Libavius (1606) is far removed from the works of rabid Paracelsists. Rather, like Guinter before him, Libavius sought a middle road which belonged to neither faction. While Libavius welcomed the practical benefits of the chemically prepared medicines and defended a limited number of the theoretical views of the chemists—especially the three principles—he coldly rejected much of the Paracelsian philosophy. Much of the latter for him was dangerous black magic, and furthermore, no less than Erastus, he was appalled by a "philosophy" that employed chemical analogies as a means of cosmological explanation and scriptural interpretation.

The English scene differed greatly from Paris. Here chemical medicines had been widely accepted through the publication of works by Brunschwig and Gesner in the mid-sixteenth century, and the contribution of Paracelsus at first seemed not too far removed from the work of authors already approved. There was, in fact, an early and fairly general acceptance of chemically prepared medicines by pharmacists and surgeons in England. As for the London College of Physicians, its members welcomed to their fellowship the well-known Paracelsian apologist Thomas Moffett in 1588 and immediately set him to work on the chemical section of their proposed pharmacopoeia.

The cosmological aspects of the chemical philosophy met with relatively little response in England in this period. Bostocke's apology (1585), surely one of the best of its sort ever written, was largely ignored. The lengthy translation from Duchesne made by Thomas Tymme (1605) was more influential, but it elicited no response comparable to the impact the same writings had made on the medical profession in Paris. These were the only two texts on the chemical philosophy of any significance to be published by Englishmen prior to the folio volumes on the macrocosm and the microcosm written by Robert Fludd in the period after 1617.

With this background it is understandable that Turquet de Mayerne—although a controversial figure in France for his views on chemistry—should not have been the center of controversy in England. His work on the official *Pharmacopeia* of the College of Physicians, finally published in 1618, helped to ensure the acceptance of the chemically prepared remedies in England by all but the most stubborn Galenists. And while some English authors continued to write of the harsh medical debate between the ancients and the Paracelsians, these same authors had seen relatively little of this conflict in their own country.

In many ways Sennert's discussion of the medical world he saw in 1619 is representative of the period. Like most chemists he strongly defended the three principles while at the same time deploring the all-encompassing claims of many Paracelsian theorists. Like Libavius he rejected the macrocosm-microcosm analogy, and like Guinter he tried to show how the Paracelsian views were similar to those of the ancients. For Sennert the humoral pathology remained a crowning glory of antiquity, an approach to medicine that could not safely be discarded. Indeed, even though during his own lifetime Sennert had been accused of founding a new Sennerto-Paracelsian medical heresy,[239] his eclecticism has made it possible for Lester King, in an important paper, to interpret his work primarily in a Galenic framework.[240]

An analysis of the period 1570-1620 indicates that the familiar characterization of Renaissance medicine in terms of a bitter Galenist-Paracelsian conflict is only partially true. To be sure, the Parisian debate was lengthy and at times venomous, but there were many influential authors who sought the best of both worlds. Although these conciliators varied on individual points, they generally sought to maintain the traditional humoral pathology while they looked on the mystical implications of the Paracelsian chemists with dismay. At the same time, they had read the Paracelsian works with care and were obviously impressed with the new chemically prepared medicines. Their answer was simple: the chemicals should be used by physicians along with the traditional Galenicals.

[239]Johann Freitag, *Novae sectae Sennerto-Paracelsicae, recens in philosophiam & medicinam introductae . . . detectio & solida refutatio . . .* (Amsterdam, 1636; 2nd ed., 1637), a volume of 1,356 pages, discussed by Partington, *History of Chemistry*, Vol. II, p. 276. Accused of heresy by Freitag, Sennert consulted the theological faculties of eight German universities in preparation for his reply, the *De origine et natura animarum in brutis sententiae Cl. theologorum in aliquot Germaniae academis* (Wittenberg, 1638). See Thorndike, *History of Magic*, Vol. VII, p. 204. On this debate see also B. Nevelthavius, *De paradoxis et erroribus novae sectae Sennertiano-Paracelsicae . . . disputatio* (1635?).

[240]Lester S. King, "The Transformation of Galenism" in *Medicine in Seventeenth-Century England*, ed. Allen G. Debus (Berkeley/Los Angeles: The University of California Press, 1974), pp. 7-31.

The sixteenth- and seventeenth-century discussions of the role of chemistry in medicine and in the interpretation of nature point to deep divisions within the ranks of the chemists themselves. These divisions are seen in most sharp relief in the decades following 1620, in a comparison of the mystical alchemical-medical system of Robert Fludd with the far different but equally chemically oriented medical system of J. B. van Helmont.

4
THE SYNTHESIS OF ROBERT FLUDD

> I will stoop a little, for your better instruction ... by some vulgar or ocular demonstrations, which may guide them and perswade them to believe that, which may otherwise seem uncredible. ... *Experience is the mother of fools.*
>
> Robert Fludd (1638)*

THE CLOSING YEARS of the sixteenth century and the opening decades of the new one witnessed a widespread interest in the chemical philosophy among scholars and physicians. This is not to say that the chemists were agreed on all points. Essentially they comprised two groups. The first of these sought a universal system of nature that linked man and God through the macrocosm-microcosm relationship. Scriptural authority played a major role in their work, and they stressed proof by analogy in the two worlds. Their interest in observational evidence most frequently was linked to their interest in chemical operations. No less active was the second group, those who rejected the macrocosm-microcosm analogy but who considered the new interest in chemistry to be a major innovation in medicine. These authors expressed far less interest in the phenomena of the great world, and while they frequently borrowed some chemical concepts from their more universally oriented colleagues, in general they were distinguished from them by their greater emphasis on medical and pharmaceutical problems. Both groups were a cause for concern to the most conservative followers of Aristotle and Galen, since these new chemists were questioning the fundamental tenets of traditional medicine and philosophy. While some confined their

---

*Robert Fludd, Esq., *Mosaicall Philosophy: Grounded upon the Essentiall Truth or Eternal Sapience* (London: Humphrey Moseley, 1659), p. 200. As the *Philosophia Moysaica* this work appeared first posthumously from the press of Petrus Rammazenius (Gouda) in 1638.

doubts to specific doctrines, there were other chemical philosophers who were calling for a blanket replacement of ancient authority with their new truths extracted from nature.

The proponents of the chemical philosophy in the early seventeenth century were confident that their system would replace that of the ancients in time. And why should Crollius and his colleagues not have expressed such confidence? Paracelsian medicine was already widely accepted in Central Europe, it was looked on with interest by the Medical College in London, while in Paris the traditionalist Medical Faculty was clearly on the defensive. Nor did this situation change rapidly. The views of the more mystical chemists were a subject of concern to Kepler in the second decade of the seventeenth century, as they were later to Mersenne and his circle in the third and the fourth.

The reasons for this continued interest in a universal chemical system are varied. Partially they may be ascribed to the uncertainty of proof in this period, but no less important was the chemists' strong call for reform in the understanding of nature. Here their emphasis on the use of observational evidence was consistently linked with their quest for truth in religion. They supported their plea with a medical trump card—the reputed cures effected by Paracelsus and his followers.

Surely there was a no more avid advocate of a universal chemical philosophy of nature in the first half of the seventeenth century than Robert Fludd. His views, again centered on a chemical interpretation of both the macrocosm and the microcosm, were to draw forth a reaction of unexpected intensity from the European intellectual community, chemists and mechanists alike.

Robert Fludd and the Rosicrucian Problem (1617)

Born in Bearsted, Kent, in 1574, Fludd came from a well-to-do family.[1] His father, Thomas, had been knighted by Queen Elizabeth, whom he had served as treasurer both of the English

---

[1]The standard biography is that of J.B. Craven, *Doctor Robert Fludd (Robertus de Fluctibus). The English Rosicrucian. Life and Writings* (Kirkwall, 1902; reprinted New York: Occult Research Press, n.d.). Although badly outdated in terms of modern scholarship, the work remains useful for many aspects of Fludd's life and for the

forces in the Low Countries and of the Cinque Ports. In 1591 Robert was entered at St. John's College, Oxford, and it was there that he obtained his B.A. (1597) and his M.A. (1598).

Turning from the arts to medicine, he left England for the Continent, where he travelled through France, Spain, Italy, and Germany for about six years. Although in England he had been engaged in some of the studies that were to be incorporated in his work on the macrocosm and microcosm, it was undoubtedly while he was on the Continent that he became aware of the rich mosaic that constituted the Hermetic and the Paracelsian systems of nature. Fludd left no itinerary of his travels, but through his later works we catch sight of him in numerous places. We find him in Paris learning of the phenomenon of palingenesis (exhibited with the blood of a chemist), disputing the claims of geomancy with Jesuits at Avignon (during the winter of 1601-1602), and being taught the powers of magnetic sympathy by Gruterius in Rome.[2]

By 1604 Fludd had returned to England. He proceeded to enter Christ Church, Oxford, where he graduated M.B. and M.D. on May 16, 1605. Moving then to London, he set up residence in Fenchurch Street and sought admission to the London College of Physicians early in 1606. Here his success was not immediate.[3] His contempt for the ancient authorities was openly displayed, and he

---

appendix on Fludd's complex bibliography. Here, however, the reader is directed to J. J. Manget, *Bibliotheca scriptorum medicorum* (Geneva, 1731), Vol. I, Pt. 2, p. 298 for the most complete list of Fludd's works. Additional points on the life of Fludd are given in the prefatory material to "Truth's Golden Harrow. An Unpublished Alchemical Treatise of Robert Fludd in the Bodleian Library," *Ambix*, 3 (1949), 91-150 by C. H. Josten. The most recent French study is that by Serge Hutin, *Robert Fludd (1574-1637): Alchimiste et philosophe rosicrucien* (Paris: Omnium Littéraire, 1971). References to recent research on Fludd will be made throughout this chapter, but for a survey of current scholarship plus a bibliographic essay on both the primary and the secondary sources see the article on Fludd by the present author in the *Dictionary of Scientific Biography*, Vol. V (New York: Scribner's, 1972), pp. 47-49.

[2]Craven, *Doctor Fludd*, pp. 24-25. C. H. Josten, "Robert Fludd's Theory of Geomancy and his Experiences at Avignon in the Winter of 1601 to 1602," *Journal of the Warburg and Courtauld Institutes*, **27** (1964), 327-335. Robert Fludd, *Doctor Fludds answer unto M. Foster, Or, The squeesing of Parson Fosters sponge, ordained by him for the wiping away of the weapon-salve* (London: Nathaniel Butter, 1631), p. 133.

[3]Fludd's difficulties with the College are discussed by Craven, *Doctor Fludd*, pp. 28-29.

failed the examination repeatedly. At least four appearances before the board (1606-1607) proved to be disastrous, and when, finally, after an examination on March 21, 1608, it was reported that he was considered worthy to be a candidate, he was turned down once more because he had conducted himself insolently and had offended everyone. Rejected one time after this, Fludd was eventually admitted as a fellow of the College on September 20, 1609. In later years he was treated as a highly respected member. He was a close friend of the royal physician, Sir William Paddy, who presented two sets of Fludd's works to St. John's College, and he was chosen to serve as censor four times, in 1618, 1627, 1633, and 1634. Fludd, in turn, was proud of his membership in the College, and he esteemed the research of its fellows. Thus, he was to discuss at great length the magnetic studies of William Gilbert (1600) and Mark Ridley (1613) and was the first to support in print the views of his colleague William Harvey on the circulation of the blood.

As a physician Fludd was highly successful. He had a secretary at hand at all times, and his interest in the proper preparation of medicines prompted him to maintain an apothecary on his premises. Still, Thomas Fuller, writing later in the century, ascribed Fludd's medical success to his familiarity with the mystical phrases that were so evident in his books. "The same phrases he used to his patients; and seeing conceit is very contributive to the well working of physic, their fancy or faith natural was much advanced by his elevated expressions."[4]

Fludd's early difficulties with the College of Physicians were the product of his temperament. Acutely self-righteous, outspoken at all times, Fludd could not easily accept criticism. Yet, it is because of these traits that his works are now of special interest to us, for all those who attacked him could be certain of being answered. It made little difference whether his opponents were famed throughout Europe—men such as Kepler, Mersenne, and Gassendi—or were little-known Englishmen—such as William Foster or Patrick Scot. The thin-skinned physician replied to all his detractors, making it possible for us to follow a number of controversies in a nearly step-by-step sequence.

Although Fludd had long been engaged in preparing his

---

[4]Thomas Fuller, *The Worthies of England*, ed. John Freeman (London: Allen & Unwin, 1952), p. 615.

PLATE XVII Robert Fludd (1574-1637). From the *Integrum morborum mysterium: sive medicinae catholicae tomi primi tractatus secundus* ... (Frankfurt: Fitzer, 1631), sig.):(1ᵛ.

manuscript on the macrocosm and the microcosm, he ventured to put nothing into print until the age of forty-two, when the publication abroad of the first two Rosicrucian documents caused a widespread reaction among European scholars. These texts, the *Fama fraternitatis* (1614) and the *Confessio* (1615), presented a manifesto to the European scholarly community. In them was exhibited the outspoken call for educational reform that was a basic part of the utopian literature of the early seventeenth century.[5] Oddly enough, there is no evidence that any order of Rosicrucians ever existed. Although many hoped to contact them, no one of any consequence claims to have been a member.

Before discussing the *Fama* and the *Confessio* it may be useful to note that the Paracelsians shared fully in a rather widespread distaste for the established system of higher education shared by intellectuals in the sixteenth century. Although it is now well known that many of the universities of that time were far less conservative than charged, this was not always evident to the outsider.[6] To many the doctrines of Aristotle and Galen seemed so deeply entrenched that little less than a total eradication of the traditional educational system would permit the foundation of a new system based on God's truth. One need only point to the Elizabethan

[5] I have discussed the relationship of the Rosicrucian manifestoes to the utopian literature in general and to Robert Fludd specifically in *Science and Education in the Seventeenth Century. The Webster-Ward Debate* (London: Macdonald; New York: American Elsevier, 1970), pp. 15-32 *passim*. Of special importance is Frances A. Yates, "The Hermetic Tradition in Renaissance Science," *Art, Science, and History in the Renaissance*, ed. Charles Singleton (Baltimore: The Johns Hopkins Press, 1968), pp. 255-274. More recently Dr. Yates has expanded this subject to a book, *The Rosicrucian Enlightenment* (London/Boston: Routledge & Kegan Paul, 1972). See also A. E. Waite, *The Real History of the Rosicrucians* (London: G. Redway, 1887), A. E. Waite, *The Brotherhood of the Rosy Cross* (London: W. Rider and Sons, 1924), Paul Arnold, *Histoire des Rose-Croix et les origines de la Franc-Maçonnerie* (Paris: Mercure de France, 1955), Will-Erich Peuckert, *Die Rosenkreutzer. Zur Geschichte einer Reformation* (Jena: E. Diederichs, 1928).

[6] See above, Ch. 1 and also the discussion in Debus, *Science and Education*, pp. 5-14. The reader is referred particularly to Eugenio Garin, *L'educazione in Europa 1400-1600, problemi e programmi* (Bari: Editori Laterza, 1957); John Herman Randall, Jr., "The Development of Scientific Method in the School of Padua," *Journal of the History of Ideas*, 1, (1940), 177-206 and "Scientific Method in the School of Padua," *Roots of Scientific Thought. A Cultural Perspective*, ed. Philip P. Wiener and Aaron Noland (New York: Basic Books, 1957), pp. 139-146.

Statutes for Cambridge (1570) or the Laudian Code for Oxford (1636)—both of which maintained the authority of the ancients in educational matters[7]—or recall the reaction of the Parisian Medical Faculty to Joseph Duchesne and Turquet de Mayerne in 1603 to understand how conservative the academic institutions must have appeared. One might then turn to as varied a group as Peter Ramus, Rabelais, Telesio, William Gilbert, René Descartes, and Francis Bacon to discover a persistent underlying theme in their writings—disillusionment with the formal training of the period and its emphasis on the infallible truths of antiquity.[8] And no matter what differences might separate these authors in their own views of scientific method, they were agreed that in the future more attention must be paid to observational evidence in the interpretation of nature.

These were the goals of the chemical philosophers as well.[9] Paracelsus had called for more observations and fewer speculations, while Severinus had encouraged his readers to abandon their books for the study of nature. Bostocke reflected the same views, but he went on to indicate the very real Paracelsian disillusionment with the universities. He complained that although Paracelsian chemistry could be shown by experience to be valid, its followers were in a desperate state since no one would give them a fair trial:

> ... in the scholes nothing may be receiued nor allowed that sauoreth not of *Aristotle, Gallen, Auicen,* and other Ethnickes, whereby the yong beginners are either not acquainted with this doctrine, or els it is brought into hatred with them. And abrode likewise the Galenists be so armed and defended by the protection, priuiledges and authoritie of Princes, that nothing can be allowed that they disalowe and nothing may bee received that agreeth not with their pleasures and doctrine ....[10]

In short, the Paracelsians called for new observations and *experientia* no less than those who today seem to be more in the

---

[7] See Phyllis Allen, "Scientific Studies in the English Universities of the Seventeenth Century," *Journal of the History of Ideas,* 10 (1949), 219-253.

[8] The views of these authors are discussed in *Science and Education,* pp. 8-11.

[9] *Ibid.,* pp. 15-32 ("The Paracelsians and Educational Reform").

[10] R. Bostocke, Esq., *The difference betwene the auncient Physicke ... and the latter Phisicke* (London: Robert Walley, 1585), sig. Fii^v.

mainstream of scientific development. And they, no less than Bacon and Descartes, rejected the formal university training they had been subjected to.

In this setting the *Fama fraternitatis* (1614) and the *Confessio* (1615) do not seem out of place.[11] The *Fama fraternitatis* called for a new learning to replace the atrophied studies of the schools, where scholars were still chained to Aristotle, Galen, and Porphyry when they should be seeking a more perfect knowledge of the "Son Jesus Christ and Nature."[12] All truly learned men agreed that the basis of natural philosophy is medicine and that this is a godly art. But the true secrets of medicine—and consequently, of all science—had been described by the founder of the order named for him, Christian Rosenkreutz. His books were studied by the brethren, who were also well aware that great truths were known by others. In Germany there were many learned physicians, philosophers, and magicians, one of the greatest having been Paracelsus, whose works alone were placed next to those of Rosenkreutz in the hidden vault of the Rosicrucians.[13]

There is little doubt, as Yates has indicated, that this was a neo-Paracelsian movement. Yet at the same time the *Fama* displays a missionary spirit: the suggestion is made that great wonders might be accomplished if the truly learned scholars of Europe united for the benefit of mankind. These scholars, it was suggested, should declare themselves in writing and join the Brotherhood in the forthcoming reformation of learning; they should examine their art and "declare their minde, either *Communicatio consilio* or *singulatim* by Print."[14] It was intended that both the *Fama* and the *Confessio* would be published simultaneously in five languages so that no one could excuse himself and say he had not seen the message. And while the Brothers refused at that time to give out their names or announce their meetings, they were willing to assure those who answered their call that they would not go unnoticed.

---

[11] Discussed by Debus in *Science and Education*, pp. 20-23 and by Yates in *The Rosicrucian Enlightenment*, pp. 41-58.

[12] *The Fame and Confession of the Fraternity of R:C: Commonly, of the Rosie Cross. With a Praeface annexed thereto, and a short Declaration of their Physicall Work*, by Eugenius Philalethes (Thomas Vaughan) (London: J.M. for Giles Calvert, 1652), pp. 1-2.

[13] *Ibid.*, pp. 36, 5, 10.

[14] *Ibid.*, p. 31.

The *Fama* was actually published in four languages in nine editions between 1614 and 1617, and an English translation appeared in 1652.[15] Indeed, the response to this appeal must have been beyond the wildest dreams of the promoters. Craven refers to a file of letters written by persons offering themselves as members dated from 1614 to 1617, now held at the library at Göttingen,[16] and Gardner lists several hundred books and tracts debating the merits of this secret group that appeared in the course of less than ten years. Major cities were visited by men who announced themselves as members of the Brotherhood and promised to show all their secrets to those who wished to become initiated. In an account published in 1619 we read:

> What a confusion among men followed the report of this thing, what a conflict among the learned, what an unrest and commotion of impostors and swindlers, it is needless to say. There is just this one thing which we would like to add, that there were some who in this blind terror wished to have their old, and out-of-date, and falsified affairs entirely retained and defended with force. Some hastened to surrender the strength of their opinions; and after they had made accusation against the severest yoke of their servitude, hastened to reach out after freedom.[17]

This "Rosicrucian Furor" declined in the following decades, but there were few intellectuals even much later in the century who did not know of the *Fama fraternitatis* and the *Confessio*.[18] We shall frequently be referring to the Rosicrucians throughout this volume.

---

[15] A listing—and discussion—of the early Rosicrucian texts is found in F. Leigh Gardner, *A Catalogue Raisonné of Works on the Occult Sciences*, Vol. I. *Rosicrucian Books*, introduction by Dr. William Wynn Westcott (2nd ed., privately printed, 1923). On pp. 4-6, 21 (items 23-29, 144) the first editions of the *Fama fraternitatis* and the *Confessio* (1614-1615) are described and discussed. Other early editions are listed in Johann Valentin Andreae, *Christianopolis. An Ideal State of the Seventeenth Century*, trans. Felix Emil Held (New York: Oxford University Press, 1916), p. 11.

[16] Craven, *Doctor Fludd*, p. 39.

[17] Andreae, *Christianopolis*, pp. 137-138.

[18] See Yates, *Rosicrucian Enlightenment*, pp. 91-102, for a discussion of the furor in Germany, and pp. 144-150, for a discussion of Andreae; and J. Kvacala, *J. V. Andreä's Antheil an geheimen Gesellschaften* (Jurjew: C. Mattiesen, 1899). Evidence of the persistent interest in the Rosicrucians in the late seventeenth century is found in the many references in the *Polyhistor* of Daniel George Morhof (1639-1691).

Among the first to discuss the Rosicrucian texts was the German mystical alchemist Michael Maier (1568-1622), who was in England shortly before Fludd published his first book, perhaps to prepare a Latin translation of Thomas Norton's *Ordinall of Alchimy*. Deeply involved in the Rosicrucian controversy, Maier published his commentary on the laws of the Rosicrucians, the *Themis aurea*, in 1618.[19] Far less complimentary to the proposed new order was the respected and renowned iatrochemist Andreas Libavius, whose *Examen philosophiae novae* paid special attention to the harmonic magic of the Brotherhood. This was immediately followed by his *Analysis confessionis fraternitatis de Rosea Cruce* and a German summary of the problem.[20] These attacks by Libavius elicited a reply from Fludd in 1616, a short *Apologia* in which he called for a new basis for all learning and expressed the hope that he might be considered worthy of membership in the Brotherhood. Although he was later to complain of not having been contacted by them, Fludd frequently referred to their worthy goals in his later publications. Surely in 1616 and 1617 Fludd was deeply affected by the Rosicrucian documents. In the latter year he expanded his twenty-three-page apology to a book of nearly two hundred pages, the *Tractatus apologeticus integritatem societatis de Rosea Cruce defendens*.[21]

Although short compared with his later works, the *Tractatus apologeticus* presents many themes which Fludd was to develop in the next decade. In company with the Rosicrucians he made a plea for reformation in education and learning. Nowhere, did he feel,

---

[19] For the life of Maier, see J. B. Craven, *Count Michael Maier* (Kirkwall, 1910; reprint London: Dawsons, 1968). Yates discusses Maier's Rosicrucian works in her *Rosicrucian Enlightenment*, pp. 70-91, 200-205, and *passim*.

[20] See *Rosicrucian Enlightenment*, pp. 51-54. See above, Ch. 3, for the views of Libavius on the Parisian medical debate.

[21] Robert Fludd, *Apologia compendiaria fraternitatem de Rosea Cruce suspicionis et infamiae maculis aspersam, veritatis quasi fluctibus abluens et abstergens* (Leiden: Godfrid Basson, 1616), enlarged the following year and retitled the *Tractatus apologeticus integritatem societatis de Rosea Cruce defendens* (Leiden: Godfrid Basson, 1617). The latter edition is the one that will be cited here as the *Tractatus apologeticus*. A German translation of this text appeared as the *Schutzschrift für die Aechtheit der Rosenkreutzergesellschaft . . . übersetzt von Ada Mah Booz* (Leipzig: Adam Friedrich Böhme, 1782). The reader is also referred to Debus, *Science and Education*, pp. 24-26 and "Mathematics and Nature in the Chemical Texts of the Renaissance," *Ambix*, 15 (1968), 1-28, 211 (15-20).

was more useless knowledge being offered than at the universities, those strongholds of the classical scholars. One could only shudder at their reliance on Aristotle in philosophy and Galen in medicine, for in the ancient world none set forth doctrines which proved to be more antithetical to Christianity than they. Aristotle and Galen had been heathens, and their followers were no better. The universities must be reformed so the divine light of Christian teachings could flourish.

In a chapter on natural philosophy, medicine, and alchemy, Fludd wrote that although innumerable authors had written on natural philosophy, they had presented only a shade of truth. They filled whole volumes with definitions, descriptions, and divisions; they droned on about the four causes; they lectured on motion, the continuum, the contiguum—of termini, loci, of space, vacuum, time, and number. This for them was the basis of physics. Then they went forth to generation, corruption, and descriptive accounts of the heavens and the universe.[22] "But, good God, how superficial and equivocal this all is."[23] For Fludd it was impossible for anyone to attain the highest knowledge of natural philosophy without a thorough training in the occult sciences.[24] It is clear that Fludd's call for a new physics was far different from that of his contemporary, Galileo. The English physician firmly rejected the study of motion and the vacuum, while he had little interest in quantification. These subjects were to be avoided as "Aristotelian."

Far more important that the natural philosophy of the Schools was medicine, the most perfect science of all. More than a simple description of diseases and the workings of the human body, it is the very basis on which natural philosophy must rest.[25] Our knowledge of the microcosm will teach us of the great world, and this in turn will lead us to our Creator. Similarly, as we progress in our understanding of the universe, the more we will be rewarded with a perfect knowledge of ourselves. The vulgar medicine of the Galenist, the astrologer, and the Paracelsist fails since it treats only of descriptive material of little value—of principles, elements, humors, the body and its parts, temperaments, virtues, and the

---

[22] *Tractatus apologeticus*, pp. 91-93.
[23] *Ibid.*, p. 93.
[24] *Ibid.*, p. 44.
[25] *Ibid.*, p. 89.

motion of the pulse. Yet, Hippocrates obscurely and mystically mentioned a single medicine which cures all ills, and from Pliny, Lull, Arnald, Rupescissa, and others we learn that this medicine unites in peace the warring elements in the body. It is this medicine which must be sought in our reformed science.[26] Once again we are told to search for hidden arcana in a mystical world system in preference to the anatomical and physiological writings of the ancients taught in the Schools.

Even the art of alchemy, from which we might expect so much, aids us little. Alchemists such as Libavius, instead of calling for reform, have described calcination, separation, conjunction, putrefaction, and all the other operations by which men are deceived and persuaded to part with their money. True chemical authors are of a different sort and write of occult secrets. We need the natural fire of the philosopher, not vulgar flame. Similarly, we need natural, not artificial, vessels for our work. No miracles are performed when true alchemy is practiced, only the work of nature.[27]

Fludd considered that real wisdom could be found in the writings of the natural magicians, men who were in truth mathematicians.[28] But if he was distressed about the state of natural philosophy, medicine, and alchemy in his day, he felt even more strongly about mathematics.[29] The arithmetic texts were filled with definitions, principles, and discussions of theoretical operations, with addition, multiplication, golden numbers, fractions, and extractions of cubes. But little heed was paid to the wisdom and doctrines of the Pythagoreans, men who, like the mystical alchemists, hid their profound mysteries from the vulgar crowd.[30] Fludd was convinced of the existence of true mathematicians who had preserved the Pythagorean secrets over the centuries no less than those selfless alchemists who had maintained their own ancient mysteries. Those who were really versed in these arcane arts were considered impious magicians. But more could be learned from the silent wisdom of the Pythagoreans than from the

[26] *Ibid.*, pp. 95-99.
[27] *Ibid.*, pp. 100-101.
[28] *Ibid.*, p. 23.
[29] *Ibid.*, p. 103.
[30] *Ibid.*, p. 105.

useless books of the philosophers. The disciples of Pythagoras reached a certainty of belief in God through their profound study of numbers and ratios; in the same way, others might learn of the unity of God in trinity, a theme continually stressed by these authors—and more, of the very fabric of the world.[31] But vainly are these secrets sought in the written records, for there are no true remains of the Pythagorean teachings; it is a science that must be wholly restored.[32]

And what of music, a subject which properly forms a division of mathematics? Are we to hear forever of harmonic systems, trite melodies, and sing *ad infinitum* the musical scales? "Good God, what does this have to do with the true and profound music of the wise, a subject through which the proportions of natural things are investigated, and the harmonic consensus and properties of the whole world are revealed . . . ?"[33] The science of music really deals with the joining of the elements, the proportions of light and weight in the stars, and their influence on our terrestrial world. Here we shall learn of the spiritual body of the sun and why the influence of Mars brings misfortune.[34] Exact knowledge of such things will come only from speculation and revelation in the occult science of music, which shows how man, here termed the microcosmic palace, agrees in celestial harmony with the macrocosm.

> Therefore happy will be he who is well versed in such mysteries of occult music since without a knowledge of these things it is impossible for anyone to know himself. And without this he will be unable to reach a perfect knowledge of God, for he who understands himself

---

[31] *Ibid.*, p. 107. Fludd would have rejoiced in the publication in the same year in Italy of a book on mystical geometry by Fortunatus, *Decas elementorum mysticae geometrae quibus praecipua divinitatis arcana explicantur* (Padua: Peter Paul Tozzi, 1617). I am indebted to Walter Pagel for this reference. For an example of Fludd's application of arithmetic to nature see his demonstration of the mystery of the world's Creation "by way of an Arithmeticall progression" in the *Mosaicall Philosophy* (1659), pp. 73-74.

[32] *Tractatus apologeticus*, p. 108.

[33] *Ibid.*, p. 109. "Sed Deus bone quid hoc ad veram & profundam sapientis musicam, quâ rerum naturalium proportiones investigantur, harmonicus consensus & totius mundi proprietates revelantur . . . ."

[34] *Ibid.*, p. 110. "Similiter cur Mars etiam pro infortunio habetur, nisi propter extensionem suae consonantiae Diapason ad elementū suae naturae contrarium, videlicet Aquam."

truly and intrinsically perceives in himself the idea of the divine Trinity. ...."[35]

Elsewhere Fludd was to describe a divine monochord which served as a key to our understanding of the harmonies of earth and heaven. This was to become a major bone of contention between him and Johannes Kepler in the coming years.

Fludd's proposals for the renovation of the remaining branches of the mathematical quadrivium may be passed over fairly rapidly since his object is already quite clear. In geometry he was aghast to find the subject dominated by the theorems, axioms, and propositions of Euclid—surely all far removed from the true and perfect geometry which is concerned with hidden secrets.[36] Like many of his contemporaries whom we today see as forerunners of modern science, Fludd called on Archimedes as the guide to a new science. Why? Because of his wondrous machines.[37] In effect, Archimedes was the archetypal natural magician. Who today, Fludd wrote, could build the masterworks of the past, the Colossus of Rhodes, the speaking birds and animals made of wood described by ancient authors, or even the bronze head of Roger Bacon? In optics we read descriptions of the membranes of the eye, the direction of rays, and the refraction of colors when we should be seeking the secrets of Archimedes' burning mirrors, which might destroy the Turks, whose continued advance posed such a threat to Christian Europe. But the state of geometry is so bad that we are faced not simply with correcting errors, but with totally restoring the lost treasures of this subject.[38]

For Fludd the mathematical science of astronomy was really astrology, but he complained that only pseudo-astrology was being cultivated because the practitioners were ignorant of the arcana of nature. While they wrote of the motions of the stars and prepared vast astronomical tables, they should learn instead of the arcane effect of the stars, for the celestial bodies are indeed the

---

[35] *Ibid.*, p. 111. "Foelix igitur erit qui in talibus Musices occultae misterijs benè est versatus, quippe sine quorum cognitione impossible est, ut quis seipsum cognoscat; quod quidem nisi fiat ad perfectam Dei cognitionem attingere non potest; nam qui seipsum verè et intrinsecus intellegit, ideam in se divinae Trinitatis percipiet ...."

[36] *Ibid.*, p. 113.

[37] *Ibid.*, p. 114.

[38] *Tractatus apologeticus*, pp. 115-116, 114.

hieroglyphics of nature.[39] Mystical alchemists embellished their works with symbolic drawings. For Fludd these were much clearer than descriptions with words or numbers devoid of occult significance.[40]

In short, Fludd believed that a total revision of man's approach to nature was in order. As a physician, however, he had a special interest in the study of man, the microcosm. His mystical views of the Creation and the relation of the greater and lesser worlds had convinced him that nothing of value could be learned of man without a true understanding of the macrocosm. To interpret correctly the book of nature and perceive the true intentions of the ancient sages in their obscure teachings we must search diligently, maintaining a definite plan with key questions forever standing in our minds. We must especially concern ourselves with the part played in Creation by the divine light of the Lord (Spirit) and further seek the relation of this light to life and motion, for it is known that whenever there is a deficiency, illness or death will follow. Special attention should be paid to the views on the invisible fire of Zoroaster and Heraclitus as well as the atomic theory of Democritus.[41]

When we turn to the microcosm we must focus our attention on the assimilation of this spirit in the body. All animated faculties require air, which contains the aetherial spirit, and its divine light hidden within. The Stoics seemed to understand this, Fludd believed, because they defined the soul as the substance of fire converted through air into water.[42] All generation is ultimately from air, which has been thickened into the form of sperm, itself an aerial spirit.[43] Among these important topics, we must single out

[39] *Ibid.*, pp. 116-118, 32-33.
[40] *Ibid.*, p. 118.
[41] *Ibid.*, pp. 187-189. See the development of this subject by the present author in "Harvey and Fludd: The Irrational Factor in the Rational Science of the Seventeenth Century," *Journal of the History of Biology*, 3 (1970), 81-105 (88-89).
[42] *Tractatus apologeticus*, pp. 188-190.
[43] The concept of all life deriving from air is Aristotelian; see *De generatione animalium* III, 11:762a, where animals and plants are described as coming into being in earth and in liquid because there is water in earth and air in water. There is vital heat in all air. This is discussed in relation to Servetus in Walter Pagel, *William Harvey's Biological Ideas* (Basel/New York: Karger, 1967), pp. 148-149.

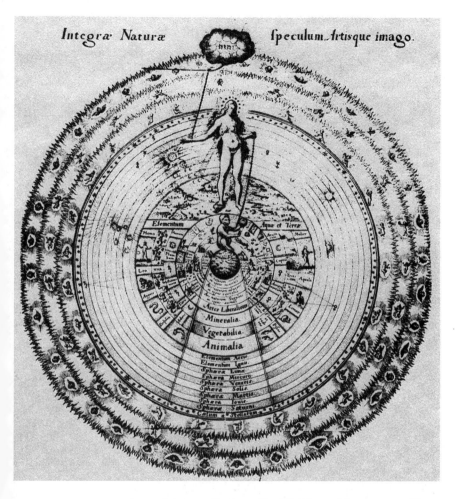

PLATE XVIII The universe of Robert Fludd. The sublunary world of the elements is separated from the lower heavenly regions, and these in turn give way to the upper celestial world beyond the sphere of the stars. The power of the Creator is shown at the top, controlling nature (symbolized by the woman), in turn connected with the ape, which signifies our poor reflection of divinity. From the *Utriusque cosmi maioris scilicet minoris metaphysica, physica atque technica historia* (Oppenheim: J. T. De Bry, 1617), pp. 4-5.

for special study the problem of how this celestial balsam or quintessence nourishes our bodies.⁴⁴ We will find that this occurs in a twofold fashion, involving the respiratory process. It seems clear that the spirit in the air is attracted through inspiration to the lungs and is then carried to the heart. The precious occult portion of the quintessence, separated from the rest, is then dispersed and dispensed as the vital spirit through the arterial system. The grosser part is mixed in the veins with the blood for its completed form; from it the more solid parts are nourished and conserved. Surely it is necessary that some of the celestial fire be united with the thickened part of the elements for the conservation of the whole body. "The ocular proof (I say) is that neither animal, nor vegetable, nor mineral could live, last, or exist for one minute of an hour without this lucid fire."⁴⁵ The section ends with a ringing defense of the interdependence of air and spirit for life.

One would hardly argue that this statement is a sophisticated anatomical or physiological study of the vascular system. It is brief and contains little original thought. On the other hand, in his search for recognition by this elusive and secret society of the Rosicrusians, Fludd had written an important work. Here he had rejected the still potent authority of Aristotle and Galen and as a corollary had called for broad curricular reform. However, his dissent had been coupled with a realization that something had to be

---

⁴⁴*Tractatus apologeticus*, p. 190. "Quomodo quodlibet animal, ab hâc quinta essentiâ invisibilitèr, duplici ratione nutriatur: Videlicet vel quatenus delitescit in aëre. Vnde fit ut attractus aër à pulmonibus per inspirationem ad cor, omni attractione, occultam quintae essentiae portionem secum rapiat, quae dispergitur & dispensatur per arterias ad nutriendos spiritus vitales in ijs contentos & congregatos. Aut quatenus colligitur & condensatur occultè in corpore composito ipsius cibi, seu nutrimento ita ut haec lux caelestis congregata, faciat ignem spissum, caelestem, visibilem facultates nutritivas & naturales, nutrientem, cujus etiam pars spiritualior per diapedisin ad cordis thalamos penetrat & transportatur atque passim per arterias dispergitur. Crassior vero ejus pars cum subtili elementorum, in venis cum sanguine, ad completam ejusdem constitutionẽ permiscetur, ut ab eo partes solidores nutriantur & conserventur. Atque necessè est ut aliquid hujus ignis coelestis cum elementorum spisso uniatur, ad conservationem ligamenti naturalis totius compositionis; licet, etiam ab arteriarum orificijs, major illius vinculi supplementi portio derivetur. Testis (inquam) ocularis fui quod nec animal, nec vegetabile, nec minerale per unicam horae minutam sine hoc igne lucido vivere, durare, aut existere potuisset."

⁴⁵*Ibid*.

proposed in its stead. Here he had outlined a new study of nature that was intimately associated with the search for divine truth. Central to this program was the proper understanding of the Creation, the universal spirit of life, and the assimilation of that spirit in the body. This, plus Fludd's emphasis on the return to the mystical truths of the ancients—here best typified by Hermes, Pythagoras, and Plato and his disciples—makes the *Tractatus apologeticus* something of a guide to his future publications.

The Fluddean Philosophy

Although the *Tractatus apologeticus* is of special interest for us, it was not Robert Fludd's only publication in 1617; his *Tractatus theologo-philosophicus* also appeared at this time.[46] Here he described the Creation, the Fall of man, the relation of God the Father to the Son, and the Resurrection. The Rosicrucian texts are explained as an allegory, and again emphasis is placed on the aerial spirit, the tabernacle of which is the sun.

Far more important was the publication of the first part of Fludd's folio works on the macrocosm and the microcosm, the *Utriusque cosmi historia*. The first part of the volume on the macrocosm, related to metaphysics, appeared in 1617 and was followed in 1618 by a second part describing eleven separate arts or sciences. The two parts of the volume on the microcosm were published together in 1619, both from the presses of John Theodore De Bry at Oppenheim.[47] These texts plus the *Tractatus apologeticus* initiated the widespread debate over Fludd's mystical and universal alchemy, but we may safely borrow from parts of his later publications as well in a discussion of his system of the world. Here the most useful general work is the posthumous *Philosophia Moysaica* (1638).[48]

---

[46]However, this text is not of major interest for an understanding of Fludd's concept of a chemically operating universe. Its content has been discussed by Craven, *Doctor Fludd*, pp. 46-61.

[47]The prominent place of De Bry in the publication of works associated with the Rosicrucian controversy is discussed by Yates in *The Rosicrucian Enlightenment*, pp. 70-90.

[48]See Allen G. Debus, *The English Paracelsians* (London: Oldbourne, 1965), pp.

Fludd, like most other Renaissance chemical philosophers, had an implacable hatred of Aristotle, even though Aristotelian influences are evident throughout his work. As his prime authority he chose to turn to God's two books of revelation: the Holy Scriptures and God's book of Creation—nature. There was no question in Fludd's mind that the first of these was the more important, for although he could appreciate the value of experimental research, he clung more tenaciously than most to the mystical approach of the neo-Platonists. He might turn to mechanical examples to demonstrate the immobility of the earth, he might invoke Harvey's dissections to demonstrate the solid structure of the septum of the heart, and he might quote the experiments of Gilbert and Ridley to show the truth of sympathetic magic, but each of these was a lesser form of truth. However, when attacked with experimental evidence, he could defend himself with like weapons instead of relying on scriptural or mystical quotations.

In his rejection of the ancients, Fludd argued that they had relied too heavily on "human or mundane wisdome," and their work had to be superseded by the "true wisdome" found in the Bible.[49] He deplored the fact that in his day the "heathnish philosophy" of the ancients was "adored" by Christians to the extent that Aristotle was worshipped "as if he were another Jesus rained down from heaven to open unto mankind the treasures of the true wisdome."[50] For Fludd, God is much more than simply the first efficient cause; He is the "generall cause of all action in this world," which He accomplishes "by his blessed Spirit, which he sent out into the world."[51] It later becomes quite clear that the sun is the temple housing this spirit.

While thus disposing of the Aristotelian corpus as a heathen philosophy not fit for Christians, Fludd finds the Bible a much more profound source than the simple human reason on which Aristotle was dependent. If one had to rely on any of the ancient Greeks it would best be Plato, who along with Hermes had the advantage of being familiar with the books of Moses. Fludd con-

---

107-127 and "Renaissance Chemistry and the Work of Robert Fludd," *Ambix*, **14** (1967), 42-59.

[49] Fludd, *Mosaicall Philosophy*, pp. 12-13.
[50] *Ibid.*, p. 28.
[51] *Ibid.*, p. 29.

cluded that what was valuable in Aristotle's works must be ascribed to the little which was absorbed from his master.[52]

Fludd's emphasis on the Holy Scriptures echoes the Hermetic and Paracelsian belief that the most important source for the study of nature is the opening chapters of Genesis. As in the *Philosophy to the Athenians*, here again in Fludd's account divine Creation is the great "spagerick act" of separation.[53] Fludd explained that

> ... it was by the Spagericall or high Chymicall virtue of the word, and working of the Spirit, that the separation of one region from another, and of the distinction of one formall virtue from another, was effected or made: of the which business the Psalmist meaneth where he saith: *By the Word of the Lord the heavens were made, and by the Spirit from his mouth each virtue thereof.*[54]

The Mosaic, or Fluddean, philosophy is based upon this mystical chemical account of the Creation. The origin of all things is the primeval dark Chaos from which arose the divine light; this, acting on the Chaos, brought forth the waters, the passive matter of all other substances. These are the three primary elements— darkness, light, and the waters, or the Spirit of the Lord—and it is from the waters that all secondary elements derive.[55]

With this knowledge, Fludd felt, it would be possible to bring order out of the widespread confusion concerning the elements. Analyzing the original texts, he showed that when Aristotle wrote of the *prima materia*, Plato of the *hyle*, Hermes of the *umbra horrenda*, Pythagoras of the "Symbolicall unity," and Hippocrates of the deformed chaos, they were all writing of the darkness or the dark abyss of Moses. Similarly by some name or another all of these

---

[52]*Ibid.*, pp. 32, 42.
[53]*Ibid.*, pp. 147-148.   [54]*Ibid.*, p. 175.
[55]On the light-dark dualism, see especially Robert Fludd, *Utriusque cosmi maioris scilicet et minoris metaphysica, physica atque technica* (Oppenheim: T. De Bry, 1617), pp. 205-206. See also the *Mosaicall Philosophy*, p. 82, and Fludd, *Tomus secundus de supernaturali, praeternaturali et contranaturali microcosmi historia, in tractatus tres distributa* (Oppenheim: T. De Bry, 1619), pp. 200-203. This Chapter 9 is titled "De principiis nostris Microcosmi, & quomodò tam cum principiis divinarum literarum, quàm cum intentionibus omnium Philosophorum & Chymicorum conveniant?" For Fludd on the *prima materia* see Walter Pagel, "Paracelsus and the Neoplatonic and Gnostic Tradition," *Ambix*, **8** (1960), 125-166 (146-147); *Das medizinische Weltbild des Paracelsus: Seine Zusammenhänge Neuplatonismus und Gnosis* (Wiesbaden: Franz Steiner, 1962), p. 94.

philosophers knew something of the Mosaic "light" and "waters." However, in their interpretations they often varied far from the truth, and it is to the works of Plato and the *Pymander* of Hermes that the true adept is urged to go for enlightenment.[56]

The three primary elements were also considered by Fludd to be the source of the ancient four qualities: hot, cold, dry, and wet. Like Paracelsus, Fludd disdained the pairing of qualities to give the elements. He explained that "cold" is "an essentiall adherent unto privative rest, and the stout of-spring and Champion unto darkness." On the other hand, heat is the "inseparable Champion or Assistant of Light."[57] Humidity arises from the "intermediate actions and passions" of the other two, while dryness is nothing other than the absence of humidity.[58]

Fludd further explained that the secondary elements, which arose from the third primary element, the waters, are none other than the ancient four elements: air, earth, fire, and water. Their creation occurred from the separation of the waters into heavenly (fiery) and earthy portions. The latter was in turn divided into the spheres of air, water, and earth.[59]

> But the world is composed only of heaven and earth, and therefore it followeth that the whole world is made and existeth of the waters, and by the waters, consisting by the word of God; Now therefore since the Starrs of heaven are esteemed nothing else but the thicker portion of their Orbes, and again every creature which is below, is said to be compacted of the Elements, it must also follow that both the Starrs in the higher heaven and the compound-Creatures, beneath in the Elementary world, be they meteorologicall, or of a more perfect mixtion, namely Animal vegetable or minerall, must in respect of their materiall part or existence proceed from waters . . . .[60]

As proof of the elementary nature of water, Fludd pointed out that earth, air, fire, and water are easily transmutable; simply by condensation and rarefaction of one element the others would come forth. Thus it was by successive heatings that earth was transformed into water, water into air, and air into fire. Recalling

---

[56]*Mosaicall Philosophy*, pp. 41-42.
[57]*Ibid.*, p. 82.
[58]Fludd, *De supernaturali*, pp. 200-203.
[59]Fludd, *Mosaicall Philosophy*, pp. 47-48.
[60]*Ibid.*, p. 48.

an early experience, he wrote of the loud noise he heard when striking rocks together when he was totally under water. "Certainly this could not have been heard if there were not air in the water since air is the medium of the sense of hearing."[61]

Similarly Fludd felt obliged to demonstrate the aqueous nature of earth, suggesting that a lump of earth or a stone should be calcined, slurried in water, and then filtered through a linen cloth. The filtrate was then to be made strongly caustic with lye, and after distillation a salt would be found in the residue. He did not consider the fact that the salt might have derived originally from the stone, but rather used this as proof that all earthy matter originates from water.[62] From these and similar observations Fludd concluded that "earth is dense water, and water is dense air, while on the other hand, air is nothing else than dense and crass fire."[63] The observable differences between them, he suggested, are due to the different amounts of light they contain. One might reason that by distilling or boiling water over a fire one would be adding heat to it. Since heat and light are inseparably connected, one could say that the operator is then adding light to the water, and on this assumption one would expect the water to be converted to air—which is just what happens.[64]

Thus the Fluddean or Mosaic philosophy teaches that there are three primary elements: darkness, light, and the waters. These are the true origin of the four qualities of the ancients, since darkness and cold, and light and heat are related, and between these extremes there is a need for a humid mean. The four elements are also closely related with this system, since in the division of the waters of the firmament by the Lord in His divine alchemy, the upper waters became the heavens (fire) and the sublunary waters were split into the spheres of air, water, and earth. But

[61]Robert Fludd, *Clavis philosophiae et alchymiae Fluddanae* (Frankfurt: Wilhelm Fitzer, 1633), p. 35.

[62]Robert Fludd, *Tom. sec. tract. sec. de praeternaturali utriusque mundi historia in sectiones tres divisa* (Frankfurt: Typis Erasmi Kempfferi, Sumptibus Joan. Theodori De Bry, 1621), p. 95.

[63]Robert Fludd, *Anatomiae amphitheatrum* (Frankfurt: Joan Theodore De Bry, 1623), p. 25. For a similar treatment see Fludd, *Mosaicall Philosophy*, pp. 69-70; *Medicina catholica seu mysticum artis medicandi sacrarium in tomos divisum duos* (Frankfurt: Wilhelm Fitzer, 1629), Vol. I, p. 107; *De praeternaturali*, p. 111.

[64]Fludd, *De supernaturali*, pp. 200-203. Fludd's work here reflects a long history of light mysticism.

because of the connection of the elements with the qualities as well as the similarity of water and air in distillation procedures, the spheres of these two are united and as a result the secondary elements must be earth, fire, and a humid sphere whose more subtle parts are air and whose more gross parts are water.[65]

Fludd had described the origin and interrelationship of the Aristotelian elements in considerable detail, and, although he seldom referred to them, he also thought it necessary to account for the three principles. He stated that Paracelsus

> ... posed three principles of all things; that is, Salt, Mercury, and Sulfur. And if we consider the matter carefully, we see that they come forth immediately from our principles. For from the center of the earth or the Darkness Salt is extracted ... ; Sulfur truly proceeds from the soul and tincture of things, that is, from Light ... ; Mercury moreover partaking partly of Salt and partly of Fire is considered not inaptly to be within the Spirit [Waters].[66]

Fludd pointed out in one place that created things are composed of the Paracelsian three principles, which he equated with the four elements of Aristotle,[67] and in another place he struck a definite Paracelsian note by insisting that there are as many different types of salt, sulfur, and mercury as there are types of substances.[68] However, he never made extensive use of these principles as an explanatory device other than in his discussion of color.[69]

Highly important throughout all of this was the divine light, which represents the active agent responsible for the divine Creation. Light and divinity are terms which are constantly related in the Fluddean writings. Fundamentally, it was the light of the Lord

---

[65] Fludd, like Duchesne (see above, Ch. 3) and other later Paracelsians, described a three-element system but also regularly utilized the four elements and the quintessence. In the *Utriusque cosmi* (1617), p. 71, he demonstrated their existence by treating a sample of wine to a series of extractions and distillations, eventually arriving at a sample with five layers or regions which for him corresponded to the four elements and the quintessence.

[66] Fludd, *De supernaturali*, p. 203.

[67] Fludd, *De praeternaturali*, p. 96. The identification of the principles with the elements was suggested by a number of authors. See above, Ch. 2, and for the position of Jacob and Gabriel Fontanus, Walter Pagel, *Paracelsus* (Basel: Karger, 1958), pp. 333, 341.

[68] Fludd, *Medicina catholica*, p. 153.

[69] *Ibid.*, pp. 147-155.

informing the chaos which was requisite for the formation of the world, and it was this same divine light arising from the Spirit which on the fourth day was formed into the sun and received into the aetherial heaven. Fludd was as much a proponent of the primacy of the sun as any Renaissance neo-Platonist. He felt that its importance was apparent in the Scriptures, especially in Psalm 18:5, where it is written, "God hath put his tabernacle in the Sun." Fludd interpreted this to mean that the Spirit of the Lord is actually in the sun, and he supported this position by theological arguments, Cabalistic analyses, and references to arithmetic, geometry, and music. Above all, the perfection of the sun indicates its connection with divinity, and, in a rhapsody to this heavenly body, Fludd wrote:

> ... the Macrocosmicall Sun's dignity and perfection is easily to be discerned, in that this Royal *Phoebus* doth sit in his chariot, even in the center or middle of the heavens, glittering with his golden hair, as the sole visible Emperour, holding the royall Scepter and government of the world, in whom all the vertue of the celestiall bodies do consist . . . .[70]

How similar this quotation is to the famous eulogy to the sun of Copernicus![71] Surely both quotations betray their neo-Platonic source, and it is of considerable interest that Fludd also spoke specifically of the centrality of the sun. Yet, in contrast to Copernicus, Fludd meant two "centralities," the first of the sun in the heavens, and the second of the earth in the universe as a whole. In a sense he must have felt that he had managed to keep the best of two worlds, for if he had no desire to depart from the traditional earth-centered universe, he still placed the sun midway between

[70]From Fludd's set of eight arguments in the *Mosaicall Philosophy*, pp. 61-64.
[71]"In the middle of all sits the Sun enthroned. In this most beautiful temple could we place this luminary in any better position from which he can illuminate the whole at once? He is rightly called the Lamp, the Mind, the Ruler of the Universe; Hermes Trismegistus names him the visible God, Sophocles' Electra calls him the All-seeing. So the Sun sits as upon a royal throne ruling his children the planets which circle round him." Nicholas Copernicus, *De revolutionibus orbium* (1543; reprint Brussels: Culture et Civilisation, 1966), fol. 9^v. Translation by Thomas S. Kuhn, *The Copernican Revolution* (Cambridge, Mass.: Harvard University Press, 1957), p. 130. For a more detailed discussion of Fludd's views see Allen G. Debus, "The Sun in the Universe of Robert Fludd," in *Le soleil à la Renaissance—sciences et mythes*, Travaux de l'Institut pour l'étude de la Renaissance et de l'Humanisme (Brussels: P.U.B./P.U.F., 1965), Vol. II, pp. 259-278.

the earth and the Creator in the heavens. In so doing, he favored two central positions, and the higher and more worthy of these was to be occupied, not unexpectedly, by the divine sun.

Fludd proceeded to support his position with mechanical analogies (which will be discussed shortly) and "mathematical" arguments based upon musical harmonies. Here he described the world in terms of a universal monochord, an approach that was to result in a bitter and long-drawn-out conflict with Kepler. He ended his discussion of the sun by noting that it was rightly termed "*the heart of heaven*, because that as in the heart doth exist the lively fountaine of blood which doth water and humect the other members of the body." But, he added, it must not be concluded that the sun is identical with God; the Scriptures attest to the fact that the sun is created and its purpose is to house the heavenly Spirit for the benefit of the created world.[72]

Beyond the sun's obvious significance for the macrocosm, Fludd was much concerned with the role of the sun as the source of light and life on earth. Its golden beams by the mercy of God are conveyed through the air and as such form a necessary aetherial nutriment for all life. There can be no doubt that "the elementarie aire is full of the influences of life, vegetation, and of the formall seeds of multiplication, forasmuch as it is a treasure-house, which aboundeth with divine beams, and heavenly gifts.'"[3] Fludd went on to interpret the Pater Noster in a fashion that the daily bread prayed for became this super-celestial aerial nutriment.[74] Nor did he consider the postulation of this substance any kind of innovation, for he believed that all true philosophers knew of this invisible fire (and here he specifically referred to Zeno, Hermes, Zoroaster, and Heraclitus).

We on earth, as the microcosm, can learn about this heavenly treasure through the proper observation of nature. Relying on scriptural references about the relationship of wheat to life and spirit, he turned to this plant for the subject of his investigation and found that the active celestial fire was a volatile salt:

> I can quickly demonstrate, that in the vegetable is a pure volatil salt, which is nothing but the essentiall aire of the specifick, which is wheat

[72] Fludd, *Mosaicall Philosophy*, p. 64.
[73] *Ibid.*, p. 163.
[74] *Ibid.*, p. 162.

or bread; this volatil salt is an unctuous liquor, as white and clear as crystall; this is inwardly neverthelesse full of vegetating fire, by which the species is multiplied *in infinitum*: for it is a magneticall vertue, by which it draweth and sucketh abundantly his life from the aire, and sunne beams, which is the principal treasure house of life, forasmuch as in it the eternall emanation of life did plant his tabernacle . . . .[75]

Fludd's actual process for this demonstration is described in his unpublished "Philosophicall Key" (c. 1618-1620), part of which he translated into Latin to form the first part of his *Anatomiae amphitheatrum* (1623).[76]

In his investigation of wheat he wrote of the putrefaction of the seed—a process which releases the elements. The actual growth of the plant results in a further elemental differentiation. Here the more spiritual quintessence (identified with fire) must rise toward the top, and through sympathetic magnetism the most spiritual part of the plant then attracts more of the aerial spirit. Later, after collection of the kernels and a lengthy distillation process that imitates the natural heat of the seasons, a "white christalline spirit" is collected. Through a series of "experiments" on this substance Fludd felt justified in identifying it as his desired universal spirit of life.

It has been mentioned earlier that the concept of an aerial spirit of life is of great antiquity. It occurred frequently also in the writings of the sixteenth-century Paracelsians,[77] and the impor-

---

[75]*Ibid.*, p. 163.

[76]This experiment has been described by C. H. Josten in his "Robert Fludd's 'Philosophicall Key' and his Alchemical Experiment on Wheat," *Ambix*, **11** (1963), 1-23. The "Philosophicall Key," Trinity College, Cambridge, Western MS 1150 (0.2.46), is being prepared for publication by the present author.

[77]For a more complete discussion see Allen G. Debus, "The Aerial Niter in the Sixteenth and Early Seventeenth Centuries," *Actes du Dixième Congrès International d'Histoire des Sciences*, 2 vols. (Paris: Hermann, 1964), Vol. II, pp. 835-839, and "The Paracelsian Aerial Niter," *Isis*, **55** (1964), 43-61. References to an active nitrous part of the air prior to John Mayow have also been studied by Henry Guerlac and J. R. Partington: Henry Guerlac, "John Mayow and the Aerial Nitre—Studies on the Chemistry of John Mayow I," *Actes du Septième Congrès International d'Histoire des Sciences, Jérusalem 4-12 Août, 1953* (Paris: Académie Internationale d'Histoire des Sciences, n.d.), pp. 332-349; Henry Guerlac, "The Poets' Nitre—Studies in the Chemistry of John Mayow II," *Isis*, **45** (1954), 243-255; J. R. Partington, "The Life and Work of John Mayow (1641-1679)," *Isis*, **47** (1956), 217-230, 405-417. Pagel has shown that the concept of a vital nitrous part of the air is Paracelsian in origin. Pagel, *Paracelsus*, p. 118, n. 324.

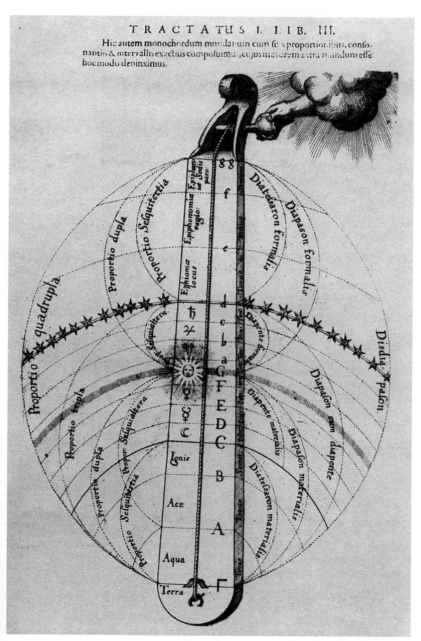

PLATE XIX The harmony of the universe illustrated in terms of Fludd's monochord. From Robert Fludd, *Utriusque cosmi maioris scilicet minoris metaphysica, physica atque technica historia* (Oppenheim: J. T. De Bry, 1617), p. 90.

tance of the *flamma vitalis* in the work of Sendivogius in the early seventeenth century has been discussed by Guerlac. For Duchesne this vital aetherial fire was identified as a vital sulfur, an aerial sulfur which was fixed in saltpeter, thus transferring vital properties to it. Fludd's "fiery" and vital air belongs to the same tradition, and for him the matter of this air was "truly saltpeter."[78]

As a physician Fludd considered it important to understand the role played by this aetherial saltpeter in our bodies. Like Galen, he insisted that the active part of the air was abstracted from the grosser part in the heart; as a Paracelsian, he called this a chemical extraction.[79] Furthermore, since man is a miniature copy of the great world, it would be enlightening, Fludd suggested, to compare our bodies with the macrocosm. Referring to Psalm 18:5 and the relationship of the sun and the Holy Spirit, Fludd concluded that in man the godly tabernacle is in the heart. Considering that the earth is stationary and the sun circles it every day, he pointed out that the sun, as the tabernacle of the Lord, affects the four cardinal winds, which contain the breath of the Lord—that vital nutriment which is breathed into our bodies, formed into the arterial blood, and then given to the body as the spiritual nourishment without which we would all perish. As the sun circles the earth daily it impresses on the winds as well a circular motion. This air is then inhaled by man, and thus the spirit of life reaches the heart and from there is carried around the body in a circular motion imitating the divine circularity. This motion impressed on the blood relates not only to the spirit of the blood in the heart but to all of the spirit of the blood in the body. As the sun rises in the east, sets in the west, and then hastens to its original position to rise again, so also in the microcosm "does it endeavor to arise anew, and it hastens through the branches of the aorta to the South, that is, the liver, and the North or the Spleen."[80] It was in

[78]Fludd, *De supernaturali*, pp. 68, 75, and *passim*. See especially p. 203, where he writes "Nam aërem universalem Macrocosmicum secundùm suam materiam esse salem petrae, superiùs probavimus, & secundùm formam suam nihil aliud esse, quàm ignem dilatatum, & ubique per orbem Macrocosmicum expansum."

[79]Fludd, *Mosaicall Philosophy*, pp. 164-165.

[80]Fludd, *Anatomiae amphitheatrum*, p. 266. In this connection see the earlier description of the vascular system by Fludd in the *Tractatus apologeticus*, p. 190 (n. 44 above). Also, in the "Philosophicall Key" (c. 1618-1620) Fludd pointed to "circularity": "Then in his midle spheare shalt thou erect a pavilion called the

this occult fashion that Fludd proposed a system of the general circulation of the spirit of the blood five years prior to Harvey.[81] These views were to become a matter of dispute between him and Pierre Gassendi.

Fludd's fundamental dichotomy of light and darkness seen first in the Creation itself was stretched by its author much further than simply to account for the origin of the world, its heavenly arrangement, and its relationship to the macrocosm and the microcosm. We have seen how Fludd equated the primary elements of light and darkness with heat and cold. The action of these in turn was apparent in condensation and rarefaction, and it was through a universal system of expansion and contraction that Fludd explained the continuing influence of God upon our world. He

---

hart, Which lik the sonne in the greater World, shall send forth the essentiall beames circularly from his centre that therby they may animat and vivify euery member of this so Well erected a Microcosme" (fol. 24^v).

[81] I have discussed Fludd's views on the circulation in more detail in "Robert Fludd and the Circulation of the Blood," *Journal of the History of Medicine and Allied Sciences*, **16** (1961), 374-393 and in "Harvey and Fludd: The Irrational Factor . . . ." The concept of the circulation of the blood as a chemical distillation has been discussed in some detail by Pagel in his "Harvey and the Purpose of the Circulation," *Isis*, **42** (1951), 25-26, *Harvey's Biological Ideas*, pp. 89-122 *passim*. See also Pagel's "William Harvey Revisited," *History of Science*, **8** (1969), 1-31 (21-29).

Duchesne, with the exception of a reference which suggests that all of the blood circulates in the body as in a chemical pelican, discusses in detail only the venous system. He states that the blood which is carried to the heart by the vena cava is there circulated and distilled over in a purer form to the brain, where it is redistilled or circulated a second time. In the translation by Thomas Tymme Duchesne writes, "the same blood being carried into the heart, by the veyne called *Vena Caua*, which is as it were the Pellican of nature, or the vessel circulatory, is yet more subtilly concocted, and obtaineth the forces as it were of quintessence, or of a Sulphurus burning Aquavita, which is the original; which is the original of naturall & unnatural heat. The same Aquavita being carried from hence by the arteries into the *Balneum Maria* of the Braine, is there exalted againe, in a wonderful maner by circulations; and is there changed into a spirit truly ethereal and heauenly, from whence the animal spirit proceedeth, the chiefe instrument of the soule . . ." [Joseph Duchesne, *The Practise of Chymicall, and Hermeticall Physicke for the preseruation of Health*, trans. Thomas Tymme, Minister (London, Thomas Creede, 1605, sig. K4^v]. Fludd was to borrow this system from Duchesne in its entirety with its implication of "local circulations." The mystical general circulation of the blood described above is a second sense of the word "circulation" used by Fludd.

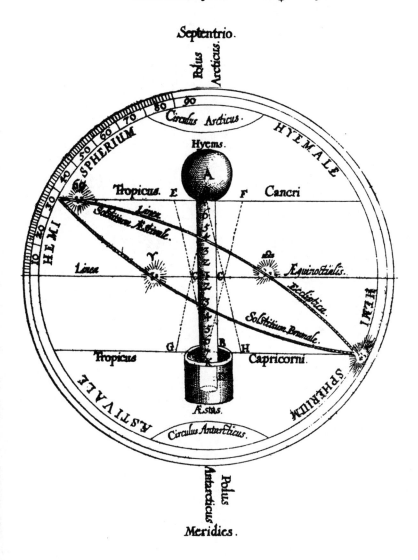

PLATE XX Robert Fludd's thermoscope and its relationship to the cosmos. From the *Integrum morborum mysterium: sive medicinae catholicae tomi primi tractatus secundus* . . . (Frankfurt: Fitzer, 1631), p. 28.

believed that the most important element involved in these processes was water and that its action could be studied observationally by means of a thermoscope, an instrument which made visible the expansion and contraction of air confined over water with temperature changes.[82] While this was really nothing more than a crude barometer like that described earlier by Galileo, for Fludd it was proof that

> ... the catholick Actor was and is the Word or Spirit of God, who acteth first in his Angelicall Organs, by the Starrs, and especially the Sun in Heaven above, and winds beneath, upon the generall sublunary Waters or Elements, according unto his volunty, altering of it after a four-fold manner, through the formall properties of the four Winds, and that either by Condensation or Subtiliation, into divers shapes and dispositions.[83]

To this English philosopher the doctrine of expansion and contraction was a universal basis for an explanation of nature, and it was the more to be praised since it could be linked with God's will. Fludd utilized this principle time and time again as observational evidence for his views. For instance, his ability to obtain the three states of matter by heating and cooling water was for him adequate proof of the interrelation of matter and evidence that water was the mother of all elements. He used the principle also as an explanation of clouds and rain, which he thought were nothing more than compressed air.[84]

The attraction and repulsion exhibited by the magnet also reflected the same universal force seen in expansion and contraction. Indeed, the relationship of man to the universe is comparable to that of the loadstone to the earth:

> The Load-stone is in comparison of its mother earth, even as man is to the whole world; wherefore Man is called the Son of the world by

---

[82]For a description of the thermoscope many of Fludd's writings might be consulted, but he devoted a considerable amount of space to its description in the *Mosaicall Philosophy*, pp. 2-9. Fludd did not claim to have invented this instrument; rather, he wrote that he learned of it through a manuscript five hundred years old. Hency C. Bolton, *Evolution of the Thermometer*, 1592-1743 (Easton, Pa., 1900), pp. 27-28, stated that this was nothing other than the Galilean thermoscope. Fludd wrote in the *Mosaicall Philosophy* that the instrument might be used for measuring the temperature.

[83]Fludd, *Mosaicall Philosophy*, p. 82.

[84]*Ibid.*, p. 100.

Hermes, as *Cardanus, Bap. Porta, D. Gilbert,* and others, have made the Load-stone the child or son of the earth. We find, I say, in the Load-stone, all the passions, as well sympatheticall as antipatheticall, which do affect his mother earth; for it hath its Poles with the earth, and it escheweth all in conformity with the earth, it flyeth from that which is contrary unto its nature. And again, doth sympathise with that which is its like, it hath its Aequinoctialls, Colures, Meridians, and Tropicks, as the earth hath; and, in conclusion, it argueth not onely a sense in motion, but a kind of reason in its action, namely, its refusing that which is contrary unto it, or embracing and desiring that, which is agreeing and conformable unto its harmony . . . .[85]

For Fludd, because the loadstone exhibited these universal forces, it was important to examine its properties, and here he relied heavily on the magnetic studies of Mark Ridley and William Gilbert. He reproduced their experiments (and Gilbert's figures) in detail, but offered his own interpretations of the phenomena. Although willing to grant the importance of Gilbert's observations, Fludd was convinced that his explanations were far from satisfactory. He could not accept a magnetism inherent in the earth as Gilbert seemed to require. The Creation account indicated that the earth derived from the waters. This granted, one could conclude that "the formall Virtue of the earth, did totally descend from heaven, and consequently the earth had no such property from it self."[86]

Nor did Fludd confine himself to magnetism in his criticism of Gilbert. Satisfied that the heavens move circularly about the immobile earth every twenty-four hours because of the divine spirit within, he paused to reject the views of Gilbert and Copernicus. One could neither accept the motion of the earth on its axis nor the motion of the earth around the sun. Gilbert, Fludd implied, may have been a fine physician, but more than medicine is necessary for a correct interpretation of the universe. Beginning with the common arguments advanced to show that the earth is stationary, Fludd turned to mechanical analogies that seemed to support his views. For example, a comparison of the force exerted in turning a heavy wheel from the center and from the circumference will give the approximate source of the motion, for surely a hundredth part of the force required from the center is sufficient to move the wheel

[85] *Ibid.,* p. 199.
[86] *Ibid.,* pp. 202-203.

DE CREATURIS COELI ÆTHEREI.
Experimentum I.

Quantò remotior est rota circumferentia à suo centro, tantò facilius & velocius movetur. Vide Regul.1.cap.2.lib.1.de motu.

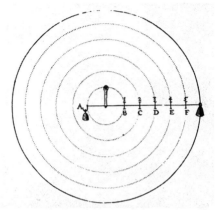

Demonstratur hoc experimentum.lib.1.de motu cap.1.& 2. Reg.1.

Experimentum II.

Multò major vis requiritur ad motum alicujus rotæ à centro (quem motum à principio seu ab interiori appellant) *quàm ad motum à superficie vel circumferentia seu ab exteriore,* qui motus in fine dicitur.

Demonstratur hæc propositio libro 1. de motu cap. 2 Reg. 4. & per regula etiam eam antecedentes.

PLATE XXI   Robert Fludd's use of mechanical analogies to indicate the most probable location of the source of motion in the universe. From the *Utriusque cosmi maioris scilicet minoris metaphysica, physica atque technica historia* (Oppenheim: J. T. De Bry, 1617), p. 155.

at the circumference.⁸⁷ Arguing from what we may term a rule of "celestial simplicity," Fludd assumed that the cause of the motion of the stars must then come from the heavens rather than the center of the earth. Again, for him this was a significant reason why the planets could not be forced around by a central sun. Other mechanical analogies were used to account for the retrograde motions of the moon and the planets.⁸⁸

All of his demonstrations were advanced to support his belief that God operated his universe from on high and that the motion of the heavenly bodies (indeed, all motion) was due to the quantity of divine spirit within. For Fludd the degree of motion signified the plane of life an object had. The greater speed of the stars showed clearly that they were much closer to divinity than the sluggish moon and the planets. Similarly, the superiority of animal life to vegetable or mineral life was indicated by their relative motion.⁸⁹

Among the more common arguments for an immobile earth which Fludd repeated was that involving parallax: if the earth is simply one of the wandering planets, we should see a change in the elevation of the pole star over the course of the year, but since this does not happen, we may conclude that it is the sun that moves rather than the earth.⁹⁰ Using another current argument, he referred to the calamities which would occur if the earth revolved on its axis: it was ridiculous for Copernicus to state that the atmosphere can revolve with the earth and thus have no effect on the sur-

⁸⁷Fludd, *Utriusque cosmi* (1617), p. 155.
⁸⁸*Ibid.*, pp. 157-160.
⁸⁹See Debus, "Sun in the Universe of Fludd," *passim*.
⁹⁰Fludd, *Utriusque cosmi* (1617), p. 154. In his examination of Fludd's philosophy, Gassendi specifically objected to Fludd's attack on the views expressed by Copernicus and Gilbert regarding the motion of the earth. Fludd, replying to Gassendi, chose this parallax argument as his chief support for an immobile earth. See Robert Fludd, *Clavis philosophiae et alchymiae Fluddanae, sive Roberti Fluddi armigeri, et medicinae doctoris, ad epistolicam Petri Gassendi theologi exercitationem responsum* (Frankfurt: Wilhelm Fitzer, 1633), p. 28. "Nam, si terra circa solem in centro fixum peripherice moueretur, tunc sequeretur, quod terra secundum latitudinem moueretur, cum hyemali tempore ipsam à Sole longinquiorem, & nonnunquam ei propinquiorem cernamus. Verum quod hoc longe aliter se habeat, exinde liquet; quoniam poli eleuatio, in omni terrae plaga, semper est una & eadem. Quibus sat liquido indicatur, quod Sol moueatur & distet secundum latitudinem per anni vices à terrae plagis, fixa manente ipsa terra . . . ."

face.⁹¹ Obviously we do suffer from terrible wind damage at times, and if this can happen at all, then surely the effects would be much greater if the earth rotated on its axis. In any case, the earth must be at the center and immobile because it is the grossest and densest material, which moves the least.⁹² We see around us the earth covered by a more tenuous water; this in turn is covered by the more tenuous air, and so on up to the heavens. Surely the sun and planets must therefore be less weighty and more easy to move than the earth. Fludd concluded that such motion as does occur must be proper to the heavens rather than the earth.

Convinced that Gilbert, his deceased fellow member of the College of Physicians, had been adequately refuted, Fludd concluded:

> Certainly the reasons of Gilbert are ridiculous—it is impossible to believe that the heavens can be carried around in the space of twenty-four hours because of their boundless magnitude; and if he would have considered with greater care both the infinite nature of the agent and the disposition of the patient, he would have discovered with great ease that it all happens far differently than he had thought.⁹³

Fludd referred those who remained unconvinced to a far more certain form of proof, Holy Scripture. Here the reader was directed to texts such as Joshua 10, II Kings 20, and Psalm 104:5.⁹⁴

For Fludd it was always possible to relate the phenomena of the great world to the phenomena associated with the earth or man, and here the existence of astral influence was connected with a fundamental question in physics, that of "action at a distance." Although there were some who argued that opaque bodies of all sorts could stop the rays of the stars or even the sun, Fludd envisioned a far more aetherial emanation. For him astral influences were divine; they must therefore be highly subtle and have infinite extension, more like the magnetic attraction of the loadstone than

---

⁹¹Fludd, *Utriusque cosmi* (1617), p. 154.
⁹²*Ibid.*, p. 156.
⁹³*Ibid.*, "Unde certè ridiculae sunt *Gilberti* rationes, impossible esse credentis, ut queant coeli propter infinitam ipsorum magnitudinem spatio 24. horarum circumduci; namq; si exactiori judicio considerasset, tum infinitam agentis naturam, tum etiam patientis dispositionem, facilimè invenisset, rem longè aliter sese habere, quàm ipse existimavit."
⁹⁴*Ibid.*, pp. 157-158.

a cloud momentarily cutting off one's view of a star.[95] Although he continually quoted Gilbert for experimental confirmation of his views, here he could not, since he felt that fundamentally all of the magnetic properties derived from astral emanations. Some believe

> ... because the Load-stone draweth the Iron but at half the distance of a table ... that the vivifying act of his vertue penetrateth or extendeth it self no farther, that at that distance in the aire, and consequently being led by their corporall eye-sight, they limit, after this externall and visible action, the Load-stones spirituall extension.

But Fludd could not agree with such an interpretation; for him the astral influences which ultimately cause the attraction of the loadstone and the iron are not limited in distance. It is simply a case of the dilution of the virtue over increasing distances. If the iron is close to the magnet, the astral beams from which the magnet draws its strength all pass through the iron, and the full strength between the magnet and the iron is evidenced in the strong attraction. On the other hand, if the iron is far away, most of the beams will pass outside the physical limits of the iron, producing a much weaker effect.[96] Above all, the immense distances involved are evidenced by the continued pointing of the compass needle to the pole star. Not only will a freely suspended loadstone turn to a north-south orientation, the same will occur with a magnetized iron or even a nonmagnetized curtain rod.[97] And although it is erroneous to believe that the compass needle points north because of magnetic rocks in the earth, even that belief would testify to the extension of this virtue over great distances rather than limited tabletop measurements.[98]

If the spirit in the loadstone is so subtle and so far-reaching, then how much more subtle and penetrating must be the spirit of life! Plants follow the influence of the sun, some closing their flowers with the approach of night, and hazel forks are used to locate metallic veins, indicating the existence of sympathetic action

---

[95] Fludd, *Mosaicall Philosophy*, pp. 221, 222.
[96] *Ibid.*, p. 224.
[97] *Ibid.*, p. 268, here citing William Gilbert, *De magnete* (London: P. Short, 1600), Book II, Ch. 33; and Mark Ridley, *A Short Treatise of Magneticall Bodies and Motions* (London: Nicolas Okes, 1613), Ch. 21.
[98] Fludd, *Mosaicall Philosophy*, p. 225.

between mineral and vegetable spirits.⁹⁹ Indeed, there is no question but that astral magnetic beams penetrate as far as the beams of a star, and the fact that man's spirit is more exalted than that of a loadstone indicates that his spirit may reach beyond the sphere of the stars to the highest throne of divinity.

For Fludd there was a direct application of this action in medical practice. A century earlier Paracelsus had described the weapon-salve, which was prepared from a number of ingredients including the blood of the injured person.¹⁰⁰ When applied to the particular weapon which had caused a wound, this salve was alleged to cure the person wounded. Although Libavius had raised his pen to attack the salve,¹⁰¹ this method of cure had excited little interest in the sixteenth century. However, a tract by Rudolph Göckel (Goclenius) in 1608 favoring the weapon-salve unleashed a minor pamphlet war that was to include tracts by Jean Roberti, Oswald Crollius, Jean Baptiste van Helmont, Daniel Sennert, and—Robert Fludd.¹⁰²

For Fludd the weapon-salve worked because of the essential unity of the world. How else might one explain the case of the

---

⁹⁹*Ibid.*, pp. 222, 225, 226. Here Fludd cites Gilbert, Book III, Ch. 6 to show that there is a magnetic effect in vegetable life. He describes Gilbert's discussion of grafting, where the definite order which is required for continued plant growth is shown to be similar to the order required for reuniting a severed magnet.

¹⁰⁰"Most men attribute this Vnguent to *Paracelsus*, or affirme it certainly to be divulged by him.

"*Paracelsus* himself, *Archidox: Magica, lib.* I giveth this description of it; Take *Scull-Mosse*, two Ounces, *Mummy*, halfe an ounce, *Mans fat*, two ounces, *Mans blood*, halfe an ounce, *Linseed Oyle*, two Drams, *Oyle of Roses*, and *Bole Armoniack*, of each one ounce.

"Mixe them together and make an Oyntment: Into the which hee puts a Stick, dipp'd in the Blood of the wounded person, and dryed, and bindeth up the wound with a rowler dipt every day in the hot Vrine of the wounded person. For the annointing of the Weapon hee addes moreover; *Honey*, one ounce, Bulls Fat, one dram."

Daniel Sennert, *The Weapon-Salves Maladie; or, A Declaration of its Insufficiencie to Performe what is Attributed to it,* trans. out of his 5th booke Pt. 4 Chap. 10 Practicae medicinae (London: John Clark, 1637), p. 2. Although the *Archidoxeos magicae* may not be genuine, it is evident that Sennert entertained no doubts about its origin.

¹⁰¹See Ch. 3.

¹⁰²The early history of this debate is sketched briefly by the present author in "Robert Fludd and the Use of Gilbert's *De Magnete* in the Weapon-Salve Controversy," *Journal of the History of Medicine and Allied Sciences,* **19** (1964), 389-417 (390-392).

... certaine Lord, or Nobleman of Italy, that by chance lost his nose in a fight or combate, this party was counselled by his Physicians to take one of his slaves, and make a wound in his arme, and immediately to ioyne his wounded nose to the wounded arme of the slave, and to binde it fast, for a season, untill the flesh of the one was united and assimulated to the other. The Noble Gentleman got one of his slaves to consent, for a large promise of liberty and reward; the double flesh was made all one, and a collop or gobbet of flesh was cut out of the slaves arme, and fashioned like a nose unto the Lord, and so handled by the Chirurgion, that it served for a naturall nose. The slave being healed and rewarded, was manumitted, or set at liberty, and away he went to Naples. It happened, that the slave fell sicke and dyed, at which instant, the Lord's nose did gangrenate and rot; whereupon the part of the nose which hee had of the dead man, was by the Doctors advice cut away, and hee being animated by the foresaid experience, followed the advice of the same Physician, which was to wound in like manner his owne arme, and to endure with patience, till all was compleate as before. He with animosity and patience, did undergoe the brunt, and so his nose continued with him untill his death.[103]

In a sympathetically operating universe such an account seems plausible. Fludd explained that as long as two bodies of the same disposition were in spiritual contact even though physically separate, they were mutually fortified by one another. However, on the death of the freedman the spiritual beams changed, and as a result the nobleman's nose began to rot.[104] For Fludd the case seemed to be a strong one for the acceptance of sympathetic action.

In this case action at a distance, or the "Aristotelicall limited spheare of Activity," posed no problem.[105] Does not the spirit of the

[103]Fludd, *Doctor Fludds Answer*, pp. 132-133. A similar story is related by van Helmont; see the *Oriatrike, or Physicke refined*, trans. John Chandler (London: Lodowick Loyd, 1662), p. 764. A well-known woodcut of arm-to-nose fixation appeared at the time when Fludd was a student. See Gaspare Tagliacozzi, *De curtorum chirurgia per insitionem libri duo* (Venice: apud Gasp. Bindonum, 1597); 2nd ed. retitled *Cheirurgia nova de narium aurium labiorumque defectu per insitionem cutis ex humero arte hactenus omnibus ignota sarciendo* (Frankfurt, 1598); re-edited by M. Troschel (Berlin: Reimer, 1831). Further details will be found in M. T. Gnudi and J. P. Webster, *The Life and Times of Gaspare Tagliacozzi Surgeon of Bologna (1545-1599)* (New York: Herbert Reichner, 1950).
[104]This example is explained primarily in terms of magnetic action in Fludd, *Mosaicall Philosophy*, p. 252.
[105]Fludd, *Doctor Fludds Answer*, p. 82.

four winds blow as far as it pleases; may not the odor of flowers be detected many miles from the source?

> What shall wee then say of the same spirit, which acteth in the little world or man, when his insensible breath or emanation tendeth affectionally toward the homogeneall place of his owne nature? I meane unto the ointment, in which the selfe same indivisible nature, either in the blood, adhering to the weapon, or having penetrated in the weapon, without any signe of external blood, is bathed? Shall wee not believe, that by his emanation, it can carry along with it in the Ayre, the occult spirit of the vegetating nature of the wounded person, included secretly in a volatile salt, to act in the oyntment?[106]

Once again we are referred to the heart as the tabernacle of the Lord in the microcosm through Psalm 18:5. The heavenly beams requisite for life, which stream down to us from the sun through the air, are chemically separated in the body and become the life spirit of the blood.

> Therefore without doubt, there is the selfe-same relation of unison betwixt this ointment with the blood in it and the wounded mans nature; as is between the string of one lute, that is proportioned unto the other in the same tone: And for this cause will be apt to evibrate & quaver forth one mutuall consent of simpatheticall harmony, if that the spirits of both, by the vertuall contact of one anothers nature, be made by conveying the individuall spirit of the one into the body of the other, that the lively balsamick vertue of the one, may comfort and stir up the deadly languishment of the other, no otherwise then the activity of one lute struck, doth stirre up the other to move, which was before still and without life.[107]

The life spirit may also be studied within the framework of the magnetic spirit—that is, in terms of polarity. Gilbert had compared magnetic polarity to grafting in trees, noting that the branch to be grafted could only be attached at one of the two ends for growth. Fludd discussed this example and then proceeded to discuss the dipolar nature of man.[108] After discussing "polarities" seen in various parts of the human body, he concluded that while there is a marked similarity, there remains a distinction between

---

[106]*Ibid.*, p. 83. As noted earlier, Fludd identified this vital aerial salt with saltpeter.

[107]*Ibid.*, pp. 98-99.

[108]See above, n. 99. Fludd comes back to this analogy again when referring to the observation that loadstones capped with steel exhibit stronger magnetic

magnetism in man and in the loadstone. The loadstone represents a lower order of life and "may in some sort be esteemed dead . . . he attracteth not by any dilation, made of a vivifying heat, through any organicall Substance or assistance, but by an earthly and Centrall, contracting disposition."[109] In short, while man has double-natured powers of attraction, magnetic attraction corresponds primarily to the northern-polar forces, a type of attraction which still resides in man after death.

Fludd's interest in the weapon-salve was clearly associated with his concern for the life spirit and its action in this world. Here it was possible to relate practical therapy with fundamental magnetic and sympathetic forces operating in the macrocosm and the microcosm. In view of his interest in alchemy one might further expect Fludd to have been one of the primary spokesmen in England favoring chemically prepared medicines. We know that he kept his own apothecary in his house, and as censor for the College of Physicians he regularly visited apothecaries to ascertain that their drugs had been prepared according to the rules of the official *Pharmacopoeia*. Yet, although Fludd recognized these preparations as a significant aspect of his universal science, he devoted relatively little space to this topic in his printed works. Because of this it is of more than average interest to note the late reference to Fludd in the rare tract *Mr. Culpeper's Ghost* (1656).[110] This work pictures the lately departed Culpeper strolling in the Elysian Fields, where he meets Robert Wright, junior, Fludd's house apothecary, who relates that

> . . . though a Trismegistian-Platonick-Rosy-crucian Doctor, [Fludd] gave his Patients the same kind of *Galenical Medicaments*, which other Physitians in the Town ordinarily appointed, and when himself was sick, he had no Chymical *Elixirs* or *Quintessential Extracts* to relie upon,

---

effects (Gilbert, Book II, Ch. 19). Here he suggests (*Mosaicall Philosophy*, p. 257) that "in like manner, if Magnetick Excresences, be grafted in the body of the Magneticall Tree, then that tree will suck and draw his like, namely the spirit of defective limbs more strongly unto it; making them to become vegetative, and to increase and grow which before did pine and wither."

[109] *Ibid.*, p. 218.

[110] Discussed by F. N. L. Poynter in "Nicholas Culpeper and the Paracelsians," *Science, Medicine and Society in the Renaissance. Essays to Honor Walter Pagel*, ed. Allen G. Debus, 2 vols. (New York: Science History Publications, 1972), Vol. I, pp. 201-220.

but after he had caused himself to be let blood (an ordinary Galenical Remedy) he sent for *Doctor Gulstone,* & relied upon his advise for the Cure of the disease, who was a *pure Galenist.*[111]

Later, when Culpeper met Fludd himself, who was conversing with the ghosts of Lull and van Helmont about the weapon-salve, the Doctor confirmed the account of Robert Wright, adding that "there were no better Medicaments in the whol world than the Galenical."[112]

Although this is an apocryphal account, it has an added interest because of the conservative nature of Fludd's approach to the theory of disease. Here he accepted both the ancient humoral theory and also the ontological viewpoint of Paracelsus. Pagel has shown that Fludd maintained the humoral theory in a close connection with astrological influences on the body.[113] However, at the same time,

> ... he maintains that the cause of disease represents the essence of disease, and taking into consideration both the ancient belief in the influence of harmful exhalations (Galen) and the emphasis laid by Paracelsus on the importance of an aerial chaos of the finest corporality, he decides that the causes of disease must be the winds.[114]

There exist in the atmosphere *spiritus mali*, which may enter the body through inhalation or the pores of the skin. These substances are under the control of four demons, identified with the four elements, and man may be pictured as being protected by four angels who help ward off these seeds of bodily destruction. Hence this essentially "modern" concept of disease is coupled with Fludd's usual mystic correspondences.[115]

---

[111]As quoted by Poynter, *ibid.*, p. 217.

[112]*Ibid.*, p. 218.

[113]See Walter Pagel, "Religious Motives in the Medical Biology of the XVIIth Century," *Bulletin of the Institute of the History of Medicine,* 3 (1935), 97-128, 213-231, 265-312 (here Pagel discusses Fludd on pp. 273-282).

[114]*Ibid.*, p. 278.

[115]Pagel has discussed Fludd's views in detail on this subject. A few specific references for various points of interest in Fludd's writings follow: Fludd discussed the humors in general in his summary to the *De supernaturali* (pp. 200-203); he specifically discussed the humors as the microcosmic elements in the *Medicina catholica* (1629), pp. 110-111, and one of his famous diagrams of man being protected in health by angels who are attacked from without may be found in the same work with an explanation, fols. 2 ff., further amplified on p. 169.

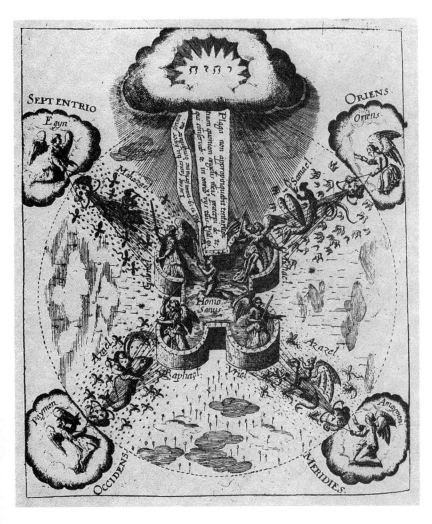

PLATE XXII Man besieged in his castle of health. From Robert Fludd, *Integrum morborum mysterium: sive medicinae catholicae tomi primi tractatus secundus* ... (Frankfurt: Fitzer, 1631), p. 338.

The Initial Reaction in England (1618-1623)

Although many of the points of Fludd's philosophy of nature were not too far removed from those of other contemporary Hermeticists and alchemists, he had presented them in far more detail and in far more sumptuous volumes than had others. His work drew additional attention since he wrote not as an alchemist from a hidden retreat, but rather as a learned physician who was at the same time an eminent member of the most prestigious society of physicians in Europe. However, it was not to be forgotten that he had published first against Libavius in defense of the Rosicrucians. And, in spite of Fludd's reference to the support of continental scholars, there were relatively few Hermeticists of the first rank who sided with him.[116] One reason for this may be that although he was outspoken against the ancients, he remained curiously tied to some aspects of ancient medical theory. So, although he had expected high praise for his labors, Fludd soon found himself attacked both at home and abroad.

In England Fludd's orthodoxy was questioned at an early date, and he hastened to prepare his "Declaratio brevis," dedicated to King James, in which he defended his works on the macrocosm and the microcosm as well as his earlier *Apologia* for the Rosicrucians. The question of his possible heresy loomed large again with a second manuscript dating from the period 1618-1620, "The Philosophicall Key," also dedicated to the King and written in English so that it might reach a wider audience. In contrast to his "Latin Philosophical discourse," "The Philosophicall Key" was to be an "experimental Treatice."[117] Here his reference was plainly to the experiment on wheat discussed earlier. However, it is to the earlier section of the "Key" that we must turn for information relating to the accusations made against him.

Already questioning the actual existence of a Society of Rosicrucians, Fludd now referred to his *Apologia* as "a certain silly

---

[116]In the "Declaratio brevis" and "The Philosophicall Key" Fludd referred to Dr. Andrews and John Selden in England and reproduced letters from Gregor Horstius, M.D. (University of Giessen) and Dr. Matthias Engelhard, "Ascaniensis." These references plus an additional one to Justus Helt are discussed by Josten in his paper on "The Philosophicall Key," pp. 12, 14-15.

[117]Fludd, "Philosophicall Key," fol. 1^r.

and poore" one; because of it he had been accused of doubt of religion, false divinity, and spurious philosophy.[118] Those who spoke against him, however, had not read his work with understanding. At no time had he subscribed to all things found in the Rosicrucian documents.[119] To those who suggested that he had written his major works with the assistance of Society members, for otherwise such details of mystical learning would not be possible, he protested, I "doe yet fly on mine owne Wings."[120] In proof he offered his "original" experiment on wheat and suggested that his friends in Germany or Dr. Andrews and Mr. Selden at home could testify that his macrocosmic history had been composed four or five years before anyone had ever heard of the Rosicrucians. In any case, "unto my greef," Fludd had never been contacted by anyone claiming to belong to that order.[121]

Fludd's folios on the macrocosm and the microcosm had been censured for dangerous innovations and "spurious philosophy." To this he replied that he had built his work upon both Sacred and philosophical authority, and in no fashion could it be considered an innovation. The ancient and true axioms he had found were subscribed to by Moses, whose concepts were later stolen by Aristotle. Beyond this, his truths were confirmed "by easy familier and oculer demonstrations, the very pointing finger of experience, which is able to instruct the rude and rusticall, yea and very fooles themselves."[122] He had simply brought forth from obscurity truths long hidden from the multitude. And if there were some who attacked him unjustly, their words were of but little weight in comparison with continental scholars of the worth of Justus Helt, Gregor Horstius, and others who had testified to the significance of his publications.[123] If both Lutherans and Jesuits could refer to his work with respect, why was it that he could be charged with heresy at home? Far from being a heretic, Fludd gladly affirmed his belief in the reformed religion "so happily celebrated heare in England!"[124]

[118] *Ibid.*, fols. $3^{r, v}$.
[119] *Ibid.*, fol. 4^{v}.
[120] *Ibid.*, fols. $12^{r, v}$.
[121] *Ibid.*, fols. $13^{r, v}$.
[122] *Ibid.*, fol. 6^{r}.
[123] *Ibid.*, fols. 7^{r}-11^{r}.
[124] *Ibid.*, fol. 5^{r}.

A third unpublished manuscript by Fludd is "Truths goulden Harrow," which was directed against *The Tillage of Light* (1623). The author of this short and undistinguished work was one Patrick Scot, who wrote of alchemy as an allegory and suggested that the many publications on the subject simply presented a guide to the gathering and perfecting of wisdom.[125] Specifically Scot had objected to a material alchemy and had then gone on to state that the search for the elixir, or philosopher's stone, was inimical to true philosophy.

In his critique of Scot, Fludd quoted numerous passages in Scripture which seemed to indicate the truth of transmutation.[126] He could not accept Scot's assertion that metals cannot propagate themselves,[127] and he went on to discuss the relation of light and spirit. After copious scriptural citations, Cabalistic analyses, and occasional references to observational evidence, Fludd concluded that "the Elixir . . . is a material, yea and an earthly substance but of an exalted nature by a supream purification."[128] Not only is the elixir real, it is true wisdom and "the lif of every thing."[129] And, to the charge that the study of alchemy and the search for the elixir were inimical to philosophy, Fludd could only reply that

> . . . he that is not a true beleever and lover of this excellent master peece, is drowned in darknes and hath an iron gate before his eyes of understandinge . . . . And to conclude, in steed of being an enimy to philosophy it is the greatest friend she hath for the onely Lady she serveth as the body doth the soule is wisdome, which is the summum bonum of the philosophers, and mayne subiect of philosophy who therfore hath her denomination from her love unto wisdome, or the earthly sonne of the philosophers which is as well the tabernacle of the divine emanation as the heavenly.[130]

---

[125] Patrick Scot, Esquire, *The Tillage of Light, or a true Discoverie of the Philosophical Elixir commonly called the Philosophers Stone Serving to enrich all true, noble and generous Spirits, as will adventure some few labours in the tillage of such a light, as is worthy the best observance of the most Wise* (London: William Lee, 1623). This work is discussed by Josten, "Truth's Golden Harrow," pp. 92-95. The text of Fludd's manuscript is given in the same paper, pp. 102-150.

[126] "Truth's Golden Harrow," p. 109.
[127] *Ibid.*, p. 140.
[128] *Ibid.*, p. 123.
[129] *Ibid.*, pp. 118-119.
[130] *Ibid.*, p. 108.

Fludd and Kepler (1619-1623)

Fludd's debate with Johannes Kepler is on a different level from that evidenced by the unpublished manuscripts in English and Latin. The *Utriusque cosmi* had appeared in 1617 and was seen by Kepler as he was preparing his *Harmonices mundi* for the press. Fludd's views of a universal harmony as expressed in his volumes on the macrocosm and the microcosm were anathema to Kepler, who hastened to prepare an appendix to the final book of the *Harmonices mundi* in which he discussed Fludd's errors. The latter replied in his *Veritatis proscenium* (1621).[131] Kepler, by this time thoroughly provoked,[132] prepared a far more detailed refutation in the *Pro suo opere harmonices mundi apologia. Adversus demonstrationem analyticam Cl.V.D. Roberti de Fluctibus medici Oxoniensis. In qua ille dicit respondere ad appendicem dicti operis* (1622). This in turn was answered by the English physician in a final *replicatio*, the *Monochordum mundi symphoniacvm, seu replicatio Roberti Flud . . . ad apologiam . . . Ioannis Kepleri*, which he appended to a major work, the *Anatomiae amphitheatrum* (1623).

The complexity of the Fludd-Kepler debate has been noted in a brilliant essay by Wolfgang Pauli.[133] As Pauli indicates there, it is perhaps all too easy to judge these works as a conflict between the true mathematician and the occultist.[134] This is the interpretation Kepler himself wished to make, but the correct application of mathematics to natural phenomena was not yet clearly understood. Fludd had emphasized in his *Tractatus apologeticus* that true

---

[131]The full title is *Veritatis proscenium; in quo aulaeum erroris tragicum dimovetur, siparium ignorantiae scenicum complicatur, ipsaque veritas à suo ministro in publicum producitur, seu demonstratio quaedam analytica, in qua cvilibet comparationis particvlae, in appendice quadam à Joanne Kepplero, nuper in fine harmoniae suae mundanae edita; facta inter harmoniam suam mundanam, & illam Roberti Fludd, ipsissimis veritatis argumentis respondetur.*

[132]Letter from Kepler to Johannes Seussius (July 15, 1622) in *Gesammelte Werke*, 18 vols. (Munich: C.H. Beck, 1937-1949), Vol. XVIII, pp. 93-94.

[133]Wolfgang Pauli, "The Influence of Archetypal Ideas on the Scientific Theories of Kepler" in C. G. Jung and W. Pauli, *The Interpretation of Nature and the Psyche*, trans. Priscilla Silz (New York: Pantheon Books, Bollingen Series 51, 1955), pp. 147-240. See also Frances A. Yates, *Giordano Bruno and the Hermetic Tradition* (Chicago: University of Chicago Press, 1964), pp. 440-444, and Robert Lenoble, *Mersenne ou la naissance du mecanisme* (Paris: Vrin, 1943), pp. 143 ff.

[134]Pauli, "Archetypal Ideas," pp. 200-201.

The Synthesis of Robert Fludd

mathematicians sought the mysteries of the Pythagoreans,[135] and if he illustrated his works with complex symbols, it was because he belonged to an alchemical tradition which considered such symbols a valid means of offering truths to all in a veiled fashion. This, he argued, was far superior to the common diagrams of "vulgar" mathematicians.[136]

As Kepler charged, Fludd did indeed discuss two mathematics: the one, concerned with "quantitative shadows," belonged to the common mathematician; the other, comprehending "the true core of natural bodies," was that of the alchemists and the Hermetic philosophers,[137] the mathematics of Paracelsus and the natural magicians. No more powerful example of mathematical analysis could be offered by Fludd than John Dee's hieroglyphic monad. He advanced this in the *Veritatis proscenium* of 1621 to prove the significance of the quaternary number in opposition to Kepler's emphasis of the Trinity. "And finally," Fludd wrote,

> ... if we consider mystic Astronomy we shall indeed perceive in it the whole power of the quaternary, and this most clearly; for its whole secret lies in the hieroglyphic monad which exhibits the symbols of sun, moon, the elements, and fire, that is to say, those four which are actively and passively at work in the universe in order to produce therein the perpetual changes whereby corruption and generation take place in it. The figure is as follows:

In this symbolic image we see, first of all, an indication of the quaternary in the cross, four lines being arranged so as to meet in a common point. Joined with the number 3, which denotes the moon, the sun, and fire, this [quaternary] will produce the number 7, which can also be demonstrated by the four elements. And yet this number 7 is in itself none other than the quaternary considered formally.[138]

---

[135]Fludd, *Tractatus apologeticus*, pp. 103-107.

[136]*Ibid.*, p. 118. See also Fludd's defense of his "delight in pictures" as quoted from the *Veritatis proscenium* by Pauli, "Archetypal Ideas," p. 195.

[137]Pauli, p. 196.

[138]As translated by Pauli, pp. 233-234. The relationship of Dee to Fludd, particularly in regard to the subjects of study described in the *De architectura* of

Fludd's approach was sharply criticized by Kepler. His own emphasis on the number three derived from a belief in the doctrine of signatures.[139] The Creator, the perfect example of three in one, had stamped this symbol throughout the universe; it was seen most clearly in the sphere where the central point, the surface, and the intervening distance indicated the Trinity.[140] Turning to chemical theory, Kepler added that the error of Fludd's adherence to four would be evident even to him if he would read with understanding the works of the Islamic chemists.[141] They, in their resolu-

---

Vitruvius, is discussed by Frances A. Yates in her *Theatre of the World* (London: Routledge & Kegan Paul, 1969), pp. 42-59. In *The Rosicrucian Enlightenment*, pp. 3-40, she goes on to picture Dee as a key figure in the Rosicrucian literature—primarily on the basis of the frequent references to the *Monas hieroglyphica*. In this context Fludd's use of the hieroglyphic monad in his debate with Kepler and his own deep interest in the Rosicrucian manifestoes are significant. However, in a broader context Yates' interpretation of the evolution of sixteenth-century Hermetic thought to a seventeenth-century chemical philosophy is open to question. In defining the "Rosicrucian Enlightenment" she states that "it is a phase in which the Renaissance Hermetic-Cabalist tradition has received the influx of another Hermetic tradition, that of alchemy. The 'Rosicrucian manifestoes' are an expression of this phase, representing, as they do, the combination of 'Magia, Cabala, and Alchymia' as the influence making for the new enlightenment." (*Rosicrucian Enlightenment*, pp. ix-x). It is the added ingredient, alchemy, then, that makes for a new force acting in the direction of the new science.

Dr. Yates' research on this subject stems from her earlier important study of Bruno, whose interests, like those of many Italian figures, did not center on alchemy. Had she started rather from Paracelsus she would have discovered the same neo-Platonic and Hermetic antecedents along with the magic and Cabala she found in the work of Bruno. However, in addition she would have found that the emphasis on chemistry, alchemy, and medicine forms part of the mainstream of Paracelsian thought from the beginning. Thus, the Dee-Rosicrucian material may be ancillary to the chemical philosophy, but it is not really an essential part of it. To be sure, in alchemical literature Dee was reasonably well known for his *Monas hieroglyphica* (it was printed in the second volume of Zetzner's *Theatrum chemicum*, 1602, 1613, and 1659), but his reputation surely was not on a level with Paracelsus or van Helmont, who are more typical of the sixteenth- and seventeenth-century attitudes toward chemistry and medicine. Dee is distinctly atypical. Among other points that might be raised, his work is for all purposes devoid of an interest in the relationship of chemistry to medicine. (See the discussion of the work of Dee above in Ch. 1.).

[139]Pauli, "Archetypal Ideas," p. 159.

[140]*Ibid.*, p. 168.

[141]Kepler, *Adversus demonstrationem analyticam . . . de Fluctibus . . .* in *Gesammelte Werke*, Vol. VI (1940), p. 439.

tion of natural bodies, recognized three principles: salt, sulfur, and mercury. They even questioned the elementarity of fire. Indeed, the quaternary number of elements, along with their corresponding qualities, had been dissipated if not destroyed by the *tria prima* and the true understanding of the elements.

As Kepler presented his case, the differences between his approach and that of Fludd were simple. His descriptions of the harmonies were "mathematical," while Fludd's were "aenigmaticos, pictos, Hermeticos."[142] The latter's works abounded with symbolic pictures, while his own instructed with true mathematical diagrams. Fludd delighted in shadowy enigmas, while Kepler sought to rescue the same phenomena from darkness and to bring them into the light.[143] Fludd borrowed fables from the ancients, while Kepler built upon the fundamentals of nature with mathematical certitude.[144] The former confused things that he did not properly understand, while the latter proceeded in an orderly fashion according to the laws of nature.[145] Again and again Kepler returned to the sharp distinction between the mathematician and the chemist, Hermeticist, and Paracelsian.

And yet, as Pauli clearly understands, the debate was not simply one of the mystic versus the mathematician. Fludd was quite capable of relying on observational evidence and laboratory equipment when he felt that such arguments might bolster his case. As for Kepler, he surely cannot be classed as a pure "modern." His development of the laws of planetary motion was the result of an interest that was deeply rooted in the search for mathematical perfection that forms part of the Pythagorean tradition. Convinced of the truth of the music of the spheres, he sought a movement of the planets in the same proportions that appear in the harmonious sounds of tones and regular polyhedra.[146] No less than Fludd did he believe in an *anima mundi*, no less than Fludd did he argue for a near-divine sun in the center of the world (although

---

[142] *Ibid.*, p. 431.

[143] Kepler, *Harmonices mundi, appendix habet comparationem huius operis cum harmonices Cl. Ptolemaei libro III cumque Roberti de Fluctibus, dicti Flud. medici oxoniensis speculationibus harmonicis, operi de macrocosmo & microcosmo insertis* in *Gesammelte Werke*, Vol. VI, p. 374.

[144] Kepler, *Adversus demonstrationem analyticam*, p. 428.

[145] Kepler, *Harmonices mundi, appendix*, p. 374.

[146] Pauli, "Archetypal Ideas," pp. 156 ff.

he disagreed heatedly with Fludd over the meaning of centrality in this case), and no less than Fludd did he believe in the stars as living entities. However, in spite of these similarities, it is quite understandable that the true meaning of mathematics for Kepler was something quite different than it was for Fludd. The latter sought mysteries in symbols according to a preconceived belief in a cosmic plan; his proportions and harmonies were forced to fit these symbols. Kepler, perhaps just as obsessed with his symbolic spherical picture of the world, insisted that his hypotheses be founded on quantitative, mathematically demonstrable premises.[147] If his hypothesis could not accommodate his observations, he was willing to alter it. These two views were so opposite that the two men could not really understand each other. Fludd felt that he could honestly call Kepler one of the worst kinds of mathematician, one of the vulgar crowd who "concern themselves with quantitative shadows; the alchemists and Hermetic philosophers, however, comprehend the true core of the natural bodies."[148]

Fludd and the French Mechanists (1623-1633)

Although the Fludd-Kepler exchange is of considerable interest, the reaction to Fludd's publications among French scholars is even more illustrative of the problems faced by chemical philosophers in the early seventeenth century. As noted in the last chapter, the claims of the chemical philosophers had been the subject of intense debate in France from the closing decades of the sixteenth century. The conflict at Paris in the early years of the new century had served to further publicize the differences between the Paracelsians and the Galenists, and ramifications of this conflict had affected scholars throughout Europe. The publications of Duchesne and his cohorts had been frequently reprinted, while Mayerne sought the firm acceptance of the chemical medicine among the fellows of the Royal College of Physicians in London after his final departure from Paris.

Nor had there been any abatement of interest in the "chemical problem" in France proper. The second and third decades of the

[147] *Ibid.*, p. 194.
[148] As quoted by Pauli, p. 196, from the *Veritatis proscenium*, p. 12.

seventeenth century saw the translation and publication of numerous alchemical classics. A French *Les douze clefs de philosophie* by Basil Valentine appeared in 1624, and other titles by this author appeared in translation in succeeding years. The numerous editions of Beguin's textbook testify to the very real contemporary need for a practical guide to chemical preparations.[149] A similar collection of recipes, but one that included the important theoretical "Admonitory Preface," was made available in *La royalle chymie* of Crollius. This was popular enough to appear in two editions at Lyon (1624, 1627) and another two at Paris (1633, 1634).

In addition to the older authorities, a new generation of French chemical physicians added their publications to a list that was already impressive. Estienne de Clave argued against the Aristotelians on behalf of the chemists long before the publication of his popular chemical textbook in 1646.[150] Pierre Jean Fabre (d. 1650), a graduate of Montpellier, published numerous chemical works of medical significance beginning with his *Palladium spagyricum* (1624). Many of these attained considerable popularity and some continued to be reprinted well into the eighteenth century. Another author, David de Planis Campy (1589 - c. 1644), published on traditional alchemical subjects as well as iatrochemistry. He insisted that "Hermetic medicine" was the "true medicine" and he wrote at length on the relationship of the macrocosm and the microcosm and the doctrine of signatures.[151] In his early *L'hydre morbifique exterminee par l'Hercule chimique* (1628) he attempted to show the similarities between the works of Hippocrates and Paracelsus,[152] but he added that the ancients no longer suffice, since the times had changed and with them had developed new illnesses unknown to Galen for which new medicines were required.[153] The success of chemists in the cure of seven

---

[149] The best survey of the French textbook tradition is still to be found in Hélène Metzger, *Les doctrines chimiques en France du début XVIIe à la fin du XVIIIe siècle,* Tome I (Paris: P.U.F., 1923), pp. 17-97.

[150] *Ibid.,* pp. 51-59.

[151] Robert Multhauf, "Medical Chemistry and 'The Paracelsians,' " *Bulletin of the History of Medicine,* **28** (1954), 101-126 (109).

[152] David de Planis Campy, *L'hydre morbifique exterminee par L'hercule chymique ou les sept maladies tenues pour incurables iusques à present rendues querissables par l'art chymique medical* (Paris: Hervé du Mesnil, 1628), sigs. ẽ iij^r-ĩ ji^r.

[153] *Ibid.,* sig. ẽ vii^r.

diseases traditionally considered incurable argued conclusively for the adoption of the chemical medicine.

> O heureuse nouueauté! puis qu'elle nous debroüille d'un cahos d'erreur & d'ignorance en laquelle la cômune opinion nous detenoit. O heureux remedes chimiques! puis que par vostre nouueauté nous voyons toutes les maladies, tenuës pour incurables du commun, totallement exterminées par vostre vsage.[154]

The Rosicrucian episode received notice also among French scholars. During the winter of 1619 René Descartes was in Germany seeking to contact the members of the Brotherhood. On Descartes' return to Paris only a visit to Marin Mersenne could still the latter's fear of his friend's conversion.[155] Again, in 1623, placards announcing the arrival of the Brethren of the Rose Cross were set up in Paris. These offered to "show and teach without books or marks how to speak all languages of the countries where we wish to be, and to draw men from error and death," according to Gabriel Naudé, who discussed the Brotherhood in his *Instructions à la France sur la verité de l'histoire des frères de la Roze-Croix*, published that same year.[156] Attacking their "vain and useless" words, he complained bitterly about the chemists and especially pointed to the malignant influence of Paracelsus, the "foundation stone of this Congregation."[157]

On August 24 and 25, 1624, fourteen alchemical theses were defended at the residence of the influential Hermeticist François de Soucy, sieur de Gerson.[158] Among them was a denial of the *materia prima* and the doctrines of substantial form and privation. The sublunary world was affirmed to be composed of the two elements earth and water, plus the three Paracelsian principles. Other theses emphasized the world soul, various aspects of the specific

---

[154]*Ibid.*, sig. vii$^{r, v}$.

[155]Yates, *Rosicrucian Enlightenment*, pp. 103-117, esp. pp. 115-116.

[156]Gabriel Naudé, *Instruction à la France sur la verité de l'histoire des frères de la Roze-Croix* (Paris: François Iulliot, 1623), p. 27.

[157]*Ibid.*, sig. c̃^{v}, p. 105.

[158]This episode is discussed in a lengthy note by the editor in the *Correspondance du P. Marin Mersenne. Religieux Minime*, ed. Cornelis de Waard, Vol. I: 1617-1627 (Paris: P.U.F., 1945), pp. 167-168. The theses are described by Mersenne in *La verite des sciences. Contre les septiques ou Pyrrhoniens* (Paris: Toussainct Du Bray, 1625; reprint Stuttgart/Bad Cannstatt: Friedrich Fromman Verlag [Günther Holzboog], 1969), pp. 79-80.

PLATE XXIII David de Planis Campy (1589–c. 1644). Portrait from *L'hydre morbifique exterminee par l'Hercule chymique* ... (Paris: Herue du Mesnil, 1628), sig. e ii^v.

elements, and the atomic nature of matter. The actual meeting was dispersed on the order of the Parlement of Paris, and a principal participant in the discussions, Estienne de Clave, was arrested. By September 6 the doctors of the Sorbonne had officially condemned the theses, and before the end of the year Jean-Baptiste Morin had published his *Réfutation des thèses*.

Clearly concerned by what appeared to be an ever-increasing interest in Rosicrucian and alchemical mysticism, Marin Mersenne (1588-1648) examined these subjects both in his *Quaestiones celeberrimae in Genesim* (1623) and his *La verite des sciences* (1625). The second volume is primarily devoted to a description of mathematics, a subject which Mersenne considered a necessity for an understanding of Holy Scripture no less than for the works of Plato and Aristotle. Without mathematics it would not be possible to interpret properly any branch of knowledge; indeed, it was as essential for the investigation of medicine as it was for astronomy.[159] Even Paracelsus (in the *De ente astrorum*) accepted this, continued Mersenne, and surely chemists required a mathematical understanding of the proportion of saltpeter in excrement, of sal ammoniac in bone, of salt, sulfur, and mercury in all things that exist.[160] These subjects for Mersenne were all basically mathematical.

And yet, although Mersenne clearly sought a mathematical basis for the understanding of nature, he viewed with alarm the approach taken by some contemporary alchemists. In the first book of the *La verite* Mersenne discussed these views in a dialogue between an alchemist, a skeptic, and a "Christian Philosopher." For the alchemist no science was more certain than his own because alchemy teaches through experience—"car vous voyez que tout ce qui y entre, est vray, palpable, & reel."[161] While he was ready to agree that alchemical charlatans existed, the emphasis that alchemy placed upon observation had resulted in a new and significant system of elements and principles. Furthermore, the works of Aristotle, admittedly filled with dangerous theological views, had been replaced by an approach more sound.[162]

[159] Mersenne, *La verite des sciences,* pp. 233-235, 242-243, 525.
[160] *Ibid.,* p. 566.
[161] *Ibid.,* p. 41.
[162] *Ibid.,* p. 97.

Mersenne's rejection of the views of the alchemist was mild but firm. For the Christian Philosopher the recent condemnation by the Sorbonne was a just one, because these doctors had rightly questioned the theological implications of the alchemical theses.[163] The Christian Philosopher looked upon the alchemical revival of atomism with special alarm.[164] As for the alchemists' much-vaunted "observationally" based system of elements and principles, Mersenne replied that the Paracelsian principles may yet be decomposed artificially.[165] Should this happen, these principles need no longer be considered elementary.

And yet, Mersenne continued, if alchemy may be faulted specifically in some areas, it must not be rejected *in toto*; rather, some method of control must be sought to avoid the pitfalls it has too often fallen into. As an answer Mersenne suggested the establishment of alchemical academies in each kingdom which would take as their goal the improvement of the health of mankind. These academies might police the field, not only by punishing charlatans, but also by actively engaging in the reform of the science. Allegorical and enigmatic terms such as "Christiano-cabalistique, Divino magique, Physico-chymique" and the like would be discarded while in their place would be substituted a clear terminology based upon chemical operations performed in the laboratory.[166]

For Mersenne a reformed alchemy would steer clear of religious, philosophical, and theological questions, which were of absolutely no concern.[167] There were some for whom the subject seemed to serve as a counter-church and who argued that the most ancient theology, magic, and pagan fables were best explained through this science. Many, indeed, held to the chemical interpretation of the Creation. These dreams and speculations must be rejected immediately if the subject was to gain the approval of the Catholic church. With these views Mersenne's pliant alchemist heartily concurred.[168]

---

[163] *Ibid.*, pp. 82-83.
[164] *Ibid.*, p. 81.
[165] *Ibid.*, p. 56.
[166] *Ibid.*, pp. 105, 106.
[167] *Ibid.*, p. 107.
[168] *Ibid.*, pp. 118, 116, 119.

But whom did Mersenne have in mind in his critique? In the *La verite* he referred to the recent *Traittez de l'harmonie et constitution generalle du vray sel, secret des philosophes, et de l'esprit universelle du monde . . . oeuvre . . . traittant de la cognoissance de la vraye medecine chimique* (1621) by Sieur de Nuisement.[169] Still, Mersenne's strictures pertained no less properly to the works of Robert Fludd. Indeed, in the *Quaestiones celeberrimae in Genesim* it was not Nuisement, but rather Francesco Zorzi, Jacques Gaffarel, Fludd, and many others who were singled out as leading proponents of this mystical approach to nature.[170] Fludd was attacked with true venom for his defense of the Rosicrucians, his belief in chiromancy (palmistry), and his interpretation of Genesis.[171] His reliance on the Cabala as a key to both medicine and astrology was similarly condemned by Mersenne, who specifically attacked the view that the tetragram of Yahweh signifies the four elements.[172] Mersenne charged Fludd with being pantheistic, and, in a fit of anger, he even termed the man a "cacomagus, foetidae et horrendae magiae doctor et propagator, haereticomagus, brevibus submergendum fluctibus aeternis."[173]

The intemperance of Mersenne's attack deeply wounded Fludd, who replied in two works, the *Sophiae cum moria certamen* (1629; "Epistola" dated Oxford, July 11, 1626), and, under the name "Joachim Frizius," the *Summum bonum* (1629). Both of these are replies to the *Quaestiones celeberrimae in Genesim* of 1623. In the *Sophiae*, a folio volume of 118 pages, Fludd defended his macrocosmic physics. Here he insisted that his views on the harmony of the two worlds were sound, and he elaborated on the relationship between the macrocosm and the microcosm.[174] In a long passage he discussed once again the spirit of life, its separation from the common air in the microcosmic chemical laboratory (left ventri-

---

[169]*Ibid.*, p. 116.

[170]See Lenoble, *Mersenne*, pp. 27-29 and *passim.*; Yates, *Giordano Bruno*, pp. 432-438; Lynn Thorndike, *A History of Magic and Experimental Science*, Vol. VII (New York/London: Columbia University Press, 1964), pp. 426-464.

[171]Mersenne, *Correspondance*, Vol. I, pp. 42-43, 61-62.

[172]Lenoble, *Mersenne*, p. 106.

[173]Mersenne, *Correspondance*, Vol. I, pp. 38, 62.

[174]Robert Fludd, *Sophiae cum moria certamen, in quo, lapis lydius a falso structore, fr. Marino Mersenno, monacho, reprobatus, celeberrima voluminis sui Babylonici (in Genesin) figmenta accurate examinat* (n.p. [Frankfurt], 1629), p. 3.

cle), and its dispersal throughout the body by the arterial system which, as the container of this spiritual substance, can never be connected with the venous blood, a substance destined for a less noble purpose.[175]

Fludd went on to defend his use of the Cabala, his views on the universal musical harmonies, and his treatment of chiromancy, which had been badly maligned by Mersenne.[176] Surely, his detractor had not understood his distinction between impious and true magic; if he had, Mersenne surely could not have charged that he was a "cacomagus." Fludd repeated once more that his opinions were not innovations and were far removed from atheism. And, although Mersenne had rejected his views in many places by name, he had attacked him in many additional places without properly citing his works.[177] Convinced of the truth of his own views and appalled by the acidity of Mersenne's attack, Fludd could only conclude that it was the nature of his opponent to be violent, indeed, insane.[178]

In the *Summum bonum* one "Joachim Frizius" discussed the *Quaestiones celeberrimae in Genesim* once again.[179] In addition to a defense of the Rosicrucian cause, this work contains lengthy sections contrasting the proper and the improper use of magic, Cabala, and alchemy. It is the last of these that is of most interest to us. To be sure, "Frizius" asserts, there are those who seek to transmute the base metals into gold and others who continually experiment with metals and salts in an unending search for new medicines. However, these practical alchemists may be clearly distinguished from those who study the true science which has as its goal the establishment of the entire chemical philosophy. This is to be accomplished according to basic truths by means of investigations and inquiries which may not always be apparent to the eye. Once again Mersenne was charged with not understanding the

[175]*Ibid.*, pp. 62-63.
[176]*Ibid.*, p. 3.
[177]*Ibid.*, p. 11.
[178]*Ibid.*, p. 109.
[179]The identity of Frizius has not been established. It would appear from the texts that Mersenne and Gassendi considered the author to be Fludd, but Yates (*Rosicrucian Enlghtenment*, p. 102) considers him most likely to have been a collaborator. In any case, the position of Frizius is identical with that of Fludd and was considered to be so at the time.

The Synthesis of Robert Fludd

subject that he presumed to criticize: the French monk had discussed only the work and goals of those who followed a spurious alchemy, while totally ignoring those who sought to explain the mysteries of man and the universe.[180]

It is clear that Fludd's understanding of an *alchemia vera* was precisely what Mersenne sought to strip away from alchemy. Above all, Fludd was disturbed by Mersenne's warning that alchemists should disassociate themselves from religious and theological matters. What of Thomas Aquinas, Roger Bacon, George Ripley, and Raymond Lull, he asked? These *were* men of religion while at the same time they were true alchemists. Must we call them fools as Mersenne would wish us to do?[181] And further, how can it be impious for the alchemist to concern himself with theology when chemists and theologians have a common subject to investigate? Fludd went further to equate mystical and occult chemistry with a part of practical theology, although he added, "I do not understand by this moral, but rather occult and mystical [theology]."[182] This subject is one that seeks to comprehend the Creation and the spirit of life with its relationship to man and the universe. Nature and supernature are clearly united, and chemistry serves as a key to the understanding of both.

Although Fludd's replies to Mersenne were not officially published until 1629, it appears that advance copies were available late in 1628, and it is evident that Mersenne was becoming ever more alarmed as he continued to read the new works of the English physician. Repetitive though they were, Fludd's publications throughout the 1620s continued to interpret the universe in terms of a mystical alchemy. Some time prior to December 2, 1628, Mersenne sent a collection of Fludd's works to his friend Pierre Gassendi, with an appeal to aid him in his attack on the alchemists.[183] The works he sent apparently included the *Sophiae cum moria certamen* and the *Summum bonum* and no doubt the newly published *De motu cordis* of Fludd's colleague William Harvey.

---

[180] Joachim Frizius (Fludd?), *Summum bonum, quod est verum [(magiae, cabalae, alchymiae: verae); fratrum Roseae Crucis verorum] subjectum* . . . (Frankfurt?, 1629), pp. 24, 25, 26.

[181] *Ibid.*, p. 28.

[182] *Ibid.*, p. 31.

[183] Mersenne, *Correspondance*, Vol. II (1945), p. 181.

Intrigued by the request of his friend, Gassendi turned immediately to his task, and the resultant work, the *Epistolica exercitatio in qua principia philosophiae Roberti Fluddi, medici, retegunter, et ad recentes illius libros R.P.F. Mersennum . . . respondetur* (Paris, 1630), was completed in manuscript form on February 4, 1629.[184]

The *Epistolica* is divided into three parts: an introductory summary of Fludd's philosophy and discussions of the *Sophiae cum moria certamen* and the *Summum bonum*. The work closes with a short letter by the Parisian philosopher Franciscus Lanovius, in support of Mersenne. Gassendi began with an examination of Fludd's elementary principles of light and darkness.[185] From here he turned to his description of the Creation, where he paused to attack the English physician's alchemical experiment on wheat as an improper application of chemistry to the interpretation of Scripture.[186] It is of interest that he accepted Fludd as an atomist, but in his description of Fludd's rejection of Copernicus and Gilbert he could only conclude that "he understands another nonvolatile earth and central sun than that commonly understood by us."[187] When Gassendi discussed the distinction made by Fludd between the true and the false alchemy in the *Summum bonum* he fully supported the views of Mersenne. He complained of the authorities cited by Fludd—Paracelsus, Pico, Khunrath, and the Cosmopolite—all chemists who wrote not only in an obscure style, but who also carried their mysteries and arcana into the sphere of religion. And, in an impassioned passage, Gassendi complained of an interpretation that would make "alchemy the sole religion, the alchemist the sole religious person, and the tyrocinium of alchemy the sole catechism of the faith."[188]

[184]In addition to the works by Yates, Thorndike, and Lenoble cited above, see also Luca Cafiero, "Robert Fludd e la Polemica con Gassendi," *Revista Critica di Storia della Filosofia*, 19 (1964), 367-410; 20 (1965), 3-15, and T. Gregory, *Scetticismo ed empirismo i studio su Gassendi* (Bari: Editori Laterza, 1961), pp. 50-62.

[185]Pierre Gassendi, *Examen philosophiae Roberti Fluddi* in *Opera omnia*, 6 vols. (Florence, 1727), Vol. III, pp. 217-267 (217). For the date of completion of this work see p. 266.

[186]*Ibid.*, p. 218.

[187]*Ibid.* pp. 219, 224: "cùm aliam Terram non volatilem, aliúmque Solem centralem, quàm vulgò à nobis intellectum intelligat."

[188]*Ibid.*, pp. 257, 259: "Quae tanta haec denique potestesse omnium impietatis, vt non stet per ipsos, quin Alchymia sola sit Religio, Alchymista solus Religiosus,

Gassendi's *Epistolica* has an added interest because it contains the first significant published abstract of Harvey's system of the circulation of the blood. Gassendi introduced this subject in his commentary on Fludd's discussion of the spirit of the blood in the *Sophiae cum moria certamen*, and he made use of the opportunity to contrast the work of Harvey with the mystical symbolism of Fludd. In sending the works of the two authors together, it had evidently been Mersenne's hope that Gassendi would set forth just such a comparison. Fludd's mystical and chemical schemes of the circulation had been outlined in his *Anatomiae amphitheatrum* (1623), where the circulation of the aetherial spirit in the microcosm (through the arterial system) was compared with the macrocosmic circulation of the sun about the earth.[189] We would hardly expect this mystical scheme to have appealed to the experimentalist Gassendi. We find, however, that although he preferred Harvey's work to that of Fludd, he rejected both schemes. Before the *Epistolica* had appeared in print, Gassendi wrote to his friend Peiresc (August 28, 1629) to give him an account of his views on the circulation. From this letter it becomes apparent that his forthcoming refutation of Fludd—insofar as the circulation of the blood is concerned—would be directed not at Fludd but at Harvey, with whom Gassendi disagreed because of his own firm belief in the intraventricular pores of the septum.[190]

Gassendi's promise to Peiresc was fulfilled in the published *Epistolica*, in which he summarized Fludd's views on the blood, centering his discussion on the question of where the arterial blood

---

tyrocinium Alchymiae solum Catechismus Fidei?"

[189] Fludd, *Anatomiae amphitheatrum*, p. 266.

[190] Nicolas-Claude Fabri de Peiresc, *Lettres de Peiresc publieés par Philippe Tamizey de Larroque . . . Tome quatrième. Lettres de Peiresc à Borilly, à Bouchard et à Gassendi. Lettres de Gassendi à Peiresc 1626-1637* (Paris, 1893), p. 208. "Le livre dont Mr. Valvois vous à parlé, Mr. du Puy en à un exemplaire pour vous envoyer. Je l'avois desja veu avant que partir pour l'Allemagne et en avois dit mon sentiment en ma lettre à P. Mersenne qui enfin se verra peut estre bientost imprimée. Son opinion de la continuelle circulation du sang par les arteres et veines est fort vraysemblable et establie mais ce que je trouve à dire en son fait est qu'il s'imagine que le sang ne sauroit passer du ventricule droit du coeur au gauche par le (septum), là où il me souvient que le sieur Payen nous à fait voir autrefois qu'il y à non seulement des pores, mais des canaux très ouverts . . . qui rendent cette doctrine evident. Vous verrez jour ce que j'en ay dit."

originated. Citing the *Anatomiae amphitheatrum*, Gassendi stated that Fludd believed that

> ... the subtle blood which is in the arteries is not seized from the veins either in that manner in which it is commonly taught, that is, by an inflow in a branch of the vena cava into the right ventricle of the heart out of which the blood, having crossed over into the left side rushes forth then into the arteries,—or in that manner in which extremes can be fashioned together, that is, the venous capillaries with the extreme arteries, and then out of the veins the blood is instilled into the arteries. For he teaches that this (the arterial) blood is simply from that aetherial spirit which is attracted through inspiration . . . and he does not admit anything from the venous blood "for this reason," he says, "if the heavy and elementary blood, destined for procreating the solid members, mixes itself with the spirits of life, confusion would follow."[191]

But, continued Gassendi, "since the common method of explanation did not please Fludd, he should have listened to Harvey, his countryman."[192] Harvey's account was based on experiment rather than analogies of the macrocosm and microcosm. Further, his insistence on the connection of the veins and the arteries through capillaries required a basic identity between the two types of blood. The Galenical system, as understood by Gassendi, required this because of the percolation of the venous blood through the pores of the septum. Gassendi felt then that the real problem was to decide between the Galenical and the Harveian positions on this question, and he went on to summarize Harvey's work in addressing Mersenne:

---

[191]Pierre Gassendi, *Epistolica exercitatio in qua principia philosophiae Roberti Fluddi, medici, reteguntur, et ad recentes illius libros adversus R.P.F. Marinum Mersennum . . . respondetur* (Paris, 1630), pp. 128-129. ". . . sanguinem illum subtilem, qui est in arteriis, non hauriri ex venis, seu eo modo, quo vulgo docetur; scilicet ramo venae cavae influente in dextrum ventriculum cordis, ex quo sanguis trajectus in laevum, in arterias deinde prorumpat; seu eo modo, quo fingi posset coire extremas, seu capillares venas cum extremis arteriis; sicque ex venis sanguinem in arterias instillari. Docet proinde hunc sanguinem esse dumtaxat ex ducto illo per inspirationem aethereo spiritu; . . . Neque vero quidpiam hic admittit ex venoso sanguine; quoniam 'hac ratione,' inquit 'sanguis grossus elementaris, & ad membra solida procreanda destinatus se permisceret cum spiritu vitae, atque sequeretur in opere humano confusio."

[192]*Ibid.*, p. 132. "Denique nisi Fluddo communis explicandi ratio placeret, audiendus forte fuerat Harveus ejus conterraneus."

If you have not seen it [Harvey's book], the argument is that the blood out of the trunk of the vena cava goes into the right ventricle of the heart through a branch, and then being sprinkled in the lungs, it is collected through the arterial vein and is then thrown together through the venous artery into the left ventricle, out of which, after having been sent into the arteries, at length it is returned into the veins through insensible openings of the arteries and junctions with the veins; and in the same way again out of the veins through the lungs and the heart into the arteries in a perpetual circuit.[193]

However, although Gassendi esteemed the work of the "learned anatomist," he had witnessed dissections at Aix at which an industrious surgeon by the name of Payanus had demonstrated the existence of the interventricular pores. "Therefore there definitely are these passages . . . and since they are really there, they ought not to be useless, and there is a purpose in readiness, the percolation of the blood out of the right vessel into the left; and I might argue that the arterial blood is derived in this manner."[194] Thus, in a detailed criticism of Fludd's philosophy, Fludd's own views on the blood flow are summarily brushed aside, and Gassendi gives instead a refutation of Harvey's thesis, which more nearly coincides with his own concept of the relation of the venous and arterial blood.

Yet, although this was a pointed attack on Harvey, it was not Harvey who answered. Prior to the appearance of Gassendi's *Epistolica*, Fludd had completed his *Pulsus* (on October 19, 1629), in which he affirmed that his view of the microcosmic circulation of the blood

. . . exactly seems to confirm the feeling and opinion of the learned William Harvey, a most skillful doctor of medicine, most clear in the art of anatomy, and yet highly versed in the mysterious profundities of

---

[193]*Ibid.*, pp. 132 ff. "Nisi videris, argumentum est, quod sanguis ex venae cavae trunco per ramum transversum in dextrum usque ventriculum cordis, & inde in pulmones per arteriosam venam respersus colligatur, atque coniiciatur per venosam arteriam in sinistrum ventriculum, ex quo in arterias immisus, per insensibilia tandem arteriarum, veniarumque juncta oscula regrediatur in venas; similique rursum modo ex venis per pulmones, & cor in arterias, circuitione perpetuā."

[194]*Ibid.*, pp. 133-136. "Hi igitur meatus sunt . . . ut cum illi reverā sint, nec frustra esse debeant, & aliunde causa sit in promptu, ipsa sanguinis e dextro sinu in laevum percolatio; sanguinem illum arterialem hac derivari arguerem."

philosophy, a man who is my esteemed compatriot and the most faithful of the College: about which he instructed the world advisedly and prudently in his little book *Exercitatio de cordis sanguinis in animalibus motu*, and he declared remarkably well with reasons produced from the treasure of philosophy as well as with demonstrations for the eye that the motion of the blood is circular.[195]

It is likely that the *Pulsus* was not published until the following year,[196] but from this quotation it is evident that for Fludd an attack on Harvey could be taken as an attack on his own views.

By 1631 Fludd had written his reply to Gassendi's criticism, a detailed point-by-point discussion which appeared as the *Clavis philosophiae et alchymiae Fluddanae* (Frankfurt, 1633).[197] Once again he defended the subjects of concern to him, but it is of special interest to follow his treatment of the blood flow. On this point he maintained his earlier position that the arterial blood could not be formed from the venous blood, but he specifically turned to Gassendi's attack on Harvey. From the standpoint of modern science, Fludd never appeared to better advantage. Surely, he wrote, Gassendi must be wrong about the interventricular pores. Either Payanus had forced the opening in the septum with his probe, or perhaps in a few abnormal individuals such openings occur naturally;

> ... but from this single case it should not be considered to be universal. For thus in one man such a single vein may be come upon, while in another there may be no spleen and with many others there may be

---

[195]Robert Fludd, *Pulsus seu nova arcana pulsuum historia, e sacro fonte radicaliter extracta, nec non medicorum ethnicorum dictis & authoritate comprobata* (Frankfurt?, n.d.). This quotation will be found on p. 11: "Hoc exactè illam viri grauissimi Gulielmi Haruei, Medicinae Doctoris peritissimi, arte Anatomica, quàm clarissimi, necnon in profundis Philosophiae mysteriis versatissimi, compatriota mihi charissimi, & collegae fidelissimi; sententiam atq; opinionem confirmare videtur; qua, idq; consultè & prudenter satis mundum in *libello quodam suo, cui titulus est: Exercitatio Anatomica, De cordis sanguinisq; in animalibus motu;* instruit, insigniterq; cùm rationibus à Philosophiae arca depromptis, tum multiformi demonstratione oculari declarat motum ipsius sanguinis esse circularem."

[196]The date of the completion of the *Pulsus*—October 19, 1629—is given on p. 93, but a large folding plate, generally missing, is dated 1630.

[197]We know that Fludd had read Gassendi's *Epistolica* by 1631. In *Doctor Fludds Answer* (1631) he compares Gassendi's scholarly refutation of his works with Foster's "ill-mannered" and "impudent" reply, and he mentions in several places his own reply to Gassendi (the *Clavis philosophiae et alchymiae Fluddanae*).

strange and unexpected deviations from the ordinary human composition.... Certainly one or even a second example does not prove the argument—neither does one swallow prove the advent of summer.

In contrast to Gassendi's one example, Fludd had personally observed Harvey search repeatedly for these pores, but, he asserted, "not in any one of the many cadavers that he examined did he find such a septum; and neither I nor any others who with most acute and almost lynx-like eyes saw this when we examined the septum of the heart." Thus, the mystical alchemist informed the mechanist Gassendi that "we know and speak as experts" when we "assert with confidence that the septum of the heart is not ordained by nature to that purpose called for by Gassendi."[198]

Although he had spoken of an exact confirmation of his views,

---

[198] Robert Fludd, *Clavis philosophiae et alchymiae Fluddanae, sive Roberti Fluddi Armigeri, ut medicinae doctoris, ad epistolicam Petri Gassendi theologi exercitatem responsum* (Frankfurt: Wilhelm Fitzer, 1633), pp. 33 ff. "Inprimis sciendum quod autor hic noster, ad statuminandum suam caussam, ad relationem experimenti cuiusdam Aquis-Sextiis peracti recurrat, vbi (ait) se Anatomica inquisitione, mediante spathula seu ferro quodam obtuso, porosos in septo cordis meatus inuenisse, quos mille in septo ianuas, licet occultas vocat; vnde concludit cum Medicis & chirurgis, qui dissectioni tali adfuerunt, esse transmissionem sanguinis à dextro sinu in laevum licet insensibilem. Idque aduersusme ... Memini ego Chirurgum quendam iactitasse, se probo aut spathula, irregulari quadam dilatione, admirabilem illum plexum retiformem ab implicatione venae & arteriae factum, qui descendit à vasis seminariis praeparantibus ad testiculos, penetrasse: quod quidem propter contexturae subtilitatem impossibile est, sine meatuum occultorum violatione; nisi raro in aliquo particulari hoc contingat ... Praeterea ista inquisitio a pluribus collegarum meorum, & praecipue à D. Harueo Anatomico expertissimo, saepius instituta est magna cum diligentia, quatenus ipse ad suam in sanguinis circulationis causam, cum fatigatione huius rei experimentum fecit; sed ne in vnico quidem, ex pluribus cadaueribus inuenit ipse tale quidpiam; nec ego, nec alii, qui oculis acutissimis & quasi lynceis cordis septum sumus scrutati. Quare concludendum est, quod, licet in vno cadauere vel alio (si ita esset sine spathula violatione) res talis Aquis-Sextiis comparauerit, hoc tamen particulare non inferat vniversale. Sic enim in vno homine, vnus inuenitur Ren tantum; in alio nullus Lien, atque per multa alia ab ordinaria compositione humana aliena atque incosueta. At anne ob eam caussam concludendum, nimirum ob aliquos partium defectus vel augmenta in vno homine, necessario omnes debere ita se habere? Vnum certè vel alterum exemplum, non probat argumentum, nec hirundo vnica probat aestatis aduentum. Quare experimento a Gassendo prolato, partim haud fidem habebimus, & partim si pro concesso illi detur, dicimus quod vnum particulare non demonstret generale. At in generali agnoscimus, & experti loquimur, istam Gassendi assertionem esse falsissimam:

it is hardly likely that Fludd could ever have accepted the Harveian thesis in its entirety. While he could speak in terms of the circulation of the spirit of the blood, he certainly was opposed to a mixing of the venous and the arterial systems. The venous blood served for the nourishment of the flesh and in no fashion could it be mixed with the spiritual content of the arterial blood.

It is significant that Harvey's work was taken up for discussion by the Mersenne circle in 1628—and in the Fluddean context of a universal mystical alchemy. It is of further importance that Fludd himself felt that Harvey's work confirmed his own mystical views of the circulation and that when Gassendi attacked him, he answered not only the arguments directly aimed at him, but also those on the constitution of the septum which had been directed primarily at his friend and colleague Harvey. This was surely the first significant controversy over the circulation of the blood, and there can be no doubt that in the mind of Mersenne the work of the two English physicians could be considered together—a position Fludd would have heartily concurred in. It is with Gassendi that we see a recognition of the basic difference between the works of the two men, but here the issue is clouded by Gassendi's own misjudgment and Fludd's remarkable reply to him.

The decade between the publication of the *Quaestiones celeberrimae in Genesim* and his own final reply to Gassendi had surely been a trying one for Fludd. Yet, if he considered Gassendi to have produced a scholarly examination of his system of the world, Fludd clearly felt deeply distressed by the works produced by his other detractors. When William Foster, the parson of Hedgly in Buckinghamshire, criticized Fludd's defense of the weapon-salve in 1631, Fludd complained bitterly of those who had attacked him:

> And who is *Mersennus*? A rayling Satyricall Babler, not able to make a reply in his owne defence, and therefore being put to a *Non plus*, hee went like a second *Job* in his greatest vexation to aske Counsell of the learnedst Doctors in *Paris*: And at last for all that, he fearing his cause, and finding himselfe insufficient, procured by much Intreatie his

---

ac proinde asseuerare haud erubescimus, quod septum cordis, non ad illud officium à Gassendo nominatum à Natura ordinetur. Nam si hoc esset pro statuto, sequeretur, quod nulla esset inter sanguinem Arterialem & Venalem differentia."

friend *Peter Gassendus* to help him, and called another of his friends unto his assistance, namely one Doctor *Lanovius*, a seminarie Priest, as immorall as himselfe, and one that professeth in his Iudiciary Letter much, but performeth little. And in good faith, I may boldly say, that for three roaring, bragging, and fresh-water Pseudophilosophers, I cannot paralell any in Europe, that are so like of a condition as are *Mersennus, Lanovius* and *Foster* . . . .[199]

The successive debates between Fludd on the one hand and Kepler, Mersenne, and Gassendi on the other became the subject of intense interest on the Continent. Fludd referred to letters of encouragement he had received from correspondents in Poland, Swabia, Prussia, Germany, Transylvania, France, and Italy,[200] while the Mersenne *Correspondance* attests to the number of those who supported him and Gassendi. Some thirty-five surviving letters between 1629 and 1633 refer to this debate. Here we see Mersenne writing at length to Nicolas de Baugy at The Hague (April 26, 1630) on the manifold errors in the works of Fludd,[201] and we find Gassendi busily sending copies of his *Epistolica* to his friends while they in turn write back the expected words of encouragement and praise. Authors representing views as divergent as those of van Helmont, Athanasius Kircher, and Galileo are included in this wealth of correspondence.[202]

When Joseph Gaultier at Aix commented (May 22, 1631) that it had been necessary to answer Fludd's "fumées chimiques, theologiques et physiques," Gassendi replied (July 9, 1631) that he had heard that his English opponent was now composing a reply to him.[203] However, it was not until the following March that Gabriel Naudé wrote to say that the *Clavis* was listed in the new Frankfurt book catalogue.[204] Finally, early in the new year, Mersenne received the work, and in a letter to Gassendi he suggested that this "key" now required its own *clavis*.[205] After he had

[199]Fludd, *Doctor Fludds Answer*, p. 18.
[200]*Ibid.*, p. 24.
[201]Mersenne, *Correspondance*, Vol. II, pp. 438-447.
[202]*Ibid.*, (Dec. 19, 1630), pp. 582-589; (Sept. 23, 1634), Vol. IV (1955), p. 364; (Jan. 15, 1633), Vol. III (2nd ed., 1969), p. 369.
[203]*Ibid.*, Vol. III, pp. 162, 174.
[204]*Ibid.* (Mar. 6, 1632), p. 266.
[205]*Ibid.* (Jan. 5, 1633), pp. 354 ff.

received a copy of the work that had been sent on by Luillier, Gassendi was relieved to find that the *Clavis* treated him less harshly than he had expected.[206] Soon we find in the *Correspondance* a number of queries regarding Fludd's new book: but on February 25 Gassendi's friend Peiresc noted that the volume was unworthy of reply, and he cautioned Gassendi not to involve himself further, since he surely had better ways to spend his time.[207]

Although Gassendi in fact played no further part in this debate, a reply to the *Clavis* was written by Jean Durelle in 1636.[208] The published work appeared in a very small edition, and it is doubtful whether it was ever seen by Fludd, who died in London on September 8, 1637. Mersenne, still unaware of Fludd's death as late as February 1640,[209] was quite distressed that Durelle's book was known to so few scholars. Accordingly, he sent a copy to Theodore Haack in London with the hope that it might be reprinted there. He suggested that if this were done, he might add a short refutation of Fludd's discussion of the weapon-salve in the final book of the *Philosophia Moysaica*[210] (published posthumously at Gouda in 1638). In a series of letters to Haack and John Pell he pressed both for their judgment of the new book by Fludd and the critique by Durelle.[211] Only in November 1640 did he learn of the unfavorable opinion of Durelle's volume taken by the London group.[212] They clearly had no plan to reprint it. Later (1642), when Mersenne questioned Durelle himself about the possibility of a

[206]*Ibid.*, p. 371.
[207]*Ibid.*, p. 379.
[208]P. Jean Durelle, *Effigies contracta Roberti Flud Medici Angli, cum naevis, appendice et relectione. In lucem producente Eusebio a S. Justo theologo Segusiano. Sanctitas Dei defensa. Agnito vultus eorum respondebit Isa. 3. Ad clariss. virum Iacobum Dazam* (Paris: Apud Guillelmum Baudry, 1636). I have not seen this book.
[209]See Mersenne's repeated queries to Theodore Haack on this point in the *Correspondance* (Jan. 15 and 20, Feb. 12, 1640), Vol. IX (1965), pp. 16-17, 45-46, 106.
[210]*Correspondance* (Mar. 20, 1640), Vol. XI (1970), p. 408.
[211]*Correspondance* (Nov. 24, 1639), Vol. VIII (1963), pp. 636-660; (Jan. 15 and 20, Feb. 12, Mar. 7, 1640), Vol. IX, pp. 16-17, 45-46, 106, 182; (Mar. 20, Nov. 16, 1640), Vol. XI, pp. 408, 419.
[212]*Correspondance* (Nov. 27/Dec. 7, 1640), Vol. X (1967), pp. 304-306 (text of letter from Joachim Hübner at London to J. A. Comenius at Leszno reporting on the unfavorable nature of their report in London on the book by Durelle). See also the letter from Mersenne to Haack (Nov. 22, 1640), Vol. XI, pp. 425-428) in which he acknowledges news of this report.

new edition, even the author had to admit that this was not likely.[213]

Fludd and the Weapon-Salve Controversy (1631-1638)

Although Robert Fludd had no need to defend himself against the members of the Mersenne circle after the publication of his *Clavis*, his remaining years were not free from controversy. William Foster had published his short tract, the *Hoplocrisma-spongus: or, A Sponge to Wipe Away the Weapon-Salve*, early in 1631. For Foster the theological implications of the cure were of overriding significance, and he pointed out that one reason he wrote was that the Jesuit Jean Roberti, in attacking the weapon-salve, had called its practitioners "Magi-Calvinists." Indeed, Foster agreed, Roberti was correct in calling the cure magical, but to suggest that it was generally accepted by Protestants had to be denied.[214] Foster wrote also to protect the poor people of England from such dangers as the weapon-salve, which might well imply making a compact with

[213]Vol. XI (Oct. 5, 1642), p. 285. Evidence of the continued interest in the "chemical problem" in France may be found in the *Recueil général des questions traictées ès conférences du Bureau d'Adresse* edited by E. Renaudot in three volumes and published in Paris between 1660 and 1666. These volumes are based upon public discussions organized by the elder Renaudot from 1633 to 1644. I have consulted the two volumes of the set that were translated into English as *A General Collection of Discourses of the Virtuosi of France, upon Questions of all Sorts of Philosophy, and other Natural Knowledg. Made in the Assembly of the Beaux Esprits at Paris, by the most Ingenious Persons of that Nation,* trans. G. Havers and (Vol. II) J. Davies (London: Thomas Dring and John Starkey, 1664, 1665). Although none of the persons present are named, the topics give a good impression of the subjects of interest to them. Typical "chemical" subjects are the following—Vol. I: "Of the First Matter," p. 18; "Of Fire," p. 31; "Of the Air," p. 38; "Of Water," p. 44; "Of the Earth," p. 51; "Sympathy and Antipathy," p. 191; "Of the Philosophers Stone," p. 256; "How Minerals Grow," p. 262; "Of Quintessence," p. 361; "Of the Magnetical Cure of Disease," p. 408; Vol. II: "Whether it be good to use Chymical Remedies?" p. 37; "Of the Generation of Metals," p. 156; "Whether there be an Elementary Fire, other than the Sun," p. 159; "Of the Fraternity of the Rosie-Cross," p. 321; "What Paracelsus meant by the Book M," p. 326; "Of the Art of Raimund Lully," p. 329; "Of Atoms," p. 454; "Of Natural Magick," p. 465.

[214]William Foster, *Hoplocrisma-spongus: or, A sponge to wipe away the weapon-salve* (London: Thomas Cotes, 1631), sig. A2. Foster's tract was registered with the

the devil. "Till of late it was little known amongst us, and therefore little or not at all inquired into," but the salve was suddenly becoming more popular as a cure, and as yet there had been nothing in English which adequately described the dangers involved.[215]

Foster thoroughly abhorred the neo-Platonists and Hermeticists, and he freely called on Aristotle in his proof that the weapon-salve was diabolical in nature:

> ... it is not of divine Institution, because it is no where registred in Scripture ... it workes not naturally, because it workes after a different manner from all naturall agents. For 'tis a rule amongst both Divines and Philosophers that; *Nullum agens agit in distans.* Whatsoever workes naturally, workes either by corporall or virtuall contact. But this workes by neither, therefore it workes not naturally.... *Paracelsus* saieth, if the weapon be annoynted, the wounded partie may be cured, though 20 miles absent.[216]

Neo-Platonists held that there was a life force in all things in the universe. But for Foster only heretics could believe such nonsense. "When *Marsilius Ficinus* can perswade mee that the Starres have the senses of seeing and hearing, and do heare mens prayers; then *Paracelsus* shall perswade me that the Loadstone hath life sense and fantasie."[217] Yet, it was the starry emanations that were essential to the defenders of the weapon-salve. They argued that "this naturall Balsame by the influence of the Starres, causeth a sympathy betwixt the weapon and the wound: and so the application of the Medicine to the one, effects the cure upon the other."[218] For an Aristotelian such as Foster sympathetic action deriving from the heavens was impossible: "as if the smearing of a Weapon

---

Company of Stationers of London on Dec. 31, 1630. *A Transcript of the Registers of the Company of Stationers of London 1554-1640 A.D.* (London: Privately printed, May 1, 1877), Vol. IV, p. 212. William Foster (1591-1643) attended St. John's College, Oxford, and became chaplain to the Earl of Carnarvon. Other than his attack on the weapon-salve, his only other printed work is *The means to keepe sinne from reigning in our mortall body. A sermon preached at Paul's Cross, May 26, 1629* (London: John Haviland, 1629).

[215] Foster, *Hoplocrisma-spongus*, sig. A3^r.
[216] *Ibid.*, p. 5.
[217] *Ibid.*, p. 26.
[218] *Ibid.*, p. 20.

The Synthesis of Robert Fludd

here below, can call the Starres above, at any time when we will, to give an influence which they gave not before, nor had not given at all, had not the Weapon been smeared at all. O inchanted Salve!"[219]

How might one be certain that such a cure is diabolical, asked Foster. Certainly the substances employed—moss taken from the skull of a hanged thief, human blood still warm, and other similar items—were highly suspect, but even more could be learned from the defenders of the cure:

> The Author of this Salve, was Philippus Aureolus Bombastus Theophrastus Paracelus. Feare not Reader, I am not conjuring, they are onely the names of a Conjurer, the first inventer of this Magicall oyntment. . . . for mine owne part, to satisfie my selfe and my Readers, I will goe no farther than to the Tract wherein the Vnguent is described, and thare to the prescription next adjoyned, which is a Receipt to cure one decayed in nature, unable to performe due benevolence. Take an horse-shooe cast from a horse, let it be wrought into a trident Forke, impresse these and these Characters on it, put a staffe of such a length into the socket for the stale of it; let the Patient take this Forke and sticke it in the bottome of a River of such a depth, and let it remaine sticking there so long as is prescribed, and he shal be restored to his former manlike abilitie. If this be not Witchcraft, I know not what is! Now when Paracelsus being a Witch, and this experiment being placed amongst his Diabolicall and Magicall conclusions, it cannot choose but be Witchcraft, and come from the grand master of Witches, the Divell, if Paracelsus were (as most repute him) the Author and Founder of it.[220]

Having thus established his major point, Foster went on to accuse, in addition to Paracelsus, Crollius, Porta, Cardan, Burgravius, Goclenius, van Helmont (perhaps the earliest reference to him in England), Fludd, and even Sir Francis Bacon of promoting this devilish cure.[221] In spite of this blanket attack, he was very specifically aiming at Fludd, the only one of the lot who was promoting the weapon-salve in England at the time. Fludd's defense of the

---

[219] *Ibid.*, p. 21.

[220] *Ibid.*, pp. 13-15. This comparison of Paracelsus' name with a diabolical invocation may have been taken by Foster directly from John Donne's *Ignatius his Conclave* (1611). See John Donne, *Complete Poetry and Selected Prose* (London: Nonesuch, 1929), pp. 366-367.

[221] Foster *Hoplocrisma-spongus*, pp. 33-34.

salve was enough to make him suspected of witchcraft by Foster, who wrote that it was a wonder that "King James (of blessed memory) should suffer such a man to live and write in his Kingdome." Summarizing Fludd's view of the theoretical basis for the cure, Foster questioned the orthodoxy of the man and then went on to suggest that any wounds that might be cured by this method were actually healed by the simple means of keeping them constantly clean.[222]

So pleased was the country parson with his witty attack that he saw to it that two of the title pages of his tract were nailed to Fludd's door in the middle of the night. This, combined with his impertinent attitude and the rather serious charge of witchcraft, was enough to ensure a reply from the distinguished physician. For Fludd, Foster's attack was a bitter pill to swallow, and immediately after the completion of his reply to Gassendi's *Epistolica* he began his reply to Foster, which was published as *Doctor Fludds answer unto M. Foster, Or, The squeesing of Parson Fosters sponge, ordained by him for the wiping away of the weapon-salve* (1631).[223] Here he complained that while his works had been largely ignored in his own country, they were now attacked by someone of little significance in a most insulting fashion. Englishmen "in stead of encouraging me in my labours . . . doe prosecute me with malice & ill speeches, which some learned *Germans* hearing of, remember mee in their letters of this our Saviour CHRIST his speech: *Nemo est Propheta in sua patria.*"[224] Foster, noting that Mersenne had accused Fludd of magic, assumed that this was "one cause why he hath printed his bookes beyond the Seas. Our Universities, and our Reverend Bishops (God bee thanked) are more cautelous than to allow the Printing of Magical books here."[225] Not so, Fludd replied; the reason had been purely financial.

> I sent them beyond the Seas, because our home-borne Printers demanded of me five hundred pounds to Print the first volume, and to find the cuts in copper; but beyond the Seas it was printed at no cost

---

[222]*Ibid.*, pp. 38, 39.

[223]*Doctor Fludds Answer*, sig. A3^V. Fludd's reply to Foster was registered with the Company of Stationers on Mar. 25, 1631. *Transcript*, Vol. IV, p. 216.

[224]*Doctor Fludds Answer*, p. 24.

[225]Foster, *Hoplocrisma-spongus*, p. 37.

of mine, and that as I would wish: And I had 16 copies sent me over with 40 pounds in Gold, as an unexpected gratuitie for it.[226]

And, far from being his enemy, King James had been his "iust and kingly Patron all the dayes of his life."[227]

Fludd was convinced that the real answer for Foster's scurrilous attack must rest in his friendship with surgeons. Foster was the son of a barber-surgeon, and surely he had been encouraged by the "Chirurgicall faction" which

> ... decryes the Weapon-Salve, fearing that few wounded persons would trouble them for their cure, being that *frustra fit per plura, quod potest fieri per pauciora*: it would bee but vaine for a wounded man, to bee tormented by slashing, eating corrosives, incisions, and dolorous rentings of Chirurgions, besides great bargaines and compacts for the cures, and perchance also little attendance, when the immediate Act of God doth operate the cure *gratis*, gently without dolorous rents or grievous incision. ... [228]

As for Foster's accusation of witchcraft, Fludd went to great lengths to summarize, once again, his system of the world and the interdependence of all its parts. It was here that he recounted the story of the transplanted nose of the Italian nobleman and discussed the purely natural cause for such cures. Thus, while witches and wizards might use such knowledge to evil effect, it did not follow that this purely natural correspondence was evil in itself.

William Foster did not venture a reply to Fludd's *Answer*, but in the following years Fludd's statement was discussed and seldom in a favorable light. In 1631 James Hart, a conservative physician of Northampton who had already spoken out against the Paracelsians in two works on the examination of urine,[229] wrote his ΚΛΙΝΙΚΗ, which was licensed by the Royal College of Physicians. In this work Hart discussed the Fludd-Foster exchange, much to the disadvantage of the former. Never mentioning Fludd by name,

---

[226] *Doctor Fludds Answer*, pp. 21-22.
[227] *Ibid.*, p. 21.
[228] *Ibid.*, p. 127.
[229] P. Forestus, *The Arraignment of Urines* . . . *Epitomized and Translated* . . . *by I. Hart* (London, 1623). James Hart, *The Anatomie of Urines* (London: James Field for Robert Mylbourne, 1625).

he referred only to the "learned Physitian of the Colledge of London" who "satisfies me not, whatsoever he doth others."[230]

For Hart the whole hypothetical basis of the weapon-salve was unsatisfactory. Not only did the prescriptions for the unguent vary from one author to the next, but some like Crollius even omitted blood as an ingredient (this for Fludd was the very essence of the cure).[231] Above all, action at a distance remained a basic problem for this traditionalist. He could only conclude that even in the case of a magnet some form of material emanation must cause contact between the attracting bodies:

> ... although some agents worke at some distance, yet is there alwaies some proportion to be observed betwixt the agent and the patient; and although there be not alwaies a natural contact, yet there is commonly some effluxe, or emanation whereby the one toucheth the other. And this is the ordinary manner of operation. And that this is the case with the loadstone may easily appeare, in that it attracteth yron, more or lesse, as it is of efficacy and power.[232]

The fragrance of flowers is also a material emanation. In short, in all cases where there seems to be action at a distance there is in reality physical contact of one sort or another. Furthermore, such attraction is exerted only over limited distances: it is false to allege that the pole star controls the loadstone, since the stars cannot have such influence, a fact that is further demonstrated by the variations of the compass needle.[233] And surely the case of the Italian noble's nose proves nothing, since the "artificiall nose might rot off about the same time the slave died, casually, or yet by reason of a like radicall temperature of the part with the whole."[234]

[230] James Hart, *KΛINIKH, or The Diet of the Diseased* (London: John Beale, 1633), p. 362. Although printed in 1633, this book was registered with the Company of Stationers on Oct. 12, 1631. *Transcript*, Vol. IV, p. 228.

[231] Hart, *KΛINIKH*, p. 365.

[232] *Ibid.*, p. 367. The problem of sympathetic action was transferred to corpuscular theory in the work of Sir Kenelm Digby. See the discussion in Lester S. King, "The Road to Scientific Therapy: 'Signatures,' 'Sympathy,' and Controlled Experiment," *JAMA*, **197**(1966), 250-256; Lester S. King, *The Road to Medical Enlightenment 1650-1695* (London: Macdonald; New York: American Elsevier, 1970), pp. 140-145.

[233] Hart, *KΛINIKH*, pp. 368, 360, 367.

[234] *Ibid.*, p. 379.

Attempting to show his impartiality, Hart willingly accepted the Paracelsian chemically prepared medicines which "produce very laudable and desired effects."[235] Furthermore, he continued with much less enthusiasm,

> I doe not deny, that many excellent and rare conclusions are by that called naturall magicke, or wisdome brought to passe. But I affirme againe, that this same hath often proved a stalking horse to cover a great deale of cacomagicall impiety, as might easily bee proved.[236]

As for Paracelsus and his followers in general, they should be studiously avoided by the prudent. Although they claim to prolong life through their metallic remedies, Paracelsus himself,

> ... this great miracle-monger (as his foolish followers would make him) died (not without tormenting arthriticall paines many times, notwithstanding all his secrets) before hee attained the 60th yeere of his age, yet will not their folly depart from them if they were braied in a morter, affirming him yet to live in his grave by vertue of *aurum potabile*, writing great voluminous bookes, and inditing many profitable precepts to his disciples. I hope the Printers shall not want work when they are ready.[237]

In any case, Paracelsus was "addicted to infamous magick," and Fludd was very much in the wrong defending it. Hart concluded that if we accept the works of those who defend the weapon-salve, we shall not be

> ... tied to the ordinary operation of agents and patients, but adhere to *Paracelsus* and his followers, and beleeve their mystical, miraculous, if not cacomagicall manner of curing, and so by this meanes must we take for current whatsoever they shall obtrude upon us, as may by the question now in hand plainely appeare. By this means also should all our rationall and methodicall proceeding by our antient Physitians so carefully prescribed, be quite overthrowne.[238]

A translation of Daniel Sennert's views on the weapon-salve added more fuel to the fire in a tract approved for publication in January 1636.[239] In the preface the anonymous translator spoke of

---

[235] *Ibid.*, p. 373.
[236] *Ibid.*, p. 380.
[237] *Ibid.*, p. 6.
[238] *Ibid.*, p. 372.
[239] See above, n. 100.

the weapon-salve as a topic of great interest, but one which had been debated far too sharply—the fault largely of Foster:

> ... they are blame worthy, who in searching for the Truth doe fly upon their Adversarie with uncivill language. In which kinde, I must needes confesse Mr. *Foster* hath exceeded the boundes of Christian charities. Hee hath in mine opinion the Truth on his side, and hath cleared it well but [he surely would have presented a stronger case if he had stated it with less hostility] ...These rules I wish had been observed in that disputation; especially considering that the Adversary whom hee opposeth, is a learned Doctor, well esteemed at home for his practicall skill in Physicke, and much honored abroad for his learned Bookes in Print.[240]

There is no need to analyze the views of Sennert here in detail. He believed that such cures as were observed occurred simply through the healing power of nature.[241] He especially stressed that this "occult" magnetism was something different from natural magnetism,[242] which worked only over short distances in a straight line: that is, its action could not be extended *in infinitum*, and it was hindered by the interposition of other substances. For Sennert there was a great difference between a loadstone, a natural body, and a weapon-salve that had been compounded by man. Indeed, for him this whole method of therapy should be avoided, since it had an aura of black magic.[243]

This final controversy dominated Fludd's last years. In the *Philosophia Moysaica* (1638) he included a lengthy defense of the weapon-salve.[244] It was all well and good, he wrote, to defend the weapon-salve on philosophical grounds and with the assertion that

---

[240] Sennert, *The Weapon-Salves Maladie*, sig. A3^V.

[241] A point stressed earlier by Hart and Foster. The former stated (*KAINIKH*, p. 366), "Wash an ordinary wound, and keepe it cleane, and I warrant it will heale without this curious ointment, which effecteth just nothing, especially as it is used."

[242] Sennert, *The Weapon-Salves Maladie*, p. 10.

[243] *Ibid.*, pp. 22-23.

[244] The section on the weapon-salve in the *Philosophia Moysaica* seems to have been prepared over a lengthy period. At one point (*Mosaicall Philosophy*, p. 232) Fludd referred to "this last yeare 1630," but later (p. 287) he referred to an event in 1637. A marginal printed "note here" and "*a third* [example] *of Mr.* Cotton's *Dog*" (both on p. 251) would lead one to believe that the text was unfinished at Fludd's death.

the magnetic virtue had been acknowledged by Christ,[245] but the time had come for a more detailed reply, one which would show the similarity of the magnetic effect in the salve and the loadstone. There was precedent for this since

> ... the well experimented Doctor, *Paracelsus*, when he writeth of the mysticall Mummies, as well corporal as spirituall, and of the attractive means or manner to extract them, as well out of the living as dead bodie; He for the better instruction of his Schollers, and such he tearmeth *filios Artis, the children of Art,* expresseth examples, drawn from the Load-stone and the Iron.[246]

Once this similarity had been established, it would be possible to apply Gilbert's experiments in proof of this medical magnetism.

Although the Paracelsian physician Rudolph Göckel (Goclenius; 1572-1621) and van Helmont had both appealed to the *De magnete* in works published in 1617 and 1621, it would appear that no early-seventeenth-century author had read Gilbert with more care than Fludd, who referred to many of the most important experiments and had significant cuts copied for the *Philosophia Moysaica*. In addition to his familiarity with the *De magnete*, a work he cited more than forty times, Fludd gives similar evidence of having studied Mark Ridley's *Treatise on magneticall bodies* (1613). In comparison, his references here to Paracelsus and the Bible (seven each) run far behind.

Once again Fludd based his work on the universal sympathy and antipathy which spring directly from the eternal life spirit and thus are directly connected with the will of God. However, in a world system dominated by correspondences between the macrocosm and the microcosm, the loadstone is important because it displays so vividly these universal forces, and for Fludd it bore the same relation to the earth as man does to the macrocosm as a whole. Because of this relationship he thought it "most fit to search out diligently . . . the dark mystery of the Loadstone's or magnet's nature, that we may with the more assurance make our ingression into the practical demonstration, of so arcane and occult a contemplation."[247]

---

[245]*Ibid.*, p. 243. John 12:32. "I shall be exalted from the earth, I will draw or attract all things unto myself."
[246]*Ibid.*, p. 244.
[247]*Ibid.*, p. 200.

Gilbert's experiments were applied to establish principles dear to Fludd's heart. It was a general Paracelsian belief that like attracts like, and in the *De magnete* (Book I, Ch. 5) Gilbert showed the effect of cutting an oblong loadstone midway between its poles: the original poles will now attract each other and the original "equatorial" regions will seek to reunite. For Fludd this was proof primarily that in the universe each of the two basic natures—"nolunty" and "volunty"—seeks its own kind.[248] Polar natures are similar and attract each other, and "southerly" equinoctial regions also tend to rejoin after separation. Similarly, Gilbert's demonstration of the reaction of a compass needle to the surface of a spherical loadstone (*terrella*) (Book 2, Ch. 2) was applied to Fludd's cosmological scheme to show primarily that the two basic natures in the universe—the equinoctial and the polar—cannot be joined.[249]

Nevertheless, Fludd's insistence on two basic types of magnetism (the polar and the equinoctial) in addition to the distinction between north and south polarities leads to some confusion for the reader. Thus, although Fludd accepted Gilbert on the difference between north and south poles in the magnet,[250] he affirmed that this "polar" magnetism is essentially one of the two basic types. The southern "dilative" magetism is of a different sort, found basically in live objects, not in a "dead" loadstone where the cold polar magnetic force dominates. The difference is seen best when he described the magnetic attraction of the living spirit into the body, something which may be accomplished only "by the animation of a dilating and vivifying heat."[251]

But is man truly like a loadstone? Yes, because the loadstone exhibits many lifelike properties in addition to its lifelike ability of attraction. Even more important is the fact that man with body and soul is dipolar in nature. This is seen anatomically as well: the spleen with its "melancholy and terrestrial juices" tending to give

---

[248] William Gilbert, *De magnete* (1600), p. 16; Fludd, *Philosophia Moysaica* (1638), fol. 131^r.

[249] Gilbert, *De magnete*, p. 76; Fludd, *Philosophia Moysaica*, fol. 106^v.

[250] *Mosaicall Philosophy*, p. 248.

[251] (*Ibid.*, p. 218). It should be added that the concept of polarity in life processes becomes of prime importance again in early-nineteenth-century *Naturphilosophie*. This has been discussed by Walter Pagel in his "Religious Motives in the Medical Biology of the 17th Century," pp. 291-292.

a saturnine virtue exhibits the property of the polar regions, while the liver with a southern or "Aequinoctial dilation disperseth the blood which it hath rubified, by veiny channels through the whole Microcosmicall earth." Indeed, the combination of the cold (north polar) nature of man with the warm spiritual (south polar) nature of the inspired life spirit results in the motion of the heart and the pulse.[252]

James Hart or others might argue that in general cure by sympathy—as in the case of the noble's nose—had little or nothing to do with a weapon-salve. This, however, seemed patently false to Fludd, since the man's blood is an essential part of the salve and will react in turn with the blood drawn from the wound with which the weapon is now impregnated. That there is blood—and consequently spirit—in such a weapon cannot be denied, since "experience teacheth us, that though no corporal blood can be found on the Iron, . . . by putting the weapon into the fire . . . the part of the sword or weapon will discover it self, which wounded the party, being that it will change colour from the rest."[253] The theory of the salve cure then simply presents itself in terms of magnetic properties. The blood in the weapon contains cold northern spirit, while the salve, prepared from fresh blood, is a storehouse of dilative southern spirit. This supply of life spirit causes the "congealed bloody spirits in the weapon" to "ripen and exspire out little by little." And "so also by little and little doth the wound heal, and mend in the wounded creature."[254]

Fludd thought that such sympathetic cures should be considered the proof of the presence of a true "cure-all," since the magnetic action of the "southern" agent will extract the deadly polar nature from the body, no matter what the affliction might be:

> . . . by it the Dropsie, Pluerisie, Gout, Vertigo, Epilepsie, French-Pox, Palsey, Cancer, Fistula, foul Ulcers, Tumors, wounds, Herniaes, Fractions of Members, superfluity and suppression of Menstrues in women, as also sterility in them; Feavors, Hecticks, Athrophy, or wastings of members, and such like, may by this naturall magicall means be cured, and that at distance, and without any immediate contact.[255]

[252]*Mosaicall Philosophy*, pp. 217,218.
[253]*Ibid.*, p. 228.
[254]*Ibid.*, p. 230.
[255]*Ibid.*, pp. 291-292.

The conflict over the weapon-salve and cure by sympathy was far from over at the time of Robert Fludd's death in 1637. We have already noted Mersenne's desire to reply to Fludd on this subject in a planned new edition of Durelle's answer to the *Clavis*. In his lengthy *Magnes sive arte magnetica* (1641) Athanasius Kircher praised Gilbert but devoted much of one chapter to a refutation of Fludd's *Philosophia Moysaica*. He concluded that the magnetic cure of wounds was false and that it implied a pact with the devil.[256] In England a translation of van Helmont's work on the magnetic cure of wounds was made first by Walter Charleton (1650) and then by John Chandler in his large edition of the *Opera* (1662).[257] Sir Kenelm Digby's oft-quoted *Discourse* appeared in 1658, and here he modestly implied that he was the one who had introduced this cure, if not specifically in England, at least to "this Quarter of the World."[258] Yet, it seems clear that not too many years earlier the cure was quite generally attributed first to Paracelsus and then to his partisans who developed it.

Conclusion

Although hardly seen as a major figure in most accounts of the scientific revolution today, Robert Fludd takes on a new significance in the light of contemporary evidence.[259] It is possible, of course, to argue that Fludd had done little that was new. He was

---

[256] Athanasius Kircher, *Magnes sive de arte magnetica* (Rome: Ludovici Grignani, 1641), sig. a7^r. Yet Kircher was quite willing to criticize Gilbert (see p. 87 and *passim*). On the *Philosophia Moysaica*, see Book III, Ch. 2, pp. 777-784.

[257] J. B. van Helmont, "The Magnetick Cure of Wounds" in *A Ternary of Paradoxes*, trans. Walter Charleton (London: J. Flesher for W. Lee, 1650); J.B. van Helmont, *Oriatrike, or Physick Refined*, trans. John Chandler (London: Lodowick Loyd, 1662).

[258] Sir Kenelm Digby, *A Late Discourse made in a Solemne Assembly of Nobles and Learned men at Montpellier in France Touching the Cure of Wounds by the Powder of Sympathy* (2nd ed., London: R. Lowndes, 1658), p. 5. "And I believe, Sirs, when you shall have understood this History, you will not accuse me of vanity, if I attribute unto my self the introducing into this Quarter of the World this way of curing." The first (French) edition of this work was printed in the same year.

[259] Arguments in favor of Fludd as a significant scientist in modern terms are given by Josten, "Truth's Golden Harrow," pp. 95-96, where pertinent references to his optical work and his use of the thermoscope and its relationship to the thermometer are mentioned.

one of many of his generation to have been fascinated with the Rosicrucian documents which called for a new, chemically oriented learning in opposition to the jaded texts of the ancients. And he, like many of his Paracelsian predecessors, sought to use alchemy as a key to the interpretation of Holy Scripture. Once again we meet here with a chemical version of the Creation and an insistence on the truth of the macrocosm-microcosm analogy. It is true that Fludd differed from some of his Paracelsian contemporaries on a number of questions: he preferred the four elements generally to the three principles, and, as Pagel has shown, he maintained the traditional humoral theory. Still, in general he falls into the pattern of the mystical alchemist of the period—clearly a "universal Hermeticist" while at the same time a man who shared many concepts with those Paracelsians and Rosicrucians who found medicine a semi-divine science.

The work of Fludd was clearly the subject of intense discussion in his own day. Both the quantity of his work and the argumentative nature of the man help to explain this fact. The record of his many debates is of great interest, for they involve some of the most important scientists and scholars of the early seventeenth century and cover a wealth of topics, only a few of which were touched upon in this chapter. Much of this interest can be traced to Fludd's insistence on the interrelation of all parts of the universe. Thus, a subject such as cure by sympathy which at first glance may seem to be a minor one could be rapidly expanded to include problems of broad significance, such as action at a distance, a subject crucial for Aristotelian physics. Again, we find the question of scientific method constantly reappearing. If Fludd preferred to argue from the basis of scriptural authority, he soon found that he was unable to do so in a way that was convincing to all others. When this occurred he changed his approach. Thus, his early "Philosophicall Key" was written as an "experimental Treatice" for those who were unable to accept the deeper arguments set forth in the *Utriusque cosmi*. Again, in his rejection of the Copernican system Fludd used mechanical examples and formulated a rule of simplicity as a basis for maintaining an earth-centered universe. In his rejection of Gassendi's insistence on the existence of the interventricular pores in the septum of the heart, it is evident that Fludd was able to argue correctly on the basis of experimental evidence when he so chose.

The reaction to Fludd in the third decade of the new century was something quite different from the late-sixteenth-century attack on Paracelsus and his followers. It is true that the accusations of diabolical magic were carried over along with the persistent medical overtones. However, Fludd had presented a chemically oriented macrocosm-microcosm universe in a broad restatement that could not be tolerated by scholars who were neither chemists nor physicians. His work clearly attracted the attention of writers who had paid little or no attention to chemists such as Duchesne or Crollius. Thus, Francis Bacon complained of those who now tried to explain all nature in terms of light "as if it were a thing half way between things divine and things natural."[260] For Kepler, Fludd's universe brought into focus the fundamental question of how mathematics should be employed in the investigation of nature. And while Mersenne sought desperately to prevent Fludd and his friends from extending alchemy into the realm of scriptural interpretation, Gassendi discussed in great detail the manifold errors in all aspects of the work of the English physician. Their arguments were all in vain, however, for Fludd wrote that he was completely unable to understand how Gassendi could believe that it was not the intent of the Scriptures to teach us physics.[261]

At one extreme there are those who would reject Fludd as a person of any consequence in the scientific revolution. At the other, there are those who make of him a major figure. The latter position is best seen in the work of Frances Yates, who pictures Fludd as a prime transmitter of Renaissance Hermetic thought, via Dee and Rosicrucianism. This for her is the direct line of development of modern science through its impact on the Royal Society and Isaac Newton. I would prefer to judge the importance of Fludd on the basis that (1) his work represents the most complete compendium of the chemical philosophy extant (although it is surely not a pure "Paracelsian" interpretation) and (2) his work became a focus for discussion and debate among the most important continental scientists of the period. Many of the questions raised in the heated Fluddean debates were fundamental for the development of a new

---

[260] Francis Bacon, *Works*, ed. James Spedding, Robert L. Ellis, and D. D. Heath, Vol. IV (London, 1870), p. 403.

[261] Fludd, *Clavis philosophiae et alchymiae Fluddanae*, p. 56.

science, and they accrued to older questions more commonly related to the acceptance of Paracelsian medicine. It was almost unavoidable that the acceptance or rejection of a chemically oriented universe—and the work of Fludd himself—should continue to be the subject of controversy in the following decades.

Volume II

5
THE BROKEN CHAIN:
The Helmontian Restatement of the Chemical Philosophy

> I praise my bountiful God, who hath called me into the Art of the fire, out of the dregs of other professions. For truly Chymistry ... prepares the understanding to pierce the secrets of nature, and causeth a further searching out in nature, than all other Sciences being put together: and it pierceth even unto the utmost depths of real truth.
>
> Jean Baptiste van Helmont*

AMONG THOSE REQUESTED by Mersenne to take sides in the Fluddean confrontation was Jean Baptiste van Helmont. The acquaintance of these two came about in a not unusual fashion. Plagued by a severe case of *herpes mordax*, Mersenne sought the medical advice of van Helmont while visiting Brussels in June 1630. Van Helmont, then at Malines, replied at length to the

---

*John Baptista van Helmont, *Oriatrike or Physick Refined. The Common Errors therein Refuted, And the whole Art Reformed & Rectified. Being A New Rise and Progress of Phylosophy and Medicine, for the Destruction of Diseases and Prolongation of Life*, trans. J(ohn) C(handler) (London: Lodowick Loyd, 1662), p. 462. Although Chandler's translation has been (correctly) accused of inaccuracy in places, he has in general interpreted the meaning of the Latin properly, and his translations will be used whenever possible. This quotation is from the "Pharmacopolium ac dispensatorium modernorum" (sect. 32); the Latin is from the *Ortus medicinae. Id est, initia physicae inaudita. Progressus medicinae novus, in morborum ultionem, ad vitam longam* (Amsterdam: Ludovicus Elsevir, 1648; reprinted Brussels: Culture et Civilisation, 1966), p. 463. The *Ortus* was issued in 1648 with a companion volume, the second edition of the *Opuscula medica inaudita*, which included the "De lithiasi," the "De febribus," the "De humoribus Galeni," and the "De peste" (Amsterdam: Ludovicus Elsevir, 1648; reprinted Brussels: Culture et Civilisation, 1966). These three editions will henceforth be cited as *Oriatrike*, *Ortus*, and *Opuscula*.

The Helmontian literature has been discussed frequently by Walter Pagel. His most recent, and most convenient, description is in his article on van Helmont in the *Dictionary of Scientific Biography*, Vol. VI (New York: Charles Scribner's Sons, 1972), pp. 253-259 (257-259). This is to be supplemented with

famed *savant*, and a rich correspondence followed.[1] Of this some fourteen letters are extant covering the period through July 1631. These touch on topics ranging from mechanics to theology, and they are explicit in indicating van Helmont's opinion of Robert Fludd. Replying to a query by Mersenne regarding Gassendi's *Epistolica*, van Helmont referred to the *fluctuantem Fluddum* and went on to comment that

> J'ay veu altrefois un tome de Flud (j'ay cognu l'homme en sa patrie pour un pauvre medicin et encor moindre alchymiste, *sed erat garrulus, stentor, superficialiter doctus, parum sui constans*), mais l'ay fueilletté vistement, ne trouvant en ses songes argument digne pour amuser mon temps longuement. *Miseror itaque rarum hominem Gassendum trivisse tot horas in refutandis neniis, deliciis et somniis viri Athei, quippe quae, succo veritatis destituta, sponte suâ erant rasura*, de sorte qu'à son livre je ne voudrois dire sinon le distichon:
>
> R habet Ausonium liber hic, R habetque Pelasgum,
> R habet Hebreum, praetereaque nihil.[2]

It is clear from the letters that Mersenne had a high opinion of the Belgian physician. Yet it is of real interest to learn that van Helmont was a member of the Mersenne circle, for he was in fact a chemical philosopher whose work was characterized by much that Mersenne had attacked in his *La verite des sciences* in 1625. For this reason it is of considerable significance to examine the views of van Helmont. Here, in contrast to Fludd, few of whose works were

---

the bibliography of secondary sources prepared by J. R. Partington, *A History of Chemistry* (London: Macmillan, 1961), Vol. II, pp. 209-210 (n. 4). A discussion of the nineteenth- and early-twentieth-century literature on van Helmont will be found in Paul Nève de Mévergnies, *Jean-Baptiste van Helmont. Philosophe par le feu* (Paris: Libraire E. Droz, 1935), pp. 7-35. Other than this—and the useful summary of earlier work on the life of van Helmont—the work of de Mévergnies cannot be recommended. For a detailed critique, see Walter Pagel, *The Religious and Philosophical Aspects of van Helmont's Science and Medicine, Supplements to the Bulletin of the History of Medicine*, No. 2 (Baltimore: Johns Hopkins Press, 1944), pp. vii-ix (n. 3).

[1] Van Helmont (Malines) to Mersenne (Brussels), June 1630, in the *Correspondance du P. Marin Mersenne, Religieux Minime*, ed. Cornelis De Waard, Vol. II (Paris: P.U.F., 1945), pp. 496-503 (497).

[2] Van Helmont (Brussels) to Mersenne (Paris), Dec. 19, 1630, *Correspondance*, Vol. II, pp. 582-599 (584). The editor of the *Correspondance* notes that the Latin couplet is explained in a marginal notation by van Helmont: "*Latinum* ER, *Graecum* RO, *Hebreum* RES."

republished after his death, we find an author of great influence whose restatement of the chemical philosophy was to become the basis of the iatrochemical school of the seventeenth century.

Van Helmont: The Early Years

Born in Brussels on January 12, 1579, Jean Baptiste van Helmont came from a family that was both well-to-do and influential.[3] His father, Christian van Helmont, who died that same year, was state counselor of Brabant who had married into the Stassart family of Brussels. Through marriage to Marguerite van Ranst in 1609, Jean Baptiste became the manorial lord of Merode, Royenborch, Oorschot, and Pellines.

Van Helmont wrote that since he was the youngest in his family he was "brought up in studies." After having completed his first course in classics and philosophy at the University of Louvain (1594), he eagerly went on to advanced subjects. It was at this time, however, that he became thoroughly disillusioned with the entire educational process. He observed that none were

> ... admitted to Examinations, but in a Gown, and masked with a Hood, as though the Garment did promise Learning; I began to know that Professors for sometime past, did expose young men that were to take their degrees in Arts, to a mock: I did admire at the certain kinde of dotage in Professors, and so in the whole World, as also the simplicity of the rash belief of young men. I drew my self into an account or reasoning, that at leastwise I might know by my own judgement, how much I was a Phylosopher, I examined whether I had gotten truth, or knowledge,
> I found for certainty, that I was blown up with the Letter, and (as it were the forbidden Apple being eaten) to be plainly naked, save, that I had learned artificially to wrangle. Then first I came to know within my self, that I knew nothing, and that I knew that which was of no worth. For the sphere in natural Phylosophy, did seem to promise something of knowledge, to which therefore I had joyned the Astrolobe, the use of the Ring or Circle, and the speculations of the Planets. Also I was diligent in the Art of *Logick*, and the Science

[3] See the important autobiographical sections in the works of van Helmont, especially the "Confessio authoris." A useful summary is found in Nève de Mévergnies, *Van Helmont*, pp. 110-148.

Mathematical, for delights sake, as often as the reading of other things had brought a wearisomness on me.

Whereto I joyned the Elements, or first Principles of *Euclide*; and this Learning, I had made sociable to my Genius or natural wit, because it contained truth; but by chance, the art of knowing the Circle of *Cornelius Gemma*, as of another Metaphysick, came to my hand. Which, seeing it onely commended *Nicholas Copernicus*, I left not off, till I had made the same familiar unto me. Whence I learned the vain excentricities, or things not having one and the same Center, another circular motion of the Heavens: and so I presumed, that whatsoever I had gotten concerning the Heavens, with great pains, was not worthy of the time bestowed about it.

Therefore the Study of Astronomy, was of little, or no account with me, because it promised little of certainty or truth, but very many vain things. Therefore having finished my Course, when as I knew nothing that was sound, nothing that was true, I refused the Title of Master of Arts; being unwilling that Professors should play the fool with me, that they should declare me Master of the seven Arts, who was not yet a Scholar. Therefore I seeking truth, and knowledge, but not their appearance, withdrew my self from the Schooles.[4]

Declining a wealthy canonship, van Helmont resumed his studies in philosophy at Louvain, since a new program with new courses was being offered by Jesuits who had previously been restrained from teaching. At this time he listened to the lectures on magic given by the famed Martinus del Rio, but "instead of a Harvest, I gathered onely empty stubbles, and most poor patcheries, void of judgement."[5] Far more satisfying were the works of the Stoics Seneca and Epictetus. Being led from these authors to Thomas à Kempis and Tauler he came to believe that through Stoicism he might gain in Christian perfection. However, his confidence was shattered by a dream which led him to understand that "in Christ Jesus, we live, move, and have our being. That no man can call even on the name of Jesus to Salvation, without the special grace of God." Thus his study of Stoicism, both Pagan and Christian, had made him "an empty and swollen Bubble, between the bottomless pit of Hell, and the necessity of imminent death."[6]

[4]"Confessio authoris" (Ch. 2, sects. 1-5), *Ortus*, p. 16; *Oriatrike*, pp. 11-12.
[5]"Confessio authoris" (Ch. 2, sect. 6), *Ortus*, p. 17; *Oriatrike*, p. 12.
[6]"Confessio authoris" (Ch. 2, sect. 8), *Ortus*, p. 17; *Oriatrike*, p. 12.

PLATE XXIV J. B. van Helmont, portrait from his *Aufgang der Artzney-Kunst* (Sulzbach: Endter, 1683).

Now, again in despair, van Helmont turned to medical subjects, since earlier, when wearied from his studies, he had relaxed through the reading of Mattioli and Dioscorides. The subject of plants and their medicinal virtues was appealing, for here could be seen the grace of the Lord in natural things. But once again his hopes were shattered: in the herbals he found pictures and names, "but nothing of their properties, virtues and uses." The confusing information given on the qualities and temperatures of herbs made him wonder at the "great darkness in applying and healing." As a result, "I inquired whether there were any Book, that delivered the Maxims and Rules of Medicine? For I supposed, Medicine might be taught, and delivered by Discipline, like other Arts and Sciences, and so to be by tradition: but not that it was a meer gift."[7]

Seeking the theorems and the chief authors of medicine, he felt certain that he would find some text that would give him the truths he so urgently sought. However, a professor of medicine told him that neither Galen nor Avicenna would give him the answers he longed for.[8] Discouraged, van Helmont temporarily considered the study of law, but when he reflected that he could barely manage his own self, he turned once again "with a singular greediness, unto the most pleasing knowledge of natural things."

> Therefore I read the Institutions of *Fuchius*, and *Fernelius*, whereby I knew that I had lookt into the whole Science of Medicine, as it were by an Epitome, and I smiled to my self. Is the knowledge of healing thus delivered, without a Theoreme and Teacher, who hath drawn the gift of healing from the Adeptist? Is the whole History of natural properties, thus shut up in Elementary qualities? Therefore I read the works of *Galen* twice, once *Hipocrates* (whose Aphorismes I almost learned by heart) and all *Avicen*, and as well the *Greeks, Arabians,* as *Moderns,* happily six hundred, I seriously, and attentively read thorow, and taking notice by common places, of whatsoever might seem singular to me in them, and worthy of the Quill. At length, reading again my collected stuffe, I knew my want, and it grieved me of my pains bestowed, and years: When as indeed I observed, that all Books, as institutions, singing the same Song, did promise nothing of soundness, nothing that might promise the knowledge of truth, or the truth of knowledge.[9]

[7]"Confessio authoris" (Ch. 2, sects. 10-11), *Ortus*, pp. 17-18; *Oriatrike*, p. 13.
[8]"Confessio authoris" (Ch. 2, sect. 13), *Ortus*, p. 18; *Oriatrike*, p. 13.
[9]"Confessio authoris" (Ch. 2, sects. 14-15), *Ortus*, p. 18; *Oriatrike*, p. 13.

Although he was no less disillusioned than before, van Helmont did not now reject the degree of doctor of medicine, which he had earned in 1599.[10]

The following years were devoted to travel. We find him in Switzerland and Italy (1600-1602) and later in France and England (c. 1602 and 1604).[11] Like Fludd he was accustomed to move in court circles, and while in London he conversed with Queen Elizabeth.[12] Esteemed both as a physician and as an adept, van Helmont was sought out by Ernest of Bavaria, the Archbishop of Cologne, and the Emperor Rudolph II, who hoped to attract him to their courts.[13] Van Helmont, however, rejected all such offers, since he had no desire to accumulate riches and to endanger his soul. After his marriage in 1609,

> I withdrew my self from the common people, to *Viluord*, that, being the lesse troubled, I might proceed diligently to view the Kingdoms of Vegetables, Animalls, and Minerals by a curious Analysis, or unfolding, by opening Bodies, and by separating all things, I went about to search for full seven years. I searched into the Books of *Paracelsus*, filled in all parts with a mocking obscurity or difficulty, and I admired that man, and too much honoured him: till at length, understanding was given, of his Works, and Errours.
>
> For not a Friend, the thief of dayes, never riotous feasting, or drinking Bouts, detained me, who then as yet, could not bear Wine: but continual labour, through extraordinary watching nights, did accompany my desires. And at length, being perfectly taught, that the corporeal faculties or powers, were bound up in their principles or beginnings; and those not worthily to be known, without an unlocking of their bolts, I sung a Hymne to my God, that it repented me not of time, pains, costs, and gain neglected, being recompenced with the sweetness of knowledge Adepticall.[14]

Clearly it was during these seven years of study and research that van Helmont became deeply influenced by the works of Paracelsus and the chemical philosophers who followed him.

[10]"Promissa authoris" (col. 3, sect. 7), *Ortus*, p. 12; *Oriatrike*, p. 7.

[11]Pagel, "Van Helmont," *DSB*, Vol. VI, p. 254.

[12]Van Helmont, "De lithiasi" (Ch. 3, sect. 13), *Opuscula*, p. 19; *Oriatrike*, p. 838.

[13]Pagel, "Van Helmont," *DSB*, Vol. VI, p. 254; Van Helmont, "Amico lectori" (F. M. Van Helmont), *Ortus*, sig. **2^r; *Oriatrike*, sig. b2^r; "Praefatio," *Ortus*, p. 630; *Oriatrike*, p. 632.

[14]"Promissa authoris" (col. 3, sects. 7-8), *Ortus*, p. 12; *Oriatrike*, p. 7.

The Tract on the Weapon-Salve (1621)

Like Fludd, van Helmont found himself drawn into the conflict over the weapon-salve. The effect of the salve had been ascribed to purely natural causes by Rudolph Goclenius in works published in 1608 and 1613. The Jesuit Jean Roberti took exception to these publications, and a literary duel between the two men produced seven attacks and counterattacks in the years 1615-1625. Perhaps seeking support for his point of view, Roberti summarized the 1613 defense of Goclenius for van Helmont. Recalling this, van Helmont wrote:

> There was very lately brought me a succinct Anatomy of the aforesaid Book, composed by a certain Divine, rather in the form of a fine or jocund censure, than of a disputation: my judgement therefore, however it should be, was desired, at least-wise in that respect, that the thing found out by *Paracelsus,* concerned himself, and me his follower.[15]

Van Helmont quickly penned his *De magnetica vulnerum curatione,* a work that was directed no less at Roberti than at Goclenius. Here he charged that Goclenius had grossly oversimplified the problem by confounding sympathy with magnetism alone, thereby considering the phenomenon to be purely natural, but Roberti was no better since he thought the true author of the cure could be none other than Satan himself.[16] Such a view could not be tolerated. It is little wonder that

> ... Nature ... called not Divines for to be her Interpreters: but desired Physitians only for her Sons; and indeęd, such only, who being instructed by the Art of the Fire, doe examine the Properties of things, by separating the impediments of their lurking Powers, to wit, their Crudity, Poysonousnesses, and Dregs, that is the Thistles and Thorns every where implanted into Virgin-Nature from the Curse: For seeing Nature doth dayly Distil, Sublime, Calcine, Ferment, Dissolve, Coagulate, Fix, &c. Certainly we also, who are the only faithful Interpreters of Nature, do by the same helps draw forth the Properties of things from Darkness into Light.[17]

One might think that it was Galileo rather than van Helmont who

---

[15] "De magnetica vulnerum curatione" (sect. 1), *Ortus,* p. 748; *Oriatrike,* p. 759.
[16] *Ibid.,* "De magnetica" (sect. 3), *Ortus,* p. 749; *Oriatrike,* p. 760.
[17] "De magnetica" (sect. 8), *Ortus,* p. 750; *Oriatrike,* p. 761.

admonished Roberti to "let the Divine enquire concerning God, but the Naturalist concerning Nature."[18]

To be sure, van Helmont continued, the magnetic cure is a natural one, but nature's magnetism stems from the heavens. No evil, no superstitious rites or chants are associated with this cure, and its only purpose is for the healing of the sick.[19] But although witchcraft is not the same as sympathy, and neither one is the same as magnetism, there are surely similarities between magnetic action and the weapon-salve. As did Fludd later, van Helmont referred his readers to William Gilbert ("by whose Industry, the variation of the Compass may be restored")[20] for information on the properties of the loadstone. However, to this he added a number of magnetic observations of medical value: the attraction of flowers to the sun was an example of natural magnetism, as was the case of the servant and the noble's new nose (as with Fludd, borrowed from Tagliacozzi).

> Thou wilt not account this to be diabolical, because there is another privie Shift at hand; to wit, that there is a Harmony of superiour Bodies with Inferiour, and an attractive Faculty, plainly Celestial, in no wise to be communicated unto sublunary things: As if indeed, the Microcosme or little World, being unworthy of a heavenly Condition, could in his Blood, and Moss, take notice of no revolution of the Stars.[21]

It is clear that "all particular things contain in them a delineation of the whole Universe"[22] because of implanted celestial virtues. As a result, "the Motion of the Heavens, because the most known, because the most common, directs the Heavens of particular things . . . according to it self."[23] As for Paracelsus, he is far from being wicked and ignorant. Rather,

> . . . he hath disclosed Magnetism, unknown to Antiquity, and in the room of that natural Study which is barrenly taught up and down in the Schooles, hath brought to us another real one; which by the

---

[18]"De magnetica" (sect. 8), *Ortus*, p. 750 ("De Deo Theologus, naturalis verò de natura inquirat"); *Oriatrike*, p. 761.

[19]"De magnetica" (sects. 11-16), *Ortus*, pp. 750-751; *Oriatrike*, p. 762.

[20]"De magnetica" (sect. 19), *Ortus*, p. 751; *Oriatrike*, p. 763.

[21]"De magnetica" (sect. 26), *Ortus*, p. 753; *Oriatrike*, p. 764.

[22]"De magnetica" (sect. 130), *Ortus*, p. 772; *Oriatrike*, p. 785.

[23]"De magnetica" (sect. 54), *Ortus*, p. 759; *Oriatrike*, p. 771.

Resolution, and Composition of Bodies is made probably to our hands, and far more plentiful in Knowledge; that from thence he hath rather by a just title, snatch'd away the Denomination of the Monarch of Secrets, from all that went before him...."[24]

In regard to the three principles, salt, sulfur, and mercury, or "Salt, Fat, and Liquor," it is from these that "every Body is composed, and again resolved into the same."[25]

By magic, "we understand the most profound inbred knowledge of things, and the most potent Power for acting, being alike natural to us with *Adam*, not extinguished by Sin, not obliterated, but as it were become drowsie, therefore wanting an Excitement."[26] A key to the revival of magic could be found in the "Art of the Cabal" through which man himself is able "to cause an excitement in himself, of so great a Power at his own Pleasure and these are called Adeptists; or Obtainers, whose Governour also, is the Spirit of God."[27] The power in the weapon-salve is real; it is nothing but a natural magic which can be utilized either for good or for evil. Indeed, van Helmont wrote, he would "assume the very works of Witches; the which although they are wickedly mischievous and detestable, yet are supported by the same root, namely a Magical power, without difference as unto good, and also unto evil."[28]

As for Aristotle and the teachings of the Schools, if they are adhered to, the Godly powers attributed to the relics of the saints would be meaningless. "So let the Sweat in the Stove of St. *Paul*, be a magnetical Unguent," but at the same time understand that the "Sweat of the Sick" exhibited in the salve is no less miraculous even though it may be regularly utilized for our benefit.[29] Van Helmont closed by protesting, "I am ... a *Roman* Catholick, whose mind hath been to ponder of nothing which may be contrary to

[24]"De magnetica" (sect. 53), *Ortus*, p. 759; *Oriatrike*, p. 771.

[25]"De magnetica" (sect. 153), *Ortus*, p. 775; (sect. 152 in the English ed.) *Oriatrike*, p. 788.

[26]"De magnetica" (sect. 126), *Ortus*, p. 771; (sect. 125 in the English ed.) *Oriatrike*, p. 784.

[27]"De magnetica" (sect. 129), *Ortus*, p. 772; (sect. 127 in the English ed.) *Oriatrike*, p. 784.

[28]"De magnetica" (sect. 106), *Ortus*, p. 769; (sect. 105 in the English ed.) *Oriatrike*, p. 781.

[29]"De magnetica" (sect. 46), *Ortus*, p. 757; *Oriatrike*, p. 769.

God, and that may be contrary to the Church."[30] Unfortunately, everyone did not agree with him.

The Letters to Mersenne (1630-1631)

The publication of van Helmont's tract on the weapon-salve could hardly have appeared at a less opportune time. These were years that had witnessed the debate over the Rosicrucians and the Paracelsians, and it was a period of retrenchment for those who represented tradition. Mersenne's attack on Fludd and the natural magicians in the *Questiones celeberrimae in Genesim* appeared in 1623. The condemnation of the alchemical theses at the Sorbonne occurred two years later. Van Helmont's attack on a prominent Jesuit, his defense of magic and Paracelsus—combined with his interpretation of the miraculous power of relics—could hardly go unnoticed.

The *De magnetica vulnerum curatione* was denounced by the Faculty of Medicine at the University of Louvain in 1623.[31] The following year van Helmont published a second work on the fountain at Spa (*Paradoxa de aquis Spadanis* and *Supplementum de Spadanis fontibus*), which compounded his problems since it was a direct attack on the popular *Spadacrene, hoc est fons Spadanus* written by Henry van Heer.[32] Here van Helmont freely turned to the macrocosm-

[30]"De magnetica" (sect. 174), *Ortus*, p. 780; (sect. 173 in the English ed.) *Oriatrike*, p. 793.

[31]The complex history of the prosecution is summarized by Nève de Mévergnies, *Van Helmont*, pp. 123-143, but this account is heavily indebted to earlier studies: C. Broeckx, *Commentaire de J.B. van Helmont sur le premier livre du Régime d'Hippocrate: Peri diaites* (Antwerp, 1849); C. Broeckx, *Interrogatoires de J.B. Van Helmont sur le magnétisme animal* (Anvers: Buschmann, 1856); Willem Rommelaere, *Études sur J.-B. Van Helmont* (Brussels: Manceaux, 1868) and A. J. J. Vandevelde, "Helmontiana," 5 parts, in *Verslagen en Mededeelingen. K. Vlaamsche Academie voor Taal-en Letterkunde*, Pt. 1 (1929), pp. 453-476; Pt. 2 (1929), pp. 715-737; Pt. 3 (1929), pp. 857-879; Pt. 4 (1932), pp. 109-122; Pt. 5 (1936), pp. 339-387.

[32]J.B. Van Helmont, *Paradoxa de aquis Spadanis* (Leodii, 1624) and *Supplementum de Spadanis fontibus* (Leodii: L. Streel, 1624). These texts are to be found in the *Ortus*, pp. 685-699 and the *Oriatrike*, pp. 687-703. Henricus van Heer's *Spadacrene, hoc est fons Spadanus; ejus singularia, bibendi modus, medicamina bibentibus necessaria* (Leodii: Arn. de Corswaremia, 1614) went through many editions in the course of the seventeenth century (the one in my possession was printed at Leiden in

microcosm analogy and argued against the commonly accepted view that internal distillations occur within the earth.[33] More important, he turned to element theory in his insistence that air and water are fundamentally different substances which cannot be converted into one another by the application of heat, cold, or pressure change.[34] For van Helmont water vapor is both

> ... materially, and formally, nothing else but a heap of the Atoms of Water lifted up on high. ... The Air therefore, whether it be received in hot, or cold Glasses, and pressed together therein, shall never afford Water, but according to how much of a Vapour, that is, of an extenuated Water, it shall contain within it.[35]

This was a direct denial of van Heer's explanation of the origin of springs, and he replied immediately to van Helmont in a *Deplementum* to the *Supplementum*, and in a personal confrontation van Helmont could report that his adversary had been struck "dumb."[36] Clearly he had made another enemy.

Alarmed by his views, his enemies cited twenty-four propositions in the weapon-salve work which they judged to be heretical.[37] These were submitted to the tribunal at Malines-Brussels, but they were quickly transferred for consideration to the Inquisition of Spain with the hope that they would receive more rapid action there. In short order the propositions were condemned for heresy

---

1685). The work is of interest because of the author's concern with analytical procedures. The text was published in a French translation in 1616 (as *Les fountaines de Spa*) that was reprinted in 1630, 1646, and 1654. Beginning with the 1630 edition there is an acute interest in color changes occurring in aqueous media, and late editions show characteristic crystal forms of various salts. The subject has been touched upon in A. G. Debus, "Solution Analyses Prior to Robert Boyle," *Chymia*, **8** (1962), 41-61 (50).

[33]J. B. Van Helmont, "De Spadanis fontibus. Paradoxum quartum" (sects. 1-2), *Ortus*, pp. 693-694; *Oriatrike*, pp. 696-697.

[34]"Aer" (sects. 3-4), *Ortus*, p. 61; *Oriatrike*, p. 58.

[35]"De Spadanis fontibus. Paradoxum secundum" (sects. 9-10), *Ortus*, p. 689; *Oriatrike*, p. 692.

[36]Henricus van Heer, *Deplementum supplementi de Spadanis fontibus, sive vindiciae pro sua Spadacrene, in quibus aroph. certissimum Paracelsi ad calculos remedium sincere explicatur* (Leodii: A. de Corswaremia, 1624). Van Helmont, "Aer" (sect. 12), *Ortus*, p. 64; *Oriatrike*, p. 60.

[37]On these developments see Nève de Mévergnies, *Van Helmont*, pp. 123-130, and the notes of the editor of Mersenne's *Correspondance*, Vol. II, pp. 499, 589.

and magic (October 16, 1625) and the tract itself impounded. Subsequently van Helmont was interrogated at Malines on these propositions plus an additional three taken from the works of Paracelsus (September 3, 1627). He duly asserted his innocence and submitted to the judgment of the Church three days later. Nevertheless, new censures were forthcoming from the Faculty of Theology of the University of Louvain, from Cologne (1628), and from the Collge of Physicians at Lyons (1629). He was forced to acknowledge his errors once again in 1630.

It was just at this time—in the midst of these problems—that the surviving letters to Mersenne were written. These letters indicate that little change had occurred in van Helmont's views. In the first (June 1630) he accepted the macrocosm-microcosm analogy as well as sympathetic action in nature.[38] Shortly thereafter, in response to Mersenne's request for his opinion of Gaffarel's *Curiositez inouyes sur la sculpture talismanique des Persans* (1629), van Helmont questioned Gaffarel's explanation of the influence of the stars.[39] The subject was fundamental for our understanding of nature since

> En verité icelle puissance est la source et fondament de la magie naturelle et de la cabale (qui, à mon jugement, n'est point fondée sur la delineature de l'alphabet hebreu, comme les Juifs nous veulent persuader, mais est la *magie celeste* ou *celeste astrologie*, qui lit les evenemens dedans les astres, et ce sans aulcune supputation astrologale, mais d'une façon cogneue aux Adeptes) et de toutes les sciences qu'on acquiert de la lation superieure, dont j'en ay faict quelque touche et petite chanterelle, au livret *de Magnetismo*, mais encore plus que cest ulceré siecle permet.[40]

He went on to attack Gaffarel for his gross description of the signatures of herbs and for his discussion of the dreams of the Old Testament prophets.[41] These, he pointed out, were correctly interpreted by Paracelsus in the *Philosophia sagax*. When asked to comment on Gassendi's examination of the philosophy of Robert

---

[38] Van Helmont (Malines) to Mersenne (Brussels), June 1630, *Correspondance*, Vol. II, pp. 496-503 (500, 502).

[39] Van Helmont (Brussels) to Mersenne (en voyage), Sept. 26, 1630, *ibid.*, 530-540.

[40] *Ibid.*, p. 533.

[41] *Ibid.*, p. 534.

Fludd, van Helmont could concur heartily in the condemnation.[42] However, on no account could he accept a blanket attack on Paracelsus by authors of the like of Goclenius, Erastus, or Libavius, who could no more judge "ce nonpareil homme" than the Sorbonne could judge a person's knowledge of nature.[43] Above all, care must be taken not to injure "ces rares personnages des Adepts, et avec le temps il no peult estre que Paracelse ne soit mis au plus hault des scavans en nature, et par là l'escrit (quoyque tres docte et judicieux) de nostre amys, moins estimé lors."[44]

The remaining letters indicate that Mersenne was a difficult task master. He questioned van Helmont on muscular fibers, and the hardness of bodies,[45] on the relationship of air to water[46] (surely a sore point in light of the van Heer debate), the force of projectiles,[47] and many other problems related to mechanics as well as a number of theological subjects. And, when asked "si tria principia sunt, prima vel ultima," his answer reflected the views of most of the chemical philosophers: "Dico, prima in synthesi, et ultima in analysi."[48]

Because of Mersenne's continual prodding—and his primary interest in problems related to the physical sciences—van Helmont's letters touch on subjects not discussed elsewhere in his *Opera*.[49] However, they surely indicate that his views had not changed significantly in the ten years between 1621 and 1631. They reveal a man who revered Paracelsus and who had no desire to criticize him. He was later to say that those who hated him

[42]Van Helmont (Brussels) to Mersenne (Paris), Dec. 19, 1630, *ibid.*, pp. 582-593 (584).

[43]*Ibid.*, p. 585.

[44]*Ibid.*, p. 586.

[45]Van Helmont (Brussels) to Mersenne (Paris), Jan. 11, 1631, *Correspondance*, Vol. III (2nd ed., 1969), pp. 10-19.

[46]Van Helmont to Mersenne, Jan. 15, 1631, *ibid.*, pp. 31-50 (32).

[47]Van Helmont to Mersenne, Feb. 6, 1631, *ibid.*, pp. 74-94.

[48]Letter of Jan. 15 cited above (n. 46), p. 32.

[49]An excellent example of this may be seen in the letter from Van Helmont to Mersenne dated Mar. 29, 1631. In writing of the barrenness of the French queen and the need for an heir, he suggested that traditional remedies should be replaced by the more effective Hermetic preparations. *Ibid.*, pp. 151-155 (153). Nève de Mévergnies has examined these letters in his "Sur les lettres de J. B. Van Helmont au P. Marin Mersenne," *Revue Belge de Philosophie et d'Histoire*, **26** (1948), 61-83.

called him a Paracelsian while those who esteemed his work considered him an adept.[50] Yet in 1631 when Mersenne addressed him in this fashion, he replied, "Je vous prie ne me donner le tiltre d'Adepte, car je ne le merite."[51]

Van Helmont's Final Years (1631-1644)

In the letters to Mersenne there are repeated references to the continual difficulties van Helmont faced with the Church and the universities. His replies to the twenty-seven condemned propositions were subjected to a most minute criticism, and a resultant condemnation was at last submitted to the Ecclesiastical Court. In a strong document van Helmont was charged with having departed from the true philosophy to espouse superstition, magic, and the diabolical art. Above all, he had followed Paracelsus and his disciples in preaching the chemical philosophy, thus having spread more than "Cimmerian darkness" over all of the world.[52]

On March 4, 1634, van Helmont was arrested and his books and papers seized. After having been imprisoned in the archbishoprical jail for four days, he was transferred—after the deposit of six thousand florins as bond—to the Minorite convent at Brussels. Once again he repudiated his errors and promised to submit humbly to the decisions of the Church court. Following a series of new interrogations van Helmont wrote to the archbishop to request permission to live at home. This was granted, but he was placed under house arrest for nearly two years, until March 16, 1636, again after posting six thousand florins as bond. It was not until 1642 that van Helmont was able to obtain an ecclesiastic imprimatur for the third work he published, the *Febrium doctrina inaudita*.[53] Later there appeared the *Opuscula medica inaudita* (1644),

[50]J. B. van Helmont, "Praefatio," *Ortus*, p. 630; *Oriatrike*, p. 632.

[51]Van Helmont to Mersenne, Feb. 14, 1631, *Correspondance*, Vol. III, pp. 95-109 (108).

[52]". . . alter pyrotechnice philosophando, has plus quam cimmerias tenebras, quibus utriusque intellectus altissime obsessus est, et pro nutu daemonis circumagitur, toti mundo effundant . . . ." As quoted by Nève de Mévergnies, *Van Helmont*, p. 133. See pp. 135-143 for a summary of the subsequent details of the case.

[53]J. B. Van Helmont, *Febrium doctrina inaudita* (Antwerp, 1642). See the *Opuscula*, p. 67 (separately paginated); *Oriatrike*, pp. 931-1011.

which included a revised text of the work on fevers, tracts on the stone and the plague, as well as an attack on the deception and ignorance of physicians who subscribed to the humoral pathology.[54]

Van Helmont was working unceasingly at this time, and in the winter of 1644 he contracted pleurisy. Recognizing his imminent death, he asked his son, Franciscus Mercurius, to

> Take all my Writings, as well those crude and uncorrected, as those that are thorowly expurged, and joyn them together; I now commit them to thy care, accomplish and digest all things according to thy own judgement: It hath so pleased the Lord Almighty, who attempts all things powerfully and directs all things sweetly.[55]

A few days later—December 30, 1644—he was dead. It was only two years after his death that his widow received a form of religious rehabilitation of her husband from the religious authorities.

The Helmontian Philosophy of Nature: Mathematics and Motion

The labors of Franciscus Mercurius van Helmont appeared in 1648 as the *Ortus medicinae*, and this was accompanied by a new edition of the *Opuscula medica inaudita*. The two works combined proved to be among the most influential medical and scientific publications of the seventeenth century. Totalling more than eleven hundred pages in the editions of 1648, the Latin texts were published again in 1651, 1655, 1667, 1682, and 1707. In addition, there were translations into English (1662, 1664), French (1671), and German (1683). A Flemish compilation, probably prepared by van Helmont himself, appeared in 1659 and 1660.[56]

---

[54] *Opuscula medica inaudita: I. De lithiasi; II. De febribus* (2nd ed. of *Febrium doctrina inaudita*); *III. Scholarum humoristarum passiva deceptio atque ignorantia; IIIa. Appendix ad tractatum de febribus sive caput XVI et XVII; IV. Tumulus pestis* (Cologne, 1644). Reprinted as a separate volume in 1648 and cited in this chapter through the 1966 Brussels edition, this work appears in the *Oriatrike*, pp. 815-1161.

[55] "Amico lectori" (F. M. Van Helmont), *Ortus*, sig. **1^V; *Oriatrike*, sig. b2^V.

[56] For a succinct discussion of the various editions of the *Opera* see Pagel, "Van Helmont," *DSB*, Vol. VI, pp. 257-258. An interesting additional item is *Ioannis Baptistae Van Helmont physices universalis doctrine et christianae fidei congrua et necessaria philosophia*, published in Wallachia in 1709 by Demetrius Cantemir. Although it was a paraphrase and selection from the *Ortus medicinae*, it made available to

Internal evidence indicates that although much of the research on which his work was based may date from his period of study at Vilvord (1609-1616), van Helmont must have completed most of his writing after 1634, since no relevant manuscripts were found when his house was searched in that year. And although his son included the work on the weapon-salve in his collection, the views expressed in most of the other tracts are far more critical of Paracelsus even when praising the value of chemistry. While it is possible to picture van Helmont as a typical chemical philosopher of the early seventeenth century, his views go far beyond those of his predecessors in terms of both experimental method and theory. It is thus necessary to be alert to tradition as well as novelty in his writings.

Van Helmont resembles many of the other chemical philosophers in his strongly worded opposition to the ancients. He considered their work to be heathenish in its origin and hardly fit for Christian consumption. Yet from his own education he could attest that this remained the basis of the university curricula, and it thus seemed clear to him that he must "destroy the whole natural Phylosophy of the Antients, and to make new the Doctrines of the Schooles of natural Phylosophy."[57]

But what specifically was to be rejected? As other Paracelsians had before him, he paid full tribute to Hippocrates. After him, however, had come Aristotle and Galen, in whose works were many intolerable statements, but whose fundamental reliance on mathematics and local motion necessitated special attention. No less than Severinus and other Paracelsians, van Helmont accused Aristotle and Galen of having introduced mathematics into the

---

Eastern Europe the views of van Helmont. I am indebted to my colleague William H. McNeill for this information. See Dan Badaru, *Filozofia lui Dimitrie Cantemir* (Bucharest: Editura Academici Republicii Populare Romîne, 1964) with a French summary, pp. 394-410. In an essay review of the recently reprinted German edition of van Helmont (1683) (Munich: Kösel-Verlag, 1971), the present author has discussed the current literature on van Helmont. "Some Comments on the Contemporary Helmontian Renaissance," *Ambix*, **19** (1972), 145-150.

[57] J. B. van Helmont, "Promissa authoris," *Ortus*, p. 6, *Oriatrike*, p. 1. See also the "Logica inutilis" (sect. 3), *Ortus*, p. 41; *Oriatrike*, p. 37. "Etenim mihi necesse fuit, in condenda nova Philosophia, omnia paenè rescindere, à prioribus tradita, multaque institui, & restitui debere, qua pari animo, non recipiet quilibet."

study of nature. Aristotle

> ... brought learning by demonstration into Nature, by a forced Interpretation, as that he would have natural causes wholly obey numbers, lines, and letters of the Alphabet, by a rashness altogether ridiculous .... And although, before the Elements of *Euclide* sprang up, he was more ignorant of the Mathematicks: yet *Aristotle* being far more skilful in this, than in nature, endeavored to subdue Nature under the Rules of that Science.[58]

Galen, in turn, seduced by the "Mathematical demonstrations" of his predecessor, proceeded to apply this method erroneously to the study of man. He "dispersed his Theoremes into a great Body, which afterwards, the prattle of the Greeks increased into a huge one, and which, the Schooles even to this very day, do superstitiously worship, because they have made themselves Trophies by others labours."[59]

Nevertheless, van Helmont's condemnation of Aristotle and Galen on this score was one of degree. All that had been done by the Scholastics had not been useless, he wrote, for indeed,

> The Heathnish Schooles ... may have an Historical knowledge, the observer of things contingent or accidental, of things regular, and necessary: which is a memorative knowledge of the thing done: they may also get Learning by demonstration, which is the knowledge of applying things unto measure. And lastly, they may promise rational knowledge, which is derived from either of these, by the fitting of discourse.[60]

If they had confined themselves to these things they might have produced satisfactory descriptive accounts and have known of "the reflux, or going back of the Starrs and Sea, that the water bends to a levelled roundness, and downward, draw divers Sequels from thence, and stablish them into maxims."[61] Yet they had gone too

---

[58] "Causae et initia naturalium" (sects. 38, 40), *Ortus*, p. 38; *Oriatrike*, p. 33. The views of van Helmont on mathematics and motion will be found (with additional points) in two papers by the present author: "Mathematics and Nature in the Chemical Texts of the Renaissance," *Ambix*, **15** (1968), 1-28, 211; "Motion in the Chemical Texts of the Renaissance," *Isis*, **64** (1973), 4-17.

[59] Van Helmont, "Promissa authoris" (col. 1, sect. 3), *Ortus*, p. 7; *Oriatrike*, p. 2.

[60] "Promissa authoris" (col. 1, sect. 14), *Ortus*, p. 9; *Oriatrike*, p. 4.

[61] *Ibid.*

far in search of causes and had drawn unwarranted conclusions from their imperfect knowledge.

In his criticism of the philosophy of the Aristotelians, van Helmont placed heavy emphasis on the theological implications of the study of motion. Aristotle had said that nature is the "*Principle, or beginning of motion, as also of rest in Bodies, in whom it is in, by it self, and not by accident.*" Not so, Helmont countered, for "I believe, that *Nature is the Command of God, whereby a thing is that which it is, and doth that which it is commanded to do or act.* This is a Christian definition taken out of Holy Scriptures."[62]

It was in Aristotle's *Physics* that there was a discussion of matter based on mathematical abstraction. Above all, this work brought in "locall motion, as it serves to Science Mathematical or Learning by demonstration, alike foolishly, and with an undistinct indiscretion, into nature."[63] And what had been the result? Following Aristotle's dictum "*that in every locall motion, a first unmoveable Mover is of necessity to be appointed,*" the Scholastics had been forced to "require an unmoveableness in the first Mover."[64] To thus confine the Creator could not be tolerated, for

> ... being altogether free, as well in his beck and motion, as in rest, he indifferently and alike powerfully moveth all things: Therefore his own unmoveable essence doth not import a necessity required by the Schooles, but the meer good pleasure of his glory. For his own word ... hath departed into nature, which afterwards is for moving of it self.[65]

Both Aristotle and Galen had based their medicine and their natural philosophy upon a premise that was fundamentally incorrect. Motion and life were clearly a result of the divine will, in reference to which van Helmont coined a new word, *blas*, a universal motive power. Present in the stars, this "is the general beginning of motion; it seemes no lesse to respect the Earth, than the Air

---

[62] "Physica Aristotelis et Galeni ignara" (sects. 2-3), *Ortus*, p. 46; *Oriatrike*, p. 42.

[63] "Physica Aristotelis et Galeni ignara" (sect. 8), *Ortus*, p. 49; *Oriatrike*, p. 45. "Ac postremo motum localem, quatenus Mathesi inseruit, in naturam aeque futiliter, atque indistincta indiscretione introducunt."

[64] "Blas humanum" (sect. 1), *Ortus*, p. 178; *Oriatrike*, p. 176.

[65] "Blas humanum" (sect. 3), *Ortus*, p. 180; *Oriatrike*, p. 176.

and Water."⁶⁶ Van Helmont's description was largely limited to meteorological effects insofar as macrocosmic phenonema were concerned: the *blas* was directly responsible through its own motion for rain, snow, hail, the winds, and the tides; its study would reveal a true understanding of heat, cold, humidity, and dryness; and the concept would not only help to explain clear and covered skies but even aid in forecasting the weather. Indeed, the *blas* made astrological and medical prognoses possible for the physician.⁶⁷

It is quite clear that van Helmont had more interest in the *blas humanum* than the *blas meteoron*. Because of the order of Creation described in Genesis he was convinced that the stars could not directly cause motion in man.⁶⁸ Plants had been created prior to the stars; thus, although animals and men were created later, it was unthinkable that their motions should be subject to the stars while those of the plants—a lower form of life—were not. For van Helmont this was a conclusive argument against the macrocosm-microcosm universe. In contrast to the tract on the weapon-salve and the letters to Mersenne, in the later works of van Helmont there are constant attacks on this analogy. For him this chain of nature had been broken.

> The name therefore of Microcosm or little World is Poetical, heathenish, and metaphorical, but not natural or true. It is likewise a phantastical, hypochondriacal, and mad thing, to have brought all the properties, and species of the Universe into man, and the art of healing: But the life of man is too serious, and also the medicine thereof, that they should play their own part of a Parable or Similitude, and metaphor with us.⁶⁹

And although he was willing to admit that there appeared to be a relationship between man and the stars, he was adamant in insisting that such imitation or correspondence was at best an indirect rather than a direct influence.

---

⁶⁶"Blas meteoron" (sect. 5), *Ortus*, pp. 81-82; *Oriatrike*, p. 79.

⁶⁷"Blas meteoron" (sect. 6, 12), *Ortus*, p. 82; *Oriatrike*, p. 79.

⁶⁸"Because the Stars were before the Creation of sensitive things; therefore it was meet, that the Blas of men should not indeed follow the guidance of the Stars, but only that it imitate the motion of those, not as of motive powers, but no otherwise than as by a free motion we do follow the foot-steps of a Coach-man or Post . . . ." "Blas humanum" (sect. 5), *Ortus*, p. 180; *Oriatrike*, p. 177.

⁶⁹"Scabies & ulcera scholarum" (sect. 33), *Ortus*, p. 328; *Oriatrike*, p. 323.

Van Helmont believed that all living things had their separate *blas* implanted in the initial seed by the Creator. Thus they all had "not onely a Principle of every Motion, but also their own limited Period of Durations proper unto every Motion."[70] These views were closely connected with his concept of biological time. Arguing against Aristotle, he rejected a definition of time in terms of motion and succession; rather, it was to be thought of as indivisible and devoid of succession,

> ... being essentially bound up with duration. This is shown in the life-span and life rhythm specific to each individual and given to the divine *semina* by the Creator. By virtue of this participation in divinity, time (*duratio*) was not different from eternity, as propounded in the Christian (Augustinian) doctrine.[71]

In man every organ has its own *blas*,

> ... readily serving for the uses of its own ends, flowing out of the Beginning of its own Essence, which are the will of the flesh, and the lust or desire of a manly will. Therefore there is in us a two-fold Blas: To wit, One which existeth by a natural motion; but the other is voluntary, which existeth as a mover to it self by an internal willing.[72]

For van Helmont there could be no doubt that

> ... there is something in sublunary things which can move it self locally, and alteratively, without the Blas of the Heavens, and an unmoveable natural mover. The will especially, is the first of that sort of movers, and moveth it self; also a seminal Being, as well in seeds, as in the things constituted of these. Moreover as God would, so all things were made: Therefore from a will they were at first moved: For whom hence whatsoever unsensitive things are moved, they are moved as it were by a certain will and pleasure or precept of nature, and have their own natural necessities, and ends; even as is seen in the beating of the Heart, Arteries, expelling of many superfluities, &c.[73]

---

[70]"De tempore" (sect. 18), *Ortus*, p. 632; *Oriatrike*, p. 636.

[71]Pagel, "Van Helmont," *DSB*, Vol. VI, p. 257. For a detailed analysis of the text (with a new translation) see Walter Pagel, "J. B. van Helmont, De Tempore, and Biological Time," *Osiris*, **8** (1949), 346-417, to be supplemented with Helene Weiss, "Notes on the Greek Ideas Referred to in Van Helmont, De Tempore," *Osiris*, **8** (1949), 418-449.

[72]"Blas humanum" (sects. 8-9), *Ortus*, p. 181; *Oriatrike*, p. 177.

[73]"Blas humanum" (sect. 9), *Ortus*, p. 181; *Oriatrike*, p. 177.

It is clear that van Helmont was primarily concerned with the problem of freeing the Creator from the shackles placed on him by the Aristotelians. The formulation of a *blas*, or motive power, made the doctrine of immovability no longer a necessity for him. Therefore he could conclude that the Aristotelian descriptive interpretation of nature

> ... is a Paganish Doctrine drawn from Science Mathematical, which necessitates the first Mover to a perpetual unmoveablenesse of himself, that without ceasing he may move all things ....[74]
>
> Therefore let the Schooles know, that the Rules of the Mathematicks, or Learning by Demonstration do ill square to Nature. For man doth not measure Nature; but she him.[75]

The Elements and the Principles

The case study of van Helmont's treatment of "mathematical demonstration" and local motion is indicative of his work in general. Rarely willing to give an inch to his opponents, he was constantly on the offensive in his dual task, the destruction of the philosophy and medicine of the Schools coupled with his own formulation of a "new Philosophy."[76] Theological considerations are in evidence everywhere, and if he believed that it was the duty of naturalists to study nature, he was equally convinced that this study was intimately connected with an understanding of the Creator. Then again, van Helmont's study of motion indicates his special interest in man and the important relationship of medicine to natural philosophy.

In a fashion reminiscent of other Paracelsians, van Helmont from the beginning turned his attention to the problem of the four elements. These were clearly the basis of the ancient philosophy and its resultant medicine; they furthermore remained the subject of interminable disputes at the universities. For these reasons it was essential to examine the four elements in detail:

> Surely it is a wonder to see, how much brawling and writing there hath been about these things: and it is to be pitied, how much these loose dreams of trifles, have hitherto circumvented or beset the World:

[74] "Blas humanum" (sect. 3), *Ortus*, p. 180; *Oriatrike*, p. 176.
[75] "Causae, et initia naturalium" (sect. 41), *Ortus*, p. 39; *Oriatrike*, p. 34.
[76] For van Helmont's use of the term *nova Philosophia*, see above, n. 57.

they have prostituted destructive vain talkings in the faires of the Schooles, instead of the knowledge of Medicine, and so, so damnable a delusion, hath thereby deceived the obedience of the sick, in healing.[77]

The truth of the matter could be established rapidly through reference to Holy Scripture rather than to the ancient philosophers. The account in Genesis clearly indicated the order of Creation, a far more valid guide to the elements than Aristotle or Galan.

> ... in the beginning, the Heaven, Earth: and Water, the matter of all Bodies that were afterwards to arise, was created. But in the Heaven were the Waters contained, but not in the Earth; hence I think the Waters to be more noble than the Earth: yea, the Water, to be more pure, simple, indivisible, firm or constant, neerer to a Principle, and more partaking of a heavenly condition, than the Earth is. Therefore the Eternall would have the Heaven to contain Waters above it, and as yet something more (by reason whereof it is called Heaven) that which we call, the Air, the Skie, or vitall Air. For therefore neither is there mention made of the creating of the Water and Air, for that, both of them, the Etymologie of the Word, Heaven, did include. Therefore, I call these two Elements Primigeniall, or first-born, in respect of the Earth.[78]

However, if these two elements might be accepted, fire could not be. Since nowhere in Holy Scripture is there reference to the creation of fire, it cannot be considered an element.

> Therefore I deny that God created four Elements; because, not the fire, the fourth. And therefore it is vain, that the fire doth materially concurre unto the mixture of bodies. Therefore the fourfold kinde of Elements, Qualities, Temperaments or Complexions, and also the foundation of Diseases, falls to the ground.[79]

As for the element earth, this too was created, but it was created from water, a fact van Helmont felt could be shown through a quantitative experiment. Although a similar suggestion had been made earlier by Nicholas of Cusa, van Helmont was the

---

[77]"Elementa" (sect. 2), *Ortus*, p. 52; *Oriatrike*, p. 47 (numbered incorrectly as 45).

[78]"Elementa" (sects. 5-8), *Ortus*, pp. 52-53; *Oriatrike*, p. 48.

[79]"Elementa" (sects. 9-10), *Ortus*, p. 53; *Oriatrike*, p. 48.

first to actually demonstrate that a willow tree grown in a measured amount of earth to which nothing is added but water consumes none of the earth.[80] For him it seemed reasonable to conclude that the tree growth (partially earth, as might be proved by fire analysis) had been formed from water alone.

> ... I took an Earthen Vessel, in which I put 200 pounds of Earth that had been dried in a Furnace, which I moystened with Rainwater, and I implanted therein the Trunk or Stem of a Willow Tree, weighing five pounds; and at length, five years being finished, the Tree sprung from thence, did weigh 169 pounds, and about three ounces: But I moystened the Earthen Vessel with Rain-water, or distilled water (alwayes when there was need) and it was large, and implanted into the Earth, and least the dust that flew about should be co-mingled with the Earth, I covered the lip or mouth of the Vessel, with an Iron-Plate covered with Tin, and easily passible with many holes. I computed not the weight of the leaves that fell off in the four Autumnes. At length, I again dried the Earth of the Vessel, and there were found the same 200 pounds, wanting about two ounces. Therefore 164 pounds of Wood, Barks, and Roots, arose out of water onely.[81]

The thesis seemed to be sustained further by the fact that earthy substances invariably could be reduced to water. Stones, gems, sand, and clay could all be changed into salt "equall in weight to its own body." This salt was its prior state, and it, in turn, could be changed into a liquor and then into a unsavory water "of equall weight with its Salt." By a similar process plant and animal matter could be reduced to its original form, and although far more difficult, it was also possible to reduce metals and the fundamental sand, or virgin earth (*quellem*), to their salts.[82]

---

[80]Cusanus suggested that seeds be planted in a weighed earthen pot, and after they had grown, the observer "woulde find the earth but very little diminished, when he came to weigh it againe: by which he might gather, that all the aforesaid herbs had their weight from the water." Nicholas Cusanus, *The Idiot in Four Books. The First and Second Wisdome. The Third of the Minde. The Fourth of Statick Experiments, Or Experiments of the Ballance* (London: William Leake, 1650), pp. 188-189. The Gnostic and neo-Clementine sources for the experiment have been discussed by H. M. Howe in "A Root of van Helmont's Tree," *Isis*, **56** (1965), 408-419.

[81]"Complexionum atque mistionum elementalium figmentum" (sect. 30), *Ortus*, pp. 108-109; *Oriatrike*, p. 109.

[82]"Elementa" (sect. 11), *Ortus*, p. 53; *Oriatrike*, pp. 48-49.

The semi-elementary nature of the *quellem* was established not only through its difficult resolution to water, but also by the evidence of the spade.[83] The variety of surface earths all rested upon this white sand, which was the substance of the center of our planet. It was true that this sand could be converted to glass, which was not altered by fire, but if the operator

> ... shall co-melt the fine powder of glasse, with more of the *Alcali*, and shall set them forth in a moyst place; he shall straightway finde all the glasse to be resolved into water: on which, if *Chrysulca* be powred, so much being added as sufficeth to the filling or satisfying of the *Alcali*, he shall presently finde in the bottom, the Sand to settle, it being of the same weight, which at first was fitted for the making of the glasse. Therefore the Earth remains unchanged . . . .[84]

Clearly there was no quaternary system of elements, but could anything more convincing be offered in support of the three principles of the Paracelsians? These could hardly be dismissed in a cavalier fashion since their influence was both wide and deep. Here van Helmont turned to a careful study of the effect of fire and heat on analysis. It was true, he began, that fire separates bodies into the three fractions commonly called salt, sulfur, and mercury. Paracelsus, however, was not to be credited with originating the *tria prima*, for he had learned of them from reading Basil Valentine, who stated that "Water, Oyle, and Salt, were to be separated by Distillation from most Bodies." Subsequently he began to call them "the first universal beginnings of corporal beings."[85] This was a concept that was quickly adopted because of its usefulness, but van Helmont did not hesitate to point out that Paracelsus also gave contradictory views on the elements and the principles.

Van Helmont concluded that although the *tria prima* were useful because they were obtained by distillation from so many things, they were "a late Invention, contrary to the truth of Nature."[86] The apparent separation by the fire was actually a delusion, since these three things do not exist in substances prior to the

---

[83]"Terra" (sect. 4), *Ortus*, p. 54; *Oriatrike*, p. 50.
[84]"Terra" (sect. 16), *Ortus*, p. 56; *Oriatrike*, p. 52.
[85]"Tria prima chymicorum principia, neque eorundem essentias de morborum exercitu esse" (sects. 3-5), *Ortus*, p. 399; *Oriatrike*, pp. 402-403.
[86]"Tria prima" (sect. 46), *Ortus*, p. 405; *Oriatrike*, p. 408.

application of heat. Instead, "*as by a Trans-mutation made by the Fire, they are there generated, as it were new Beings, and there is made that, which was not before.*"[87] Since analyses show that there are variable proportions of oil and ashes in vegetable matter at different times, it can be concluded that they were not in the substance from the beginning and therefore cannot be called primary principles. Then again, analyses reveal some substances from which only one or two fractions can be distilled, others from which three things are never separated, and a few which are completely unchangeable in the fire.[88]

The constancy of gold was explained by van Helmont by the predominance within the metal of mercury, which derives from the truly elementary water and the mercurial seed from which neither salt nor sulfur can be extracted.[89] A similar case is sand, from which neither sulfur nor mercury can be obtained by fire. Here the seed is "content with a stonyfying coagulation of water, without an appointment of fatnesses, or Mercuries."[90] But the chemist need only turn to tartar to show the complex action of fire. In the distillation of sixteen ounces of this substance thirteen ounces of oil are obtained, yet tartar does not have an oily nature, and, if one attempts to reconstitute tartar from its distilled oil and the residuous salt, only a soap will be recovered.[91]

Van Helmont's discussion of the principles is strikingly similar to that of Robert Boyle in *The Sceptical Chymist* (1661). In short, van Helmont thought it possible to speak generally of mercury, sulfur, and salt, since they were so commonly obtained from distillation processes, but he also advised the chemist to always remember that these substances originated from the application of heat.[92] While here van Helmont could play the part of the "Sceptical Chymist" himself, his readers had to be constantly on guard for lapses, for, as he admitted himself, he occasionally ascribed the three principles to water, "which in it self being defiled with no Seed, hath on every side, a co-like simplicity, and impossibility of

[87]"Tria prima" (sect. 47), *Ortus*, p. 405; *Oriatrike*, p. 408.
[88]"Tria prima" (sects. 48-50), *Ortus*, p. 405; *Oriatrike*, p. 408.
[89]"Tria prima" (sect. 53), *Ortus*, p. 407; *Oriatrike*, p. 410.
[90]"Tria prima" (sect. 64), *Ortus*, p. 409; *Oriatrike*, p. 411.
[91]"Tria prima" (sect. 68), *Ortus*, p. 410, *Oriatrike*, p. 412.
[92]"Tria prima" (sect. 63), *Ortus*, pp. 408-409; *Oriatrike*, p. 411.

separation." Such statements, he advised the reader, were to be considered only as analogies.[93]

Chemistry as the Key to Nature

One cannot proceed far in the works of van Helmont without realizing that his approach to nature is indecipherable without chemistry. If his ultimate goal was a new medicine and the understanding of disease, this understanding could not be granted to anyone unskilled in the art of the fire. Van Helmont thought that Lullius had been correct in referring to the universal nature of alchemy;[94] he could even add that "as the art of the fire, unlocks Bodyes before our eyes; so it opens the Gate unto Natural Philosophy."[95] Indeed,

> I praise my bountiful God, who hath called me into the Art of the fire, out of the dregs of other professions. For truly, Chymistry, hath its principles not gotten by discourses, but those which are known by nature, and evident by the fire: and it prepares the understanding to pierce the secrets of nature, and causeth a further searching out in nature, than all other Sciences being put together: and it pierceth even unto the utmost depths of real truth: Because it sends or lets in the Operator unto the first roots of those things, with a pointing out the operations of nature, and powers of Art: together also, with the ripening of seminal virtues. For the thrice glorious Highest, is also to be praised, who hath freely given this knowledge unto little ones.[96]

The wonders of chemistry were so very real that van Helmont found it easy to slip into a familiar alchemical dream sequence in their description:

> ... sliding as it were into a dream, I saw my self in a certain Kingly Pallace, excelling humane artifices. But there was a high Throne, encompassed with an unaccessable light of Spirits. But he who sate in the Seat of the Throne, is called [He is] And the foot-stool of his feet

---

[93]"Tria prima" (sect. 54), *Ortus*, p. 407; *Oriatrike*, p. 410.

[94]"De lithiasi" (Ch. 3, sect. 1), *Opuscula*, p. 21; *Oriatrike*, pp. 839-840. The pertinent quotation is transcribed at length above in Ch. 1.

[95]"De lithiasi (Ch. 5, sect. 4), *Opuscula*, p. 40; *Oriatrike*, p. 859.

[96]"Pharmacopolium ac dispensatorium modernorum" (sect. 32), *Ortus*, p. 463; *Oriatrike*, p. 462.

PLATE XXV The chemical philosopher as imitator and interpreter of natural phenomena. From the frontispiece to J. B. van Helmont, *Opera omnia* (Frankfurt: J. J. Erythropilus, 1682).

The Broken Chain

[*Nature*] The Porter of the Court, was called [*Understanding*] who without speech, reached unto me a little Book, a choice out of darkness, the name whereof was [*The bud of a Rose not yet opened*] And although the Porter uttered no voice, yet I knew, that little Book was to be devoured by me. I stretched forth my hand, and ate it up. And it was of an harsh and earthy taste, as it would stop up my winde-pipe; so as I swallowed it with a great slowness of labour. From whence, afterwards, my whole head, seemed to be transparent. Then, afterwards, another spirit of a superior order, gave me a bottle, wherein was [*Fire-water*] as being in one word: A name altogether simple, singular, undeclinable, unseparable, unchangeable, and immortal. But I knew not what my business was with it: Neither, heard I any thing more of it; and by reason of the fear of its greatness, my jawes were shut up, and my voice clave to my jawes. At length, having performed due worship before the Throne, I endeavoured diversly to experience, what the bottle might contain. Behold, before the doors of the Court, there was the Art of the Fire, a cheerful old Woman being the Turn-Key, who did not open the locks without, unless the Porter had first withdrawn the bolt within; the which he did not attempt, unless, from a sign given him by the light of the Throne. But unto those that knocked at the doors, the Porter answered, the Key-keeper holding her peace: I know you not. But they who tryed to look in thorow the lattices of the windows, being smitten with darkness, forthwith fell down mad, many wandred up and down, promising great things without a foundation. I stood a good while silent; and then afterwards, a hand (the rest of whose Body I saw not) led me aside unto a pleasant Garden: where on a sudden, all Simples worshipped me, as though every one had been singular by themselves. In which assault, I felt or perceived all the Simples of the world, not indeed, as if their qualities did act in me (for I being but one, had not been sufficient for the bearing of them all) as it were, their object: but they all were seen, as on a Theatre, to represent in me their Tragedies. And I wish, I may well declare them with my pen![97]

The following pages offer a host of chemical observations related to medicine, but the chapter closes with a longing for the new lost bottle—and praise for the Lord who made these truths known.[98]

There is little doubt that van Helmont had a sincere belief in alchemy. He confessed that he had transmuted mercury to gold on

---

[97] "Potestas medicaminum" (sects. 3-4), *Ortus*, p. 471; *Oriatrike*, pp. 170-171.
[98] "Potestas medicaminum" (sect. 66), *Ortus*, p. 483; *Oriatrike*, p. 482.

more than one occasion, and in a circumstantial account he wrote that

> I therefore contemplate of the New-birth or renewing of those that are to be saved, to be made in a sublunary and earthly Nature, just, even as in the Projection of the Stone which maketh Gold: For truly, I have divers times seen it, and handled it with my hands: but it was of colour, such as is in Saffron in its Powder, yet weighty, and shining like unto powdered Glass: There was once given unto me one fourth part of one Grain: But I call a Grain the six hundreth part of one Ounce: This quarter of one Grain therefore, being rouled up in Paper, I projected upon eight Ounces of Quick-silver made hot in a Crucible, with a certain degree of Noise, stood still from flowing, and being congealed, setled like unto a yellow Lump: but after pouring it out, the Bellows blowing, there were found eight Ounces, and a little less than eleven Grains of the purest Gold: Therefore one onely Grain of that Powder, had transchanged 19186 Parts of Quick-silver, equal to it self, into the best Gold. The aforesaid Powder therefore, among earthly things, is found to be after some sort like them, the which transchangeth almost an infinite quantity of impure Mettal into the best Gold, and by uniting it to it self, doth defend it from cankering, rust, rottenness, and Death, and makes it to be as it were Immortal, against all the torture of the Fire, and Art, and translates it into the virgin-Purity of Gold; only it requires Heat.[99]

For van Helmont one of the wonders of chemistry was the *alkahest*, a liquor which "resolves every visible Body into its first matter, the power of the Seeds being reserved."[100] This marvel he ascribed to Paracelsus.[101] But little less remarkable was the universal remedy of the Irish alchemist Butler, who employed a small stone (called *drif* by van Helmont) which through proper application was capable of curing any illness.[102] Describing it as a metallic body that could be prepared only by a true adept, van Helmont insisted on its spirituality which necessitated repeated sublimations

---

[99]"Vita aeterna," *Ortus*, p. 743; *Oriatrike*, pp. 751-752. At the opening of the "Arbor vitae" van Helmont stated that he had transmuted mercury to gold "distinctis vicibus." *Ortus*, p. 793; *Oriatrike*, p. 807.

[100]"Explicatio aliquot verborum artis," *Opuscula*, sig. A4^V; *Oriatrike*, sig. Nnnnn 4^V

[101]"Progymnasma meteori" (sect. 7), *Ortus*, p. 68; *Oriatrike*, p. 65.

[102]"Butler," *Ortus*, p. 587; *Oriatrike*, p. 587.

in its preparation.[103] Here again his description is closely related to the alchemical tradition.

Quantification: A New Chemical Tool

Van Helmont's chemistry had close ties with both traditional alchemy and Paracelsian iatrochemistry, but it would be misleading to judge him only in this light. His work went far beyond that of his predecessors in almost all areas, and he was consistently able to bring forth observational evidence based upon his own laboratory research in support of his views.

Of special significance is van Helmont's interest in quantification, a tool rarely employed by earlier chemists except in mining technology and in checking weights in cases of alleged transmutations. To those instances of van Helmont's careful attention to weights in chemical reactions which have already been noted,[104] three additional examples will be added at this time, the first being his interest in specific gravity. In a letter to Mersenne in 1631 van Helmont rejected the views of a chemist who had asserted that copper is heavier than gold. Referring to his own observations, he wrote, "I have found these proportions: one pound of tin occupies the same place in water as $1\frac{3}{8}$ pounds of silver, $1\frac{5}{8}$ pounds of lead, $1\frac{7}{8}$ pounds of mercury and $2\frac{2}{3}$ pounds of gold."[105] Elsewhere van Helmont observed that water was an eighth part heavier than the same volume of ice,[106] and he applied the same technique to the examination of urine. Hardly satisfied with the traditional method of uroscopy, van Helmont found the distillation procedure described

---

[103]Van Helmont's method of preparing the *drif* is found in "Butler," *Ortus*, pp. 595-596; *Oriatrike*, pp. 595-596.

[104]Van Helmont's reaction against the "Mathematical" (logical) reasoning of the Schools, discussed earlier, must be contrasted with his approval of weight determinations.

[105]Van Helmont (Brussels) to Mersenne (Paris), Jan. 30, 1631, *Correspondance*, Vol. III, pp. 56-57. Here again, as in the willow tree experiment, one finds a background in *The Idiot* of Cusanus. In the fourth book of "Static experiments" it is suggested that "by the difference of weights, I thinke wee may more truly come to the secret of things, and that many things may be known by a more probable conjecture." *The Idiot in Four Books*, p. 172. He goes on to suggest that the weights of metals of the same "bulke" or "Bignesse" might be compared (p. 182).

[106]"Gas aquae (sect. 35), *Ortus*, p. 72; *Oriatrike*, p. 75.

by Thurneisser little better. "Being willing, that man should not onely be a *Microcosme* or little World but also, that his urine should rejoyce in the same prerogative," Thurneisser had suggested that the distilled vapors and drops would indicate to the chemist both the places affected and the disease matter.[107] But van Helmont rejected this purely chemical alternative and replaced it with quantitative comparisons; for there was, he wrote,

> ... a safe method of examining urines by their weight; To wit, an ounce weigheth 600 grains. But I had a glassen vessel, of a narrow neck, weighing 1354 grains: But it was filled with rain water, weighing besides, 4670 grains: the urine of an old man, was found to weigh in the same vessel, 4720 grains; or to exceed the weight of the rain water, 50 grains. ...[108]

Another six cases follow. Besides representing the origin of modern urinalysis, this is significant in the present context as an indication of van Helmont's frequent use of quantitative evidence.

A second case of van Helmont's interest in quantification is related to degrees of heat. Like Fludd, van Helmont described an air thermoscope which he used to determine temperatures.[109] Again like Fludd, he fitted this instrument with a scale for greater accuracy. Elsewhere he rejected the four degrees of heat used by previous alchemists in favor of a fifteen-point scale ranging from "the greatest cold" (first) and melting ice (second) to boiling water (tenth), dark-red heat (thirteenth), and the "ultimate vigour of the Bellows and *Reverbery*" (fifteenth).[110]

---

[107] "De lithiasi" (Ch. 3, sect. 20), *Opuscula*, p. 25; *Oriatrike*, p. 843.

[108] "Scholarum humoristorum passiva deceptio atque ignorantia" (Ch. 4, sect. 31), *Opuscula*, p. 108; *Oriatrike*, p. 1056. Once again in *The Idiot*, Cusanus suggested that a physician might give a more valid judgment of urine "by the weight and colour both together then by the deceitfull colour alone" (p. 174).

[109] Described and illustrated in the tract "Aer" (sects. 11, 12), *Ortus*, p. 64; *Oriatrike*, pp. 60-61. In addition to the thermoscope and a new catheter, van Helmont used a pendulum for the measurement of motion. As Pagel has noted ["J. B. van Helmont, De Tempore, and Biological Time," *Osiris*, **8** (1949), 398], "he is fully aware that the length of the pendulum and not its weight determines the period of its movements." The pertinent texts will be found in the "De tempore" (sects. 49-51), *Ortus*, pp. 639-640; *Oriatrike*, p. 642.

[110] "Calor efficienter non digerit, sed tantum excitative" (sect. 35), *Ortus*, p. 206; *Oriatrike*, p. 202. Van Helmont's views on the thermoscope and the degrees of heat are discussed by Partington in *A History of Chemistry*, Vol. II, p. 220.

Third, and perhaps most important, van Helmont believed implicitly that matter could not be destroyed and that the truth of this could be examined quantitatively. In the sand glass experiment all the sand could be recovered through precipitation with the addition of *aqua fortis*.[111] Many similar examples can be culled from his writings: pearls will dissolve in vinegar, but their powdered substance can be recovered upon the addition of salt of tartar;[112] gold can be converted to a red oil, but it can be recovered in its original weight with little trouble.[113] "Whence I collect it into a new position for the Schooles. That no substance is to be annihilated by the force of nature, or art."[114]

Again, in his insistence that all earths derive from an elementary water, van Helmont was faced with the problem of accounting for increased specific gravity—and here he referred in particular to the metals gold and lead. His explanation was that in these cases the water is pressed together or condensed, because "of nothing, nothing is made. Therefore weight is made of another body weighing even so much; in which there is made a transmutation as of the matter, so also of the whole essence."[115] In short, not only did van Helmont speak in general terms of the indestructibility of matter, he insisted as well upon the permanence of weight in chemical change.

The Vacuum and the *Magnal*

Van Helmont's insistence on the ability of a body to be compressed was closely related to the problem of a vacuum. The existence or nonexistence of a vacuum was a key question for seventeenth-century scientists. It arrested the attention of Galileo, Torricelli, Pascal, and Boyle, and its ultimate acceptance was closely related to the rise of the mechanical philosophy. For this reason alone van Helmont's speculations are of considerable interest. That this subject gives us an additional opportunity to com-

[111] "Terra" (sect. 14), *Ortus*, p. 56; *Oriatrike*, p. 52.
[112] "De febribus" (Ch. 8, sects. 7-10), *Opuscula*, pp. 40-41; *Oriatrike*, p. 971.
[113] "Progymnasma meteori" (sect. 6), *Ortus*, p. 68; *Oriatrike*, 64.
[114] "Formarum ortus" (sect. 68), *Ortus*, p. 115; *Oriatrike*, p. 143.
[115] "Progymnasma meteori" (sect. 18), *Ortus*, p. 71; *Oriatrike*, p. 67.

pare the approach of van Helmont with that of Robert Fludd further enhances its interest.

For van Helmont the Aristotelians had incorrectly postulated that nature abhors a vacuum. Experience teaches that

> ... air may be pressed together in an Iron Pipe of an ell, about the length of fifteen fingers, at the expansion or enlarging of which copressed air, the sending forth of a small Bullet thorow a Board or Plank, should happen, no lesse than if it were driven out of a Handgun. ... from thence have I first of all learned the matter and conditions of the air; that it should sometimes most easily sustain a pressing together, and enlarging of it self, as the sight doth shew. From whence I consequently have supposed, that by all meanes there must needes be in the air enlarged, some free space and vacuum, according to the double extension of it.[116]

Proof of the vacuum's existence was to be found in the case of a burning candle standing in water and covered by a glass bulb. The observation that the water rises as the candle burns had been known to Philo of Byzantium, and more recently Robert Fludd had described the phenomenon several times.[117] Van Helmont began his proof of the vacuum as follows:

> Let a piece of Candle be placed in the midst of the bottom of a dish, being fastened to its melted Tallow in the bottom: Let it burn, and let water be powred round about it, to two or three fingers space; but let a deep Cupping-glasse be set over the flame, the flame appearing three fingers space out of the water, so that the mouth of the Glasse set over it, may stand upon the bottom of the dish: Thou shalt straightway see the place of the air in the aforesaid free Glasse, but the water by a certain sucking to be drawn upwards, and to ascend into the Glasse in the place of diminished air: and at length the flame to be smothered ....[118]

But why should the volume of the air be diminished? For van Helmont the fire and smoke were essentially "gas," while the tallow of the candle could easily be expanded to ten thousand times its original volume in the course of the combustion. Additionally, the heat of the flame should be expected to result in expansion rather

---

[116]"Vacuum naturae" (sects. 4, 5), *Ortus*, p. 84; *Oriatrike*, p. 82.

[117]See the discussion later in this chapter and Allen G. Debus, *The English Paracelsians* (London: Oldbourne, 1965), pp. 116-118.

[118]"Vacuum naturae" (sect. 7), *Ortus*, p. 84; *Oriatrike*, p. 82.

than contraction. "From whence I conclude, that the place of the air, ought not to be lessened by the flame, but necessarily to be increased, unless some place in the air were empty, which is lessened."[119]

For van Helmont the observed contraction of the air must be explained in terms of pores or holes in the air. In the example of the pipe there is a simple contraction due to pressure, but in the case of the candle and the dish these pores are actually annihilated.[120] Thus, even though new matter may be added, the resultant volume is still less. But, he asked, what if there should be an inflammable substance in the air "which is consumed by the flame of the Candle?" He answered that it was absurd to assume that a body can be annihilated by the flame and not be enlarged; there can only be something in the air

> ... that is lesse than a body, which fills up the emptinesses of the air, and which is wholly annihilated by the fire. Nor that indeed, as if also it were the nourishment of the fire it self: For although that thing be impertinent to this Question and place; yet that which is not truly a body, can nourish nothing. And then, seeing it is neither a body, nor a fat thing, it cannot be inflamed, kindled, or wasted or consumed by the fire.[121]

In addition, because "Air is an Element, and a simple thing, it cannot admit of composition, or a conjoyning of divers things or Beings in its own nature." As a result, although it is true that the flame through its heat increases the volume of the air, the smoke at the same time fills the pores:

> ... this is the one onely cause of the diminished space in the air, whence the flame is also consequently smothered. For the heat that is externall to the Glasse, seemes to inlarge the air in the Glasse: but the fire within, by reason of its smoakes, doth actually stir up a stifling and pressing together of the air. Therefore the heat doth by it self enlarge the air, as appeareth by the Engine meating out the degrees of the encompassing air: but the fire by reason of its smoakes, presseth it together.[122]

Air is clearly a receptacle for exhalations, and to receive them

[119] *Ibid.*
[120] "Vacuum naturae" (sect. 10), *Ortus*, p. 85; *Oriatrike*, p. 83.
[121] "Vacuum naturae" (sect. 11), *Ortus*, p. 85; *Oriatrike*, p. 83.
[122] "Vacuum naturae" (sects. 11, 12), *Ortus*, pp. 85-86; *Oriatrike*, p. 83.

its pores must be empty. Still, it is evident that any given amount of air has a limited ability to support combustion:

> ... it receiveth those exhalations, by its set and due proportion: and where it hath its emptinesses filled up to a just measure, the air fleeth away, and in its flight, it forceth or gathereth all the flame into a Pyramide or Spire. But if the air being detained from its flight, be loaded with too much smoak, it straightens it self, and extinguisheth the fire, which fills it self with smoak above due measure.[123]

The fire dies "for want of room, which cannot contain the smoak, by the pressing together whereof, the fire being stifled, is extinguished."[124]

Nevertheless, the aerial porosities under normal conditions, while void of matter, do contain a being which is midway between matter and spirit. This is the *magnal*, or "sheath of the air," through which the *blas* of the stars exerts its effect on us.[125] It is through the *magnal* that rarefaction of the air can be explained, for

> ... as oft as rarefying doth appear in the air, it must needes by all meanes happen through an increase of the Magnall: Which sounds, that a vacuum being increased in the air, the pores of the air are enlarged and extended; and so, so far it is, that by reason of heat, the air by it self, and in its own body doth sustain a rarefying, and that the body of the Element is changed: that rather it is coagulated, at least is pressed together, and that the little holes of the vacuum, do extend themselves, or that the Magnall it self is multiplied in the air.[126]

Similarly the *magnal* may be contracted in the presence of cold. Given these observations, it can be concluded that the "Magnall is like light, and is easily made, and easily brought to nothing. For that which is in it self the vacuum of the air, is almost nothing in respect to bodies."[127]

Van Helmont's difficulty in explaining the contraction of air in the case of the enclosed candle was magnified by his insistence on

[123]"Vacuum naturae" (sect. 15), *Ortus*, p. 86; *Oriatrike*, p. 84.
[124]"Vacuum naturae" (sect. 16), *Ortus*, p. 86; *Oriatrike*, p. 84.
[125]"Vacuum naturae" (sects. 19, 20), *Ortus*, p. 87; *Oriatrike*, p. 85.
[126]"Vacuum naturae" (sect. 27), *Ortus*, p. 88; *Oriatrike*, p. 86.
[127]"Vacuum naturae" (sect. 27), *Ortus*, p. 89; *Oriatrike*, p. 86.

the existence of the vacuum plus the truly elementary nature of air. The introduction of the *magnal* added new difficulties for Mersenne, who questioned him in regard to the relation of this to the universal soul and asked further "*an sancti plus magnalis habituri quàm nos modo?*"[128]

Van Helmont's explanation of the candle experiment can be compared to that of Fludd, who had postulated an active component in the air which he identified with the breath of the Lord, fire or saltpeter.[129] If he had equated his divine nitro-aerial spirit with a material component in the atmosphere, his experiment with the candle could well have led to important conclusions. However, this was not the case: in 1617 Fludd explained that the rise of the water is proportional to the amount of air in the flask, because "air nourishes fire, and in nourishing it, it is consumed, and lest there be a vacuum admitted, water, which is the third element, takes over the place of the consumed air."[130] Two years later he used the same experiment to illustrate the relationship between the soul and body.[131] Finally, in 1631, he referred to the experiment a third time in a refutation of the Aristotelian explanation. By this time the experiment was well known, for Fludd complained that the Peripatetics were employing it to show that heat attracts water. To him this was quite impossible, for when the bulb of his thermoscope was warmed, the heated air inside depressed the water, and therefore the opposite was true. According to Fludd the water rises in the bulb not because of the attractive power of the flame (heat), but because the air becomes rarefied. Consequently, with the cold (water) entering, the flame is extinguished.[132] These explanations are all interesting, but Fludd's purpose in the experi-

---

[128]Van Helmont's discussion of the *magnal* occurs in his letter to Mersenne dated Feb. 21, 1631, *Correspondance*, Vol. III, pp. 110-130 (111-112; quotation on p. 112).

[129]See above, Ch. 4, and Robert Fludd, *Tomus secundus de supernaturali, praeternaturali et contranaturali microcosmi historia, in tractatus tres distributa* (Oppenheim: T. De Bry, 1619), p. 203, and Robert Fludd, *Clavis philosophiae et alchymiae Fluddanae* (Frankfurt: Wilhelm Fitzer, 1633), p. 33.

[130]Robert Fludd, *Tractatus secundus de naturae simia seu technica macrocosmi historia* (2nd ed., Frankfurt: J. T. De Bry, 1624), pp. 471-472.

[131]Fludd, *De supernaturali*, p. 138.

[132]Robert Fludd, *Integrum morborum mysterium: Sive medicinae catholicae tomi primi tractatus secundus* (Frankfurt: Wilhelm Fitzer, 1631), pp. 456-457, 497-498.

ment was (1) to show that a vacuum was *not* created, (2) to show the relationship between the soul and the body, and (3) to explain the relationship of heat and water on the basis of expansion and contraction. He was not interested in what was consumed, nor would he have considered it worthwhile to examine the properties of the confined air before or after the combustion.

Thus, although the interpretation of van Helmont was not correct, his discussion suggests that he was more aware of the problems posed by the candle experiment than was Fludd. The widespread influence of his work—and that of Fludd as well—later in the century ensured the further discussion of this experiment and the problems that it presented.

A Model for the Geocosm

Prior to his study of man and his search for the cause of disease, van Helmont examined the earth itself. Since Holy Scripture teaches that medicines are created from the earth, it follows that the physician who does not understand natural philosophy as a whole cannot be truly learned in medicine.[133] Yet, although van Helmont's works reflect his wide knowledge of the current literature, his views of the "macrocosmic" world in many cases are not far in advance of those defended by Fludd. No less than his English counterpart did he reject the Copernican hypothesis as a young man. Later he was to note that the decree of the Church had forbidden any consideration of a circular motion for the earth, and the work of the Polish astronomer could be dismissed as "fiction."[134] He was well aware of Galileo's discoveries with the telescope, and he referred to the "stars" that circled Jupiter; but although the Copernican theory was defended by learned champions, van Helmont feared that should this opinion "break forth," it "will ruine all apparitions in the Heaven, and Predictions."[135] The result would be tragic, said van Helmont, for in spite of his rejection of the macrocosm-microcosm analogy, there were few who believed more firmly than he that the stars indicate times and

---

[133]Van Helmont, "Tumulus pestis" (Ch. 1), *Opuscula*, p. 5; *Oriatrike*, p. 1074.
[134]"Terrae tremor" (sect. 2), *Ortus*, p. 92; *Oriatrike*, p. 92.
[135]"Astra necessitant; non inclinant, nec significant de vita, corpore vel fortunis nati" (sect. 48), *Ortus*, p. 127; *Oriatrike*, p. 126.

PLATE XXVI Robert Fludd's illustration of the candle experiment. From the *Integrum morborum mysterium: sive medicinae catholicae tomi primi tractatus secundus* ... (Frankfurt: Fitzer, 1631), p. 457.

seasons and so implicitly "contain" all that happens in the future.

> Yea truly, I have been bold to attribute more Authority to the Heaven, than what hath wont to be given unto it by the Holy Scriptures: To wit, that the *Stars* are to us for foreshewing Signes, Seasons, or changes of the Air; lastly, for dayes, and years: wherefore the Text takes away all power of causes, besides in the abovesaid revolutions of seasons, dayes, and years: Neither do they act, I say, but by a motive and alterative Blas.[136]

This *blas* in turn impresses earthly things by means of the *magnal*. It is possible to foretell plagues through a study of the firmament,[137] and even the human seed is said to frame its directions according to the general motion of the heavens.[138] Indeed, "the stars do necessitate, not incline, nor signifie, of the life, the body, or fortunes of him that is born." Were this not true it would not be written that "*The Heavens declare the glory of God, and the Firmament sheweth his handy works.*"[139]

The universe itself is a vast edifice with the stars many thousands of earth diameters away; so far, indeed, that if they generated pernicious air, it would take many years to reach us.[140] As for the sun, van Helmont gave its diameter as 160 times that of the earth, and with the aid of a magnifying glass he identified its beams with true fire.[141] Our earth, an oval or "egg-shaped" object inclined to the poles, is formed basically of flint (*quellem*), which fell from heaven to the center of the universe at the time of the Creation.[142] The earth's shape is determined through a study of major river systems,

> Because the Waters do slide with a more swift course from *North* to *South*, than otherwise from *East* to *West*. For very many Waters do always descend by Rivers from the *North*, which do never run back unto the *North*. So the River *Danubius*, with many others, doth slide

[136] "Astra necessitant" (sect. 13), *Ortus*, p. 120; *Oriatrike*, p. 121.
[137] "Tumulus pestis" (Ch. 15), *Opuscula*, p. 62; *Oriatrike*, p. 1136.
[138] "Archeus faber" (sect. 9), *Ortus*, p. 41; *Oriatrike*, p. 36 (incorrectly numbered 26).
[139] "Astra necessitant" (title and sect. 2), *Ortus*, pp. 117-118; *Oriatrike*, pp. 118-119.
[140] "Tumulus pestis" (Ch. 3), *Opuscula*, p. 18; *Oriatrike*, p. 1086.
[141] "In sole tabernaculum," *Ortus*, p. 780, *Oriatrike*, p. 794.
[142] "Aqua" (sects. 13-15), *Ortus*, pp. 59-60; *Oriatrike*, pp. 55-56.

thorow the *Hellespont* or *Greek* Sea, into the Archi-pelago or the chief Sea: the Waters descend, neither doth any thing return from the Mediterranean Sea. Whatsoever doth once descend into the Mediterranean, is never spread into the Ocean. For the River *Nilus*, always descending in a right line from the Mountains of the Moon, is wholly plunged into the *Zebunutican* Sea with its dresses: neither doth the Mediterranean Sea in the mean time increase, nor become the salter. Which thing notwithstanding should be altogether needful so to be, if in manner of a naked vapour, the waters powred into it, should exhale out of the Sea.[143]

The shape of the earth and the continual flow of its rivers and streams testify to the wisdom of the Creator, who thus provided a continual supply of fresh water. There is a constant circulation of water in nature as the rivers flow into the seas, which in their deepest parts rest on the virgin earth, where the water with its salt is greedily absorbed by the *quellem*, which in time releases it again

> ... into Springs, Streams, Rivers, and the Sea, to moysten the Earth, and appointed to enrich it with Minerals. Whither again, the waters being driven, they are supt up partly by the *Quellem*, and partly do snatch the Air. So indeed doth the Universe distribute its waters, and lay them aside for divers fruits.[144]

Van Helmont's model of the interior of the earth was clearly in conflict with the traditional belief in a central fire which resulted in "distilled" mountain streams. In his view the attraction of the *quellem* for sea water was of great strength, resulting in some places in the formation of great whirlpools which rushed the water through stony channels, after which it gradually deposited its salt in earthy veins. In addition, the force of the attractions could be great enough to carry fish along, and the largest may become trapped within the channels and gradually turn to stone.[145] Nor should this phenomenon be wondered at, since the existence of natural springs with "stonifying" properties is well known. However, this is not the only explanation for fossilized substances, for over the passage of time the surface of the globe had shifted. Thus,

[143] "Aqua" (sect. 13), *Ortus*, p. 59; *Oriatrike*, p. 55.

[144] "Terra" (sect. 8), *Ortus*, p. 55; *Oriatrike*, p. 51.

[145] "Aqua" (sects. 5-9), *Ortus*, p. 55; *Oriatrike*, p. 51. The somewhat similar model of Seneca (in the *Naturales quaestiones*) is described by R. J. Forbes in *Studies in Ancient Technology* (Leiden: E. J. Brill, 1963), Vol. VII, pp. 8-9.

The Broken Chain

In *Hingsen* nigh *Scalds*, twelve foot under the Horizon, in a moyst Meadow, was found an *Elephants* Tooth, with the whole Cheek-bone, whose third part, being two foot long, I keep with me. And so living *Elephants* were once in this Countrey. But, very lately, *Groenland* hath ceased to be found subverted by the Sea, whence the Centre of the Earth ought necessarily to be changed or removed.[146]

Van Helmont's views on the origin of springs and mountain streams had first been formulated in the course of his dispute with Henricus van Heer, and they were clearly associated with his element theory. Van Heer had postulated that mountain streams derive from air thickened in the interior regions of the earth, wrote van Helmont, because

> ... springs in the tops of Mountains, were not seen to proceed from the Sea, whither they at length do rush. Therefore Springs have been hitherto falsely judged by the Schooles to take their Beginnings and Causes, from Air condensed or co-thickned by the force of cold, between the hollow places of Mountains, ready to fall upon each other. The which, I, in a little Book concerning the Fountains of the Spaw, printed in the year 1624 at *Leidon*, have shewne ....[147]

Here van Helmont had argued that this could not be true since both water and air are true elements. An experiment would show that no matter what force is brought to bear on the air in a pipe, no moisture is ever found. Furthermore, temperature seemed to have no effect, since this experiment was done both in mid-winter and in the heat of summer. "What if therefore the Air being pressed together by force in a Pipe, and cold season, be not changed into water; by what authority shall the Schooles confirm their fictions, touching the co-thickning of the Air, for the springing up or overflowing, and the continuance of Fountains?"[148]

The Mineral Kingdom

A thorough-going vitalist, van Helmont proceeded to develop an explanation for the existence of all things based upon his system of elements and their own life cycles. Here he discussed life sources

[146] "Terra" (sect. 17), *Ortus*, p. 56; *Oriatrike*, p. 52.
[147] "Aqua" (sect. 12), *Ortus*, p. 58; *Oriatrike*, p. 54.
[148] "Aer" (sects. 3, 4), *Ortus*, p. 61; *Oriatrike*, p. 58.

and semina, which were responsible for results as diverse as the minerals and human disease. It is of some importance to discuss the Helmontian *archeus*, the *ferment*, and the *seed* first in terms of inorganic nature, not only because this is the way they were introduced by him, but also because van Helmont's views on metals and minerals were to carry considerable influence later in the century.

Van Helmont was convinced that all earthly things were derived from a material base of elementary water. This was true of the *quellem* as much as other substances, for the *quellem* had only a semi-elementary nature and could not be considered the basis of other earths. Referring to Paracelsus, van Helmont compared the earth to a

> Womb that was great with Child with the seeds of minerals, wherein the *Lord* implanted Reasons or Respects, Endowments, and seeds that were to be sufficient for Ages. For so indeed, the wealthy seed of Rocky stones and minerals is implanted in the Water, that it may receive its determination and Ferment in the womb of the Earth.[149]

Thus, as minerals and earths could be resolved into a water, they originated from the water and a seed—and both stem from the original Creation.

However, with all live beings there is a governing or life force, the *archeus*, which is an inherent part of all seeds be they animal, vegetable, or mineral. Given this view of the universality of the *archeus*, it is understandable that van Helmont should compare the generation and life process of minerals to the human analogy. The waters are described as being

> ... with Child through the desire of the seeds, and the Almighty hath disposed the Idea's of his pleasure, or Precept, through the Water. Yet these seminal desires of the water, do not fructifie through a successive propagation of one thing by another, after the manner of plants: but a seminal vertue lurking in the Treasures of the water, doth peculiarly stir up its own Of-springs from it self, and successively perfect them. For a seed or seminal and mineral Idea, is included in the water, which never goes out of it: but locks up and incloseth it self in that matter, until at length, under the maturity of dayes, that be made thereof, which was born to be made of it.[150]

[149]"De lithiasi" (Ch. 1, sect. 6), *Opuscula*, pp. 11–12; *Oriatrike*, p. 830.
[150]*Ibid.*

The Broken Chain

The gift of generation had thus been given to the earth by means of a seminal "spirit" (*archeus*) acting on water. Unfortunately van Helmont suffered from the inconsistency he objected to in Paracelsus: occasionally it is the ferment that is the seminal beginning, while this same term serves double duty as the instrument of the seed and its spirit. Thus the ferments are defined first as "gifts, and Roots stablished by the Creator the Lord, for the finishing of Ages, sufficient, and durable, by continual increase, which of water, can stir up and make Seeds proper to themselves."[151] Later we read that the seminal spirit

> I name the *Archeus* or Master-workman; call thou it as thou wilt. Be it sufficient to know, that nothing doth arise anew in Nature, without a seed. In the next place, Every seed operates by dispositions its Handmaids, which it propagates in the matter for its intended desires. But the mediating Instruments, whereby seeds do dispose of their matters, I call *Ferments* . . . .[152]

If there appears to be some confusion in terminology, it is not of major importance, for in van Helmont's vitalistic system all things were produced from water and seed, the latter operating through an inborn spirit to produce its specific form.

The search for specificity in nature is closely associated with van Helmont's concept of *gas*, a term which along with *blas* he took special pride in having coined. For van Helmont the *gas* of a substance is a "wild spirit" which is released if a body is forced to give up its fixed state.[153] If, for instance, a substance is heated—and if it does not revert to elementary water—it becomes a *gas* characteristic of the original substance. Everything thus contains its specific *gas*, which is in essence a spiritual form of matter that still contains the seed (and therefore the *archeus*) of that substance.[154] In the course of his work van Helmont described numerous gases,[155]

---

[151] "Causae, et initia naturalium" (sect. 25), *Ortus*, p. 36; *Oriatrike*, p. 31.

[152] "De lithiasi" (Ch. 5, sect. 5), *Ortus*, pp. 40-41; *Oriatrike*, p. 859.

[153] See Walter Pagel, "The 'Wild Spirit' (Gas) of John Baptist van Helmont (1579-1644) and Paracelsus," *Ambix*, 10 (1962), 1-13, *Religious and Philosophical Aspects*, pp. 16-22.

[154] "I call this Spirit, unknown hitherto, by the new name of Gas, which can neither be constrained by Vessels, nor reduced into a visible body, unless the seed being first extinguished." "Complexionum atque mistionum elementalium figmentum" (sect. 14), *Ortus*, p. 106; *Oriatrike*, p. 106.

[155] The best discussion is to be found in Partington, *History of Chemistry*, Vol. II,

which by definition had to be characteristic of the substances they were derived from. These gases could not be changed into water without the loss of the original seeds that gave specificity to them.

At last, with an armory of new definitions of forces and fundamental substances, it seemed possible to explain natural phenomena with a detail that was not possible earlier. A subject of traditional importance for chemists was the origin and life cycle of metals, which van Helmont proceeded to describe as follows:

> ... the Roots or Ferments of Mineralls, do sit immediately in place, and do breath without disdain, for fulness of dayes: The which, when it hath compleated a seed, then the Gas environing the water in the same place, receiveth a seed from the place, which afterwards begets the Sulphur of the water with childe, condenseth the water, and by degrees transplants it into a Minerall water. For it oft-times happeneth, that a digger of Metalls in Mines breaking great Stones asunder, the Wall cleaves or gapes, and affords a chink, from whence a small quantity of water of a whitish-green color hath sprung, which hath presently grown together like to liquid Sope, (I call it *Bur*) and afterwards its greenish paleness being changed, it waxeth yellow, or growes white, or becomes more fully green: For thus that is seen, which else without the wound of the stone, comes to pass within: because that juyce is perfected by an inward efficient. Therefore the first life of a metallick seed, is in the Buttery or Cellar of the place, plainly unknown to man. But when as the seed comes forth to light cloathed with a Liquor, and Gas hath begun to defile the Sulphur of the water, there is the middle life of the seed: But the last life is when it now waxeth hard: But the last life of the metallick seed, is the first life of the Metalls, or at leastwise very nearly conjoyned to it. But while that Masse doth breath Sulphur, and shuts up its Mercury within; then I say, is the middle life of Metalls: But their last life is, when it hath attained a fixedness, and the proper stability of a vein. ... Wherefore there is a more manifest progress of a life, and seed in Metallick Bodies, than in the two fellow Monarchies .... [156]

The life process of the mineral kingdom could be properly compared with that of the vegetable and animal kingdoms. Thus, the concept of a middle life was to play a significant part in van Hel-

---

pp. 227-232. Here are described fifteen different gases mentioned by van Helmont.

[156] "Magnum oportet" (sects. 39, 40), *Ortus*, p. 157; *Oriatrike*, pp. 155-156.

mont's disease theory. A second analogy is found in the liquid *bur*, or proto-mineral, which corresponded to the *leffas* in vegetable life:

> ... the Earth stinking with metally ferments, doth make out of water, a metally or Mineral Bur. But the water being elsewhere shut up in the Earth, if it be nigh the Air, and stirred by a little heat, it putrifieth by continuance, which is no more water, but the juyce *Leffas* or of Plants. ... From whence ariseth every kind of Plant wanting a visible seed, and from whence seeds that are sown, are promoted into their appointments: therefore there are as many rank or stinking smells of putrefactions by continuance, as there are proper savours of things: for that, odours are not onely the messengers of savours, but also their promiscuous parents. The smoak *Leffas* being now gathered together, doth at first wax pale, afterwards wax yellowish, straightway it waxeth a little whitishly green: And at length it is fully green. And the power of the Species or particular kinde being unfolded, it assumeth divers Colours and Signates .... but *Leffas* is constrained to perfect the Tragedy of the conceived seed. Therefore Rain conceiving a hoary ferment, and being made *Leffas*, is drawn into the lustfull roots by a certain sucking. And it is experienced, that within this Kitchin, there is a new hoary putrefaction of the *Ferment* the Tenant: by and by, it is brought from thence to the Bark or Liver, where it is enriched with a new *ferment* of that bowel, and is made an Herby or woody juyce, and at length, a ripeness being conceived, it becommeth Wood, becometh an Herb, or departs into *fruit*: but the Trunk or Stem, if it sooner putrifies under the Earth than the Bark or Rhine becomes dry, it cleaves asunder by its own ferment, sends forth a smoak thorow the Bark, which in its beginning is spongie, and at length hardens into a true root; and so planted branches become Trees by the abridgement of art.[157]

Through a similar fermentation nature produces worms and gnats in the water of fountains, and frogs, shellfish, and snails from fen water. Thus, we need no longer speak in terms of the mixture of four elements when we talk of the composition of bodies. All things arise from water and seed. They arise from the action of the internal *archeus* and resultant ferments. And, their life cycles played out, all substances return in time to their primal state, water.[158]

[157] "Imago fermenti impraegnat massum semine" (sects. 31, 32), *Ortus*, pp. 116-117; *Oriatrike*, p. 116.
[158] "Imago fermenti" (sect. 33), *Ortus*, p. 117; *Oriatrike*, p. 117.

The Chemical Geocosm of Edward Jorden (1631)[159]

Van Helmont's model of the earth and the forces inherent in it are in many cases borrowed or modified from earlier or contemporary sources. His *archeus* and seed are fundamentally Paracelsian in origin. Similarly, the concept of fermentation had been traditionally significant for alchemists, who compared the action of yeast to that of the philosopher's stone. Martin Ruland (1612) quoted Hermes, Geber, Richardus Anglicus, and Raymond Lull in his description of the relation of fermentation to the great work.[160] For Bonus of Ferrara (c. 1330) the ferment was the great secret of the art of alchemy, and as such it was one that the sages spoke of only in the most obscure terms. Its action is "not quite analogous to that of leaven. For leaven changes the whole lump of dough into a kind of leaven, but our Stone, instead of converting metals into the Tincture transmutes them only into gold."[161] With Paracelsus we find an interest in fermentation directed more at an understanding of generation in the earth. He suggested that the reaction of salt and sulfur in the earth produces an internal fermentation that results in a liquid that subsequently solidifies into sharp rocks on exposure to air.[162] And although van Helmont

---

[159]To a large extent the following material has been taken from Allen G. Debus, "Edward Jorden and the Fermentation of the Metals: An Iatrochemical Study of Terrestrial Phenomena," *Toward a History of Geology: Proceedings of the New Hampshire Inter-Disciplinary Conference on the History of Geology, September 7-12, 1967*, ed. Cecil J. Schneer (Cambridge, Mass.: M.I.T. Press, 1969), pp. 100-121.

[160]Martin Ruland, *Lexicon alchemiae* (Frankfurt, 1612; Hildesheim: Georg Olms, 1964), pp. 211-213.

[161]Bonus of Ferrara, *The New Pearl of Great Price*, trans. and abridged by A. E. Waite (London: Vincent Stuart, 1963), p. 254. For a late medieval alchemical discussion of fermentation see also Sir George Ripley, "The Compound of Alchymie" in the *Theatrum chemicum Britannicum*, collected and annotated by Elias Ashmole, introduction by Allen G. Debus (New York/London: Johnson Reprint, 1967), pp. 107-193 (173-178). Walter Pagel has discussed the role of fermentation in alchemy in his *Paracelsus*, pp. 261-263, and in "Van Helmont's Ideas on Gastric Digestion and the Gastric Acid," *Bulletin of the History of Medicine*, **30** (1956), 524-536.

[162]Paracelsus, "Philosophia de generationibus et fructibus elementorum" in *Opera Bücher und Schrifften*, ed. J. Huser, 2 vols. (Strasbourg: Zetzner, 1616), Vol. II, p. 57. "Dann das Salz / und der Sulphur, die da gestanden seind / haben sich nicht mögen vergleichen in der Globul mit der Erden / sondern durch die Erden in das widerspiel gebracht / uund in ein Wüten kommen / unnd (sich) gebläet /

had a special interest in medicine, as did most other sixteenth- and seventeenth-century chemists, his genuine concern for essentially geological problems was certainly not unique. Here Paracelsus may again be brought forth as his mentor, but it is even more interesting to show the influence of these themes on an English contemporary, Edward Jorden (1569-1632), whose *Discourse of Naturall Bathes* (1631) exhibits striking similarities—as well as differences—in comparison with van Helmont's work.

Jorden was educated first at Oxford and then at Cambridge, where he took his M.A. in 1586.[163] Like so many other sixteenth-century English students who were interested in medicine, he then went to the Continent, where he obtained an M.D. at Padua (c. 1591). On his return to England he became first a licentiate (1595) and then a fellow (1597) of the London College of Physicians. Here Jorden was in regular contact with the most noted English scientists and physicians of his day, many of whom took an active part in discussions of the chemical innovations of the Paracelsians. Thomas Moffett had been made a fellow seven years earlier than Jorden, and he must have been present when the decision was made by the College in 1594 to license the chemical enthusiast John Banister to practice medicine. Jorden was to have the opportunity to witness the election to membership of a number of fellows we have referred to earlier: Matthew Gwinne in 1599, Thomas Rawlins in 1605, Robert Fludd in 1609, and Sir Theodore Turquet de Mayerne in 1616.

Jorden must have been at ease among these chemically oriented members of the College of Physicians. He had surely read widely in all fields of natural philosophy, his writings show that he was well acquainted with the important chemical authors of the period, and among metallurgists he frequently quoted Biringuccio, Agricola, and Ercker. Even more interesting for our purpose is the

---

als sein schaum oder gest von einem Fleisch oder Wein." Partington noted this passage in the Latin translation in *History of Chemistry*, Vol. II, p. 147. More traditional alchemical thought on fermentation is reflected in the pseudo-Paracelsian *Aurora philosophorum*. See Paracelsus, *Chirurgische Bücher und Schrifften*, ed. J. Huser (Strasbourg: Zetzner, 1618), p. 772.

[163]On the life of Jorden see the biographical sketch by Joseph Frank Payne in the *Dictionary of National Biography*, and the note in William Munk, *The Roll of the Royal College of Physicians of London* (2nd ed.; London: Published for the College, 1878), Vol. I, p. 113.

fact that Jorden represents a direct English link with the continental iatrochemical tradition. Although his detailed knowledge of the refutation of Paracelsus by Thomas Erastus is evident throughout, and he spoke with some distaste of the "Platonicall grandiloquence" of Peter Severinus,[164] Jorden referred often to Penotus, Thurneisser, Moffet, Dorn, and Crollius, as well as Paracelsus. Among chemical authors, however, it is evident that he leaned most heavily on Andreas Libavius, who had evidently befriended him while he was a student on the Continent. In Libavius' most important work, the *Alchemia* (1597), we find Jorden listed first among a group of friends—among them Tycho Brahe—who had offered him information for his book.[165] It was in the *Alchemia* that Libavius included a lengthy and important chemical discussion of mineral waters, and it is for his investigation of the baths at Bath that Jorden was best known in the seventeenth century. Similarly, there is a rejection of the works of the mystical alchemists that is common to both Libavius and Jorden.

Like many Paracelsians and natural magicians Jorden consistently rejected the supernatural as a valid form of explanation. He wrote: "we must not imagine that the gouernment and ordering of the world and nature in a constant course, is performed by miracle, but that naturall effects have naturall causes, and must be both under the same *genus*."[166] This approach is clearly evident in Jorden's first publication. Having been called in to investigate a case of demonic possession, he proceeded to reject that explanation and to attribute the attack to natural causes.[167]

Nor did Jorden share the belief of many of his contemporaries

[164]Edward Jorden, Dr. in Physick, *A Discourse of Naturall Bathes, and Minerall Waters* (2nd ed. "in many points enlarged"; London: Thomas Harper, 1632), p. 69. Unless otherwise noted, all citations are from this edition, which is considerably expanded from the first edition of 1631. All later editions (1633, 1669, 1673) are based upon the revision of 1632.

[165]Andreas Libavius, *Alchemia* . . . (Frankfurt: Saurius, 1597), sig. c.

[166]Jorden, *Discourse of Naturall Bathes*, pp. 109-110.

[167]Edward Jorden, *A briefe discourse of a disease called the Suffocation of the Mother, written upon occasion which hath beene of late taken thereby, to suspect possession of an evill spirit or some such like supernaturall power. Wherein is declared that divers strange actions and passions of the body of man. which . . . are imputed to the Divell, have their true natural causes, and do accompanie this disease* (London, 1603). This short work, which was never reprinted, and the *Discourse of Naturall Bathes* are the only two works published by Jorden.

that there had been an appreciable decay of the pristine knowledge of Adam. Instead, he felt that men of the present could not help but know more than the ancients, since "*Dies Diem docet: Alpham Beta corrigit.*"[168] Still, even the great minds of his age were subject to error; and, as they felt free to publish their views, Jorden thought that he too might be excused for presenting the results of his research. He wrote, "I trauell where no path is to be seene of any learned foot that here hath beene,"[169] yet his work constantly reveals the mark of a cautious scholar who diligently sought the advice and criticism of others before publishing his views.

Jorden is an attractive figure to the twentieth-century mind for a number of reasons. We find him highly skeptical of the evidence offered for transmutation in nature. Like van Helmont, he rejected the belief that metals are changed into one another in certain mineral water springs. In his discussion of vitriol, he referred to such springs in Cyprus, in Transylvania, and the famous Hungarian waters at Smolnicum, places in which

> Copper is ordinarily made out of iron by infusing it in these waters. I will not determine whether this be transmutation of one species into another, as some doe hold, or rather a precipitation of the Copper which was formerly dissolved in the water by means of the sharpe Vitriol; which meeting with Iron, corrodes it, and imbibeth it, rather then the Copper, and so lets the Copper fall, and imbraceth the Iron in place of it. We daily see the like in Aqua fortis, which having imbibed one metall, will readily embrace another that is more familiar to it and let fall the first.[170]

Here Jorden cited the earlier discussions of Libavius and Ercker. The latter, noting that iron precipitates copper from coppery solutions, was first convinced that "it is only the copper in such solutions that is precipitated by the iron and the iron itself does not change in copper." Later, however, he observed that where copper ore is being mined, iron objects appear to become copper. Noting further that in solutions the precipitated copper was greater in weight than the metal in solution originally, he reversed his earlier opinion. "I have therefore been forced to the conclusion that iron turns into copper." Libavius, concerned whether or not this was a

[168] Jorden, *Discourse of Naturall Bathes*, p. 6.
[169] *Ibid.*, p. 103.
[170] *Ibid.*, pp. 49-50.

true transmutation, came to the same conclusion and cited Ercker in support of this view.[171]

No less susceptible to criticism was the contemporary use of sympathy and antipathy to explain natural phenomena. On this Jorden commented,

> ... if we examine aright what this sympathy and antipathy is, we shall finde it to bee nothing but a refuge of ignorance, when not being able to conceiue the true reasons of such actions and passions in naturall things, wee fly sometimes to indefinite generalities, and sometimes to this inexplicable sympathy and antipathy: attributing voluntary, and sensitiue actions and passions to insensible substances.[172]

For Jorden heat could generally be thought of as a form of motion of the "seeds or formes" of a substance. Thus he explained that a substance could be heated or set on fire through various forms of motion, such as collision, concussion, dilation, compression, putrefaction, fermentation, and reflection. However, some results led Jorden to question the universality of this rule. Anticipating the procedure of Joule two centuries later, Jorden studied water in violent motion, at the cataracts of the Rhine near Splug, and found that "the agitation and fall of water upon rocks is most violent, and make a hideous noise; yet it heats not water."[173] His incorrect conclusion was that water was an exception to the general rule; in short, air and liquids are not heated but cooled by attrition.

His view of heat as a form of motion clearly affected his concept of the system of the universe. Concerned with the origin of the heat of some mineral water spas, he referred to theories which held that the sun was the source of the heat. If this is so, Jorden continued, it must be a result of the motion of the sun—either on its own axis or circumgyrating on a fixed sphere. The latter, or "common

---

[171] See the chapter titled "Whether Iron Turns into Copper," in Lazarus Ercker, *Treatise on Ores and Assaying*, trans. from the German ed. of 1580 by A. G. Sisco and Cyril Stanley Smith (Chicago: University of Chicago Press, 1951), p. 223. Libavius referred to Ercker on this point in the *Syntagma selectorum undiquaque et perspicue traditorum alchymiae arcanorum* (Frankfurt: Excudebat Nicolaus Hoffmannus, Impensis Petri Kopffi, 1611), p. 280.

[172] Jorden, *Discourse of Naturall Bathes*, p. 121.

[173] *Ibid.*, p. 90.

opinion," could not be true, for "to haue such a bullet as the globe of the Sunne, which is held to be 166 times bigger then the globe of the earth, to bee carried in a swifter course [than the bullet], and that perpetually, is a monstrous, furious, and mad agitation, *insanus motus*, as one termeth it."[174] Unwilling to discuss this subject in greater detail, Jorden referred his readers to the sixth book of Gilbert's *De magnete* for a more detailed confutation of the motion of the spheres.

Jorden's interest in mineral water spas led him directly to the analysis of these springs. Van Helmont also gives evidence of having analyzed spa waters, but Jorden emerges as one of the outstanding analytical chemists of the early seventeenth century. Well aware of the earlier literature on this subject, he praised Agricola, Fallopius, Baccius, Solenander, Thurneisser, and Libavius for their contributions. Because of their work in describing detailed analytical procedures, Jorden felt confident of the inadequacy of others who gave results of the contents of springs simply by distilling a sample to dryness and then tasting the residue.[175] He suggested that the first check for impurities in a water should be a comparison of the weight of a sample with a sample of pure water, though he complained that "it is hard to have great balances so exact, as a small difference between them may be discerned by them."[176] Jorden's relative lack of emphasis on quantification by weight, in comparison with van Helmont, seems to stem from the inadequacy of the available balances.

In the course of his analytical procedure Jorden took a special interest in the identification of different types of salts. Here his procedure was based primarily on distillation followed by the recrystallization of the residue. Upon completion of this step he was able to identify specific crystal forms. He noted that "the shooting or roching of concrete iuyces, is worthy to bee observed, seeing euery kinde hath his seurerall manner or fashion of shooting, whereby a man may see the perfection of each kinde."[177] The method was simple: after a stick had been placed in the liquor, it was allowed to evaporate, and

[174]*Ibid.*, p. 91.
[175]*Ibid.*, p. 122.
[176]*Ibid.*, p. 13.
[177]*Ibid.*, pp. 37-38.

... within a few dayes, the concrete iuyces will shoot upon the wood, in Needles, if it bee Niter; in Squares, if it be Salt; and in Clods and Lumps, if it be Allum, or Coperose; and the other minerall substance which the waters have receiued, will either incorporate a tincture with them, or if it be more terrestriall, will settle and separate from it, and by drying it at a gentle fire will shew from what house it comes, either by colour, taste, smell, or vertue ...[178]

Crystal form had been used for identification earlier than this, and, in more limited scope, it was to be used by van Heer and van Helmont.[179] Still no one had considered the test as valuable as did Jorden. He suggested that saltpeter be tested for purity by crystallization techniques, for only when it is pure will it shoot into needles; when impure, it will appear in "squares or angles or lumps." He also saw that when saltpeter is mixed with salt of ashes and then allowed to crystallize, the salt of ashes precipitates first as squares and the saltpeter appears last in needles.[180] Jorden found it difficult to find a reason for the characteristic crystal forms. It would be expected, he said, that when different salts are dissolved

[178] *Ibid.*, p. 124.

[179] Van Helmont distinguished sea salt from the salt of urine on the basis of their different crystalline shapes. "De lithiasi" (Ch. 3, sect. 22), *Opuscula*, p. 25; *Oriatrike*, p. 844. Late editions of van Heer's *Spadacrene* include a folding plate of crystal forms for the purpose of identification.

[180] Jorden, *Discourse of Naturall Bathes*, p. 38. A less detailed method of analysis including crystallization was suggested for English miners at approximately the same time. Gabriel Plattes wrote, in *A Discovery of Subterraneall Treasure* (London, 1639), pp. 9-10, "The nexte worke is to finde out the Springs of Water issuing out of the said Mountaynes; and those being found, a quantity of the sayd water is to be boiled in a new cleane pipkin, to the consistency of a thinne Oyle, but no so thicke as a Syrrup, and when it is almost cold, then put it into a Urinall, and to set it in the coldest place that can be found for 3 daies, then to play the Physitian, and to observe it exquisitely what residence it yeeldeth: if nothing settle but a black earth or mudde, it is a signe of Coales: if some part thereof shoot into Ice, or a substance like Ice or Vitrioll, then to observe the colour thereof; if it be greene or blewish, it is an evident signe of copper; if whitish then it may signifie any other Metall with out exception."

It appears that Jorden was in error in his discussion of the precipitation of saltpeter and salt of ashes. Based on the solubilities of K_2CO_3 and KNO_3, one might expect the saltpeter to precipitate first. Indeed, in describing a similar process, Nicholas Lemery concluded that "the long Crystals that we see *Salt-peter* shoot into, do proceed from its volatile part, for that which is Crystallized last, is fixt like sea-salt, and looks just like it." *A Course of Chymistry*, trans. Walter Harris (2nd ed. from the 5th French ed.; London, 1686), p. 292.

The Broken Chain

in the same water, the resultant crystals would be uniform, but this is not the case, since such a "mixt water will yeeld his seurerall salts distinctly, and all at once."[181] The reason can only rest in the fact that nature has distinguished these species no less definitely than plants. We need not look for varying proportions of elements or qualities for our answer,

> ... but only from the forme, *anima*, seed, &c. which frames euery species to his owne figure, order, number, quantity, colour, taste, smell, &c. according to the science, as *Seuerinus* termes it, which euery seed hath of his owne forme. So also it is in minerals, which haue their seuerall and distinct species in nature, and their seeds to maintaine and perpetuate the Species.[182]

As with van Helmont, for Jorden it is the seed or the *anima* associated with it that confers specificity. He went on to say that in the earth these salts may not always appear in their characteristic crystalline forms because of mixing with other minerals, because of lack of space, or because of lack of water to dissolve them. Yet chemical crystallization techniques are able to uncover the characteristic forms.

Jorden suggested also that if the analyst did not want to wait for the salts to crystallize, they might be isolated by precipitation. This is done with the aid of either two classes of substance: (1) "Salts, as Tartar, Soape, Ashes, Kelps, Vrine" (bases) or (2) "sowre iuyces, as Vinegar, Lymons, Oyle of Vitrioll" (acids).

> ... I haue observed that the Salts are proper to blew colours, and the other to red: for example, take a piece of scarlet cloath, and wet it in Oyle of Tartar (the strongest of that kinde) and it presently becomes blew: dip it againe in Oyle of Vitriol, and it becomes red again.[183]

While this is an important early instance of the use of an acid-base indicator, there is no evidence that Jorden recognized a neutral state, as did van Helmont.

[181] Jorden, *Discourse of Naturall Bathes*, p. 39.
[182] *Ibid.*
[183] *Ibid.*, p. 124, and also the 1631 ed., p. 76. One cannot be certain just what dye was in the cloth Jorden used. However, Sidney M. Edelstein has pointed out (in a letter to the author) that the common scarlet dye in Jorden's lifetime was cochineal or kermes and that this "would have acted exactly as Jorden describes the reactions with alkali and acid in his text."

The subjects most important to Jorden in his *Discourse of Naturall Bathes*, however, were not these, but his search for the true explanation of the growth of minerals and the source of the heat of mineral water springs. In his solution to these problems Jorden drew upon a varied background of current thought and offered a fermentation hypothesis to account for all observed phenomena. This theory was based upon Jorden's conviction that there is a life force in metals. He felt that the book of Genesis could be used as a guide, for the Lord taught that seeds were created before the plants. By analogy, therefore, we may assume that minerals

> ... were not at first created perfect, but disposed of in such sort, as they should perpetuate themselues in their seuerall kindes. Wherefore it hath euer beene a receiued *Axiome*, among the best Philosophers, that minerals are generated, and experience hath confirmed it in all kindes.[184]

The influential forces behind this growth or increase were either celestial or terrestrial, and the first seemed unlikely. The most Jorden was willing to concede was that the sun might help further the natural process, but he would not consider either the sun or the stars to be the principal cause of growth of metals or their ores.[185] At that point he sought a terrestrial explanation,[186] examining in particular the role of the elements in metallic generation. Jorden reflected the current dispute over the elements in his impatience with those who attributed nearly all things to them. Some authors insisted that souls are elementary, and others derived from them tastes, colors, and a host of other properties.[187] All these views Jorden summarily rejected, for surely the elements could not give properties to other things when they did not possess them themselves.

Among the elements only fire could be a possible answer, since the other three were cold and therefore could not cause generation of any sort. But the authorities who stated that fire is responsible for the generation of metals did so because they believed that heat must assemble all elements together and then proceed to "ranke

---

[184]*Ibid.* (1632), p. 67.

[185]*Ibid.*, p. 71. The arguments against astral influences are developed at greater length in Debus, "Edward Jorden," pp. 113-114.

[186]Jorden, *Discourse of Naturall Bathes*, pp. 72-78.

[187]*Ibid.*, p. 75.

them." Not so, Jorden replied, "heat can but heat," and all chemists know that the result is the separation of different substances and the congregation of similar ones.[188] Indeed, while none would deny the existence of common fire, the existence of the element is a different matter. Jorden rejected the latter, as did many of his contemporaries:

> As for the element of fire, which should bee pure, not shining, and therefore invisible, and subsisting without a subject or fewell: let them finde it who know where to seeke for it. For my part I know no element of fire, vnlesse we should make it to be that which is naturall to all creatures and their seeds, causing their fermenting heat . . . .[189]

Turning to the alchemists and their sulfur-mercury theory of the metals, Jorden thought that if they meant common sulfur and common mercury, they were obviously mad. But it seemed that most of them understood by these words "some parts of the seminary of metals which haue some analogye with these: and so their opinion may be allowed. For the spirit, which is efficient in these generations, doth reside in a materiall substance, which may be resembled to Sulphur or Oyle, as some other part may be resembled to Mercury."[190] The chemists' views are permissible, then, if they are understood in terms of the seed-sperm concept of generation.

For Jorden real truth is found in an application of two axioms of Hippocrates: (1) "euerything is dissolved into that whereof it was made"; and (2) "wee are nourished by such things as we consist of." We find that neither natural nor artificial dissolution gives us the four elements. Hippocrates, speaking of natural dissolution or putrefaction, observed three distinct substances: *calidum, humidum* or *fluidum*, and *siccum* or *solidum*. This agrees well with the artificial distillation analyses of the chemists, who obtain *vaporosum, inflammabile,* and *fixum*—that is, mercury, sulfur, and salt. Thus, if all things are dissolved into their original components, these cannot be the four elements. Furthermore—again following the maxims of Hippocrates—if things do consist of the four elements, they are not nourished by them. Jorden therefore felt it safe

---

[188] *Ibid.*, p. 72.
[189] *Ibid.*, p. 96.
[190] *Ibid.*, p. 73.

to assert that metallic growth was not nourished by the four elements. At best the traditional elements could be considered as matrices or wombs within which the growth process occurs from "the seeds of things, which are agreeable to their species."[191]

Since Jorden found none of the earlier concepts to be satisfactory, he felt free to give his own views on the generation of metals. Here he adapted to his own purposes the process of fermentation:

> There is a Seminarie Spirit of all minerals in the bowels of the earth, which meeting with conuenient matter, and adiuuant causes, is not idle, but doth proceed to produce minerals, according to the nature of it, and the matter which it meets withall: which matter it workes vpon like a ferment, and by his motion procures an actuall heate, as an instrument to further his work; which actuall heate is increased by the fermentation of the matter.[192]

Indeed, this may properly be likened to the making of malt,

> ... where the graynes of Barley being moystened with water, the generatiue Spirit in them, is dilated, and put in action; and the superfluity of water being remoued, which might choake it, and the Barly laid vp in heapes; the Seeds gather heat, which is increased by the contiguity of many graines lying one vpon another. In this worke natures intent is to produce moe indiuiduals, according to the nature of the Seede, and therefore it shoots forth in spyres: but the Artist abuses the intention of nature, and conuerts it to his end, that is, to increase the spirits of his Malt. The like we find in minerall substances, where this spirit or ferment is resident, as in Allum and Copperas mynes, which being broken, exposed, and moystened, will gather an actuall heat, and produce much more of those minerals, then else the myne would yeeld; as *Agricola* and *Thurneiser* doe affirme, and is proued by common experience.[193]

Thus, Jorden continued, the heat experienced in deep mines need not be ascribed to the sun, which clouds can remove, or to an internal subterranean fire. The earth abounds with matter "which attends vpon the species of things: and oftentimes for want of fit opportunity and adiuvant causes, lies idle, without producing any species."[194] Again rejecting any active role of the elements, Jorden

[191] *Ibid.*, pp. 77-80.
[192] *Ibid.*, p. 84.
[193] *Ibid.*, pp. 84-85.
[194] *Ibid.*, p. 86.

The Broken Chain

insisted that "this matter is not the Elements themselves, but subterraneall seeds placed in the Elements, which not being able to liue to themselues, do liue to others." Here, then, we have the proper explanation of the nutrition of natural things. It is only in this way that we may properly account for colors, tastes, numbers, proportions, distempers, and all the other observed differences of the varying substances of the earth.

Jorden's seminary hypothesis may be pushed somewhat further with additional analogies between the organic and the inorganic worlds. Transplantations occur in both the animal and vegetable kingdoms: there are variations from the norm in the issue of dogs and wolves, partridges and hens, and similar changes in plants, since when species are mixed the first seeds "doe oftentimes bring forth other fruits then their owne."

> In Minerals we find the like transplantations: as Salt into Niter; Copperasse into Allum, lead into Tinne, Iron into Copper; Copper into Iron, &c. And this is the transplantation whereupon the Alchymists ground their Philosophers stone.[195]

Thus, although common explanations of transmutation are to be frowned upon, the fermentation concept makes it possible to think in terms of the transplantation of mineral seeds. Jorden's position is not far removed from that of van Helmont, who wrote that "the ferment is the parent of transmutation."[196] For Jorden this law of natural growth seemed basic and indisputable. The idea of a seminary spirit acting as a ferment had the backing of philosophical authority (Aristotle), observational evidence, and even Holy Scripture, which required the growth of all things from the first benediction, "*crescite & multiplicamini.*"[197] Jorden thought that the generation of minerals formed only part of a greater divine law, one that was founded no less firmly than the laws of mathematics and logic.

The answer to Jorden's second major question—the source of the heat of mineral springs—now seems obvious. The heat was surely not due to the friction of the wind, air, or other exhalations rising from the internal regions, since friction seems to cool rather

---
[195] *Ibid.*, p. 87.
[196] Van Helmont, "Imago fermenti" (sect. 23), *Ortus*, p. 115; *Oriatrike*, p. 115.
[197] Jorden, *Discourse of Naturall Bathes*, pp. 87-88.

than heat air and liquids. Nor can the sun heat the internal waters of the earth any more than it does the metallic ores in the mines. Those who apply the doctrine of antiperistasis are equally wrong, since this would require that the waters be hotter in the winter than the summer. And those who postulate within the earth large deposits of quicklime, which are heated on contact with water, may be refuted, since quicklime is a substance that does not exist in the natural state and can only be made in the laboratory.[198]

Nor did Jorden believe the most common explanation, which presumed the existence of a vast subterranean fire causing heated waters, volcanos, and all sorts of other surface effects. If the fire existed, it must be assumed that the center of the earth is hollow, not a very likely situation, since all heavy things descend toward the center, implying that it is more compact than the surface. Again, if the earth were hollow and filled with a gigantic fire, we could think in terms of an immense distillation system: vast receptacles below the surface acting in the form of chemical vessels filled with water which are then heated to boiling so that the surface springs form the end product of a macrocosmic distillation system. However, laboratory experience shows that in such systems almost all the heat is lost on condensation, whereas in the hot natural springs the temperatures remain very high.[199]

And what of the subterranean fire itself? If it is like the common kitchen fire, it is difficult to explain how it is kindled and how it continues to exist. Any fire requires a tremendous amount of air, a fact obvious to anyone who has confined a flame with a cupping glass and seen how it is immediately put out, since "fuliginous vapours . . . choake it if there were not vent for them into the ayre."[200] Surely the required abundance of air cannot reach the central regions under normal circumstances. Those who further insist that a limitless supply of fuel exists in deposits of bitumen and sulfur do not consider that the earth would sink in the areas of these deposits as they were consumed.[201] Furthermore, experience teaches us that not all hot baths are sulfureous or bituminous, nor all bituminous or sulfureous springs hot.[202] To those who argued

[198]*Ibid.*, 89-95.
[199]*Ibid.*, pp. 99-100.
[200]*Ibid.*, pp. 97, 101.
[201]*Ibid.*, p. 112.
[202]*Ibid.*, p. 102.

that the existence of volcanos overrides all objections, Jorden countered that volcanic action could easily be confined to regions close to the surface, where deposits of sulfur or other combustible material could be kindled by lightning and thence proceed because of the proximity of the atmosphere.[203] In short, a subterranean central fire could not be maintained; it was an outmoded hypothesis which should be dropped entirely.

We must instead return to the fermentation hypothesis of the generation of the metals, which accounts for the heat by natural means as the result of the motion between the agent spirit and the passive matter. Such a heat is natural rather than violent and will result in generation rather than the familiar destruction of the ordinary fire. And since it is a slow process rather than a rapid combustion, it will account satisfactorily for the heating of springs over long periods of time. If we accept this explanation, we need not exhaust our wits to locate vast sources of fuel nor need we rack our minds to think of ways for air to be admitted to the bowels of the earth. And what more natural explanation could there possibly be for the presence of minerals in water, since their generation was the very cause of the heating process?[204]

For Jorden fermentation had become the basis of a new system of the earth. His conclusions differed from those of van Helmont on numerous points, but it is important to note that both authors were affected by the current disputes over the origin of mineral water springs and the growth of metals. Strongly influenced by alchemical and iatrochemical tradition, both found it easy to reject Aristotelian elementary theory and to find a new basis for natural phenomena in the concepts of active semina and ferments.[205]

Helmontian Medicine:
The Divine Office of the Physician

Van Helmont's search for a replacement for ancient and Scholastic philosophy is apparent throughout his discussion of natural phenomena. However, it was in keeping with Renaissance

---

[203] *Ibid.*, p. 101.
[204] *Ibid.*, pp. 104-107.
[205] Jorden's work was widely known in the seventeenth century. See Debus, *English Paracelsians*, p. 174 (n. 101).

neo-Platonic and Paracelsian tradition that he should take as his goal the reformation of medicine. And here again we see the impact of his belief in divine revelation.[206] For van Helmont the thirty-eighth chapter of Ecclesiasticus provided indisputable proof of the divine office of the physician. Reflecting on his own long period of study, he recalled that when

> ... *the Art of Medicine fell into disesteem with me, I lighted on a Text of holy Scripture, having been often read, yet never understood; To wit,* That the most high had created the Physitian, and had commanded him to be honoured, by reason of the necessity there is of him. *Wherein I presently discerned,*
> 1. That he who created all things, doth notwithstanding, particularly glory, that he is the Creator of the Physitian.
> 2. That for his own glory sake, for the issuing forth of his goodness for the necessities, helping, and succours of the Sick, and so by the Physitian, the Almighty will be appeased, in restoring health that was taken away.
> 3. That he to whom all honour and glory is due, hath commanded, that Parents, and Physitians onely created by him, be honoured; as if a Physitian had something of a fatherly Nature.
> 4. And then, in my manhood, I not a little carefully inquired day and night, what happy Man he should be, whom the Almightly from Eternity ordained, chose, and created for a Physitian, and from hence also, commanded to be honoured.[207]

---

[206] For general accounts of van Helmont's medicine and its relationship to religion and philosophy see Walter Pagel, "The Debt of Science and Medicine to a Devout Belief in God. Illustrated by the Work of J. B. Van Helmont," *Journal of the Transactions of the Victoria Institute,* **74** (1942), 99-115; "Religious Motives in the Medical Biology of the 17th Century," *Bulletin of the Institute of the History of Medicine,* **3** (1935), 97-128, 213-231, 265-312; *Religious and Philosophical Aspects;* "Helmont, Leibniz, Stahl," *Sudhoffs Archiv für Geschichte der Medizin,* **24** (1931), 19-59; "The Position of Harvey and Van Helmont in the History of European Thought," *Journal of the History of Medicine and Allied Sciences,* **13** (1958), 186-198; "William Harvey Revisited," *History of Science,* **8** (1969), 1-31 and **9** (1970), 1-41 (esp. 8, 18-21); "The Reaction to Aristotle in Seventeenth-Century Biological Thought," *Science, Medicine and History. Essays in Honour of Charles Singer,* ed. E. A. Underwood, 2 vols. (London: Oxford University Press, 1953), Vol. I, pp. 489-509.

[207] Van Helmont, "De lithiasi" ("Philiatro lectori"), *Opuscula,* p. 4; *Oriatrike,* sig. Nnnnn 2^V. This is van Helmont's most thorough treatment of the text of Eccelsiasticus, yet he mentions it throughout the *Opera.* see also "Tumulus pestis" (Ch. 1), *Opuscula,* pp. 5, 8; *Oriatrike,* pp. 1074, 1076: "Promissa authoris" (col. 2, sect. 6), *Opuscula,* p. 10; *Oriatrike,* p. 6.

However, continued van Helmont, in contrast to what it should have become, the art of medicine had been cultivated by Greeks, Arabs, and Jews, men given to gain who at the same time were enemies of Christianity. Although they called themselves physicians, they had not been called to this profession by the Lord and for this reason were not true healers nor were they able to cure difficult diseases.[208] For all their detailed texts, Vesalius had proved in 106 places that the ancients had not even dissected humans.[209] In recent past God has sent to us Paracelsus, "*a rich forerunner in the resolutive knowledge of Bodies.*" It is now to be hoped that God would "*over-add the knowledg of an Adeptist, which that other* [Paracelsus] *wanted.*"[210]

The Theory of Disease[211]

Not willing to consider himself an adept, van Helmont nevertheless strove to destroy the entire framework of ancient philosophy and to suggest a new approach.

> ... I have sufficiently demonstrated, that there are not four Elements in nature, and by consequence, if there are only three, that four cannot go together or encounter: Therefore that the squadron being broken, cannot cause four unlike Elementary combates, temperatures, mixtures, contrarieties, hatred, strifes, &c. For I have taught cleerly enough, that the fruites which antiquity hath believed to be mixt bodies, and those composed from a concurrence of four Elements, are materially of one onely Element.[212]

---

[208]"De lithiasi" ("Philiatro lectori"), *Opuscula*, p. 5; *Oriatrike*, sig. Nnnnn 3ʳ.
[209]"De febribus" (Ch. 3, sect. 10), *Opuscula*, p. 16; *Oriatrike*, p. 947.
[210]"De lithiasi" ("Philiatro lectori"), *Opuscula*, p. 6; *Oriatrike*, sig. Nnnnn 3ʳ.
[211]A recent and most useful account is that of Lester S. King, *The Road to Medical Enlightenment 1650-1695* (London: Macdonald; New York: American Elsevier, 1970), pp. 37-62. See also Peter H. Niebyl, "Sennert, Van Helmont, and Medical Ontology," *Bulletin of the History of Medicine*, 45 (1971), 115-137. Of major importance is Walter Pagel's "Van Helmont's Concept of Disease—To Be or Not to Be? The Influence of Paracelsus," *Bulletin of the History of Medicine*, 46 (1972), 419-454, which I was privileged to read in galley proof.
[212]"Scholarum humoristarum passiva deceptio atque ignorantia" (Ch. 1, sect. 3), *Opuscula*, p. 70; *Oriatrike*, p. 1017. See also the "Natura contrarium nescia" (sect. 1), *Ortus*, p. 164; *Oriatrike*, p. 161.

Devastating as the denial of the four elements seemed to be, it was necessary that much more be examined. A student of anatomy, physiology, and pathology, van Helmont proceeded to reject the concept of "catarrh" which was generally employed to account for disease.[213] Here vapors were alleged to ascend from the stomach to the brain, where they were condensed to a mucus that then flowed down from the base of the skull to all parts of the body. In the lungs the result was pneumonia and consumption, in the joints, rheumatism and gout, and in the legs, ulceration. This was a doctrine of a general disease due to a surplus of a corroding fluid, and it was closely linked with humoral pathology.

Van Helmont argued that the anatomy of the body makes it impossible for vapors to rise from the stomach to the head, so there can be no production of catarrhs, and thus disease does not flow from the head. The physician must look eleswhere for the cause— to the broader truths of natural philosophy. Diseases do not derive from a corroding catarrh flowing down from the brain; they are localized phenomena which develop from their own specific seeds[214] and have unique causes and life cycles. The growth process itself involves primarily an interaction of the *archeus* of the intruding seed with that of its host.[215]

Man is dominated by his vital principle, the *archeus influus*, situated in the *duumvirate* (stomach and spleen) and governing the whole body. In addition, each local organ possesses its own *archeus insitus*. A bodily upset, such as a fever or convulsion, will occur when either the main *archeus* or the local *archei* have reacted to an irritant. The result may be the departure or the neutralization of the offending irritant, bringing a return to health. Should this not happen, the disease will take root and grow, and the parent *archeus* then "runs away," leaving the intruder in possession of the

[213]This is the primary concern of Walter Pagel's *Jo. Bapt. Van Helmont. Einführung in die philosophische Medizin des Barock* (Berlin: Julius Springer, 1930). See also Pagel, *Paracelsus*, pp. 134-137. The major text is van Helmont, "Catarrhi deliramenta," *Ortus*, pp. 426-447; *Oriatrike*, pp. 429-450. A second English translation is that of Walter Charleton, *Deliramenti catarrhi: or the Incongruities, Imposibilities and Absurdities couched under the Vulgar Opinion of Defluxions* . . . (London, 1650).

[214]"Natura contrarium nescia" (sect. 15), *Ortus*, p. 168; *Oriatrike*, p. 164. "Ignotus hospes morbus" (sect. 40), *Ortus*, p. 492; *Oriatrike*, p. 492.

[215]"De lithiasi" (Ch. 5, sect. 5), *Opuscula*, p. 30; *Oriatrike*, p. 859.

organ.[216] Needless to say, the resultant damage is potentially much greater if the disease becomes lodged in the domain of the *duumvirate* rather than one of the lesser organs.

The origin and actual growth of the disease was explained by van Helmont both in spiritual and in materialistic terms. For him the *semen* was a substance made by the parent who conceives an image of himself through desire. Again, a seed may be produced through the odor of a ferment that disposes matter toward the idea of an object.[217] The concept of odor is ever recurrent in the Helmontian *Opera*. The ferment of the intruder may produce an odor that actually disposes the host *archeus* to terror.[218] This power may indeed be so great that should the alien seed rest in an area with no local *archeus*, something akin to one may be formed—the result being spontaneous generation.[219]

The seed of the disease having been hatched in the body, it begins its material growth and soon becomes independent. Thus, although initially a case of spiritual generation involving the interaction of inherent forces (*archei*), the result is soon evident in a material form.[220] In the Helmontian system exogenous agents are important as a source of irritation; however, the morbid ideas (*entia morborum*) are images that form driving forces within the disease *semina*. Still, no disease could develop if there were not a specific sympathy between an alien *archeus* and that of the host. Indeed, the concept of sympathetic action pervades van Helmont's works from his earliest tract on the weapon-salve to his last.

The introduction of alien *archei* is accounted for by van Helmont in his tract *Magnum oportet*, in which he outlined his concept of the "middle life."[221] Once again he turned to Genesis for authority. The Fall of man resulted in the loss of his immortal soul and the new ascendency of the sensitive soul which now controls his body. The perfection of the immortal soul had ensured the

---

[216]"Ortus imaginis morbosae" (sect. 2), *Ortus,* pp. 552-553; *Oriatrike,* p. 552; "Tumulis pestis" (Ch. 13), *Opuscula,* p. 53; *Oriatrike,* p. 1128.

[217]"Imago fermenti" (sect. 11), *Ortus,* p. 113; *Oriatrike,* p. 113.

[218]"Imago fermenti" (sect 19), *Ortus,* p. 114; *Oriatrike,* p. 114.

[219]"Imago fermenti" (sects. 8, 9), *Ortus,* pp. 112-113; *Oriatrike,* pp. 112-113.

[220]As discussed by Pagel in "Van Helmont's Concept of Disease." See "Ignotus hospes morbus" (sect. 34), *Ortus,* p. 491; *Oriatrike,* p. 491.

[221]Van Helmont, "Magnum oportet," *Ortus,* pp. 149-163; *Oriatrike,* pp. 148-159.

elimination of all alien bodies that might enter the body, but the imperfection of the sensitive soul made it impossible to eliminate anything completely. Thus pork maintains a fishy smell when pigs have been fed seafood, and an infant's urine exudes a characteristic odor if its wet nurse has eaten asparagus.[222] Although the original form and substance have been altered, the odor attests to the persistence of the original ferment and *archeus*. This, then, is a way in which exogenous bodies can be introduced to a new system and bring about a chain leading from irritation to disease. In short, van Helmont not only discarded the concept of catarrh, he also erected a new edifice for the explanation of disease based upon odors, ferments, *archei*, and semina.

Tartaric Disease

A special problem presented itself in the case of the Paracelsian tartaric disease. Here van Helmont rejected the views of Paracelsus rather than the ancients. Here also chemistry played an essential part in his understanding of medicine. Turning first to the origin of the concept of tartaric disease, he summarized the views of Paracelsus:

> Nature being at first a beautiful Virgin, was defiled by sin; not indeed by her own, neither therefore for a punishment to her self; but seeing she was created for the use of ungrateful man, she was as it were defiled with the fault of her inhabitant, that even by the defect of nature, he might in some sort purge the guilt. It after some sort repented the Creator, that he had commanded nature to obey the disobedient: Therefore he appointed, that the Earth should henceforward bring forth Thistles and Thorns: under the allegory whereof, the curse and rise of *Tartarers* are designed unto us; To wit, their matter which should exceeding sharply prick us: For the words do shew the progeny of the Earth, by the use whereof they do signifie, that Diseases should at length be incorporated in us: for first of all, the hostile *Tartarers* do trayterously enter with meats and drinks, they pierce into the bottom, are radically co-mingled, and shut up with a hidden Seal: Therefore some of them do even presently separate themselves within, from the pure nourishment; but others do remain

---

[222]"Magnum oportet" (sects. 4, 48), *Ortus*, pp. 150, 158; *Oriatrike*, pp. 149-150, 157

together with the nourishment, which being wasted away, the surviving *Tartarers* are coagulated under the form of Mucilage, Clay or Bole, next, of Sand, or a Stone, which then, are not onely uncapable of receiving the breath of life; but moreover, they keeping their wild Thorn, have become as the most inward immediate causes of all Diseases, the daily Nurses of the calamity of mortals: For as soon as the bloud is converted into the substance of the thing nourished, and afterwards consumed, this off-spring of Thorns doth often remain, surely inconvenient through a forreign coagulation, if not also through acrimonies or sharpnesses: For it waxeth more hard daily, and bespotteth its own Inn with a 1000 Hostilities: But a *Tartarer* or tartarous humour, differs from the humane excrements of meats in that, because these do putrifie, but that is coagulated: Therefore that stomach, and Liver is onely happy, which have known how to banish the sweepings of Tartar from the stinking excrements, in the beginning. As these Thorns are procured unto us by our antient Tartarous enemy; So the Stone that adhereth to the Joynts or Ribs of the Wine-Hogs-heads, giving by reason of its manifest Prerogative, a name to the other Ranks of coagulable vices, is called *Tartar:* For truly the Wine in the Vessel is on every side incrusted with a Stony bark, which is *Tartar,* diverse from the Lees: For this falls down to the bottom, knowing no coagulation: but that being extended round about, doth arm the Vessel, and preserve it within, for ever from corruption: But that guest being through nourishments, a stranger, is called a forreign *Tartar,* to distinguish it from that which groweth together within us, with a fatall Spectacle, by a Microcosmical Law: For whereby any violent thing doth rush into us, for that very cause the nourishable humours being destitute of life do appear hostile, are coagulated, and called the *Tartar* of the venal bloud: whence are Apostemes, stoppages, and other Calamities, according to the delighted property, and pleasure of every *Tartarer*: And so *Tartar* insinuating it self from the mouth, even unto the ultimate Coasts of the Pipes, is also the principal cause of all Diseases.[223]

Even though he admitted that many Galenists accepted the "Tartaric disease," van Helmont could not. All disease occurs from specific seeds and their *archei,* and in the case of tartar the inherent force drives toward a stonifying condition.[224] The tartar of wine forms no part of our food or drink, so it is hardly surprising to find that human tartar is different from the tartar that forms in

---

[223] "Tartari historia" (sect. 7), *Ortus,* pp. 234-235; *Oriatrike,* p. 231.
[224] "Alimenta tartari incontia" (sect. 12), *Ortus,* p. 247; *Oriatrike,* pp. 242-243.

wine casks.[225] The latter is a salt rather that a stone. Furthermore, processes leading to stonification in nature all differ from the formation of the urinary calculus (*Duelech*) in man which is caused by the action of the spirit of urine on a volatile earth procreated by a seed.[226] This stirs up a spirit of wine in the urine, and the entire process is comparable to the formation of ochre in spa waters with an "iron spirit." Here a solid is deposited and the water reverts to its original condition.

Van Helmont considered the treatment of the stone to be primarily a chemical problem. But it was not enough to find a solvent; the means of introducing it into the bladder had to be considered. Since the path from the mouth is a long one, subjecting substances to both heat and chemical change, a simpler method might be the direct introduction of the solvent into the bladder with the aid of a catheter. Van Helmont's long and arduous search was referred to in his treatise on the stone:

> ... at first, I ran through the Monarchy of Vegetables: but I found not that which could dissolve *Duelech* in the Bladder; But whatsoever of those would make *Duelech* to melt in a Glass, was either hostile, or at leastwise it came not with those qualities unto the Bladder; but if it might seem to be cast in by a Syringe, it was not by the Bladder to be endured. Therefore Vegetables being distilled and decocted; and likewise their ashes, *Calx's* or Limes, pouders, and all things being extracted, I learned but vain and slender Remedies against so great an Enemy. The more sharp ones indeed did diminish *Duelech* in his entireness; but being taken in at the mouth, they entred not under that power unto the Bladder; but being cast in from without, however they seemed mild, like unto Wine, yet they imitated bright burning Iron in the sense of pain.[227]

Van Helmont's extensive experiments on bladder injection made evident to him the need for an improved catheter. The traditional "little silvered Horn, wherewith Chirurgians do with the greatest Tourments, fetch out the urine, is cruel and bloudy; and therefore it hath altogether displeased me." In its place he invented a catheter made of thin leather with a plain seam which

---

[225]"Tartari vini historia" (sects. 10-19), *Ortus*, pp. 237-238; *Oriatrike*, pp. 233-234.
[226]"De lithiasi" (Ch. 4, sects. 1-3), *Opuscula*, pp. 32-33; *Oriatrike*, p. 851.
[227]"De lithiasi" (Ch. 7, sect. 28), *Opuscula*, p. 65; *Oriatrike*, p. 884.

proved to be practically painless. Therefore "Let praise Eternal be unto God in the Highest, and let it please him, to bedew, and make my services and desires fruitfull, which are offered for the help of mortals."[228]

When van Helmont's own search for a suitable solvent appeared fruitless, he turned to the *Ludus lapis* of Paracelsus—reputedly a specific for bladder stones. Now speaking in the fashion of an alchemist, he confessed that "at length, God taking pity on the anguishes of the Complainer, gave me the knowledge of the *Ludus*."[229] But "although the world be worthy of Compassion" so that

> ... its preparation may in a more manifest sense be described; yet the manifold Contemners of secret things are unworthy, that those things should be manifested now, which God for most weighty reasons, would have to remain among a few, and the little ones of this world, in the possession of the treasures of his own dispensation, until that nothing be hidden, which shall not be revealed in its own fulness of dayes: in which fulness of time, Wo to the world, and to its confusion![230]

Yet, if van Helmont then proceeded to "speak a little plainer," his directions are far from explicit and purposely intended only for those "who are skilful in the Phylosophy of the art of the fire."

As in so many other cases, van Helmont's work on the stone is a blend of mysticism and deep insight based upon experimental observation. There is little doubt that this treatise which emphasizes the chemical search for a solvent for the stone was of great influence and did much to unite chemistry and medicine in the course of the following century.

Chemical Inquiries: The Search for the Vital Spirit[231]

Throughout the works of the Paracelsian authors there is a persistent interest in the blood and the vital spirit. The background to

---

[228]"De lithiasi" (Ch. 7, sect. 34), *Opuscula*, pp. 67-68; *Oriatrike*, p. 886.
[229]"De lithiasi" (Ch. 7, sect. 28), *Opuscula*, p. 66; *Oriatrike*, p. 885.
[230]"De lithiasi" (Ch. 7, sect. 22), *Opuscula*, p. 62; *Oriatrike*, p. 881.
[231]As mentioned earlier, the most important recent treatment of the chemical work of van Helmont has been Partington's. His chapter in *A History of Chemistry*

this concept has already been touched on, and we have followed the early references to a vital aerial saltpeter in the Paracelsian tradition through the work of Robert Fludd, whose attempt to isolate this spirit formed the major portion of his "Philosophicall Key."

Here again the research of van Helmont indicates his keen awareness of a subject of great contemporary interest. The significance of the blood could not be denied. According to Holy Scripture "life"—which he compared with fire and light—is contained in the arterial blood.[232] Van Helmont's understanding of the blood flow was basically Galenic, and there is nothing in his work that indicates an acquaintance with the work of Harvey.[233] The chyle is the origin of both the urine and the venal blood, and the latter through contact with an indrawn aerial spirit is converted to the arterial blood.[234] One could then say that "the vitall Spirit *Is arterial Blood resolved by the force of the Ferment and Motion of the Heart, into a salt Air being vitally enlightened.*"[235]

However, it was not enough for van Helmont "to have known that the venal Blood doth ascend into arterial Blood" and that the "arterial Blood passeth over, partly into vital Spirit, and partly departeth into the Nourishment of the solid Parts"; it also "behoves us to have known the Marrow of the vital Spirit."[236] To find this essence, van Helmont, like Fludd, turned to chemical (distillation) analysis. However, rather than isolating the "quintessence" of wheat as Fludd did, van Helmont turned to human blood, which he may have been the first to distill in a search for its components.[237] Noting the common origin of urine and venous blood, he examined each in turn. From urine a volatile salt spirit is

---

(Vol. II, pp. 209-243) was based upon an earlier paper, "Joan Baptist van Helmont," *Annals of Science*, 1 (1936), 359-384. The reader is referred to Kurd Lasswitz, *Geschichte der Atomistik vom Mittelalter bis Newton*, 2 vols. (Hamburg/Leipzig: Leopold Voss, 1890), Vol. I, pp. 343-351, for van Helmont's limited use of the corpuscular hypothesis.

[232] Van Helmont, "Custos errans" (sect. 2), *Ortus*, p. 259; *Oriatrike*, p. 255.

[233] Evident throughout, but see as an example the "Aura vitalis," *Ortus*, p. 727; *Oriatrike*, p. 734-735. Here the vital spirit is discussed in terms of a Galenic framework.

[234] "Spiritus vitae" (sect. 12), *Ortus*, pp. 197-198; *Oriatrike*, p. 194.

[235] "Aura vitalis," *Ortus*, p. 728; *Oriatrike*, p. 734.

[236] "Aura vitalis," *Ortus*, p. 726; *Oriatrike*, p. 733.

[237] *Ibid.* See also "Spiritus vitae" (sects. 15-16), *Ortus*, pp. 198-199; *Oriatrike*, p.

obtained which when fermented with the earth produces saltpeter. The venous blood also contains a volatile salt spirit, which is the same as the spirit of urine except that "the Spirit of the Salt of Venal Blood cureth the Falling-sickness, but the Spirit of the Salt of Urin not so."[238] Noting the relationship of venous to arterial blood and the vital spirit, he concluded that the vital spirit must also be saltlike as well as airy; while not identical with saltpeter, it is similar to the spirit of urine which gives rise to saltpeter in the earth. The vital spirit is "plainly Salt, therefore Balsamical, and a Preserver from corruption."[239] Yet, although we know of the existence of this salt spirit, we do not know just how the whole venal blood may be homogeneally transchanged by the ferment of the heart. Elsewhere van Helmont suggested that

> ... the salt Spirits, and Sulphur of the arterial Blood, do by the Pulse, rub themselves together in the Sheath of the Heart, and a formal Light together with Heat, is kindled in the vital Spirit; from the Light I say, of the most inward, and implanted sunny Spirit, in which is the Tabernacle of the specifical Sun, even unto the Worlds end.[240]

Van Helmont's very typical insistence on the existence of "life" in the blood was partially responsible for his very unorthodox rejection of bloodletting.[241] Any loss of the vital principle could be equated with a corresponding loss in the strength or the vitality of the body. This theme was adopted quickly by his followers and was to become one of the most characteristic aspects of Helmontian iatrochemistry. Thus, Noah Biggs in 1651 wrote that "the *soul* or *vital* strength, rides in the Chariot of the bloud," and, as a result, bloodletting can only lead to "the fall or losse to the whole ocean of

---

195. The influence of this work on Robert Boyle has been touched upon by A. R. Hall in "Medicine and the Royal Society," *Medicine in Seventeenth Century England*, ed. Allen G. Debus (Los Angeles/Berkeley: University of California Press, 1974).

[238] "Aura vitalis," *Ortus*, p. 726; *Oriatrike*, p. 733.

[239] "Aura vitalis," *Ortus*, p. 727; *Oriatrike*, p. 733.

[240] "In sole tabernaculum," *Ortus*, p. 782; *Oriatrike*, p. 796. The process is here compared with the analogous heating of spa waters by the interaction of salt and sulfur in the earth.

[241] "De febribus" (Ch. 4), *Opuscula*, pp. 17-25; *Oriatrike*, pp. 949-957. The subject has been covered intensively by P. H. Niebyl in "Galen, Van Helmont and Blood Letting," *Science, Medicine and Society in the Renaissance*, ed. Allen G. Debus, 2 vols. (New York: Science History Publications; London: Heinemann, 1972), Vol. II, pp. 13-23.

strength."²⁴² Similar considerations led the Harvard graduate George Starkey to observe in 1657 in a volume titled *Natures Explication and Helmont's Vindication* that the "life is in the bloud ... and by how much of it is taken away, by so much is the vital balsam wasted, and therefore very unwisely taken away, if the disease may be cured without . . . ."²⁴³

A New Concept of Digestion

Van Helmont's insight and ability as a chemist is also seen to good advantage in his investigation of the process of digestion.²⁴⁴ The Galenic tradition taught in the Schools attributed to digestion to heat, suggesting further that the process was similar to cooking. However, as noted in the case of his search for a solvent for the stone, van Helmont was well aware that the action of bodily forces on ingested substances was highly complex. It was a simple matter for him to show that the results of heating a substance were considerably different from the action of the digestive process on that same substance.²⁴⁵ He further showed that the digestive process varied from one animal to another, thus permitting different species to digest different foods.²⁴⁶

Once again involved in a search for specificity, van Helmont explained that digestion in the stomach results from a specific acid ferment. This he had learned at an early age:

²⁴²Noah Biggs, Chymiatrophilos, *Mataeotechnia medicinae praxeωs. The Vanity of the Craft of Physick* . . . (London: Giles Calvert, 1651), p. 140. This work is discussed by the present author in "Paracelsian Medicine: Noah Biggs and the Problem of Medical Reform," *Medicine in Seventeenth Century England*, pp. 33-48.

²⁴³George Starkey, *Natures Explication and Helmont's Vindication* (London, 1657), p. 265.

²⁴⁴This subject has received considerable attention in recent years. The reader is particularly directed to Robert P. Multhauf, "J. B. Van Helmont's Reformation of the Galenic Doctrine of Digestion," *Bulletin of the History of Medicine*, **29** (1955), 154-163; Walter Pagel, "J. B. Van Helmont's Reformation of the Galenic Doctrine of Digestion—and Paracelsus," *Bulletin of the History of Medicine*, **29** (1955), 563-568; Walter Pagel, "Van Helmont's Ideas on Gastric Digestion and the Gastric Acid," *Bulletin of the History of Medicine*, **30**, 524-536; King, *The Road to Medical Enlightenment*, pp. 49-54.

²⁴⁵Van Helmont, "Calor efficienter non digerit, sed tantum excitative" (sect. 2), *Ortus*, p. 201; *Oriatrike*, p. 198.

²⁴⁶"Calor efficienter" (sect. 7), *Ortus*, p. 202; *Oriatrike*, p. 199.

... while being a Boy, I nourished Sparrows; I oft-times thrust out my tongue, which the Sparrow laid hold of by biting, and endeavoured to swallow to himself, and then I perceived a great sharpness to be in the throat of the Sparrow, whence from that time I knew why they are so devouring and digesting.

And then I saw that the sharp distilled Liquor of *Sulphur* had seasoned my Glove, and that it did presently resolve it into a juice, in the part which it had moistened; which thing confirmed to me a young Beginner, that meats are transchanged by a sharp or soure thing, and so that a ferment doth inhabit in the stomack, which should change all things cast into it, although sweet, presently into a sowreness: . . .[247]

Acidity alone is not enough to bring about the change, since many acids alone do not cause the alteration. "Therefore the Digestive Ferment *is an essential property, consisting in a certain vital sharpness or soureness, mighty for transmutations*; and therefore of a specifical property."[248] Pagel has shown that van Helmont came close to identifying this acid with our hydrochloric acid.[249]

In his discussion of the acid ferment as the effective digestive agent, van Helmont accused Paracelsus of being "deluded by a digestive heat, and ignorant of the Ferment of the stomack."[250] Pagel has further shown that this charge was not wholly just, since Paracelsus was aware of "appetising acids" (*acetosa esurina*) which he considered to be the normal digestive factor for certain animals.[251] Fabius Violet (1635) had gone further to insist that the *acetosa esurina* were the agents of digestion for humans as well as a limited number of animals.[252] Thus, important and original as van Helmont's contributions were on this subject, we find him also working within the general framework of the Paracelsian tradition.

[247]"Calor efficienter" (sects. 23, 24), *Ortus*, p. 204; *Oriatrike*, p. 201.
[248]"Calor efficienter" (sect. 28), *Ortus*, p. 205; *Oriatrike*, p. 201.
[249]Pagel, "Ideas on Gastric Digestion," pp. 532-533. The identification is based primarily upon the Flemish and German passages in the *Dageraad* and the *Aufgang der Artzneykunst*.
[250]"Calor efficienter" (sect. 20), *Ortus*, p. 204; *Oriatrike*, p. 200.
[251]Pagel, "Ideas on Gastric Digestion," pp. 529-530.
[252]Fabius Violet, *La parfaicte et entiere cognoissance de toutes maladies du corps humain, causées par obstruction* (Paris: Pierre Billaine, 1635), p. 142. This passage is discussed by Pagel in his *Paracelsus*, pp. 161-163.

After his discussion of the acid ferment of the stomach, van Helmont proceeded to attack the traditional threefold process of digestion in (1) the stomach and the intestines (to produce chyle), (2) the liver (to produce urine and the venous blood), and (3) the blood (by which it is changed to nourishment).[253] In his replacement of this system he postulated a series of six digestive steps. The first five took place in the stomach, the small intestine, the mesenteric veins, and the heart, where venous blood is changed to arterial blood and where the vital spirits are produced. The sixth digestion involved the nourishment of the various parts of the body under the assumption that each part had its own "kitchen" or "stomach" with the proper ferment to bring about the required activity and transformation.[254] Indeed, in all cases local ferments were required for the processes involved. His system was a complex one encompassing many processes that we would not call "digestive," but in the course of his discussion van Helmont again offered a number of significant observations that shed light on the relationship of his physiology to his chemical research. Two examples will illustrate his approach. The first derives again from van Helmont's reading of Paracelsus. Following an observation made by his predecessor, he dissolved salt in a bladder that was "hanged up in a Brasse Kettle of hot water." In time,

> I found the water in the Kettle to be not much less salt than that which was in the Bladder, whose neck was tied fast to the handle of the Kettle appearing above the water; from whence I knew, that the water did pierce within and without the Bladder; to wit, That the Bladder was passable by Salt, and hot water, but not by air.[255]

This observation of osmosis made it possible for him to explain that the "cream" [*cremor*] entering the small intestine from the stomach passed through the bowels at least "partly by its imbibing of them [the cream], even as Salt water doth a Bladder . . . ."

Van Helmont's consideration of the digestive process led him to observe the neutralization that occurs in the small intestine

---

[253] "Triplex scholarum digestio," *Ortus*, pp. 207-208; *Oriatrike*, pp. 203-204.
[254] "Sextuplex digestio alimenti humani," *Ortus*, pp. 208-225; *Oriatrike*, pp. 205-221.
[255] "Sextuplex digestio" (sect. 48), *Ortus*, p. 218; *Oriatrike*, pp. 214-215.

through the action of the bile.

> The Cream sliding out of the *Pylorus* or neather mouth of the Stomack, into the *Duodenum*, being straight-way snatched within the Sphear of activity, by the in-breathing of the Gaul, doth exchange its sourness into Salt, and its more watery part is made severable from its more pure or un-mixt part, which is drawn by the Reins.[256]

The concept of acid-base neutralization, so important for the development of chemistry later in the century, derives principally from van Helmont. It was a process he discussed in contexts besides the physiology of digestion. Thus he noted that diuretics are generally alkaline because "in every Wound, a Tartnesse or Acidity, the Betokener and Companion of all putrefaction in the flesh, doth arise: the which *Alcalies* do easily sup up into themselves, and consume."[257] And again, writing on the effect of fire on substances, he observed that new things may be produced that were not present originally: either acids (spirit of salt [HCl] from salt or oil of vitriol [H_2SO_4] from vitriol) or alkalis, while "now and then, the things themselves, together with their adjuncts, are diversly transchanged by the Fire, and become neutral."[258]

The Chemical Remedies

Van Helmont's rejection of cure by contraries was a direct consequence of his denial of the humors. The Galenic system had essentially been founded upon a "warfare" of the elements in our bodies, and cure by contraries arose as a necessity. However, for van Helmont it was a "mathematical" error of the ancients to assume that "every excess . . . might be reduced into a mediocrity or mean." Those who subscribed to this doctrine accepted a perpetual strife and conflict in nature when instead they should have emphasized its harmony and integrity.[259]

---

[256] "Sextuplex digestio" (sect. 56), *Ortus*, p. 220; *Oriatrike*, p. 217.
[257] "De lithiasi" (Ch. 5, sect. 17), *Opuscula*, p. 45; *Oriatrike*, p. 863.
[258] "Potestas medicaminum" (sects. 36, 37), *Ortus*, p. 479; *Oriatrike*, pp. 477-478.
[259] "Natura contrarium nescia" (sects. 11 *seq.*) *Ortus*, p. 167; *Oriatrike*, pp. 163-164.

So convinced was van Helmont of the error of the Galenic position that he carried his argument beyond the realm of medical practice to that of the physics of motion. The natural philosophers of the Schools affirm that "every patient or sufferer doth likewise of necessity re-act, and for that cause likewise every agent or acter doth re-suffer; neither also that it is any other way weakened."[260] But it is incorrect to charge that any action results in an equal reaction.

> For truly, after that I with-drew contraries out of Nature, I could not afterwards, in sound judgment, find out any re-acting in the patient, as neither could I admit of hostilities in nature, elsewhere than among soulified or living creatures: For contariety is in those things alone, wherein there is an actual defence in the will of the patient against the injuries brought on it, and felt from the Agent: Wherefore there is never a re-acting of the patient on the agent, unlesse there is a contrariety conceived in the soul.[261]

In a lengthy section van Helmont proceeded to offer numerous examples designed to show that under no circumstances is there any relation of action to reaction in the case of collisions. Thus,

> If an impressive force of strength doth act indeed by it self, but in the mean-time be limited by space of place, duration, or be weakned by impediments, or lastly, if it act measuringly, by reason of figure or hardnesse; at leastwise, there is never in these, any re-acting of the patient, or re-suffering of the Agent.[262]

In the case of a man who strikes an anvil with his fist,

> ... and thereby receives a wound, or bruise, there is not in that stroak any re-acting of the Anvil .... the fist doth act simply on the Anvil,

---

[260]"Ignota actio regiminis" (sect. 3), *Ortus*, p. 329; *Oriatrike*, p. 325. The Latin for this important passage reads "Itaque stabilivere omne patiens, vicissim reagere necessario, atque eatenus similiter omne Agens repati, nec etiam aliunde debilitari." This is a far cry from Newton's succinct statement of the third law of motion, "Actioni contrariam semper & aequalem esse reactionem: sive corporum duorum actiones in se mutuo semper esse aequales & in partes contarias dirigi." Isaac Newton, *Philosophiae naturalis principia mathematica* (London: Jussu Societatis Regiae ac Typis Joseph Streater, 1687), p. 13. A paper on the relationship of van Helmont to Newton and the third law is forthcoming.

[261]Van Helmont, "Ignota actio regiminis" (sect. 4), *Ortus*, p. 329; *Oriatrike*, p. 325.

[262]"Ignota actio regiminis" (sect. 6), *Ortus*, p. 332; *Oriatrike*, p. 327.

and the Anvil suffers simply, although it took no offence thereby: but the fist suffers by accident, if it do the more strongly strike: the Agent of which suffering is notwithstanding, not the Anvil, but the fist it self: Because there is one only and single action of the stroak, and hurt, which I therefore call a rebounding one: . . .[263]

On the level of medical practice the denial of contraries by Paracelsus and his followers had been replaced by a doctrine of cure by similitude, with which van Helmont could not agree.

For it is not requisite that the Remedy and external Cause of a Disease should have a co-resemblance, how ever notwithstanding *Paracelsus* hath so commanded . . . . For Poysons are not overcome by a co-resemblance of the Venome, but by that which conquers the Venome. For those medicinal Powers are the gifts of God, which do neither bear a contrariety, or character of hostility, mutually towards themselves, nor towards Diseases. But every thing acteth from a gift, that which it is commanded to act. And moreover, bodies being freed from their lump, enclosure, filths, and impediments, do unfold most noble gifts and most excellent Powers or faculties.[264]

For van Helmont the long search for the inherent virtues of substances was made all the more difficult by the state of the medical profession, burdened with quacks, Jews, and fugitive chemists. The latter "are Idiots, being fugitive Apostates from Chymical furnaces."[265] But the Schools are most to blame for deception. Their professors know nothing of the cause of disease and they perpetuate their ignorance by promoting their own scholars:

. . . this man, because he is a Latinist, and hath his father a Chyrurgion, or an Apothecary; or another, because he was made Master of Arts, and hath heard some Lectures of Professours; another lastly, because he in part, brags of *Euclide*, or hath learned to dispute, from *Aristotle*.[266]

The learned physician will not deign to soil his hands and thus relegates the preparation of his medicines to his apothecary.[267] And

[263]"Ignota actio regiminis" (sect. 7), *Ortus*, p. 332; *Oriatrike*, p. 327.
[264]"Tria prima chymoricum principia" (sects. 42, 43), *Ortus*, p. 404; *Oriatrike*, p. 407.
[265]"De lithiasi" (Ch. 7, sect. 2), *Opuscula*, p. 54; *Oriatrike*, p. 873.
[266]*Ibid.*
[267]"De febribus" (Ch. 15, sect. 1), *Opuscula*, p. 53; *Oriatrike*, p. 989.

if the distillers of oils are a shade better than the apothecaries, they have produced little if anything new since the time of Dioscorides. Nor do those like Duchesne who seek the virtues in the form of divine signatures aid in the advance of medicine. That chemist's observations on the ashes of nettles proved nothing; without a doubt the properties of herbs—as all other substances—are inherent in their seeds.[268] "I believe by Faith, that man was not of nature, and therefore likewise, that nature is not the Image, likeness, or engravement of man."[269]

In order to seek out the true virtues of medicinal substances, van Helmont concluded, it is necessary to turn once again to chemistry, for this is the study that "opens the Gate unto Natural Philosophy."[270]

> That science therefore which teacheth how to look, into bodies shut up, by a re-solution of themselves, and to extract their hidden virtues, is not the servant of the practick preparatory part of medicine (as the reproaches of the ignorant do sound) but it is the chief interpretation of the history of nature.[271]

Although van Helmont could proudly give himself the title "Philosophus per Ignem,"[272] he was not forgetful that others disagreed with his views. Listing the common arguments against the chemicals, he methodically proceeded to demolish them in order.[273]

Van Helmont's espousal of chemical medicines did not blind him to their frequent misuse and possible danger. He was aware that leaf gold and powdered pearls were simply not digested and thus had no effect on the body at all.[274] Mercury and antimony provoked violent purges—a treatment that was to be avoided no

---

[268] "Pharmacopolium ac dispensatorium modernum" (sects. 13, 14), *Ortus*, p. 459; *Oriatrike*, p. 459.

[269] "Pharmacopolium" (sect. 5), *Ortus*, p. 458; *Oriatrike*, p. 457.

[270] "De lithiasi" (Ch. 5, sect. 4), *Opuscula*, p. 40; *Oriatrike*, p. 859.

[271] "De febribus" (Ch. 15, sect. 8), *Opuscula*, p. 55; *Oriatrike*, p. 991 (incorrectly numbered 971).

[272] See the "Tumulus pestis" ("Chare lector"), *Opuscula*, p. 4.

[273] "De febribus" (Ch. 15, sects. 1-25), *Opuscula*, pp. 53-58; *Oriatrike*, pp. 989-992 (the latter page incorrectly numbered 1002).

[274] "De febribus" (Ch. 8, sects. 4-10), *Opuscula*, p. 38; *Oriatrike*, pp. 972-973. "Pharmacopolium" (sect. 55), *Ortus*, p. 468; *Oriatrike*, p. 467.

less than bloodletting.[275] The common preparation of extracts might at first ease pain, "but Magisteries, I willingly lay up in the place of extracts, whereby the whole substance of a thing is reduced into its primitive juice."[276] Like earlier Paracelsians, he commented on the necessity of small dosages, since the virtues of these remedies "are narrowed in a smal quantity."[277]

Throughout the works of van Helmont there are references to chemical preparations, but nowhere did he collect a group of chemical remedies in the fashion of Crollius, Duchesne, or other iatrochemists. Thus, he referred (in the "Duumviratus") to the sulfur of vitriol (without mentioning Paracelsus as his source) because of the "sweet" sleep it induces,[278] he gave elaborate directions (in the "De lithiasi") for the preparation of copper sulfate (vitriol of copper) because of the healing properties mentioned by Paracelsus,[279] and he discussed various diuretics in relation to his experiments to dissolve the stone.[280] Chemical compounds are referred to throughout, and the methods of preparation he described are at times of considerable significance. Nevertheless, his "Pharmacopolium ac dispensatorium modernorum" is more a general defense of the use of chemistry in medicine than it is a pharmacopoeia. There it also becomes clear that the physician in his chemical search for medicinal virtues in plants and minerals is at the same time performing a pious duty by showing to mankind the infinite love of the Creator.

> At length, I by Chymistry, beholding all things more clearly, it repented me of my former rashness, and blockish ignorance; For

[275]"De febribus" (Ch. 8, sect. 6), *Opuscula*, p. 38; *Oriatrike*, p. 971.

[276]"Pharmacopolium" (sect. 22), *Ortus*, p. 461; *Oriatrike*, p. 461.

[277]"De febribus" (Ch. 15, sect. 5), *Opuscula*, p. 55; *Oriatrike*, p. 991 (incorrectly numbered 971).

[278]"Duumviratus" (sect. 9), *Ortus*, pp. 346-347; *Oriatrike*, p. 339. See also the "De lithiasi" (Ch. 9, sect. 86), *Opuscula*, p. 99; *Oriatrike*, p. 918. In a number of places Pagel has discussed the use of ether mixtures by Paracelsus to induce sleep in chickens. See especially "Paracelsus' ätherähnliche Substanzen und ihre pharmakologische Auswertung an Hühnern. Sprachgebrauch (henbane) und Konrad von Megenbergs 'Buch der Natur' als mögliche Quellen," *Gesnerus*, **21** (1964), 113-125. Partington is the authority to consult for a listing of significant compounds described by van Helmont.

[279]"De lithiasi" (Ch. 8, sects. 16-22), *Opuscula*, pp. 71-73; *Oriatrike*, pp. 890-892.

[280]"De lithiasi" (Ch. 7), *Opuscula*, pp. 52-68; *Oriatrike*, pp. 871-886.

truly. I did on every side, humbly adore, with admiration, the vast Clemency and Wisdom of the Master-Builder. For he would not have Poysons, to be Poysons, or hurtful unto us: For he neither made Death, nor a Medicine of Destruction in the Earth; but rather that by a little labour of ours, they might be changed into the great pledges of his Love, for the use of Mortals, against the cruelty of future Diseases.[281]

A Challenge for the Future

Any final assessment of van Helmont is difficult. At one extreme he has been interpreted basically as an alchemist and a Hermeticist whose work was harmful for the advance of medicine.[282] At the other, we have been told that his main aim was the development of a concept of disease, "not simply or even primarily . . . 'iatrochemical.' "[283]

Neither of these interpretations is completely satisfactory. Van Helmont was a chemical philosopher who took as his goal the understanding of all nature. While he sought the complete overthrow of ancient medicine and philosophy as taught in the universities, he hoped to see it replaced by a new vitalistic philosophy based upon theological and natural truths. However, deeply influenced by the Paracelsian tradition, van Helmont believed that the most noble science was medicine, a subject unknown to those who were not true natural philosophers. But no one could claim a knowledge of natural philosophy or of medicine who did not possess the key of chemistry. Only this could open the gates. Thus, although critical of Paracelsus, van Helmont may still be safely placed in the Paracelsian tradition. And, on the basis of his accomplishments and his influence, he may rightly be termed the most important chemical philosopher of the first half of the seventeenth century.

Complex and diffuse as his writings occasionally are, their

---

[281]"Pharmacopolium" (sect. 46), *Ortus*, p. 466; *Oriatrike*, p. 465.

[282]Nève de Mévergnies, *Van Helmont*, p. 220. "L'*Ortus Medicinae* a peut-être rendu des services signalés à la science médicale. Mais, par l'esprit qui l'anime, et dont le rayonnement semble avoir été considérable, il n'en a pas moins été, malgré le génie de son auteur, un obstacle au progrès de la médecine." This statement can hardly be upheld in the light of recent research.

[283]Niebyl, "Sennert, Van Helmont, and Medical Ontology," p. 115.

effect was to infuse new life into the chemical philosophy. His work, more than that of any other, was responsible for the tremendous influence of chemistry in the decades following the publication of the *Ortus medicinae* in 1648. With van Helmont there was a new emphasis on observational and experimental data, coupled with an interest in the development of new instruments and in the application of quantification to chemical and medical problems. For him all earlier writings—be they Galenist, Aristotelian, or Paracelsian—were subject to review, and, in contrast with his more mystical colleagues, the macrocosm-microcosm analogy was reduced to the status of metaphor. Though, under certain circumstances, van Helmont could still argue for the proper reading of the stars and could give a circumstantial account of a transmutation or write in awe of alchemical adepts, he nevertheless had broken an important chain which held many to an older worldview.

However, beyond this, van Helmont seemed genuinely to excite those who read his works by his frontal attack on authority. Flinging down the gauntlet to the Galenists, he wrote

> Oh ye Schooles. . . . Let us take out of the hospitals, out of the Camps, or from elsewhere, 200, or 500 poor People, that have Fevers, Pleurisies, &c. Let us divide them into halfes, let us cast lots, that one halfe of them may fall to my share, and the other to yours; . . . we shall see how many Funerals both of us shall have: But let the reward of the contention or wager, be 300 Florens, deposited on both sides: Here your business is decided. Oh ye Magistrates, unto whom the health of the People is dear! it shall be contested for a publique good, for the knowledge of truth, for your Life, and Soul, for the health of your Sons, Widows, Orphans, and the health of your whole People: And finally, for a method of curing, disputed in an actual contradictory, superadd ye a reward, instead of a titular Honour from your Office: compel ye those that are unwilling to enter into the combate, or those that are Dumb in the place of exercise, to yeild; let them then shew that which they now boast of by brawling: For thus Charters from Princes are to be shewn: Let words and brawling cease, let us act friendly, and by mutual experiences, that it may be known hence forward, whether of our two methods are true: For truly, in contradictories, not indeed both propositions, but one of them only is true.[284]

---

[284]Van Helmont, "Respondet author" (sect. 9), *Ortus*, pp. 526-527; *Oriatrike*, p. 526.

This was an offer that was to be repeated time and time again by his followers.

As for the universities, their reform was mandatory, wrote van Helmont. No longer should young men waste their youth on Aristotelian and Galenic texts. In a seven-year program they should devote three years to arithmetic, mathematical science, the *Elements* of Euclid, and geography (which is to include the study of seas, rivers, springs, mountains, and minerals). During this same three-year period they were to study the properties and customs of nations, waters, plants, living creatures, and the astrolabe. Only then, van Helmont continued,

> ... let them come to the Study of Nature, let them learn to know and seperate the first Beginnings of Bodies. I say, by working, to have known their fixedness, volatility or swiftness, with their seperation, life, death, interchangeable course, defects, alteration, weakness, corruption, transplanting, solution, coagulation or co-thickning, resolving. Let the History of extractions, dividings, conjoynings, ripenesses, promotions, hinderances, consequences, lastly, of losse and profit, be added. Let them also be taught, the Beginnings of Seeds, Ferments, Spirits, and Tinctures, with every flowing, digesting, changing, motion, and disturbance of things to be altered.
>
> And all those things, not indeed by a naked description of discourse, but by handicraft demonstration of the fire. For truly nature measureth her works by distilling, moystening, drying, calcining, resolving, plainly by the same meanes, whereby glasses do accomplish those same operations, And so the Artificer, by changing the operations of nature, obtains the properties and knowledge of the same. For however natural a wit, and sharpness of judgement the Philosopher may have, yet he is never admitted to the Root, or radical knowledge of natural things, without the fire. And so every one is deluded with a thousand thoughts or doubts, the which he unfoldeth not to himself, but by the help of the fire. Therefore I confess, nothing doth more fully bring a man that is greedy of knowing, to the knowledges of all things knowable, than the fire. Therefore a young man at length, returning out of those Schooles, truly it is a wonder to see, how much he shall ascend above the Phylosophers of the University, and the vain reasoning of the Schooles.[285]

---

[285]"Physica Aristotelis et Galeni ignara" (sects. 9-11), *Ortus*, pp. 49-50; *Oriatrike*, p. 45.

This was the Helmontian-iatrochemical program for educational reform: students were to spend the bulk of their time learning the true facts of nature through observation and chemical analysis with fire. No longer was there a call for compromise, a plea to accept the best of both the old and the new systems. Van Helmont no less than Paracelsus claimed both truth and proper method. His claim was one that was to be heard by many who were ready to listen.

6
THE CHEMICAL PHILOSOPHY IN TRANSITION:
Nature, Education, and State

> ... Alchymy *points out to us the way, (she being the Instrument of the true praise-worthy Philosophy) and opens the Gate, whereby we may search into the inmost bowels of every thing. He who well knows the fire, and the use thereof, will not be distressed with want.*
>
> For my part I am verily of this Opinion, that there is such a wonderfull time at hand, the like of which hath neither been seen or heard of from the time of the Floud even to these days.
>
> Johann Rudolph Glauber (1656, 1661)*

VAN HELMONT'S PLEA reached many who recognized the need for educational reform. The chemists were convinced that the time was right for change. Wars and revolutions had shaken the old order in both politics and religion, while an increasing number of scholars throughout Europe were working to alter the traditional educational system. In the eyes of these chemical philosophers the heathen university curricula based upon Aristotle and Galen had to be replaced by the Christian philosophy of nature set forth by the Paracelsians and the Helmontians, and until this was done it was best for those seeking a true education to stay far away from any university. However, after the necessary educational reforms had been accomplished, scholars could expect to reap both pure and practical benefits from a greater knowledge of nature. Not only the inestimable medical benefits offered by the chemists, but much more might be expected. The application of Paracelsian

---

*Johann Rudolph Glauber, *The Prosperity of Germany* in *The Works of The Highly Experienced and Famous Chymist, John Rudolph Glauber . . .*, trans. Christopher Packe (London: Thomas Milbourn, 1689), Part I, pp. 297, 440. Here the italicized quotation is from the preface to the first part of the work, the other from the sixth part. The German was available to me only for the first two parts, which are included in J. R. Glauber, *Des Teutschlandts Wolfahrt* in *Opera Chymica, Bücher und Schrifften . . .* (Frankfurt am Mayn: Thomae-Matthiae Götzens, 1658-1659), Part II, p. 348.

speculation on plant growth, which had been directly applied to agricultural theory in the late sixteenth century, became a theme pursued in the new century as well. In Germany Johann Rudolph Glauber examined this subject in detail and included it as part of a plan for both the economic and the military prosperity of his homeland—a plan based upon chemical research.

A study of both the educational debates and the expected benefits to be derived from the chemical philosophy are subjects of considerable interest and importance. The present chapter will concentrate on the former primarily through the English conflict between the chemist John Webster and the mechanists Seth Ward and John Wilkins. The suggested reform of agriculture will be followed through the work of Bernard Palissy, Hugh Plat, and Samuel Hartlib, while a glance at Glauber's *Prosperity of Germany* will give some indication of the great breadth of the dreams envisaged by some mid-seventeenth-century chemists.

Educational Reform: Background

There are few more persistent themes in Renaissance intellectual history than that of educational reform.[1] Peter Ramus (1515-

---

[1] The following section reproduces with minor modifications much of the material that appeared in the following studies: Allen G. Debus, "The Webster-Ward Debate of 1654: The New Philosophy and the Problem of Educational Reform," *The Proceedings of the Institute of Medicine of Chicago*, **27** (1969), 248-249; "John Webster and the Educational Dilemma of the Seventeenth Century," *Actes du XXIe Congrès International d'histoire des Sciences*, 12 vols. (Paris: A. Blanchard, 1970), Vol. IIIB, pp. 15-23; "The Webster-Ward Debate of 1654: The New Philosophy and the Problem of Educational Reform," *L'univers à la Renaissance: Microcosme et macrocosme*, Travaux de l'Institut pour l'étude de la Renaissance et de l'Humanisme (Brussels: P.U.B./P.U.F., 1970), Vol. IV, pp. 33-51; *Science and Education in the Seventeenth Century. The Webster-Ward Debate* (London: Macdonald; New York: American Elsevier, 1970), pp. 1-64; "Paracelsian Medicine: Noah Biggs and the Problem of Medical Reform," in *Medicine in Seventeenth-Century England*, ed. Allen G. Debus (Berkeley/Los Angeles: University of California Press, 1974). For a general discussion of the pertinent literature on the history of education see *Science and Education in the Seventeenth Century*, pp. 11-12 (especially nn. 1 and 11). Social problems not discussed here are assessed in Hugh Kearny, *Scholars and Gentlemen. Universities and Society in Pre-Industrial Britain* (Ithaca: Cornell University Press, 1970).

1572) wrote in despair of his training, while Rabelais (c. 1495-1553) implored his readers to study nature at first hand rather than through books.[2] With Telesio, Campanella, Patrizi, and Bruno one finds a militant rejection of the classical authors and their medieval commentators joined with a conscious search for a new scientific method.[3] In the new century William Gilbert and William Harvey expressed their independence of tradition,[4] while René Descartes recalled that after completing the entire course of study at one of the most celebrated schools of Europe, "I found myself embarrassed with so many doubts and errors that it seemed to me that the effort to instruct myself had no effect other than the increasing of my own ignorance." Perhaps, he added, the whole body of the sciences need not be reformed, "but as regards all the opinions which up to this time I had embraced, I thought I could not do better than endeavor once and for all to sweep them away, so that they might later be replaced . . . ."[5]

If we turn to Francis Bacon (1561-1626), we find in the midst of his search for a new scientific method a critical reappraisal of the university system. After complaining of the low salaries allotted to university teachers, and after questioning the wisdom of concentrating students on specific professions rather than the arts and sciences, he went on:

> Again, in the customs and institutions of schools, academies, colleges and similar bodies destined for the abode of learned men and the cultivation of learning, everything is found adverse to the progress of science. For the lectures and exercises there are so ordered, that to think or speculate on anything out of the common way can hardly oc-

---

[2]Frank Pierrepont Graves, *Peter Ramus and the Educational Reformation of the Sixteenth Century* (New York: Macmillan, 1912), pp. 23-24, 169. The significant passage is quoted at length above, Ch. 1. Rabelais, *Les oeuvres*, ed. Ch. Marty-Laveaux, 6 vols. (Paris: Lemerre, 1870-1903), Vol. I, p. 256.

[3]Paul Oskar Kristeller, *Eight Philosophers of the Italian Renaissance* (Stanford: Stanford University Press, 1964), pp. 39-138 *passim*.

[4]William Gilbert, *On the Loadstone and Magnetic Bodies, and on The Great Magnet the Earth*, trans. P. Fleury Mottelay (London: Bernard Quaritch, 1893), p. 1. William Harvey, *The Works of William Harvey, M.D.*, trans. and with a life by Robert Willis (London: Sydenham Society, 1847), p. 7.

[5]René Descartes, *Discours de la méthode*, introduction and notes by Etienne Gilson (Paris: J. Vrin, 1964), pp. 49, 62.

cur to any man. And if one or two have the boldness to use any liberty of judgment, they must undertake the task all by themselves, they can have no advantage from the company of others. And if they can endure this also, they will find their industry and largeness of mind no slight hindrance to their fortune. For the studies of men in these places are confined and as it were imprisoned in the writings of certain authors, from whom if any man dissent he is straightway arraigned as a turbulent person and an innovator . . . [The] arts and sciences should be like mines, where the noise of new works and further advances is heard on every side. But though the matter be so according to right reason, it is not so acted on in practice; and the points above mentioned in the administration and government of learning put a severe restraint upon the advancement of the sciences.[6]

In his utopian *New Atlantis* (1627), Bacon went on to describe "Solomon's House" with its provision for workshops, scientific instruments, and laboratories—all of the requirements for the proper collection of scientific data according to Bacon's scheme for a new science.

This general disenchantment with Scholasticism forms an integral part of the Paracelsian tradition as well. We have already noted that the Rosicrucian *Fama fraternitatis* and the *Confessio* emphasize the need for educational reform and point to Paracelsus as the prime example of the true Christian scholar. These works glorify the calling of the physician and the learned alchemist, precisely the points to which Robert Fludd addressed his Rosicrucian apology in 1617. The utopian format of Bacon's *New Atlantis* found a parallel in the *Fama fraternitatis* as well, and this genre was to be employed by others who had an interest in the introduction of the chemical philosophy into the traditional curriculum. An example is the *Christianopolis* (1619) of Johann Valentin Andreae (1587-1654), who may have been the author of the *Fama*. The *Christianopolis* bears striking similarities to the *New Atlantis*, and its importance for seventeenth-century science merits greater attention than it has received in the past. Andreae's close association

---

[6]*The Philosophical Works of Francis Bacon . . . Reprinted from the Texts and Translations, notes and prefaces of Ellis and Spedding*, ed. John M. Robertson (London: Routledge and Sons; New York: Dutton, 1905), p. 286, from the *Novum organum*. See also *On the Advancement of Learning*, p. 76, for a similar passage.

with Comenius, who was in contact with Samuel Hartlib (d. 1670), John Dury (1596-1680), and William Petty (1623-1687), connects him with the background which culminated in the foundation of the Royal Society of London.[7]

In the *Christianopolis* Andreae deplored the decay of learning and religion and suggested that a proper community be formed, open to all of good intent and character.[8] We need not discuss this model in detail, but it is significant to recognize again the reaction against tradition in the spirit of the contemporary chemical philosophers. The citizens of *Christianopolis*, for instance, have a large library but use it little, concentrating only on the most thorough books. The meaning is clear: "The highest authority among them is that of sacred literature, that is, of the Divine Book; and this is the prize which they recognize as concealed by divine gift to men and of inexhaustible mysteries; almost everything else they consider of comparatively little value . . . ."[9] The scholars here prefer to get their knowledge directly from the book of nature, for a "close examination of the earth will bring about a proper appreciation of the heavens, and when the value of the heavens has been found, there will be a contempt of the earth."[10] Accordingly it is more than justified to look at the laboratory—the center for such studies. As might be expected, it is a chemistry laboratory, fitted

---

[7]Here see Allen G. Debus, *The Chemical Dream of the Renaissance* (Cambridge: Heffer, 1968; Indianapolis: Bobbs Merrill, 1972), pp. 18-20. For the literature on Andreae, see Ch. 4 above, n. 18. Johann Valentin Andreae, *Christianopolis. An Ideal State of the Seventeenth Century*, trans. with an historical introduction by Felix Emil Held (New York: Oxford University Press, 1916), pp. 53 ff. The influence of the *Christianopolis* has been touched on briefly by W. H. G. Armytage, "The Early Utopists and Science in England," *Annals of Science*, **12** (1956), 247-254. Held has discussed the relation of the *Christianopolis* to the founding of the Royal Society in his edition of the text, pp. 100-125. Considerable attention is paid to this work by Margery Purver, *The Royal Society: Concept and Creation* (London: Routledge and Kegan Paul, 1967); this should be read alongside the essay review by Charles Webster, "The Origins of the Royal Society," *History of Science*, **6** (1967), 106-127. See also G. H. Turnbull, *Hartlib, Dury and Comenius. Gleanings from Hartlib's Papers* (Liverpool: University of Liverpool Press, 1947), and H. R. Trevor-Roper, "Three Foreigners and the Philosophy of the English Revolution," *Encounter*, **14** (1960), 3-20.

[8]Andreae, *Christianopolis*, p. 29.
[9]*Ibid.*, p. 191.
[10]*Ibid.*, p. 187.

out with the most complete equipment, where the "properties of metals, minerals, and vegetables, and even the life of animals are examined, purified, increased, and united, for the use of the human race and in the interests of health." More important, it is here that the "sky and the earth are married together," and it is here that the "divine mysteries impressed upon the land are discovered."[11] These are clearly references to the macrocosm-microcosm analogy and the doctrine of signatures.

The importance of chemistry for Andreae becomes even more evident when it is compared with his treatment of other sciences. His hall of physics contains, not basic studies of motion, but natural history scenes painted on the walls—views of the sky, the planets, animals and plants. The visitor may examine rare gems and minerals, poisons and antidotes, and all sorts of things beneficial and injurious to man's body. Here the study of mathematics rises above vulgar arithmetic and geometry to the mystical numerical harmonies of the heavens, a subject known to the Pythagoreans of old.[12] The relation of heaven and earth is everywhere stressed, and for this reason astrology is raised to its proper place: "he who does not know the value of astrology in human affairs, or who foolishly denies it, I would wish that he would have to dig in the earth, cultivate and work the fields, for as long a time as possible, in unfavourable weather."[13] For Andreae a different system of learning is required, and if it cannot be accommodated to the current university system, a separate academy or college must be founded. Andreae's proposals could have been seconded by any of the chemical philosophers.

The search for a new learning based on chemistry and the occult sciences reached its peak in the middle decades of the seventeenth century. England in the Civil War period was particularly susceptible to such speculations. Indicative of the increased interest is an astrological work written by Christopher Heydon in 1608 but withheld from the printer until 1650: the preface now announces that writers need no longer be afraid to set forth such views, for these opinions are received favorably by a large segment of the scholarly public and are well defended by the

[11]*Ibid.*, pp. 196-197.
[12]*Ibid.*, pp. 200, 221-223.
[13]*Ibid.*, p. 228.

members of the learned London Society of Astrologers.[14] Paracelsian and Helmontian chemistry shared in the growing interest in all schemes for a new science. In the decade of the 1650s more Paracelsian and mystical chemical works were translated than in the entire century before 1650. At the same time there was a greater interest in the Rosicrucian movement. John Heydon (fl. 1667)[15] gave his books such titles as *A New Method of Rosie Crucian Physicke* (1658), while Eugenius Philalethes (Thomas Vaughan) (1622-1666) prepared a lengthy introduction to a translation of *The Fame and Confession of R:C:* (1652). The link was quite clear for George Hakewill (1578-1649) in his survey of scholarship: he praised the "*Chimiques, Hermetiques*, or Paracelsians (and a branch of them as I conceive is the order Roseae Crucis)."[16]

At the beginning of the century, almost fifteen years before the

[14]Christopher Heydon, *An Astrological Discourse, manifestly proving The Powerful Influence of Planets and Fixed Stars upon Elementary Bodies, in Justification of the Verity of Astrology, Together with an Astrological Judgement Upon the great Conjunction of Saturn and Jupiter 1603* (London: John Macock, for Nathaniel Brooke, 1650). See the note by William Lilly and the prefatory "To the Reader" by Nicholas Fisk. Although the London Society of Astrologers published none of its regular proceedings, there is ample evidence of its existence over a number of years. Elias Ashmole noted his attendance at the Society Feasts in the years 1649 (Aug. and Oct. 31), 1650 (Aug. 8, when he was chosen Steward), 1651 (Aug. 14), 1653 (Mar. 18), 1655 (Aug. 29) and 1658 (Nov. 2). After a lapse of a number of years the Feasts were resumed in 1682 and Ashmole attended the meeting in that year (July 13) and another one in 1683 (Jan. 29). *Elias Ashmole (1617-1692). His Autobiographical and Historical Notes, his Correspondence, and Other Contemporary Sources Relating to his Life and Work*, ed. C. H. Josten, 5 vols. (Oxford: Clarendon Press, 1966), Vol. II, pp. 492 f., 539, 580, 640, 679, 751; Vol. IV, pp. 1705, 1712. At least three of the addresses given at the Feasts were published: Robert Gell, *Stella Nova A New Starre, Leading wisemen unto Christ* ... (London: Samuel Satterthwaite, 1649); Robert Gell, ΑΓΓΕΛΟΚΡΑΤΙΑ ΘΕΟΎ *Or a Sermon Touching Gods government of the World by Angels* ... (London: John Legatt for Nath. Webb and William Grantham, 1650); Edmund Reeve, *The New Jerusalem: The Perfection of Beauty: The Joy of the Whole Earth* (London: J. G. for Nath: Brooks, 1652).

[15]John Heydon (b. 1629) married the widow of the chemist and physician Nicholas Culpeper (1656) and was active in London astrological circles c. 1650 - c. 1667, when he was imprisoned for "treasonable practices in sowing sedition in the navy, and engaging persons in a conspiracy to seize the tower" (*DNB*). There is no evidence that he was related to Sir Christopher Heydon (c. 1623), who was active in a bitter astrological debate in the first decade of the century.

[16]George Hakewill, *An Apologie or Declaration of the Power and Providence of God in the Government of the World* (3rd ed., Oxford: William Turner, 1635), p. 276.

Christianopolis, in England Joseph Hall (1574-1656) had written *Mundus alter et idem* (c. 1605), which described mythical newly found lands in the Southern Hemisphere. In "Fooliana" a university exists which is composed of a college of Skeptikes and a college of Gewgawiasters. The provost of the latter is none other than Paracelsus, who has invented a "Supermonicall" language for the fellows "who give themselves wholly to the invention of novelties, in games, buildings, garments, and governments."[17] Hall's conservative approach to Paracelsian thought and innovation is not unusual for the period, but how different it is from the search for reform only a few decades later. Bacon's *New Atlantis* and his attitude toward the universities has already been mentioned. Samuel Gott (1614-1671) discussed life in "Nova Solyma" (1648),[18] questioning the validity of the pursuance of knowledge for its own sake and suggesting that the study of the chemistry of nature be made not for personal fame but for the glory of God.

Samuel Hartlib dedicated to Parliament his tract on the ideal kingdom of Macaria (1641). Here we read of a "college of experience" in which scholars develop medicines for the benefit of mankind and are justly rewarded by the populace. When it is suggested that this is contrary to traditional medical practice, the narrator points out the divine nature of medicine:

> In Macaria the parson of every parish is a good physician, and doth execute both functions; to wit, *cura animarum, & cura corporum;* and they think it is as absurd for a divine to be without the skill of physick, as it is to put new wine into old bottles; and the physicians, being true naturalists, may as well become good divines, as the divines do become good physicians.[19]

---

[17]Joseph Hall, *The Discovery of a New World (Mundus alter et idem) c. 1605,* ed. Huntington Brown with a foreword by Richard E. Byrd (Cambridge, Mass.: Harvard University Press, 1937), p. 87. Hall's *Mundus* has recently been discussed by Nell Eurich in her *Science in Utopia. A Mighty Design* (Cambridge, Mass.: Harvard University Press, 1967), pp. 82-83.

[18]See Eurich, *Science in Utopia,* pp. 83-86, and (Samuel Gott), *Nova Solyma The Ideal City: Or Jerusalem Regained. An anonymous romance written in the time of Charles I. Now first drawn from obscurity, and attributed to the illustrious John Milton,* introduction, translation, essays, and bibliography by the Rev. Walter Bayley, 2 vols. (London: John Murray, 1900). This translation is somewhat abridged.

[19]Samuel Hartlib, *A Description of the Famous Kingdom of Macaria: shewing the excellent Government, wherein the Inhabitants live in great Prosperity, Health and Happiness;*

Hartlib was deeply convinced of the need for an experimental college, and he was closely connected with the "Invisible College" of London which began meeting in the mid-1640s. Furthermore, we find the young Robert Boyle (1627-1691) corresponding with Hartlib on the various utopian texts. In a letter dated April 8, 1647, Boyle specifically suggested that Campanella's *Civitas solis* and Andreae's *Christianopolis* be translated for the benefit of English readers.[20]

It was Hartlib also who turned to the great Czech educational reformer Jan Comenius for aid. Comenius had prepared a reformed system of learning, his encyclopedic "pansophia," in which universal knowledge was to be gathered. Comenius admired Francis Bacon, but Bacon's method seemed deficient since it aimed only at uncovering the secrets of nature.[21] Similarly the works of Jungius (1587-1657) were lauded, while

> If any be uncertain if all things can be placed before the senses in this way, even things spiritual and things absent (things in heaven, or in hell, or beyond the sea), let him remember that all things have been harmoniously arranged by God in such a manner that the higher in the scale of existence can be represented by the lower, the absent by the present, and the invisible by the visible. This can be seen in the *Macromicrocosmus* of Robert Flutt, in which the origin of winds, or rain, and of thunder is described in such a way that the reader can visualize it. Nor is there any doubt that even greater concreteness and ease of demonstration than is here displayed might be attained.[22]

Yet, in preference to all other authorities, Comenius referred to Andreae, "who in his golden writings has laid bare the diseases not

---

the King obeyed, the Nobles honoured, and all good men respected; Vice punished, and Virtue rewarded. An Example to other Nations: In a Dialogue between a Scholar and a Traveller printed 1641 in *The Harleian Miscellany: A Collection of Scarce, Curious, and Entertaining Pamphlets and Tracts, as well in Manuscript as in Print. Selected from the Library of Edward Harley, Second Earl of Oxford*, Vol. I (London: John White, John Murray and John Harding, 1808), pp. 580-585 (582).

[20]Thomas Birch, *The Life of the Honourable Robert Boyle* in *The Works of the Honourable Robert Boyle*, 5 vols. (London: A. Millar, 1744), Vol. I, pp. 22-23.

[21]John Amos Comenius, *A Reformation of Schooles* . . ., trans. Samuel Hartlib (London: Michael Sparke, 1642), p. 35.

[22]John Amos Comenius, *The Great Didactic*, translated and with an introduction by M. W. Keatinge (London: Adam and Charles Black, 1896), p. 339.

only of the Church and the state, but also of the Schools, and has pointed out the remedies . . . ."[23]

These views bear a marked similarity to those of the chemical philosophers mentioned earlier. Comenius wanted the heretical works of the Aristotelians discarded and replaced by the two-book concept of knowledge.[24] The truth of the Mosaic account of the Creation was not questioned, and the importance of the evidence of the senses in the study of nature was continually emphasized.[25] All this coupled with his plans for universal reform made him a scholar of great interest to Hartlib, who arranged for him to visit England in 1641.[26] However, Hartlib's hopes for the establishment of a Baconian Universal College failed to materialize at a time when the government faced collapse. The outbreak of the Civil War in 1642 resulted in the departure of Comenius for Sweden.

Although Comenius' stay in England was brief, it appears to have catalyzed the reformers to greater efforts. The decade from 1642 to 1652 produced a host of tracts dedicated to the educational problem, to the need for a new science. Although it has been shown that real changes were already underway at the universities, few of the tract authors indicated this.[27] Most of them wrote of the universities dominated by a rigid Aristotelianism, unwilling and unable to appreciate the new chemical philosophy or the mechanical philosophy, where the Sedleian Lecturer of Natural Philosophy at Oxford was confined to discussing "Aristotle's *Physics*, or the books concerning the heavens and the world, or concerning meteoric bodies, or the small Natural Phenomena of the same author, or the books which treat of the soul, and also those on

[23]*Ibid.*, p. 159.
[24]*Ibid.*, p. 26. Keatinge emphasizes the influence of Bacon and Campanella on Comenius, while August Nebe prepared a monograph on *Vives, Alsted, Comenius in ihrem Verhältnis zu einander* (Elberfeld, 1891). There is little doubt that Comenius was well read in the literature on educational reform and the utopian tracts of the sixteenth and early seventeenth centuries.
[25]Comenius, *Great Didactic*, pp. 27, 337.
[26]Eurich, *Science in Utopia*, p. 148.
[27]Robert G. Frank has recently discussed the widespread introduction of the new science in seventeenth-century Oxford and Cambridge ("Early Modern English Universities," *History of Science*, **11** (1973), 194-216, 239-269). Still, the science that Frank discusses would have given little consolation to the chemical philosophers.

generation and corruption."[28] At Cambridge the lecturer in medicine was admonished to "read Hippocrates and Galen," and he was not permitted officially to lecture on recent authors.[29] Medical students did not perform dissections themselves. In this period English medical students frequently studied abroad. Those who did so incorporated their foreign degrees at Oxford or Cambridge on their return—a procedure so prevalent in the period from 1559 to 1642 that 37 per cent of the members of the Royal College of Physicians received their medical education abroad.[30]

The situation at the universities remained a thorn for the chemists. In his *Art of Distillation* (1650) John French (1616-1657) spoke of alchemy,

> ... which is more noble than all the other six Arts and Sciences.... This is that true natural philosophy which most accurately anatomizeth Nature and natural things, and ocularly demonstrates the principles and operations of them: that empty natural philosophy which is read in the Universities, is scarce the meanest hand-maid to this Queen of Arts. It is a pity there is such great encouragement for many empty, and unprofitable Arts, and none for this, and such like ingenuities, which if promoted would render an University far more flourishing, than the former. I once read or heard, of a famous University beyond Sea, that was faln into decay, through what cause I know not: but there was a general counsel held by the learned, how to restore it to its primitive glory: the *Medium* at last agreed upon, was the promoting of Alchymie, and encouraging the Artists themselves: But I never expect to see such rational actings in this nation till shadows vanish, substances flourish, and truth prevail ....[31]

French may have almost given up hope for such rational behavior in England, but not others. The victory of the parliamentary forces signalled for some an unparalleled opportunity for reform in education as well as religion. Addressing *An Humble Motion to the*

---

[28]Phyllis Allen, "Scientific Studies in the English Universities of the Seventeenth Century," *Journal of the History of Ideas*, **10** (1949), 219-253 (227).

[29]Phyllis Allen, "Medical Education in 17th Century England," *Journal of the History of Medicine and Allied Sciences*, **1** (1946), 115-143 (119).

[30]Mark H. Curtis, *Oxford and Cambridge in Transition 1558-1642* (Oxford: Clarendon Press, 1959), p. 154.

[31]John French, *The Art of Distillation* (4th ed., London: T. Williams, 1667), sig. A3r. From the dedication to Tobias Garband dated London, Nov. 25, 1650.

Parliament of England Concerning the Advancement of Learning: and Reformation of the Universities (1649), John Hall asked: "Where have we any thing to do with Chimistry, which hath snatcht the keyes of Nature from the other Sects of Philosophy by her multiplied real experiences?"[32]

Even more important is Noah Biggs' *Vanity of the Craft of Physick* (1651), a work clearly based upon the Helmontian blueprint for reform in medicine and natural philosophy. Once again there is a plea to Parliament for action. Biggs argued that Cromwell's victory over the Scottish troops led by the pretender Charles at Evesham had at last freed the commonwealth from the threat of the Stuarts.[33] Now, with peace, it would be possible to bring about a thorough reformation of the abuses in all professions. It had been in England, Biggs stated, that there had "*sounded forth the first tidings and Trumpet of Reformation to all Europe.*" But the reformation of the Church was clearly not sufficient. Now Parliament must go further to accomplish the reformation of "the stupendious body of Universal Learning, Languages, Arts and Sciences."[34]

In his plan for educational reform Biggs can be viewed as a typical seventeenth-century iatrochemist. Since the basis of the study of natural philosophy and medicine in the schools is the output of antiquity, the "learned" physicians in reality know little, even though they claim broad privileges and a knowledge of all things.[35] How could the educational system that spawned such men and such a profession continue to be tolerated?

> Wherein is our Universities reformed, or what amendment of her Fundamental Constitutions? . . . Or wherein do they contribute to the promotion or discovery of Truth? . . . Where have we any thing to do

[32] J(ohn) H(all), *An Humble Motion To the PARLIAMENT of ENGLAND Concerning the ADVANCEMENT of Learning and Reformation of the Universities* (London: John Walker, 1649), p. 27.

[33] Noah Biggs, Chymiatrophilos, *Mataeotechnia Medicinae Praxeωs. The Vanity of the Craft of Physick. Or, A New Dispensatory. Wherein is dissected the Errors, Ignorance, Impostures and Supinities of the SCHOOLS, in their main Pillars of Purges, Blood-letting, Fontanels or Issues, and Diet, &c. and the particular Medicines of the Shops. With an humble Motion for the Reformation of the Universities, And the whole Landscap of Physick, And discovering the Terra incognita of Chymistrie. To the Parliament of England* (London: Giles Calvert, 1651), sig. a2^r.

[34] *Ibid.*, sig. a3^v.

[35] *Ibid.*, pp. 19-28.

with Mechanick *Chymistrie* the handmaid of Nature, that hath outstript the other Sects of Philosophy by her multiplied real experiences? Where is there an examination and consecution of Experiments? encouragements to a new world of Knowledge, promoting, compleating, and actuating some new Inventions? Where have we constant reading upon either quick or dead *Anatomies*, or an ocular demonstration of *Herbs*? Where a Review of the old Experiments and Traditions, and casting out the rubbish that has pestered the Temple of Knowledge?[36]

Those in charge waver "twixt negligence and uncertainty" and suspend all further inquiry while they snore "in the *Lethargy* of their idleness like *drones* in the hive of their *pedantick* Brotherhoods." These academics, in opposition to the present republican government, will hardly resign the keys to the Temple of Knowledge voluntarily; change must be forced upon them through parliamentary authority.[37]

Biggs was not hesitant to outline an orderly procedure for academic reform.[38] A meeting of "prudent and well instructed men" would be called to investigate the universities, especially the state of natural philosophy and medicine therein. The committee was to examine all medical factions and sects—the writings of Galen, Aristotle, and Hippocrates as thoroughly as those of the chemists. Only then would any recommendations or decisions be made. Although Biggs strove to appear impartial, he felt that he knew what the outcome would be, since the "unsuccessful nature of Galenical Physick" indicated that the "most excellent and natural Art of Chymistry" must needs be "called from her *Ostracism*."[39]

John Webster and the *Academiarum examen* (1654)

The advocates of reform in higher education came into direct conflict with the proponents of the mechanical philosophy, as symbolized in the debate between John Webster and Seth Ward in 1654. John Webster (1610-1682) was a typical representative of the

[36]*Ibid.*, sig. b1^r.
[37]*Ibid.*, p. 229, 3.
[38]*Ibid.*, sig. b3^r-b4^r.
[39]*Ibid.*, sig. b4^r.

reforming chemists of the period,[40] men who passionately sought a new science and a reformed educational system that would develop it. For them the works of Fludd, van Helmont, and the Paracelsians seemed to offer a Christian answer to the stale learning taught at the universities. Webster was attracted early to the study of both nature and religion; he had studied chemistry (c. 1632) under the Hungarian alchemist John Hunyades (1576-1650)[41] and had been ordained as a minister shortly afterward. With his Puritan sympathies he had served both as a surgeon and chaplain with the parliamentary army during the Civil War. By 1648 his reaction against the established Church had forced him to become a nonconformist, and in later years he supported himself as a "practitioner in Physick and Chirurgery." Although most of his writings are on religious topics, his important *Metallographia* (1671) and his *Displaying of Supposed Witchcraft* (1677) are of considerable interest to the intellectual historian as well as the historian of science. They will both be referred to again in this chapter.

It was Webster's concern with the educational training of the minister that inspired him to write both *The Saints Guide* and the *Academiarum examen* in 1653. In the former he decried the schools which taught worldly wisdom through books and disputations. This learning was of no use toward any true understanding of the Gospel truths, for this required God's Spirit through the grace of the Holy Ghost.[42] Webster returned to the same subject a few

[40]See the article on John Webster by Bertha Potter in the *Dictionary of National Biography*.
[41]The work of Hunyades has been discussed by F. Sherwood Taylor and C. H. Josten in "Johannes Banfi Hunyades 1576-1650," *Ambix*, **5** (1953), 44-52, and "Johannes Banfi Hunyades. A Supplementary Note," *Ambix*, **5** (1956), 115, where (p. 52) it is suggested that Hunyades arrived in London sometime between 1623 and 1633.
[42]John Webster, *The Saints Guide, or Christ the Rule and Ruler of Saints. Manifested by way of Positions, Consectaries, and Queries* . . . (London: Giles Calvert, 1654), pp. 1-2. Regarding the mid-century dispute over educational reform, R. F. Jones has written in his "The Humanistic Defence of Learning in the Mid-Seventeenth Century" that "the central issue, though frequently lost sight of, was the employment of learning and reason in the preparation of a preacher and interpretation of the Gospel" [in *Reason and the Imagination: Studies in the History of Ideas, 1600-1800*, ed. J. A. Mazzeo (New York: Columbia University Press; London: Routledge and Kegan Paul, 1962), pp. 71-92 (73)].

months later in the *Academiarum examen*. Again it was the problem of the training of the ministry that first attracted his attention. The insufficiency of the universities was clear—surely the student should study God's own Creation rather than the books of ancient authors who knew nothing of God's revelation.

Webster's chemical approach is evident throughout his attack on the universities. In a typical Paracelsian fashion he dismissed Aristotle as a heathen whose work could not be the basis of Christian education. He went on to detail further failings in the university curricula in a method that clearly parallels Fludd's *Tractatus apologeticus* of 1617. The emphasis on deductive logic was detestable, especially since van Helmont had pointed out glaring errors in the use of this subject and Bacon had rightly shifted the attention of scholars to inductive logic.[43] Similarly, Aristotelian physics was to be condemned because of its concentration on the study of moving bodies. "Is there no further end nor consideration in *Physicks* but onely to search, discuss, understand, and dispute of a natural movable body, with all the affections, accidents and circumstances thereto belonging?" If this is so, then "*our Philosophy is made Philologie, from whence we teach to dispute, not to live.*"[44]

Little more could be said for the mathematical sciences, where Webster found the important work of John Napier (1550-1617), Henry Briggs (1561-1630), and William Oughtred (1575-1660) ignored and geometry decayed to valueless verbal disputes. Music had become little more than the singing of songs and the playing of instruments, while Ptolemaic astronomy was still taught even though the truth of the Copernican system was well known.[45] But above all, the Aristotelian approach emphasized mathematics in an incorrect fashion as the basic guide to truth in nature. This was no small error, "for though the Mathematicks be exceedingly helpful to Natural *Philosophy*, yet is confusion of terms very hurtful; for if a mathematical point or superficies be urged in a Physicall argu-

---

[43]John Webster, *Academiarum examen, or the Examination of Academies. Wherein is discussed and examined the Matter, Method and Customs of Academick and Scholastick Learning, and the insufficiency thereof discovered and laid open; As also some Expedients proposed for the Reforming of Schools, and the perfecting and promoting of all kind of Science. Offered to the judgements of all those that love the proficiencie of Arts and Sciences, and the advancement of learning* (London: Giles Calvert, 1654), pp. 32-40.

[44]*Ibid.*, p. 18.

[45]*Ibid.*, pp. 40-49.

ment it will conclude nothing, but only obfuscate, and disorder the intellect."[46]

What was needed was a new science based less on logic and mathematical abstractions and more on reality determined through observation and experiment. Deductive logic can be safely discarded and inductive reasoning developed instead.[47] Mathematics must advance beyond arithmetic and geometry to the harmonies which exist between the great and the small worlds. In astronomy the "learning of *Copernicus, Kepler, Ticho Brahe, Galilaeus, Ballialdus,* and such like, might be introduced, and the rotten and ruinous Fabrick of *Aristotle* and *Ptolemy* rejected and laid aside." Aristotle's observations on minerals might well be retained, but the whole of his natural philosophy and his astronomy—and the accumulated commentaries on them—must be eradicated.[48]

For Webster even the conventional study of grammar and language was of little value, although recent developments gave hope of a breakthrough toward a universal or natural language.[49] The symbols of the chemists and the new mathematical notations indicated the possibility of such a written language which might be understood by all. But the emblems and hieroglyphics of the alchemists and the occult philosophers offered far more hope, and one could scarcely deny the existence of a natural language stamped by the Creator on his Creation through divine signatures and hidden virtues. In the celestial regions we must learn to read the signs of the stars, while in man and in earthly substances we must also seek the hidden signs.

> Many do superficially and by way of *Analogy* . . . acknowledge the Macrocosm to be the great unsealed book of God, and every creature as a Capital letter of character, and all put together make up that one word or sentence of his immense wisdom, glory and power; but alas! who spells them aright, or conjoyns them so together that they may perfectly read all that is therein contained? Alas! we all study, and read too much upon the dead paper idols of creaturely invented letters, but do not, nor cannot read the legible characters that are onely written and impressed by the finger of the Almighty . . . .[50]

[46]*Ibid.*, p. 68.
[47]*Ibid.*, p. 102.
[48]*Ibid.*, pp. 103, 104.
[49]*Ibid.*, pp. 18-32.
[50]*Ibid.*, p. 28.

Indeed, it was to the credit of Jacob Boehme (1575-1624) and the Rosicrucians that they have sought to decipher this language of nature.[51]

Surely, Webster continued, the Scholastic philosophy taught at the universities was grounded on little of value. Lecturers continued to emphasize the work of Aristotle when they might well have followed the advice of Cicero and Quintilian, who judged Plato to be a more worthy author than his student.[52] The preference for Aristotle and for the commentators on his works was all the more to be wondered at since there were some who thought that the books ascribed to Aristotle might be apocryphal. Even among his staunchest defenders there was little agreement as to interpretation. But there was good reason for his. In his letters to Alexander, Aristotle had admitted that his style of writing was obscure and difficult to understand, which is borne out in the *Physics*, where his position is not properly defined, and also in his confusion over the doctrine of the matter of the heavens. The great van Helmont found no fewer than thirteen errors in Aristotle's definition of nature.[53]

As a replacement for Aristotle Webster suggested the philosophy of Plato as methodized by Ficino and also

> That of *Democritus* cleared, and in some measure demonstrated, by *Renatus des Cartes, Regius, Phocylides Holwarda,* and some others; That of *Epicurus,* illustrated by *Petrus Gassendus*; That of *Philolaus, Empedocles,* and *Parmenides,* resuscitated by *Telesius, Campanella,* and some besides; and that excellent *Magnetical Philosophy* found out by Doctor *Gilbert*; That of *Hermes,* revived by the *Paracelsian* School, may be brought into examination and practice, that whatsoever in any one of them, or others of what sort whatsoever, may be found agreeable to truth and demonstration, may be imbraced, and received ...[54]

Indeed, demonstrations, observations, and experiments are

> ... the only certain means, and instruments to discover, and anatomize nature's occult and central operations; which are found out by laborious tryals, manual operations, assiduous observations, and

[51]*Ibid.,* p. 26.
[52]*Ibid.,* pp. 56-57.
[53]*Ibid.,* pp. 60-65.
[54]*Ibid.,* p. 106.

the like, and not by poring continually upon a few paper Idols, and unexperienced authors.[55]

The essential key is chemistry, which teaches the unfolding of nature's secrets through the use of manual operations. Van Helmont is noted for his statement that no philosopher is admitted to the root of science without a knowledge of the fire.[56] Surely one year of work in a chemical laboratory will prove more beneficial than centuries of disputes over the texts of Aristotle. Yet for Webster this is simply the method and basis of natural magic, "whereby the wonderful works of the Creator are discovered, and innumerable benefits produced to the poor Creatures."[57] The subject of magic is, as Giovanni Pico della Mirandola (1463-1494) observed, "to marry the world . . . that thereby nature may act out of her hidden and latent power."[58] There is nothing diabolical about magic; it is the proper sphere of the greatest philosophers and physicians.

It is quite evident, then, why Webster believed that the universities should turn to the writings of Francis Bacon and Robert Fludd in a search for the basis of a new philosophy. On the surface we are entitled to call Webster a Baconian. We find Bacon quoted more frequently than any other modern author, and the *Academiarum examen* is keynoted by a quotation from the *Novum organum* on the title page.[59] Furthermore, in his dedication Webster exhorted Major-General John Lambert (1619-1683) to reform the universities for the "*advancement of* Learning,"[60] while he offered four "rules" in his rejection of ancient authority that clearly reflect the influence of Bacon's four "Idols."[61] But it was of more interest to Webster that Bacon had defined magic as that "which leadeth cognition of occult forms into wonderful works, and by conjoyning actives to passives, doth manifest the grand secrets of nature."[62] In

[55]*Ibid.*, p. 68.
[56]*Ibid.*, p. 71.
[57]*Ibid.*, p. 69.
[58]*Ibid.*, p. 70.
[59]This is the attack on the universities cited above in n. 6.
[60]Webster, *Academiarum examen*, sig. A2^V.
[61]John Webster, *The Displaying of Supposed Witchcraft* (London: J.M., 1677), pp. 13-17.
[62]*Academiarum examen*, p. 69. Here Webster is quoting Bacon's discussion of magic in the *De augmentis scientiarum* (Book 3, Ch. 5).

addition, Bacon had shown the significance of induction and experiment as a basis of a new science.

> It cannot be expected that *Physical* Science will arrive at any wished perfection, unlesse the way and means, so judiciously laid down by our learned Countreyman the Lord *Bacon*, be observed, and introduced into exact practice. And therefore I shall humbly desire, and earnestly presse, that his way and method may be imbraced, and set up for a rule and pattern: that no *Axioms* may be received but what are evidently proved and made good by diligent observation, and luciferous experiments; that such may be recorded in a general history of natural things, that so every age and generation, proceeding in the same way, and upon the same principles, may dayly go on with the work, to the building up of a well-grounded and lasting Fabrick, which indeed is the only true way for the instauration and advance of learning and knowledge.[63]

Webster was a Baconian because Bacon sought to replace the Aristotelianism of the schools with an observational-experimental natural philosophy or natural magic.

For Webster this position need not conflict with the work of Robert Fludd. Webster's critique of traditional learning was closely modelled on Fludd's *Tractatus apologeticus*, and he emphasized that true Christians must seek a knowledge of nature

> ... that is grounded upon sensible, rational, experimental, and Scripture principles: and such a compleat piece in the most particulars of all human learning (though many vainly and falsely imagine there is no such perfect work to be found) is the elaborate writings of that profoundly learned man Dr. *Fludd*, than which for all the particulars before mentioned (notwithstanding the ignorance and envy of all opposers) the world never had a more rare, experimental and perfect piece.[64]

This is a theme which Webster returned to years later in his work on witchcraft when he spoke of the initial resistance to Harvey's doctrine of the circulation of the blood and compared this lack of appreciation with the reaction to the work of his colleague.

> Our Countreyman Dr. *Fludd*, a man acquainted with all kinds of learning and one of the most Christian Philosophers that ever writ,

[63] *Academiarum examen*, p. 105.
[64] *Ibid.*

yet wanted not those snarling Animals, such as *Marsennus, Lanovius, Foster,* and *Gassendus,* as also our *Casaubon* (as mad as any) to accuse him vainly and falsely of Diabolical Magick, from which the strength of his own Pen and Arguments did discharge him without possibility of replies.[65]

John Webster emerges neither as an "ancient" nor a "modern." Instead, he represents the chemical philosophers of the mid-seventeenth century—scholars who properly belong in neither camp. Natural magic was the goal of their new philosophy, and this was defined as the search for a true understanding of the secrets of nature through observation and experiment. The macrocosm-microcosm analogy was implicit in Webster's work, and it is readily understandable how Robert Fludd could be one of his idols. It is equally understandable how he could point to Francis Bacon—the natural magician—as a guide. Yet, if the pairing of these authors caused little soul searching for a mid-century Paracelso-Helmontian, this was far from the case for the new mechanical philosophers.

The *Vindiciae academiarum* of John Wilkins and Seth Ward (1654)

The *Academiarum examen* brought an immediate reaction from the Oxford proponents of the new philosophy. This was a period when Oxford was rapidly becoming a center for the new science. Here were gathered Robert Boyle, Thomas Willis (1621-1675), Jonathan Goddard (1617-1675), and John Wallis (1616-1703). As a group they were to form the nucleus of the Philosophical Society of Oxford, a forerunner of the Royal Society of London. Two distinguished members of the group were Seth Ward (1617-1689) and John Wilkins (1614-1672). Wilkins had been the warden of Wadham College since 1648 and Ward had been the Savilian Professor of Astronomy since 1649. Wilkins was already widely known for his defense of the Copernican system (1640), for his *Discovery of a World in the Moone* (1638), and his *Mathematical Magick* (1648). Ward had as yet published little of his research. In 1653 he

---

[65]Webster, *Displaying of Supposed Witchcraft,* p. 9.

had written a treatise questioning the astronomical system of Ismael Bullialdus (1605-1694), and he was currently engaged in the preparation of his *Astronomia geometrica*, a work which appeared in 1656 and defended the orbital system of Kepler.

For both Wilkins and Ward the recent publications on educational reform were a matter of concern, and it was to these that they turned in their *Vindiciae academiarum* (1654). In his *Leviathan* (1651) Thomas Hobbes (1588-1679) had sketched a picture of the universities that, Ward charged, conformed more to the turn of the century than to the current state of affairs.[66] Recent research tends to uphold Ward's statement: Mark Curtis, pointing to the fact that the conservative statutes of Oxford and Cambridge were not rigidly enforced and also to the growth of the tutorial system which made possible the study of recent texts, has emphasized that the actual state of science was not as bleak as it has been commonly pictured.[67]

But if Hobbes had presented an inaccurate view of the universities, William Dell (d. 1664), the master of Gonville and Caius College, Cambridge, had written an even more disturbing work. Dell had suggested that new universities should be established in the growing urban centers, and further, that young students should work part time so that they might learn a trade while they pursued their regular course of study. Such a plan was far from acceptable to Seth Ward, who wrote: "I am much assured, there is not a Learned man in all the world who hath not found by experience, that skill in any Faculty . . . is not to be attained, without a timely beginning, a constancy and assiduity in study, especially while they are young . . . . "[68]

The writings of Hobbes and Dell were viewed as dangerous,

---

[66] [Set]H [War]D, *Vindiciae academiarum containing, Some briefe Animadversions upon Mr. Websters Book, Stiled The Examination of Academies. Together with an Appendix concerning what M. Hobbs, and M. Dell have published on this Argument* (Oxford: Leonard Lichfield Printer to the University for Thomas Robinson, 1654) (with an introduction by [Joh]N [Wilkin]S, pp. 58-59. On Wilkins see Barbara Shapiro, *John Wilkins 1614-1672. An Intellectual Biography* (Berkeley/Los Angeles: University of California Press, 1969).

[67] Curtis, *Oxford and Cambridge in Transition*, pp. 227-260. The progressive nature of the universities in this respect has been more recently discussed by Frank, "Early Modern English Universities."

[68] Ward, *Vindiciae academiarum*, pp. 64-65.

but there is no question that Wilkins and Ward considered the tract of John Webster to be the most alarming of them all. Wilkins, in his introduction to Ward's detailed reply, accused him not only of ignorance of the arts and sciences he hoped to advance, but also of an almost total lack of information of the current state of the universities.[69] Logic served as an excellent example of Webster's misunderstanding. Surely, Wilkins wrote, logic is essential for any sound reasoning, but Webster had complained that the theology of the universities had become a "confused Chaos" through the application of "strict *Logicall Method*." How could logical order possibly result in chaos? For Wilkins the points made by Webster were plagiarized from Bacon and van Helmont and beyond his comprehension.[70] In addition, if Webster had visited the universities in recent years he would have seen that induction as well as syllogism was being taught.

Webster's views on the study of grammar and languages seemed even more dangerous. Ward, noting Webster's plea for the study of hieroglyphic emblems and cryptographs, replied that these forms "were invented for *concealment* of things" in contrast to grammar and languages, which serve for their explication. The new mathematical symbolism of Vieta, Harriot, Oughtred, and Descartes had been invented to avoid confusion over the understanding of words; this should not be confused with the emblems and symbols of the alchemists.[71]

The question of a true universal language—a subject both Wilkins and Ward had studied in detail—was something quite different. Ward was willing to agree that enough symbols might be found to produce a universal language "wherein all Natives might communicate together, just as they do in number and in species." But while this was possibile, it was highly impractical, for in reality it would be necessary to have an almost infinite number of characters at one's command. Sarcastically Ward added that his arguments might be easier for Webster to understand if presented in the obscure style favored by mystical alchemists.[72]

As Savilian Professor of Astronomy, Ward had a special in-

[69] *Ibid.*, p. 1.
[70] *Ibid.*, pp. 5, 23.
[71] *Ibid.*, pp. 18, 19.
[72] *Ibid.*, pp. 21-23.

terest in Webster's suggestions for the reform of the mathematical sciences. Ward admitted that like Webster he had often complained "against the neglect of Mathematics in our method of study," but nevertheless he was appalled by what he had read in the *Academiarum examen*. To have suggested that the universities had ignored the work of Napier, Briggs, and Oughtred was patently false. Oughtred had been a fellow of King's College at Cambridge while Briggs had been Professor of Geometry at Gresham College. Webster's assertion that arithmetic and geometry had been neglected was not true, while the phenomena of optics were now being studied by competent scholars and at long last removed from the realm of magic.[73] Webster's plea for a more profound study of music as a key to the nature of universal harmonies was quite lost on Ward, who commented that our instruments have "been lately out of tune, and our harpes hanged up, but if such men as he should please to come among us, and put us to an examen, without doubt we should then have a fit of Mirth &c."[74]

"But of all things," Ward continued, "the *Astronomy* Schooles he is most offended at, as maintaining with Rigour the *Ptolemaick System*." In reality, "The Method here observed in our Schooles is, first to exhibit the *Phenomena*, and shew the way of their observation, then to give an account of the various Hypotheses, how those Phenomena have been salved . . . ."[75] While it is true that the Ptolemaic system ranks as one of the major systems which have "salved the *Phenomena*" of planetary motion, the Copernican system is taught at Oxford "as it was left by him, or as improved by *Kepler, Bullialdus*, our own Professor and others of the *Ellipticall* way."[76] Indeed, while the teaching of the mathematical sciences may not have reached perfection at Oxford, these subjects were taught well by those who know their importance. How different it was from Webster's plea that music should be made a major key for the mathematical interpretation of the heavens and that astrology should be considered a worthy goal for true mathematicians: the latter is a subject fit only for those educated in the "Academy of *Bethlem*," and is based, not upon observation and ex-

[73]*Ibid.*, pp. 27-29.
[74]*Ibid.*, p. 29.
[75]*Ibid.*, pp. 29-30.
[76]*Ibid.*, p. 29.

periment, but the Aristotelian system which Webster professed to abhor.⁷⁷

Ward saw very little new in Webster's attack on Aristotle. A comparison of the *Examen* with the work of Gassendi and van Helmont showed that he had closely followed these sources without satisfactorily citing them.⁷⁸ Webster had charged that in the schools natural magic was prosecuted, that chemistry was neglected, that medicine had declined, that anatomy was sterile, that surgery was defective, and that the professors remained ignorant of celestial signatures, the three principles, and the magnetic and atomic philosophies. But, Ward countered, if we have little respect for the magic of Agrippa, Porta, and Wecker, it is because of their deceit rather than their use of the word "magic." "The discoveries of the Symphonies of nature, and the rules of applying agent and materiall causes to produce effects is the true naturall Magic, and the generall humane ends of all Phylosophicall enquiries; but M. Webster knew not this . . . ."⁷⁹ As for chemistry, it is not neglected at Oxford; rather, there has been

> . . . a conjunction of both the Purses and endeavours of severall persons towards discoveries of that kind, such as may serve either to the discovery of light or profit, either to Naturall Philosophy or Physick. But Mr. *Webster* expects we should tell him, that we have found the Elixar, (surely we are wiser then to say so) . . . .⁸⁰

One need say little more. Medicine, now firmly based upon experiment and observation, produces great new discoveries. Similarly the magnetic and atomic philosophies are not neglected. Indeed, a whole school could be furnished with the scientific instruments which are used at Oxford.⁸¹

And what may be said of Webster's two main remedies for the reformation of learning? It is true that he has stated that *"my L. Bacons way may be embraced. That Axioms be evidently proved by observations, and no other be admitted &c."* and with this Ward is in agreement: "It cannot be denied but this is the way, and the only way to perfect Naturall Philosophy and Medicine: so that whosoever in-

⁷⁷*Ibid.*, p. 31.
⁷⁸*Ibid.*, pp. 32-33.
⁷⁹*Ibid.*, p. 34.
⁸⁰*Ibid.*, p. 35.
⁸¹*Ibid.*, p. 36.

tend to professe the one or the other, are to take that course, and I have not neglected occasionally to tell the World, that this way is pursued amongst us."[82] Here Webster appears to clear away the shades of the astrologers and the Rosicrucians, and perhaps if he had stopped at this point he would have merited some praise. Unfortunately Webster followed directly with the statement "*That some Physicall Learning may be brought into the Schooles, that is grounded upon sensible, Rationall, Experimentall, and Scripture Principles, and such an Author is* Dr. Fludd; *then which for all the particulars, the World never had a more perfect piece.*" For Ward this is too much.

> How little trust there is in villainous man! he that even now was for the way of strict and accurate induction, is fallen into the mysticall way of the *Cabala*, and numbers formall: there are not two waies in the whole World more opposite, then those of the L. *Verulam* and D. *Fludd*, the one founded upon experiment, the other upon mysticall Ideal reasons; even now he was for him, now he is for this, and all this in the twinkling of an eye. O the celerity of the change and motion of the Wind.[83]

Webster has suggested further that the works of Descartes should be studied by scholars at the universities. Surely these must be empty words: Ward was convinced that Webster had no greater understanding of Descartes than he had of Bacon. And as for the plea for a deeper examination of the philosophies of Plato, Democritus, Epicurus, Philolaus, Hermes, and Gilbert, Ward could only reply: "If *De Fluctibus* be so perfect, what need we go any farther?"[84]

Ward suggested that Webster might best be dismissed as an impertinent pamphleteer, one who called for a new philosophy but who did not understand the authors he had recommended. Webster's evaluation of Bacon and Descartes was surely wrong, and even his condemnation of Aristotle seemed unjustified. And if Webster would make natural philosophy the basis of educational reform, his was a dream which could not be fulfilled. Ward pointed out that the nobility and the gentry sent their sons to the Inns of Court, while not one student in a hundred desired to proceed in

[82]*Ibid.*, pp. 46, 49-50.
[83]*Ibid.*, p. 46.
[84]*Ibid.*

natural philosophy. For this very reason the universities could never be oriented in this direction.[85]

Thomas Hall's *Whip for Webster* (1654)

The exchange between John Webster on the one hand and Seth Ward and John Wilkins on the other has unexpected overtones for the twentieth-century observer. It was the "occultist" Webster who called for observation and experiment as a basis for educational reform, while those who defended the existing system were the "moderns," the "mechanical philosophers." Against this background it is illuminating to note that an unyielding Aristotelian, Thomas Hall (1610-1665), thought his views were closely in accord with those of Wilkins and Ward rather than with Webster.

Thomas Hall was an Oxford graduate (1629), and although at first a conformist, he later became a Presbyterian, a move which resulted in the cancellation of his Church benefice after the passage of the Uniformity Act (1662). He was the author of many books and tracts, but the one of special interest to us is his *Vindiciae literarum* (1654), to which is appended his *Histrio-Mastix. A whip for Webster (as 'tis conceived) the Quondam Player* (1654).[86] From the title it seems obvious that Hall assumed that the author of the *Academiarum examen* was none other than the Jacobean playwright of the same name.[87]

In his *Vindiciae literarum* Hall spoke at length of the need for

[85] *Ibid.*, pp. 49, 50.

[86] See the short biography of Thomas Hall by Alexander Gordon in the *Dictionary of National Biography*. Gordon lists eighteen works by Hall printed between 1651 and 1661. Although John Wilkins signed the preface "N. S." and Seth Ward signed the main text "H. D.," Hall correctly identified the authors of the *Vindiciae academiarum* in his "To the Reader," dated Sept. 4, 1654. Thomas Hall, *Vindiciae literarum, The Schools Guarded: Or the Excellency and Usefulnesse of Humane Learning in Subordination to Divinity, and preparation to the Ministry . . . Whereunto is added An Examination of John Websters delusive Examen of Academies* (London: W.H. for Nathaniel Webb and William Grantham, 1655). The *Examination* is separately titled *Histrio-Mastix. A Whip for Webster*. Here see sig. O2^v.

[87] *Histrio-Mastix*, title page (sig. O1^r). This is also made clear in the anonymous work against Webster's logic appended to Hall's *Histrio-Mastix*. "This Mr. Webster (as I suppose) is that Poet, whose Glory was once to be the Author of Stage-plaies" (*ibid.*, p. 217).

maintaining the established course of study for the training of ministers. Surely, he argued, Godly learning was not obtained from divine revelation alone. Was not Moses learned in the wisdom of the Egyptians? Had not Solomon excelled the Egyptians in his knowledge of their secrets?[88] While some fields could be improperly applied by their practitioners—no better example existed than those mathematicians who used their art to cast horoscopes[89]—one need not condemn whole areas of knowledge because of the misdeeds of a few men.

Hall had barely completed his lengthy defense of the traditional educational system when he chanced upon a copy of the *Academiarum examen* and the reply to it by Wilkins and Ward. Webster's denial of the value of "humane learning" stirred him to anger, but the work of Wilkins and Ward seemed so complete that Hall was sorely tempted to give up his own defense of established logic and philosophy. Only after receiving a learned essay favoring Aristotelian logic did he decide to proceed with his original plan.[90]

Writing in a state of emotional fury, Hall branded Webster a "Herculean-Leveller, ... a dissembling Fryer ..., [a] professed friend to Judicial Astrology and Astrologers, ... [and] A great stickler for the fire and Furnace of Chymestry, for Magick and Physiognomy &c."[91] This man who spoke of the proper training of ministers should have defended languages, arts, and sciences, as well as recognize that true philosophy and reason can only be an aid to religion. As for philosophy, unquestionably Aristotle was the "Prince of Philosophers," not a blind pagan who wrote by diabolical instinct as Webster suggested.[92] But instead of the divine works of antiquity, Webster amazingly suggested magic,

> ... that noble, and almost divine science (as he cals it) of naturall Magick. This key (if you will believe him) will better unlock natures Cabinet, then syllogismes; yet he complaines, that this is neglected by the Schooles, yea hated and abhorred, and the very name seems nauseous and execrable to them.[93]

---

[88] *Vindiciae literarum*, p. 11.
[89] *Ibid.*, p. 53.
[90] *Histrio-Mastix*, sig. O2^V.
[91] *Ibid.*, p. 198.
[92] *Ibid.*, pp. 199 ff., 203-204.
[93] *Ibid.*, pp. 204-205.

Further, Webster commended the study of judicial astrology, even though in Scripture its practitioners "are oft joyned with Witches, wizards, and Sorcerers...."[94] Indeed, this ignorant man admired

> ... not only *Lilly* and *Booker*, but also Fryar *Bacon* [*sic mulus mulum*, it becomes one Fryar to claw another] and *Paracelsus*, a Libertine, a Drunkard, a man of little learning, and lesse Latine; he was not only skilled in naturall Magick ... but is charged also to converse constantly with Familiars, and to have the Devill for his Pursebearer; yet this is one of Mr. *Websters* society.[95]

Hall complained further of Webster's praise of physiognomy and of his attack on professional physicians and more specifically on Galen, "the Father of Physicians." No less detestable was his

> ... extolling of Chymistry, and preferring it before *Aristotelian Philosophy*, and advising schollars to leave their Libraries, and fall to Laboratories, putting their hands to the coales and Furnace ... this is Mr. *Websters* short cut, a quick way to bring men to the Devill, or the Devill to them.[96]

In short, according to Hall, John Webster was "against learning, against *Aristotle*, against Magistracie, against Ministrie, against Physitians, and against all that is truly good...." And in place of Aristotle and true learning he would recommend mathematics, optics, geometry, geography, astrology, arithmetic, physiognomy, magic, pyrotechny, dactylogy, stenography, architecture, and "the soule ravishing study of Salt, Sulphure and Mercury [a medicine for a Horse]."[97] He concluded with a blast at Giles Calvert, who had printed the *Academiarum examen*,[98] and then proceeded to append an anonymous defense of Aristotelian logic reiterating that Webster and his fellow astrologers were enemies of the study of logic.[99]

[94] *Ibid.*, p. 205.
[95] *Ibid.*, p. 209. I am indebted to Professor P. M. Rattansi for pointing out to me that this passage is quoted by Hall—without identification—from Thomas Fuller's short and uncomplimentary life of Paracelsus in *The Holy State* (2nd ed., Cambridge: John Williams, 1648), p. 53. (1st ed., 1642.)
[96] Hall, *Histrio-Mastix*, pp. 209-210.
[97] *Ibid.*, p. 214.
[98] *Ibid.*, p. 215. Calvert was known for publishing texts of this sort.
[99] *Ibid.*, p. 238, in the *Examen Examinis. An Examination of Mr. Websters Illogicall Logick, and Reasoning even against Reason* (pp. 217-239).

The work of Hall epitomizes the approach of those who desired no change in the educational system. Aristotle, Galen, and indeed all of the traditional authors were to be praised, while Webster's plan to replace the ancients with a new curriculum based on natural magic, chemistry, and astrology was anathema: not only were these subjects useless, they were diabolical in origin. This Aristotelian found John Wilkins and Seth Ward allies in the struggle against a new philosophy based on observation and experiment, scholars who would contribute to the great crusade to save the universities for "humane leaning."

It would be erroneous to think that the tracts of Webster, Hall, Wilkins, and Ward represent the total spectrum of thought on educational reform published in mid-seventeenth-century England. These particular tracts have been chosen to point out that the educational problem—fundamental for the Paracelsians for over a century—remained one of major importance in the 1650s. As Charles Webster has recently noted, Wilkins, Ward, and their colleagues did not oppose the study of many subjects proposed by the Paracelso-Helmontians.[100] However, there can be no doubt that their approach to these subjects often differed on basic issues. Furthermore, the mechanical philosophers favored the development of the sciences primarily through private organizations rather than through university reform. The universities, they felt, should properly be retained as centers for more traditional learning. The foundation of the Royal Society at the Restoration offered them an opportunity to realize their hopes on a scale they had not thought possible earlier.[101] Oddly, Elias Ashmole and John Webster saw this society at least as a partial fulfillment of their desire for a new learning. Ashmole was to become one of the earliest elected members while Webster, referring to the Royal Society as "one of the happy fruits of His Majesties blessed and miraculous Restauration, and that which will speak him glorious to all succeeding Generations, beyond all his Royal Progenitors," marvelled at the learning of its members,

---

[100]Charles Webster, "Science and the Challenge to the Scholastic Curriculum 1640-1660," *The Changing Curriculum* (published by the History of Education Society) (London: Methuen, 1971), pp. 21-35 (p. 31).
[101]*Ibid.*, p. 32.

whose efforts were gradually dispelling the "wonder" of the unknown.[102]

Chemistry and the State: The Agricultural Problem

Underlying all schemes for a "new philosophy" was the firm conviction that the resultant truth could lead to the betterment of man's lot. A new medicine based upon chemical knowledge promised the eventual control of disease while at the same time it benefited the scholar through a true understanding of the works of the Lord. This is a familiar theme that derives from the Renaissance natural magic tradition in which it was accepted that the *magus* could learn to control nature. The practical aspect of this becomes ever more important during the course of the seventeenth century. In the work of Francis Bacon there is a conscious effort to utilize the newly acquired learning for the betterment of practical processes. It was suggested that "Histories" of the various trades be made so that the *virtuosi* might learn, and consequently be able to improve on, the work of the artisans. If the compilation of this information was a favorite pursuit of the early members of the Royal Society of London, the practical significance of the studies of scientists was hardly less appreciated across the Channel. In Paris the members of the newly founded French Academy of Sciences were paid by the state with the expectation that they would apply themselves to the solution of national problems.

If the medical benefits of the chemical philosophy were obvious, the metallurgical implications were hardly less so, and there was some conscious effort to apply this aspect of the "new philosophy" to the improvement of industry and the economy of the state as a whole. Here the subject of agriculture is of special concern, for the concept of growth was of persistent interest to sixteenth and seventeenth-century chemists. The origins of this may be traced to Paracelsus, but once again it is instructive to take as our vantage point the English scene, where a number of familiar themes converge in the mid-seventeenth century. Samuel Hartlib

[102]John Webster, *Metallographia: or An History of Metals* . . . (London: A.C. for Walter Kettilby, 1671), sig. A2^v; Webster, *Displaying of Supposed Witchcraft*, p. 268.

is again a central figure, but of even more interest is Sir Hugh Plat (1552-1608), whose *Jewell House of Art and Nature* reflected Paracelsian views through the author's personal acquaintance with the London distiller John Hester and through his deep study of the French works of Bernard Palissy (1499 or 1510-1589).[103] *The Jewell House of Art and Nature*, not widely known at the time of its initial publication in 1594, became a much-quoted text after it was twice reprinted in 1653.

A Cambridge graduate (1572), Plat resided at Bethel Green and elsewhere in the London area while he devoted himself to numerous experiments.[104] The results were printed in a series of books of the genre best described as "books of secrets." There he listed new inventions relating to mechanics, chemistry, and agriculture. He described the brewing of beer without hops, the making of wine from English grapes, and the preservation of food in hot weather and at sea. Noteworthy is his invention of a cheap fuel prepared by the kneading of coal and clay into balls, described in his *Of Coal-Balls for Fewell wherein Seacoal is* (1603).

From his books it is evident that Hugh Plat had a deep interest in chemistry. However, like most other English chemists of the sixteenth century, he expressed his interest in a practical form. Well aware of the medical importance of chemistry through his friendship with John Hester, Plat also had a particular concern for ways in which to improve barren soil. This is evident in the second book of *The Jewell House of Art and Nature*. Titled *Diuerse new sorts of Soyle not yet brought into any publique use, for manuring both of pasture and arable ground, with sundrie concepted practises belonging thereunto*, this work was considered important enough to be issued separately in 1594.[105] Here the first sentence refers to Palissy as an authority for the fact that "all sorts and kinds of Marl, or soyl whatsoever, either known or used already for the manuring or bettering of all hungry

[103]The following sections on chemical agricultural theory derive, with minor modifications, from my paper "Palissy, Plat, and English Agricultural Chemistry in the 16th and 17th Centuries," *Archives internationale d'histoire des sciences,* 21 (1968), 67-88. The best general background is G. E. Fussell's *Crop Nutrition. Science and Practice before Liebig* (Lawrence, Kansas: Coronado Press, 1971).

[104]See the biography of Plat by Sidney Lee in the *Dictionary of National Biography*.

[105]Hugh Plat, *Diuerse new Sorts of Soyle not yet brought into any publique use, for manuring both of pasture and arable ground, with sundrie concepted practises belonging therunto* (London: Peter Short, 1594). This is also in Hugh Plat, *The Jewell House of Art and*

and barren grounds . . . draw their fructifying vertue from that vegetative salt."[106] The experimental treatises "by that learned husbandman Master *Bernard Palissy*, whereof the one is intituled, *Des sels diverses*, and the other *De la marn*, will ensure that the "true infants of Art, may receive a full light into Nature," while "ignorant Farmers" may glean "a few loose and scattered ears, to make so much bread of, as may relieve their hungry bellies."[107]

For Plat, Palissy's tracts on salts and marl form a logical whole because they present salt as the true key to all growth. So important was the subject of salt to Plat that he was willing to reproduce most of the theoretical parts of Palissy's discourse. We are told that there are so many kinds of salt that no man can number them all. Indeed,

> . . .there is not any one thing in the world, which doth not participate of this salt, whether it be man, beast, tree, plant, or any other kind of vegetable, yea, even the mettals themselves; and that which is more, there is not any kind of vegetable whatsoever, that could grow or flourish, without the action of salt, which lies hid in every seed . . . .[108]

The significance of this is so great that

> . . . if the salt were divided from the body of any living man, or from stones which are wrought up into strong buildings, or from the principal posts, the beams, and rafters of any house, they would all fal to powder in the twinkling of an eye. The like may be said both of Iron, Steel, Gold, and Silver, and all other mettals.[109]

Although Palissy does not cite Paracelsus here, surely this is his source. Sixteenth-century chemists insisted on "salt" as a principle

---

Nature. Conteining diuers rare and profitable Inuentions, together with sundry new Experiments in the Art of Husbandry, Distillation and Moulding (London: Peter Short, 1594). The edition cited here is Sir Hugh Plat, *The Jewell House of Art and Nature: Containing Divers Rare and Profitable Inventions, together with sundry new Experiments in the Art of Husbandry. With Divers Chymical Conclusions concerning the Art of Distillation, and the rare practices and uses thereof. Faithfully and familiarly set down, according to the Authours own Experience . . . Whereunto is added, A rare and excellent Discourse of Minerals, Stones, Gums and Rosins; with the vertues and use thereof, By D. B. Gent.* (London: Elizabeth Alsop, 1653). A second 1653 issue with the imprint of B. Alsop exists.

[106]Plat, *Jewell House*, p. 91.
[107]*Ibid.*, pp. 97-98.
[108]*Ibid.*, p. 98.
[109]*Ibid.*

of all things, but at the same time they cited the *De mineralibus* of Paracelsus to the effect that there are as many different types of salt as there are objects.[110] And for the Paracelsians the principle of salt conferred the property of solidity which bound together the extreme principles mercury and sulfur. Palissy would not have argued with Nicholas Lefèvre's explanation of natural putrefaction (1660) in terms of the release of sulfureous and mercurial vapors after their binding salt had been dissolved.[111]

The powers of salt can be seen in many practical processes which disclose its presence in unexpected substances. Saltpeter is extracted from earth through a leaching process, and linens are whitened through boiling with ashes. Similarly the tanning process is carried out in vats with the bark of oak trees, while the taste of cinnamon is found most strongly in its bark.[112] In all these cases, the bark, the ashes, and the earth lose their power after the extraction. One may assume that the active salt has caused the observed phenomena. Salt also acts to harden substances and preserve them from putrefaction. It is responsible for sound, and "it giveth beauty to all reasonable creatures." More important, salt aids in the procreation of all living things.[113] Not only does it stir lust in man and beast, it is the dissolved salt from dung which causes a more vigorous growth in the fields.

> But if thou wilt not give credit unto my speech, yet mark how the labouring Hind, when he carries his dung to the field, how in dis-

---

[110]Paracelsus, *Sämtliche Werke*, ed. Karl Sudhoff and Wilhelm Matthiessen, 15 vols. (Munich/Berlin, 1922-1933), Vol. III, pp. 42-43. Paracelsus discussed salt as the balsam of life in the *De naturalibus rebus*, in *Opera omnia* (Geneva: Jean. Antonii & Samuelis De Tournes, 1658), Vol. II, p. 188. See also p. 191: "In capitis huius initio dixi, Naturam etiam in Liquore Terrae salem habere, naturaliter incorporatum. Ab hoc sale crescentia omnia saliuntur, estque Balsamus salis, de quo mentionem feci." The source of Palissy's knowledge of Paracelsus is open to question. In a personal communication to the author, Aurèle La Rocque agreed that Palissy's ultimate source on the subject was probably Paracelsus, but he added, "I wonder if Palissy knew him at first hand? It seems more likely that he either heard of Paracelsus' opinions or read them in translation (perhaps manuscript ones) or in the works of Paracelsians writing in French."

[111]Nicasius Le Febure, *A Complete Body of Chymistry*, trans. P. D. C. Esq. (London: O. Pulleyn for John Wright, 1670), p. 21.

[112]Plat, *Jewell House*, pp. 99-101.

[113]*Ibid.*, p. 102.

charging of his loads he leaves it in certain heaps together, and a while after, he commeth to spread it all over the ground, and layeth the same in equal level; and afterward when the field happens to be sowed with corn, thou shalt alwayes find the corn to be more green and rank in those places where the same heaps were first laid (after they have lain there some reasonable time) then in any other place in all the ground besides; and this comes to passe, by reason that the rain which fell upon them hath carried even the salt through them, and conveyed it into the earth that was under them. Whereby thou mayest easily gather, that it is not the dung it self which causes fruitfulnesse; but the salt which the seed hath sucked out of the ground.

And hereupon it commeth to passe, that all excrements as well of man as beast, serve to fatten and enrich the earth. But if any man will plow and sow his ground yearly without dunging the same, the hungry seed in time will drink up all the salt of the earth, whereby the earth being robbed of her salt, can bring forth no more fruit, until it be dunged again, or suffer to lie fallow a certain time, to the end that it may gather a new saltnesse from the clouds, and rain that falleth upon it.[114]

Here again Palissy's source seems to have been Paracelsus, who discussed the relationship of growth to water, salt, and dung in his *Von den Natürlichen Wassern*.[115]

Plat turned next to Palissy's treatise *De la marn*, in which he discussed the similar effect of marl on growth. Asking specifically what the English farmer should know about this substance, Plat answered that the farmer should know what it is, how it is found, and why it benefits the earth. Palissy had written that marl is an earth—found in digging pits—generally white, but often colored gray, black, or yellow. When dispersed on barren ground it gives the same benefits as manure; indeed, a thorough job of marling can be so effective that the soil will be enriched for ten to thirty years.[116] The virtue of marl derives from a generative water it contains. Whereas common water maintains a regular circulation in its ascent from the earth through the attraction of the sun and in its

---

[114]*Ibid.*, p. 103.

[115]Paracelsus, *Opera, Bücher und Schrifften*, ed. J. Huser, 2 vols. (Strassburg, 1616), Vol. II, p. 150; see also the Latin translation in the *Opera omnia* (1658), Vol. II, pp. 359-360. More is said of the production of niter from urine in the *De naturalibus rebus* (Vol. II, p. 190). These passages are of importance in relation to the Paracelsian concept of an aerial salt.

[116]Plat, *Jewell House*, pp. 108-109.

subsequent condensation, the generative waters—a true fifth element—are different. Although they are carried with the others

> ... along the valleys, whether they be flouds, rivers, or springs ... they do always frame some one thing or other, and most commonly either great stones, rocks and quarries of stone, according to the grossness of the matter which is stayed with it, and carries the form of his mold wherein it rests, and this being so congealed, that common water is sometimes drunk up in the earth, and descends lower, or else it is drawn upwards, and doth vanish away in vapours and clouds, leaving his companion behind, which he is not able to carry any longer.[117]

It is this generative water which enters into the substance of plants and animals. We see, for instance, that

> ... corn and other seeds do keep themselves green untill their maturity, and when they are ripe, and that their root ceases to draw up or to drink up any more thereof, the exhalative water flieth away, and the generative remains; and as the decoction in plants begins to perfect it self, so the colour also changes, as it comes to pass also in stones and all kinds of mettal.[118]

The alchemists had generally distinguished between common water and the water of the philosophers.[119] Similarly they had spoken of a generative nature of water, as in Martin Ruland's statement that "auss dem Wasser wächst alles/hat auch die Nahrung darauss."[120] Palissy found in this concept the proper ex-

[117]*Ibid.*, p. 111.
[118]*Ibid.*, pp. 112-113.
[119]As did J. J. Becher in the *Oedipus chymicus*, when he spoke of two types of water: "Elementus Aquae duplex est, datur enim Aqua Communis & Philosophorum." "Philosophorum Aqua vocatur humidum radicale seu Aqua primordialis corporum . . ." J. J. Manget, *Bibliotheca chemica curiosa*, 2 vols. (Geneva: Sumpt. Chouet, G. De Tournes, Cramer, Perachon, Ritter & S. De Tournes, 1702), Vol. I, p. 325.
[120]Martin Ruland, *Lexicon alchemiae* (1612) (Hildesheim: Georg Olms, 1964), p. 46. Here again a Paracelsian influence seems likely. In his discussion of the role of water in the Paracelsian corpus, Pagel has studied Paracelsus' views of the generative nature of "growing water." When the salt of the earth (niter) is added to it, this water is "enabled to form the 'flesh', i.e. the full substance of a plant. It is special water—as against ordinary rain or brook water—for it contains all constituents of the plant. Rain can become grass, but only when it is mixed with this potent 'growing water'." Walter Pagel, *Paracelsus: An Introduction to Philosophical*

planation of marl. Originally it was a simple earth in which both waters had entered; the common water in time passed away while the generative water remained behind.[121] The action of the latter both hardened and whitened the earth. And now seeds, in the presence of common water, partake of the fixed generative water from the marl. In the same way that the vegetative salt from dung can be used up in the process of growth, so too, the generative water can be consumed "by often sowing of the ground," thus requiring fresh marl.

> Mr. *Bernard* concludes thus, That Marl is a natural, and yet a divine soyl, being an enemy to all weeds that spring up of themselves, and gives a generative vertue to all seeds that are sown upon the ground by the labour of man. And here ends Mr. *Bernard*.[122]

To Plat it was incredible that such a useful work should have gone unnoticed in England for so long. It seemed likely that English farmers would continue with their outmoded methods until "some studious Scholler, or other, will step forth, and take our idle Farmers by the hand, and either lead them over shooes into one of Master *Bernards* muck-heaps, or else by violence thrust them into one of his Marl-pits."[123]

And yet, Plat noted, Palissy's work was not perfect, for it had been based upon observations made in France, where phenomena might well vary from those in England. For instance, there might be colors of marl in England unknown to Palissy. It was thus likely that English authors could contribute to Palissy's text, making suggestions as innovative as Palissy's idea of using Fuller's earth for marling. Why not try the soil of the streets and the "grounds of all Channels, Ponds, Pools, Rivers, and Ditches?" Here Plat suggested that the guide should be the concept of the generative waters, since any place where rain water had settled might be rich in the congealative parts of those waters which are "full of the vegetative salt of nature, as Mr. *Bernard* notes." The importance of

---

Medicine in the Era of the Renaissance (Basel: S. Karger, 1958), p. 97. See also Pagel, *Das medizinische Weltbild des Paracelsus seine Zusammenhänge mit Neuplatonismus und Gnosis* (Wiesbaden: Franz Steiner, 1962), pp. 76-79.

[121] Plat, *Jewell House*, p. 112.
[122] *Ibid.*, p. 116.
[123] *Ibid.*

the vegetative salt should further guide the farmer in his method of storing dung. Palissy had shown that manure left standing in the fields would lose its salt value by being dissolved in the rain. Carefully prepared brick receptacles would permit the virtues of the dung to be preserved until it was spread upon the fields.[124]

Palissy—and to a lesser extent Franciscus Valetius[125] had written at length on the virtues of vegetative salt, but neither had gone far enough for Plat. Their theories and observations were splendid, he thought, but the farmer was still faced with the problem of finding a cheap and abundant source for this salt. On this point Palissy's work might actually prove to be the source of confusion, for he had stated that he did not speak of common salt, only vegetative salt. For Plat it seemed unlikely that Palissy could properly explain the generative cause of all vegetable life and still not know the relation of common and vegetative salt.[126] To solve this question it must be determined whether any essential difference exists between the four elements earth, water, air, and fire,[127] a point that would be illuminated less by a study of Aristotle, Velcurius, and Garceus than by observing the backside of the Moor fields

> ... where by undoubted arguments, I did hear it maintained, that all those elements do onely differ in attenuation and condensation: so as earth being attenuated becomes water, and water condensate becomes earth; water attenuated becomes ayr, and ayr condensate becomes water; and so likewise ayr attenuated becomes fire, and fire condensate becomes ayr: and thus all of them spring from one root: which being admitted, is a manifest proof, that there is a great and near affinity between the land and the sea, where we shall find salt water enough for our purpose. And yet further we see that of the earth and water together are made one globe, so as a small matter will make them friends, being so nearly united together.[128]

---

[124]*Ibid.*, pp. 117-121.

[125]Plat refers to the *De sacra philosophia, sive de his quae scripta sunt physice in libris sacris* of François Vallès, Valesio, or Franciscus Valetius, professor of medicine at Alcalá de Henares and physician to Philip II of Spain. This was printed at Turin in 1587, and other editions appeared in 1588, 1590, 1592, 1595, 1608 and 1622.

[126]Plat, *Jewell House*, p. 126.

[127]*Ibid.*, p. 127.

[128]*Ibid.*, p. 128.

Since theory thus explains the essential sameness of land and sea, the salt in the sea must have a close relation to Palissy's vegetative salt.[129] While some object and argue that an excess of common salt will make water unfit for drinking, these doubters should turn to the work of Valetius, who has pointed to the Biblical miracle of the purification of a well with salt.[130] In reality, salt must protect life, since it prevents putrefaction and fosters more forms of life in the seas than anywhere else. Only the degree of saltiness need be of concern, for in moderation the effect of this substance is not only beneficial but essential to life. For Plat it was quite understandable and proper for Palissy to have called salt the fifth element.[131]

Agricultural experiments seemed to provide indisputable support for Plat's contention. Seed corn soaked in sea water proved to be more fruitful than corn not submitted to this treatment, and experiments at Clapham had shown that one bushel of sea salt had a more beneficial effect on barren land than two loads of the best manure.[132] This evidence explained the west country practice of conveying saltish lands from the coast as far as five miles inland. Fields so treated remained enriched for years.[133] Even lands which had been inundated by the sea and subsequently drained were not found to be ruined, since the earth, left to her own handiwork, and "by her inward heat and transmuting nature will in some reasonable time, by way of putrefaction, convert that which was before a common salt, into a vegetative salt."[134] From all of this Plat concluded that common salt was the source of a cheap and abundant supply of vegetative salt.

But given the supply, Plat continued, care must be employed in learning the proper application of the substances. Only long experience will reveal the most effective quantity of marl for a given field, the best season of the year to apply it, and the type of grain it benefits most.[135] With the more potent salt even more care must be

---

[129] *Ibid.*, pp. 128 ff.
[130] *Ibid.*, pp. 92-95.
[131] *Ibid.*, pp. 111, 126.
[132] *Ibid.*, p. 129.
[133] *Ibid.*, p. 130.
[134] *Ibid.*, pp. 132-133.
[135] *Ibid.*, pp. 117-118.

employed. Plat added that farmers must read and reread his words,

> ... lest peradventure they take a sword by the point, and so hurt themselves by the weapon which was given them to defend their persons. And let this be a general caveat unto them, that they begin with small practises, and first upon arable grounds, before they proceed to pasture or meadow: and so being carefull in those former circumstances, which I have at large handled in the title of Marl, they shal no way indanger their estates nor hazard any great losse before they attain their desires.[136]

Plat did not recommend that farmers overflow land which had already been sown or was to be sown shortly. This practice he suggested only for waste land that was to stand fallow for some time, since "in what time the earth will sufficiently putrifie and transmute the salt, before it will be serviceable in this kind, I will not here determine." In this regard Plat again called attention to the significance of the quantitative factor by pointing out that the brine of salt pits is far more concentrated than sea water.[137]

In short, seeking improvements in agricultural practices in England, Hugh Plat found what seemed to be the key in the *Admirables discours* of Bernard Palissy. He was certain that Palissy's discussion of the vegetative salt in dung and the generative water in marl could be applied to English farming. And yet it seemed that Palissy had not told his readers enough—a defect Plat proposed to remedy. Starting from Palissy's concept of the generative waters, he endeavored to show by philosophical argument and observation that common salt was closely related to Palissy's vegetative salt and that it could be changed into it by a natural transmutation. The use of sea salt could then be considered the ultimate answer for the improvement of English farm lands.

---

[136]*Ibid.*, p. 133.

[137]*Ibid.*, pp. 133, 134: "it shall not be amiss for them to know the difference, that the brine of some of these salt-pits doth hold one third, or one fourth part of salt, whereas the sea water doth not for the most part contain above one eighteenth or twentieth part of salt, which would make a great difference betwixt them . . . ."

Agricultural Chemistry in Seventeenth-Century England

In 1699 John Woodward wrote that "the ancients generally have ascribed to the earth the production of animals, vegetables, and other bodies; ... several of the moderns have given their suffrage in behalf of water."[138] Yet neither Palissy nor Plat would have wished to have been categorized simply as Aristotelians or proto-Helmontians. Through their observations on the beneficial effects of salts on plant growth they both had contributed important evidence linking the plant with its environment. After their time we find an ever-increasing interest in the problems of plant growth. This is illustrated most effectively in van Helmont's willow tree experiment, but it is also evident in the *Silva sylvarum* of Francis Bacon. Here, in a series of experiments on the acceleration of germination, Bacon showed the effect of marl, chalk, sea sand, pond earth, salt, and dung on different seeds.[139] He attributed the beneficial effect he obtained from sea sand to its salt, "for *Salt* is the first Rudiment of life."[140] The similarity of this with Plat's interests is striking, and it fits in well with the alchemical tradition.

The Jewell House of Art and Nature appeared in print again in 1653. The new editor, one D. B. Gent., has been identified as Arnold de Boot (or Boate),[141] who, along with his brother Gerard, was in regular communication with Samuel Hartlib. There is no question that Hartlib had a considerable personal interest in schemes for agricultural reform, and it is in relation to this that he

---

[138] John Woodward, "Thoughts and Experiments on Vegetation" (1699) in *The Philosophical Transactions of the Royal Society of London from their Commencement, in 1665, to the Year 1800; Abridged ...*, notes by Charles Hutton, George Shaw, and Richard Pearson, 18 vols. (London: C. and R. Baldwin, 1809), Vol. IV, p. 382.

[139] Francis Bacon, *Sylva sylvarum* (8th ed., London: W. Lee, 1664), pp. 89-90.

[140] *Ibid.*, p. 123.

[141] By Sidney Lee in his biography of Sir Hugh Plat in the *Dictionary of National Biography*. J. R. Partington's suggestion that D.B. stands for Anselmus de Boodt, who wrote the *Gemmarum et lapidum historia* (1609), is most certainly incorrect [*A History of Chemistry*, Vol. II (London: Macmillan, 1961), p. 102]. On the other hand, references to Arnold de Boot in the Hartlib papers between 1650 and 1652 make no mention of the forthcoming *Jewel House*. I am indebted to Charles Webster for this information. Thus, while D.B. Gent. could be Arnold de Boot, it is at present impossible to be certain.

cited Hugh Plat.[142] Hartlib's friend Gabriel Plattes had written a work to encourage the improvement of agricultural practices, while he authored a series of works dedicated to the advancement of husbandry himself.[143] Chief among these was his publication of Sir Richard Weston's letter on the well-developed Flemish system of agriculture (1650). Publishing it originally under his own name, he acknowledged Weston as his source the following year in *His Legacy* (1651). This work with its appendix (1652) went through several editions and in abridged form was still being printed in the mid-eighteenth century.

Like Plat, Hartlib spoke at length of the use of marl in farming, and he was very much interested in the application of chemistry to practical agricultural problems. He spoke knowingly of fermentation,[144] and he reproduced a letter by Robert Child insisting on more accurate identification of the substances taken out of the earth. Perhaps husbandmen need not be naturalists, he said, but it would be helpful if they were "Petty-Phylosophers."[145] To this end it is important that there should be prepared a natural history of sands, earth, stones, mines, and minerals. This interest is reflected in Gerard de Boot's *Ireland's Naturall History*, a work which Hartlib and Arnold de Boot had completed after the death of Gerard in 1650. Here there was included an account of the "Metals,

---

[142]Samuel Hartlib, *His Legacy of Husbandry* . . . (3rd ed., London: J.M. for Richard Wodnothe, 1655), p. 6. Plat's appreciation of the advanced farming practices in the Low Countries would have been of special interest to Hartlib: "the Low Country men of *Flanders*, . . . are generally accounted the most skilfull and painfull husbandmen of all *Europe*." Plat, *Jewell House*, p. 137. G. E. Fussell has not mentioned Plat in his "Low Countries Influence on English Farming," *English Historical Review*, 74 (1959), 612-622. This subject is discussed further in John J. Murray, "The Cultural Impact of the Flemish Low Countries on Sixteenth- and Seventeenth-Century England," *American Historical Review*, 62 (1957), 837-854 (851-852).

[143]Gabriel Plattes, *A Discovery of Infinite Treasure* (London, 1639). Samuel Hartlib, *A Discoverie for division or setting out of Land as to the best form* . . . (London, 1653); *An Essay for advancement of Husbandry-Learning or Propositions for the errecting a College of Husbandry* (London, 1651); *The Reformed Husband-man* . . . (London, 1651); *His Legacy* (London, 1651); *An Appendix to the Legacie of Husbandrie* (London, 1652). (The last two were printed together in 1652 and 1655 and an abridgement appeared in 1742.) *The Compleat Husband-man* (London, 1659).

[144]Hartlib, *Legacy* (1655), pp. 6 ff.

[145]*Ibid.*, pp. 72-73.

Minerals, Freestone, Marble, Sea-Coal, Turf and other things that are taken out of the Ground."[146] Hartlib himself pointed to the results of the chemical analysts who had studied spa waters as the best examples of this type of investigation published to date, and he cited with special warmth Edmund Deane's study of the waters at Knaresborough.[147]

The *Legacy* is studded with references to the two leading mid-seventeenth-century authorities in chemistry—van Helmont and Glauber. Questioned on this point, Hartlib rose immediately to the defense of these authors: in his circle, he said, the chemists were in good repute, and

> ... they have laboured and experimented much, as all men know, to find out the truth, and to advance the Commonwealth of learning: Some imperfections I impute to humane frailties, old age, &c. And I should be glad to see a solid answer to *Helmont*, who hath thrown down both *Aristotles* Philosophy, and *Galens* Physick; and as yet I have not seen any man who hath in any measure vindicated their old Masters. Further, I should rejoyce to see an experimental Philosopher confute *Glaubers* Experiments by experience; till this be done, I cannot but account them ingenious men, and well deserving of the Commonwealth of Learning ....[148]

Hartlib was acutely aware of the importance of understanding plant growth, and he pointed to the need for a competent study of salts in his "Philosophical Letter concerning Vegetation or Causes of Fruitfulness." Such a study he considered essential because it was accepted that "Salt is a seat of life and vegetation, and so the subject of nutrition."[149] Since we know that rain water nourishes plants, we can be certain that it contains a vital salt. Still, we remain

> ... very ignorant of the true causes of Fertility, and know not what Chalk, Ashes, Dung, Marle, Water, Air, Earth, Sun, &c. do con-

---

[146]Charles Webster, "The College of Physicians: 'Solomon's House' in Commonwealth England," *Bulletin of the History of Medicine*, 41 (1967), 393-412 (405).

[147]Hartlib, *Legacy* (1655), pp. 72-73. On Edmund Deane and the importance of the chemical analyses made at the English mineral water spas, see Allen G. Debus, "Solution Analyses Prior to Robert Boyle," *Chymia*, 8 (1962), 41-61 (53).

[148]Hartlib, *Legacy* (1655), pp. 134-135. Hartlib's defense is in answer to Dr. Arnold Beati (A. de Boot?).

[149]*Ibid.*, pp. 217-219.

tribute: whether something Essential, or Accidental; Material, or Immaterial; Corporal, or Spiritual; Principal, or Instrumental; Visible, or Invisible? whether Saline, Sulphureous, or Mercurial; or Watry, Earthy, Fiery, Aereal? or whether all things are nourished by Vapours, Fumes, Atoms, Effluvia? or by Salt, as Urine, Embrionate, or *Non specificate*? or by Ferments, Odours, Acidities? or from a *Chaos*, or inconfused, indigested, and unspecificated lump? or from a Spermatick, dampish Vapour, which ascendeth from the Centre of the Earth, or from the Influence of Heaven? or from Water onely impregnated, corrupted, or fermented? or whether the Earth, by reason of the divine Benediction hath an Infinite, multiplicative Vertue, as Fire, and the Seeds of all things have? or whether the multiplicity of Opinions of learned *Philosophers* (as *Aristotle, Rupesc. Sendivog. Norton, Helmont, Des Cartes, Digby, White, Plat, Glauber*), concerning this Subject sheweth the great difficulty of this Question, which they at leisure may peruse.[150]

The inclusion of Plat and the strong predominance of chemists in this list of authors is noticeable. On his part, Hartlib did not wish to venture a personal answer to the question. He preferred rather to send his readers to these authors and to Francis Bacon, who "hath gathered stubble . . . for the bricks of this foundation."[151]

Hartlib and his friends were greatly influenced by Bacon's call for natural histories. In a long list of required "histories" Bacon had specifically mentioned the need for a history of agriculture.[152] And, with the foundation of the Royal Society of London, this need was translated into a Committee on the History and Improvement of Agriculture. Led by Daniel Coxe, this committee proceeded to prepare a list of "Enquiries concerning Agriculture" (1665).[153] Here an interest was expressed in the determination of the relative value of different solid composts and marl in the growth of plants, and once again it was recognized that "it is impossible to compose an exact history of vegetation, till we understand the nature of the ground, as the matrix, wherein all plants are conceived, and

[150]*Ibid.*, p. 38.
[151]*Ibid.*
[152]Francis Bacon, *Catalog of Particular Histories by Titles*, in *The Works of Francis Bacon*, ed. James Spedding, M.A., Robert Leslie, M.A., and Douglas Denon Heath (new ed., 7 vols., London, 1870), Vol. IV, pp. 265-270 (270).
[153]Daniel Coxe, "Enquiries concerning Agriculture," *Philosophical Transactions of the Royal Society of London*, **1** (1665), 91-94.

whence they derive their nourishment."[154] While no answers to these queries were printed in the *Transactions*, Coxe later investigated the effect of sea sand on plant growth in his paper on Cornish farming practices (1675).[155] Here he observed that dry sand was nearly useless, and that if it was to be used at all, it should be taken from the shore while still damp. Then, after it had been applied to the fields there would be seen an almost immediate improvement in the land, again an indication of the possible importance of sea salt. This, at least, had been Plat's conclusion. Noting this practice in the west country, Plat had written that "the whole strength and vertue thereof consists in the saltness, for otherwise we might happily find some other sorts of sand that would also be equivolent unto this."[156] For Coxe, however, it was primarily the sand which appeared to be the significant agent for bettering the soil, an effect which probably could be explained through chemistry: "It may be worth while for some ingenious *Chymist, to open* the body of *Sand*, thereby to discern its several principles that are most prevalent: And then for some good Naturalist, to consider how it becomes so advantageous to *Vegetation*, and especially as to that part which concerns *prolifique Seed.*"[157]

The same problems which Hugh Plat had discussed eighty years earlier remained of interest to the members of the Royal Society. The book reviews in the *Transactions* bear this out, for here we find specific recognition of Plat's contribution. The same issue containing Coxe's paper on Cornish agriculture has a review of a new edition of Plat's *Garden of Eden*.[158] There was appended a discussion of other recent reprints of the work of Plat, among them the 1653 edition of *The Jewell House of Art and Nature*. We are told that Plat "shew'd great respects for honest Chymistry, and was

[154]Thomas Birch, *The History of the Royal Society of London*, 4 vols. (London: A. Millar, 1756), Vol. II, p. 32. Paper read Apr. 19, 1665.

[155]Daniel Coxe, "The improvement of Cornwall of Sea Sand, communicated by an Intelligent Gentleman well acquainted in those parts to Dr. Daniel Coxe," *Philosophical Transactions of the Royal Society of London*, **9** (1675), 293-296.

[156]Plat, *Jewell House*, p. 130.

[157]Coxe (1675), p. 296.

[158]Review of "*The Garden of Eden, or an Accompt of the Culture of Flowers and Fruits now growing in England . . .* In 2 parts in 8°, written by Sir Hugh Plat Kt.; newly reprinted," *Philosophical Transactions of the Royal Society of London*, **9** (1675), 302-304.

careful in directing Distillations, for Salts, Spirits, Oyls, and shewing various uses of them: But he was cautious against the cheating Alchymist." This, however, was not the most important feature of the book; above all, Plat

> ... advanced the *Agriculture of England* by Marle, Saline materials, as far as the Seas extend, which encompass these Islands; and by other Soyles, c. 104; chiefly by Lime, and the way of Denshiring whereby the most barren lands, hills and wasts may be converted to bear the richest burthens of corn, hay, and grass.[159]

Surely, for this reviewer, Plat was one of the most important authors in print on the improvement of agriculture. For Hartlib he was one of a select group of "learned philosophers" who had published on the causes of plant growth and fertility. And yet, on examination we find this work basically to be a practical extension of the chemical philosophy. A discussion of Plat's views on agriculture necessitates a comparison with his source, the work of Bernard Palissy. And, if Plat was dependent on Palissy, Palissy in turn can be linked with the earlier alchemical-Paracelsian literature. The concepts of a nutritive water and a life-giving salt belong to this tradition. Even Palissy's greatest "innovation"—the belief that fertilizers derive their importance from water-soluble salts—can be linked to the work of Paracelsus. In short, the mid- and late-seventeenth-century English work on plant growth and agricultural reform constitute an integral part of the tradition of the chemical philosophy.

Chemistry and Economic Policy: Johann Rudolph Glauber

For the mid-century natural philosopher there were two recent chemical authors whose works had to be mastered. For Hartlib (1665) these two pillars were the "ingenious" van Helmont and the German Johann Rudolf Glauber. Hartlib's views were echoed by the chemical instructor to King Charles II, Nicholas Lefèvre (c. 1610-1669): for him van Helmont and Glauber were "as the two Beacons and Lights which we are to follow in the Theory of

[159]*Ibid.*, p. 304.

Chymistry, and the best practice of it."[160] Indeed, the works of these two, plus that of Paracelsus, were the freely acknowledged bases of Lefèvre's *Cours de cymie* (1660), which was to become one of the most popular of the chemical "textbooks" of the late seventeenth century.

Glauber's acute understanding of the potential of the chemical philosophy for economic prosperity and national power makes him a subject of special interest at the present point.[161] Although he lived much of his productive life in the Netherlands, he was a patriotic German born in Karlstadt (c. 1603), who retained a deep love for his homeland. To a large extent self-educated, he excelled in the development of chemical equipment. With new furnaces he was able to obtain higher temperatures than ever reached before, and as a result he was able to extend the range of distillable substances. In his *Furni novi philosophici* (1646-1650) he described clearly the production of concentrated mineral acids. From these he made their corresponding salt derivatives. Glauber's works indicate that he attempted to distil or decompose almost everything readily available to him. His descriptions—with their notes on practical applications—made his *Furni* extremely popular, and it

[160] Le Febure, *Complete Body of Chymistry* (1670), p. 3.

[161] A valuable recent account is the biography by Kathleen Ahonen included in the *Dictionary of Scientific Biography*, Vol. V (New York: Scribner's, 1972), pp. 419-423. This is to be supplemented with the important chapter by Partington (*History of Chemistry*, Vol. II, pp. 341-361) and by J. W. van Spronson, "Glauber Grondlegger van Chemishe Industrie," *Overdruk uit Nederlandse Chemische Industrie* (Mar. 3, 1970), 3-11. These works all discuss the work of Glauber and include detailed bibliographies. Also of value are Kurt F. Gugel, *Johann Rudolph Glauber 1604-1670, Leben und Werk* (Würzburg: Freunde Mainfränkischer Kunst und Geschichte, 1955); Erich Pietsch, *Johann Rudolph Glauber. Der Mensch, sein Werk und seine Zeit*, Deutsches Museum Abhandlungen und Berichte (Munich: R. Oldenbourg, 1956); and P. Walden, "Glauber," in Günther Bugge, ed., *Das Buch der grossen Chemiker* (Berlin, 1929), Vol. I, pp. 151-172. Somewhat less valuable is H. M. E. de Jong, "Glauber und die Weltanschauung der Rosenkreuzer," *Janus*, **56** (1969), 278-304. The editions of Glauber's works cited here are those referred to at the beginning of the present chapter. These will henceforth be noted as Glauber (English, 1689) or (German, 1658/1659). The reader is reminded that while all six parts of *Des Teutschlandts Wolfahrt* appear in the English translation, only the first two appear in the German *Opera chymica*. Because of this, Parts III - VI will only be cited through the English edition.

PLATE XXVII Johann Rudolph Glauber (1603-1670). Portrait from A. D. Clement and J. W. S. Johnsson, "Briefwechsel zwischen J. R. Glauber und Otto Sperling," *Janus*, 29 (1925), 210. A discussion of this portrait is found in Erich Pietsch, *Johann Rudolph Glauber: Der Mensch, sein Werk und seine Zeit* (Munich: R. Oldenbourg; Düsseldorf: VDI-Verlag, 1956), pp. 25-26.

went through many editions in Latin, German, French, and English.[162]

Renowned for his chemical preparations, Glauber was able to build a large laboratory in Amsterdam in 1648, after leaving Germany in 1646. The end of the Thirty Years War was probably influential in his decision to return to his homeland (c. 1650), where he established a new laboratory at Kissingen. Here he employed a number of assistants, among them one Farner, who broke an oath of secrecy and openly sold Glauber's secret preparations. The latter, deeply wounded, defended himself in print[163] and published a number of his preparations for the first time. However, he left for Amsterdam (c. 1655) once again, where he established a third major laboratory. In failing health after 1660, he died an exile some ten years later.

It is not difficult to place Glauber in the mainstream of the chemical philosophy. He is similar to van Helmont in his rejection of ancient authority and in his reliance on chemistry as a true basis for the understanding of natural phenomena. He complained that those who were trained at the universities condemned the chemists unjustly, "for whosoever they be that are Ignorant as to the Fire, and that know not its wonderful efficacy, tho' they may be most skilful Proficients in foreign Tongues or Languages (which in the more-secret Philosophy makes not all to any purpose); yet nevertheless they will not be able to accomplish any the least matter in things of so great a moment." The title of doctor in the universities was both empty and vain and, indeed,

> I confess ingeniously, that I never frequented the Universities, nor ever had a mind so to do; for should I have so done, haply I should never have arrived to that knowledge of Nature, which I mention without boasting, as I now possess; neither doth it ever repent me, that I have

---

[162] *Furni novi philosophici oder Beschreibung einer Newerfundener Destillirkunst*, 5 parts (Amsterdam, 1646, 1647, 1648, 1648, 1649). Partington (*History of Chemistry*, pp. 344-345) notes numerous editions, the latest published at Prague in 1700. An English version, made by John French, *A Description of New Philosophical Furnaces, Or a New Art of Distilling . . .*, was published in London in 1651. The work is also included in the Packe translation of the *Works* of 1689.

[163] See the *Apology of John Rudolph Glauber, Against the Lying Calumnies of Christopher Farnner* (English, 1689), Part I, pp. 147-161.

put my hands to the Coals, and have by the help of them penetrated into the knowledge of the Secrets of Nature . . . .[164]

The significance of theory is apparent throughout his work. He stressed the importance of a knowledge of the growth of metals, the composition of the earth, and the origin of hot water springs and mountain streams. As for the importance of salt, he who was not aware of its value was not a philosopher, "but a proud Ass."[165]

Glauber's authorities included the traditional alchemists and metallurgists: references to Basil Valentine, Lazarus Ercker, and Georgius Agricola are frequently encountered. Nevertheless, it was Paracelsus above all upon whom Glauber relied, "the Most Potent Lion, and Monarch of the North; to whom none in the World may be compared, nor *did* ever any excel him in Glory and Power, or *shall be* like unto him."[166] Glauber proceeded to incorporate several of the works of his predecessor into his own publications, adding to them his own commentary by way of explanatory passages and digressions.[167]

Glauber's almost unqualified praise for Paracelsus may be contrasted with the more critical approach of van Helmont. The difference between the two, however, is more fundamental than this. Like many chemical philosophers—and like van Helmont himself—Glauber had a deep interest in the preparation of medicines, especially metallic salts. Yet in contrast with the learned Belgian physician, he had little interest in the application of chemistry as a tool for the explanation of physiological processes. Rather, his concern was primarily with metals, salts, and chemical compounds along with an understanding of earthly phenomena which might have a practical significance both for individuals and for the state.

---

[164]Glauber, *The Prosperity of Germany* (Part I), (English, 1689), Part I, p. 307; (German, 1659), Vol. II, p. 373.

[165]Glauber, *Of the Nature of Salts* (English, 1689), Part I, p. 256; (German, 1658), Vol. I, p. 462.

[166]Glauber, *Prosperity* (Part III), (English, 1689), Part I, p. 351.

[167]Glauber reproduced the prophetic works of Paracelsus with his own commentary; (English, 1689), Part I, pp. 351-352. He also proceeded to incorporate Paracelsus' *Tincture of Natural Things* into the second part of his *Miraculum mundi* [(English, 1689), Part I, pp. 231-242], and the third part of the *Mineral Work* is composed of a commentary on *The Heaven of the Philosophers, or a Book of Vexations*

In his general approach to the cosmos Glauber did not differ much from other Paracelsians. With them he turned to Genesis for its account of the Creation of the elements.[168] He insisted on the validity of the macrocosm-microcosm analogy, and he compared the hairs of the body with the growth of plants on the earth.[169]. However, because of the necessity of explaining surface phenomena and generation, he considered with special care the composition of the earth. Once again we read that the metals are living things and that we must know their true method of generation if we are to perfect them in the laboratory.[170] Glauber's objective was a practical one, for he was convinced that gold could—and should—be made by the true artist.

He believed that the seminal virtues of all things were placed in the stars at the time of the Creation. These virtues are subsequently carried unceasingly by the air to the earth, where they proceed to the center cavity of the earth with its fire. Here, as by an optical mirror, they are repulsed in all directions and at last come to rest near the surface.[171] There the earth, now impregnated by the seed, gradually brings forth its offspring:

> ... the Earth (embraceing it) doth cherish, nourish, and augment from form to form, until it comes to be a perfect Metal, which it (at length) brings forth into the light, as a Mother doth her mature young one; which Conception and Generation of the Metals, taking its Original at the very beginning of the World, will alwaies continue even unto its Dissolution.[172]

The process is a complex one for Glauber. Like van Helmont, he postulated an intermediate proto-metal, *Gur*.[173] He argued further that the particular matrix would influence the final product; that is, that the same seed might result in different metals under

---

by Paracelsus [(English, 1689), Part I, pp. 125-147].

[168]Glauber, *Of the Nature of Salts* (English, 1689), Part I, p. 255; (German, 1658), Vol. I, pp. 459-460.

[169]Glauber, *The First Century, or Wealthy Store-House of Treasures* (English, 1689), Part II, pp. 30-31. (This treatise is not in the German *Opera chymica*.)

[170]Glauber, *Mineral Work* (Part II), (English, 1689), Part I, pp. 115-116; (German, 1658), Vol. I, p. 338.

[171]*Ibid.* (English, 1689), p. 116; (German, 1658), p. 340.

[172]*Ibid.* (English, 1689), p. 115; (German, 1658), p. 339.

[173]*Ibid.* (English, 1689), pp. 122-123; (German, 1658), pp. 357, 361.

different conditions.[174] The heat of the central fire also played an important role in the process: should this heat be withdrawn during growth, the resultant metal could only be imperfect.[175] The alchemist must learn to apply his art to the perfecting of such metals or minerals.

Like other Paracelsians, Glauber was convinced that air was required for all forms of life. The spontaneous generation of live maggots did not occur in glasses that were hermetically sealed;[176] similarly he did not think that metals could live without the spirit of life. Like Fludd and van Helmont he sought to identify and to isolate this spirit. He assumed a heavenly source for the spirit and thought that it originated in the sun. By exposing flat dishes filled with a "magnetic" solution to the sun he claimed to have isolated this precious substance.[177] The absorbed sunbeams produced a concentrated solution which upon subsequent evaporation yielded a salt. This proved to be saltpeter, or sal nitrum, which Glauber called the *miraculum mundi* and the universal solvent.[178] It contained fire or sulfur within itself, but saltpeter could be found in everything, including the elements. "It is for this reason that it may be called the *Materia Universalis* since no one can live without the elements."[179] Glauber emphasized the importance of aerial niter, which he believed originated in the stars and then, through the medium of the air, impregnated rain, snow, and dew, which in turn fructified the earth.[180] Indeed, sal niter was so important that Glauber felt that the Emerald Table of Hermes was no more than a commentary on this substance. Our very lives and spirits were preserved through the hidden virtues of this salt. But since that which maintains life is the aerial *spiritus universalis*, without which

[174]*Ibid.* (English, 1689), p. 121; (German, 1658), p. 354.

[175]Glauber, *Prosperity* (Part II), (English, 1689), Part I, p. 324; (German, 1659), Vol. II, p. 411.

[176]Glauber, *Prosperity* (Part IV), (English, 1689), Part I, p. 389.

[177]*Ibid.*, p. 390.

[178]Glauber, *Of the Nature of Salts* (English, 1689), Part I, p. 248; (German, 1658), Vol. I, p. 442. *Spagyrical pharmacopoea* (German, 1658), Vol. I, p. 97. This will be found in the preface to Part III, which is not in the English *Works*.

[179]Glauber, *Continuation of the Miraculum mundi* (English, 1689), Part I, p. 207; (German, 1658), Vol. I, p. 262.

[180]Glauber, *Miraculum mundi* (English, 1689), Part I, p. 169; (German, 1658), Vol. I, p. 163.

the sun could not shine nor fire burn, the vital aerial spirit must be the familiar aerial niter.[181]

If this spiritous salt was freely available to man and beast through the inhaled air, there was no reason why an analagous circulatory system would not benefit the living metals in the earth. These too required the life-giving salt.

> Seeing therefore, that the constant Circulation of Blood in the *Microcosm*, can be in no wise deny'd, why should not also such a Circulation in the *macrocosm* be admitted as true? For as the Blood of the Human Body arising from the Liver, diffuseth it self through all the Passages and Veins of the Body, as well small as great, and Conserveth the life of the whole, nourisheth all the parts, and augmenteth the good juices, which are changed into Flesh, Bones, Skin, and Hairs in the Members themselves, and leaving the unprofitable Phlegm to be expelled by the Pores of the skin: So also is it with the Nutriment and Universal Aliment of the great World, while the Salt water without intermission, of the great Sea, or Ocean, encompassing the whole Globe of the Earth, by many small and great passages of Veins, passeth through all the parts of the Earth, and nourisheth and sustaineth them with its Salt, that Minerals, Metals, Stones, Sand, Clay, Shrubs, Trees, and Grass may be nourished and grow, and in growing take their encrease. The rest of the Water being freed from all Saltness, is exterminated as a superfluity in the Superficies, and being diffused into various Springs, as well small as great, is expelled, no otherwise than the superfluous sweat of the Blood in the *Microcosm*, by innumerable passages and pores.[182]

Thus, arguing from the now generally accepted microcosmic circulation of the blood, Glauber proceeded to accept a similar circulation of the life spirit within the earth. This circulation, essential for the birth of minerals, also produced life-giving salts at the earth's surface. No less than Palissy or Plat, Glauber believed that salt might be employed efficiently in place of manure by farmers, an agricultural method that he predicted would eventually come into general usage.[183] At the same time Glauber argued that the

---

[181]*Ibid.* (English, 1689), p. 170; (German, 1658), pp. 164-165.
[182]Glauber, *Of the Nature of Salts* (English, 1689), Part I, pp. 248-249; (German, 1658), Vol. I, pp. 443-444.
[183]Glauber, *Continuation of the Miraculum mundi* (English, 1689), Part I, p. 189; (German, 1658), Vol. I, p. 210.

earthly circulation resulted in the never-ending replenishment of sulfureous substances near the surface, which were regularly burned off through the eruption of volcanoes.[184] For him, no less than others, the chemical philosophy was expected to account for a vast variety of phenomena.

The Prosperity of Germany (1656-1661)

While the *Furni novi philosophici* made it possible for Glauber to detail numerous practical chemical processes, the publication of the first part of the *Miraculum mundi* (1653) gave ample evidence of his desire to benefit his common man and aid his homeland. Not only did he promise a new technique for enrichment of the earth without manure, he also promised a true method for the recovery of health and an explanation of how to augment gold and silver and to make medicines at small cost. Glauber's plan was developed in the *Miraculum mundi*, its continuation, and in a number of shorter works and was discussed at greatest length in *Des Teutschlandts Wolfahrt*, which appeared in six parts (with a lengthy appendix to the fifth part) between 1656 and 1661. Here especially can be seen the deep impact of the Thirty Years War on Glauber. How to avoid the ravages of future wars and how to solve the post-war economic problems were subjects of pressing concern which he hoped to solve with the secrets of the chemical philosophy.

To be sure, Glauber began, war is an unavoidable calamity. The malice of men provokes God to "avenge himself on the most Impious living of this perverse World, by bringing scarcity of Food, (War, and the Plague).''[185] However, it is evident that the pious suffer as well as the wicked, and it was to save them in time of disaster that he took pen in hand. Also, convinced that Germany was enriched in natural treasures beyond all other European countries, Glauber hoped to encourage Germans to cease wasting these riches.

Glauber had observed that the constant moving of armies resulted in the destruction of livestock of all kinds, with soldiers

---

[184]Glauber, *Mineral Work* (Part II), (English, 1689), Part I, p. 118; (German, 1658), Vol. I, pp. 345-346.

[185]Glauber, *Prosperity* (Part I), (English, 1689), Part I, p. 294. From the Dedication, which is not in the German *Opera chymica*.

The Chemical Philosophy in Transition, I 435

commandeering horses, oxen, sheep, and whatever else they could find. Much of the land had lain fallow for years, while the loss of the animals prevented the proper enrichment of the soil by manuring.[186] All these factors had resulted in a scarcity of food and a rise in its prices—problems compounded by the Peace Treaty of 1648:

> After the making of Peace, and sending away the Souldiers, many Commanders being weary of warfare, and abounding with Money and Horses, bought (or rather squeez'd out) of the poor Inhabitants, everywhere destitute of Money, most notable Farms, for a very mean price, and did set themselves with the utmost of their Industry to till the Earth, which had lain fallow a long time; which Lands became so fruitfull that it even amazed all men. The poor Inhabitants too, who now had gotten a little Money, they also set their hands to the Plough, and used even the utmost of their endeavours in the tillage of the Earth: From hence it came to pass, that they were so furnished with such a vast deal of Corn in a few years space, that they knew not at all what to do therewith. As for carrying it into far distant Countries, the troublesomeness and charges of the Carriage were too great an hindrance: To lay it up in the Granaries, that the Air will not suffer them to do, tho' in the time of *Joseph* such a thing was done in *Egypt*, where the Air being Nitrous, dry, and not so easily subject to corrupting, as ours is, preserved the Corn from Corruption.[187]

The answers to all the problems were to be found in the chemical philosophy, Glauber promised. And while all nations would benefit from what he had to write, it was Germany, with its hidden wealth, which was destined to become the "Monarch of the World."

> He who well knows the fire, and the use thereof, will not be distressed with want. And he that has no knowledge of the same, neither will he throughly search into Natures Treasures: From these things it is evident what Treasures we *Germans* do unknowingly possess, and yet convert them not to our use; And were not foreign parts as ignorant too of these things as we our selves are, they would readily have upbraided us *Germans*, that we spend more time in eating and drinking, than on good *Arts* and *Sciences*.[188]

---

[186]Glauber, *Continuation of the Miraculum mundi* (English, 1689), Part I, pp. 189, 193; (German, 1658), Vol. I, pp. 210, 223.

[187]Glauber, *Prosperity* (Part I), (English, 1689), Part I, p. 303; (German, 1659), Vol. II, pp. 362-363.

[188]*Ibid.* (English, 1689), p. 297; (German, 1659), p. 348.

A primary problem was the overabundance of grapes and wheat. While the production of wine and beer was customary, the excess product could not easily be sold profitably elsewhere because of the high cost of transportation. Glauber suggested the solution to the problem was chemical concentration. In the case of wine the must was to be evaporated to the consistency of honey, a form in which it could be preserved as long as needed and then reconstituted with water prior to the fermentation stage. Not only would carriage costs be saved, the vintner might withhold his product from the market until a time of need and thus reap an appreciably greater profit.[189]

This process could be even more important for the wheat farmer. Even now, Glauber noted, a huge wheat surplus in Franconia had forced the farmers to pile their yield in heaps in the open where it was rotting. The cost of carriage was so high that it was an impossibility to ship the wheat to areas of the country where it might be needed. In this case he suggested as the answer a concentrate of malt. As an example, Glauber explained that eight barrels of wheat should be brought to a malt, then boiled into ale and condensed to a juice that occupied one barrel. This concentrate was not eaten by rats and, unlike wine and ale, could be shipped as "dry goods." Not only were the carriage rates cheaper, the single barrel could be reconstituted to eight barrels of good ale, or even ten or twelve if a weaker brew was desired. The final product was actually better than that usually bought from the brewer; indeed, Glauber planned to supply his own family with this ale.[190]

Not only could the farmer and the vintner protect their investment in times of surplus by storing their product indefinitely, the concentration process would aid the state in other ways as well. The lords of castles were advised to lay in a store of the concentrates for use in time of siege.[191] In a separate work, *The Consolation of Navigators* (1657), Glauber claimed that the malt concentrate could be reconstituted not only with water to beer, but also with wheat to bread.[192]

---

[189]*Ibid.* (English, 1689), pp. 300-301; (German, 1659), pp. 356-358.
[190]*Ibid.* (English, 1689), pp. 303, 304, 305; (German, 1659), pp. 362, 366, 368.
[191]*Ibid.* (English, 1689), p. 306; (German, 1659), p. 369.
[192]Glauber, *The Consolation of Navigators* (English, 1689), Part I, p. 280; (German, 1658), Vol. I, pp. 534-535.

The great fertility of German soil, which had resulted in a surplus of grain, also produced the vast forests which covered much of Germany, and Glauber had a plan for harvesting these for the benefit of the state. In the process of making potash, logs are normally burned and the valuable salt subsequently worked up from the ashes. Here Glauber reminded his reader that all vegetable matter contains both niter, the essential salt which is derived from the universal spirit of the air, and tartar. Both are valuable commodities, while niter is the source of medicines and even gold and silver.[193] For the army a plentiful and inexpensive supply of saltpeter is a necessity (for gunpowder), for the sailor the spirit of salt could be used as a sure cure for scurvy,[194] and for the farmer niter could serve as a substitute for manure.[195] For all of these substances the forest could be a cheap source, the secret being a "wood press" which Glauber claimed to have invented. This would rapidly squeeze out the essential juices from logs of all sizes, and these juices could subsequenty be worked up for their salts. However, the construction of the press and the extraction process were secrets too valuable to publish.[196] Those who sought them would have to negotiate with Glauber in person.

Germany's riches extended also to its mineral resources, and here Glauber wrote that the true alchemist would learn from him that minerals and base metals might be purified with the judicious use of saltpeter, which forces out unnecessary volatile sulfur.[197] It was thus possible to prepare the precious metals profitably by alchemical means. But although Glauber—as most alchemists before him—promised to discuss his processes openly, he also stated that they were not for the unskilled. As for his most noble

---

[193]Glauber, *Prosperity* (Part I), (English, 1689), Part I, p. 308; (German, 1659), Vol. II, p. 374.

[194]Glauber, *The Consolation of Navigators* (English, 1689), Part I, p. 284; (German, 1658), Vol. I, p. 546.

[195]Glauber, *Continuation of the Miraculum mundi* (English, 1689), Part I, p. 189; (German, 1658), Vol. I, p. 210.

[196]Glauber, *Prosperity* (Part I), (English, 1689), Part I, p. 311; (German, 1659), Vol. II, p. 383. Glauber ascribed the practical part of his process to Lazarus Ercker. On this see also J. R. Partington, *A History of Greek Fire and Gunpowder* (Cambridge: Heffer, 1950), p. 318.

[197]Glauber, *Prosperity* (Part II), (English, 1689), Part I, p. 320; (German, 1659), Vol. II, p. 404.

secret, it was clearly not to be imparted to the public, for in the past when he had taken pity on those ruined by the war and had divulged secrets to them, he had always been betrayed.[198]

The Prosperity of Germany closed with a plea that steps be taken for the coming struggle with the Turks. Tempests, storms, and earthquakes in 1660 had warned of God's wrath, while an abnormal comet in 1662 had indicated the same divine displeasure. Only God knew what the great conjunction in Sagittarius in 1663 would bring.

> *Sagitary* is a martial sign, and deadly, and portends nothing but Dissentions, Seditions, and uproars of War; and therefore prophesieth unto us nothing but bloudy Wars, insomuch that it is much to be feared that this ungratefull World will be consumed, and blotted out as 'twere by Fire, and the Sword, and the anger of God: which evil, God of his mercy turn away from us . . . . That old Proverb may very likely prove true, which the Ancients have pronounced, *viz.* that it will come to pass in the year 1660, that Alchemy will begin to flourish, but will bring along with it such changes for [some] years following, and such dreadfull changes too, that the third part of Men will perish with Famine and Pestilence, the third part will perish with Fire and Sword, and there will be but a third part onely left.[199]

Once again Germans might take some hope for the future. Not only should they look to their natural resources, they should also reflect upon their alchemical heritage, for theirs was the land of Basil Valentine, Jacob Boehme, and the "Monarch of the North," Paracelsus. It should not be forgotten that it was a German Monk, Barthold, who had discovered gunpowder in 1380.

> Neither is there any one ignorant, what great Victories they that well knew its use, obtained over their Enemies afore it was made manifest. But when it became publickly known, then it served both Parties, so that neither part had more advantage by it than the other, and then the deciding the Controversie consisted (next the blessing of God) in the multitude of the Men, of which their power was composed, and not in Art.[200]

Barthold himself had made known the preparation and use of

---

[198] *Ibid.* (Part III), (English, 1689), Part I, p. 380.
[199] *Ibid.* (Part VI), (English, 1689), Part I, p. 432.
[200] *Ibid.*, p. 433.

gunpowder—a potent secret that he surely should have kept. Now a new danger has appeared, for the "*Turks* doe begin more and more to persecute and vex the Christians by their Tyrannical Persecutions, and endeavour even thoroughly to root them out, as they have already made a beginning in *Hungary* of their detestable attempts." But through his chemical skills Glauber had found new weapons by which the Turks could be halted. Should these secrets be openly divulged? Although his friends urged him to make this information known, care still had to be taken, for

> I dare not here trust my Pen too much, though I could disclose them in a few words, and so, as to bring every one to acknowledge the truth hereof, and to feel it as it were with his hands. But I must deal warily, lest such sharp, and yet not killing weapons fall into the Enemies hands.[201]

From the text it is evident that Glauber's new weapons were based upon a liquid acid that could be projected over the enemy forces in the form of a mist or rain. It was only necessary to produce large quantities of "Spirits of Salt," which might be transported with no greater difficulty than water, and the "Canes" through which these acids were to be projected.[202] The result could only be total defeat for any foe.

> Suppose I am in a City or Castle besieged by the Enemies, and that I have by me some of those kind of Instruments which are accustomed to this use for the moist Fires. If now the Enemy should set upon the Trench, Wall, or a Bul-Wark, and I were furnished with that defensive Water of mine, should stand behind the Wall or Bul-wark in that place which the Enemy sets upon, he could not possibly come at me; for those moist Fires may be cast by the said Instruments far without the City, like a fiery showr of Rain, or like a Cloud, which Rain being forced out of but one onely Instrument would dilate it self far and wide upon many hundreds of Men, and whatsoever it touched it would hurt and burn like common natural Fire, and it would especially blind their sight, . . . how would they be able being blinded to get over the Walls or Bul-warks?[203]

These forces would be lucky to rejoin their units, while any that

[201] *Ibid.*
[202] *Ibid*, p 431
[203] *Ibid.*, pp. 437-438.

had made their way into the town could be easily debilitated by women or children furnished with the same weapons. Those blinded would abandon their arms in a desperate attempt to remove the fire from their eyes. Yet this could hardly be done

> ... in half a days time, nay if it should touch their Eyes in pretty quantity, it would scarse be quencht in two days time, and if they should go to wipe their Eyes with their hands, then would they make their torments much more intolerable, and would add more Fire to their Eyes, because their hands are likewise moistened and plagued with that continual shour. And now may not such unbidden Guests be overwhelmed and slain with Stones cast upon them out of the adjoyning Houses? And may not one single Woman with such a little Instrument onely filled with these moist Fires defend her house against an hundred Soldiers.[204]

Glauber suggested also that small "Granadoes" made of copper, iron, or earth and filled with these liquids might be employed by Christian forces for offensive actions. Such a hand grenade "may of its own accord (without any benefit of Gun-powder) leap assunder and fill all the whole room with his blind making Dew." Advance units of soldiers could throw this "into those Watch houses where they keep Guard, and so make all that are there present unfit to fight."[205] The city gates might then be opened for the entrance of the Christian army.

How much better this chemical weapon was than gunpowder, "by which such a multitude of Men are destroyed and slain," for "by this Invention of mine, no man is slain, and yet the victory wrested out of the Enemies hands. And the Enemies being taken alive and made Captives, may be constrained to work, and in my opinion may bring more benefit than if they were slain."[206] In any case, Glauber concluded, "is it not lawfull for us to smite our Capital Enemies the *Turks* with blindness, and to defend our selves, our Wives and Children?"[207]

He foresaw the possibility that some of these new weapons might be sold by traitors or fall into the hands of the enemy in the course of battle. Since it seemed almost inevitable that the secret

[204] *Ibid.*, p. 438.
[205] *Ibid.*, p. 439.
[206] *Ibid.*, p. 433.
[207] *Ibid.*, p. 438.

would somehow be lost in time, it was essential that "Men of a quick piercing Wit" constantly seek to improve existing weapons as well as invent new ones. Should such a program of research be instituted, "I do not question but that hereafter Wars will be waged after another manner than hath hitherto been done, and force must give place to Art. For Art doth sometimes overcome strength." Although Glauber gives enough information here to reconstruct the general outline of his invention, he does not go into detail about the mechanism of his projection devices or the fashion in which the grenades were to be constructed. These secrets—along with others—were only to be obtained from him in person.[208]

Conclusion

The works of John Webster, Hugh Plat, and Johann Rudolph Glauber stand as examples of the breadth of interests expressed by early and mid-seventeenth-century chemical philosophers. Their work is permeated by chemical influences and by a strong belief that mankind's hope for a better future lies in the study and application of chemistry. As van Helmont and his predecessors advocated, a fundamental reform had to be made in education before students could ever be exposed to the real truths of nature and religion—two fields which the chemists insisted were indissolubly united. Only when this educational reform had been accomplished would the full practical benefits of knowledge—especially chemical knowledge—be reaped.

The rapid push to apply the new theory to agricultural practice is seen in the late-sixteenth-century work of Palissy and Plat, who combined the views of Paracelsus on salt and plant growth with their own observations and experiments. Plat was surely motivated by patriotic fervor. That his work was widely appreciated is indicated by the fact that the Royal Society referred to him in company with authorities such as Bacon, van Helmont, and Glauber as they continued this essentially chemical search for a means to improve agriculture.

Of all these men, Glauber may be the most interesting. His theoretical views mark him as both a Paracelsian and a chemical

---

[208]*Ibid.*, pp. 440, 439. " 'Tis not expedient to make any larger discourse concerning these matters, and to take up any more room by revealing more."

philosopher. He is distinctive, however, in his realization that the chemical philosophy should not only provide a true new philosophy of nature but also should become the basis for the economic and military reform of the state. Living in a predominantly agricultural economy, Glauber sought to provide prosperity for the farmer no less than had Palissy and Plat. He suggested the application of salts to the soil in times when manure was in short supply and when prices were high. He urged that in times of plenty, surplus wheat products be converted to liquid concentrates, which were resistant to spoilage and which might later be reconstituted to liquid or solid food products. He further sought a cheap source of saltpeter for the state through the harvesting of forests—a process that was to be coupled with a reliable and efficient method for the transmutation of the base metals to gold and silver. Economic prosperity seemed inevitable for Germany. The state itself was to be protected with a dramatic new military technology—again based upon chemical discoveries—which would make it possible to hold off the advancing Turkish hordes. A permanent research program in this field would ensure military superiority.

It should not be thought that Glauber was unique. The mining and metallurgical texts of the period are practical examples of chemical technology, while Glauber's love of his homeland was perhaps no less than that of the mystical alchemist Michael Maier, who glorified his country through the enumeration of the inventions of his compatriots in his *Verum inventum*.[209] Here Maier praised the works of the German chemists and especially lauded Paracelsus. Johann Kunckel (1630/1638-1703) may also be advanced as an author deeply influenced by current chemical theory who was at the same time concerned with practical problems.[210] His *Ars vitraria experimentalis* (1679) reprinted the earlier work of Antonio Neri (1612) with the notes of Christopher Merrett (1662),

---

[209]The work is discussed by J. B. Craven, *Count Michael Maier* (Kirkwall, 1910; reprinted London: Dawsons, 1968), and by Alex Keller, "Mathematical Technologies and the Growth of the Idea of Technical Progress in the Sixteenth Century," *Science, Medicine and Society in the Renaissance*, ed. Allen G. Debus, 2 vols. (New York: Science History Publications; London: Heinemann, 1972), Vol. I, pp. 11-27 (24).

[210]Partington, *History of Chemistry*, Vol. II, pp. 361-377.

PLATE XXVIII Johann Joachim Becher (1635-1682). W.P. Kilian after anonymous drawing, 1675.

but Kunckel also added numerous experiments of his own which made his work the most valuable account of glassmaking to be produced for over a century.

Finally, note must be taken of Johann Joachim Becher (1635-1682).[211] Trained in medicine, he coupled this interest with one in chemistry. His theoretical views will be touched on in the next chapter, but here it is of interest to note that his marriage to the daughter of an Imperial Councilor in 1662 suddenly opened to him the prospect of an extraordinary career. Named first as *Hofmedicus und Mathematicus* to the Elector of Bavaria, he soon found it profitable to move to Vienna as Imperial Commercial Councilor to Leopold I. Here he built and organized an Imperial arts and crafts center that included a glassworks plus facilities for the manufacture of textiles. It is noteworthy that this center included a chemistry laboratory. Aware also of the importance of technical education for the advancement of the domestic economy, he proposed important educational reforms such as the institution of schools to provide practical instruction in civil and military engineering and statics. Concurrently, eager to increase foreign trade, Becher organized the Eastern Trading Company and proposed colonial settlements in South America. Nor did he neglect a role as alchemical advisor to the Emperor. His important *Physica subterranea* appeared in 1669, followed by supplements in 1671 and 1675. The failure of Becher's mercantilist policies resulted in his dismissal and a short imprisonment. However, by 1678 he had made his way to Holland, where he sold plans to the city of Haarlem for a machine that would spool silk cocoons and then submitted a plan to the Dutch Assembly for the extraction of gold from sea sand through smelting. Although an early test of this process proved encouraging, he soon left for England to study the mines of Scotland and Cornwall. He died only a few years later.

Although Becher's vast economic schemes failed, his work should be thoroughly reexamined. To date it has been appraised

---

[211] Recently discussed by Allen G. Debus in the *Dictionary of Scientific Biography*, Vol. I, pp. 548-551. Becher's life and his socio-economic plans have been most thoroughly discussed by Herbert Hassinger in *Johann Joachim Becher 1635-1682. Ein Beitrag zur Geschichte des Merkantilismus*, Veröffentlichungen der Kommission für Neuere Geschichte Österreichs, No. 38 (Vienna, 1951) and by F. A. Steinhüser in *Johann Joachim Becher und die Einzelwirtschaft* (Nuremberg, 1931).

for the most part either by economic historians intrigued by his mercantilist policies or by historians of chemistry seeking in his books the origin of the phlogiston theory. However, his interest in chemistry and medicine as well as his plans for the economic prosperity and the educational reform of the empire may all be viewed as elements of the current chemical philosophy. A more thorough study of Becher in this context should prove illuminating. And if the work of Becher might be better understood when placed in its proper setting, so too the chemical philosophy might be used by historians as a setting for a reinterpretation of the interests and activities of a number of other chemists and physicians of the period of the scientific revolution.

7
THE CHEMICAL PHILOSOPHY IN TRANSITION:
Toward a New Chemistry and Medicine

[The writings of van Helmont] are so much read, and studied, that now an *Helmontian* seems to overtop a common *Chymist, Paracelsian,* and *Galenist.*

John Webster (1671)*

BY THE MID-SEVENTEENTH CENTURY some argued that only academic upheaval would accomplish the desired establishment of the chemical philosophy and thus ensure the practical benefits expected from a truly Christian understanding of both nature and man. But while these writers sought basic educational reform, agricultural improvements, and a new economic policy tied to chemical research, they never lost sight of a fundamental goal—the uncovering of the still unknown secrets of chemistry and medicine. These two subjects ever had been inseparable for them.

Any evaluation of the chemical work of the last half of the seventeenth century reveals a continuing influence of the chemical philosophy. We see this in the roots of the phlogiston theory of chemistry, which was founded by Georg Ernst Stahl (1660-1734) on the earlier work of Johann Joachim Becher. We further see a clear influence of the Paracelso-Helmontian heritage in the writings of the "corpuscularians." Themes fundamental to an earlier period emerge not only in the work of Robert Boyle and John Mayow, but even in the writings of Isaac Newton. Similarly, mid- and late-seventeenth-century medicine reflects the impact of

*John Webster, Practitioner in Physick and Chirurgery, *Metallographia; or an History of Metals. Wherein is declared the signs of Ores and Minerals both before and after digging, the causes and manner of their generations, their kinds, sorts, and differences; with the description of sundry new Metals, or Semi Metals, and many other things pertaining to Mineral knowledge. As also, The handling and shewing of their Vegetability, and the discussion of the most difficult Questions belonging to Mystical Chymistry, as of the Philosophers Gold, their Mercury, the* Liquor Alkahest, Aurum potabile, *and such like* . . . (London: A.C. for Walter Kettilby, 1671), p. 34.

more than a century of chemical debates. This is evident in the influential iatrochemical texts of Thomas Willis (1621-1675) and Franciscus de le Boë, called Sylvius (1614-1672).

The present chapter will center on these three subjects: the background of the phlogiston theory, specific links that bind the corpuscularians with earlier chemical philosophers, and medical themes in the iatrochemistry of the late seventeenth century. No attempt will be made to discuss these topics exhaustively. Rather, selected subjects will be examined which indicate connections bridging the mid-seventeenth-century scientific watershed and thus establish a continuity of chemical and medical thought.

Chemistry in Mid-Century: Lefèvre (1660) and Rhumelius (1648)

The mid-seventeenth century was a period of intellectual ferment affecting all scholarly subjects. The impressive number of editions of Bacon, Descartes, and van Helmont dating from these years gives evidence that these authors were eagerly read by a large audience. This has been most frequently interpreted in terms of the growth of "modern" science, but there is little doubt that this same intellectual activity resulted in a new interest in the work of the chemical philosophers. The extent of this interest may be dramatically seen by a glance at the English scene. Although van Helmont's collected works had appeared as late as 1648, several of his shorter pieces were already for sale in English by 1650, and the complete *Opera* was available in translation by 1662. As for Paracelsus, the only translations prior to 1650 were a few spurious recipe collections which had been made by John Hester in the closing decades of the previous century. One of these was reprinted in 1633, but between 1653 and 1661 six new works were printed in English, including his important *Archidoxies* and the *Philosophy to the Athenians*. Besides these, there were a host of new translations from continental alchemists and iatrochemical authors, among them Basil Valentine, Glauber, Crollius, Maier, Sendivogius, and d'Espagnet. Elias Ashmole collected all the alchemical texts he could find, and his *Theatrum chemicum Britannicum* (1652), although unfinished, remains one of the fundamental collected editions. Other English authors include Thomas Vaughan, Nicholas

Culpeper, and George Starkey; and it can be argued that both Thomas Willis and Robert Boyle belong in this group. James Howell, responsible for the translations of the *Archidoxies* and the *Triumphal Chariot of Antimony*, announced his intention of making available in English versions of other works by Paracelsus, Isaac Hollandus, Bernard Trevisan, Glauber, Raymond Lull, and the *Rosary of the Philosophers*.[1] Richard Russell, after already having published English versions of the *Basilica chymica* of Oswald Crollius, the *Tyrocinium chymicum* of Jean Beguin, and the works of Geber, promised to translate the whole of the new Latin version of the *Opera omnia* of Paracelsus.[2] Indeed, the flood of chemical works was so great that the publisher William Cooper considered it important to make available *A Catalogue of Chymical Books which have been written Originally, or Translated into English* (1673).[3]

And yet the English scene is but a reflection of the chemical activity elsewhere. With the exception of a few medieval alchemists, only Robert Fludd can be cited as an English chemist of international renown prior to 1650. Most of the major iatrochemical texts had been written on the Continent, and it was at this time that the fundamental edition of the works of Paracelsus prepared by Huser (1589-1591) was finally translated into Latin (1658). Similarly, the works of van Helmont and Glauber went through many editions both in their original language and in translations. Nor were the traditional alchemical texts in a state of noticeable decline. A sixth volume was added to the massive *Theatrum chemicum* in 1661 (earlier editions: four volumes in 1602; four in 1613; a separate fifth volume in 1622; five volumes in 1659-1660), while the smaller—but still authoritative—*Musaeum Hermeticum* appeared in enlarged form in 1678 (first edition, 1625). The number

---

[1] See Howell's introductory section to his translation of Paracelsus' *His Aurora & Treasure of the Philosophers. As Also the Water-Stone of the Wise Men; Describing the Matter of, and manner how to attain the universal Tincture*, trans. J. H. (James Howell) (London, 1659), p. 328.

[2] Richard Russell's translations, and planned translations, are listed in his preface to *The Works of Geber* (London: William Cooper, 1686), sig. A2$^{r, v}$. This is dated May 3, 1678, corresponding with the first edition of this work.

[3] This appeared first as an appendix to *The Philosophical Epitaph of W.C. Esquire* ... (London: William Cooper, 1673). It was reissued separately in 1675 and other lists, supplementary in nature, were appended to his other alchemical publications.

of shorter works was now so large that there was a very real demand for guides to this literature. One result was the 250-page *Bibliotheca chemica* of Petrus Borellius (1656); another was the lengthy history prepared by Olaus Borrichius, the *De ortu et progressu chemiae dissertatio* (1668).

The persistence of traditional alchemy is readily apparent, but it is clear that the iatrochemical work of the preceding century had deeply influenced this subject. All were familiar with the chemical trinity, Paracelsus, van Helmont, and Glauber, but one finds more than references to these giants. This is noticeable in the family of chemical textbooks[4] which had derived from the initial *Tyrocinium chymicum* (1610) of Beguin. This was a work based upon the *Alchymia* of Libavius (1597), and later efforts were no less dependent upon iatrochemical sources. A series of talented chemical demonstrators at the Jardin du Roi in Paris continued to lecture on chemical preparations of medicinal value throughout the late seventeenth and into the eighteenth century. The first of these, William Davidson, appointed in 1648, is no less noted for his *Philosophia pyrotechnia seu curriculus chymiatricus* (1635) than for his new edition (with a commentary that dwarfs the original text) of the work of Peter Severinus (1660).[5] He was followed by Nicholas Lefèvre (1651), Christopher Glaser (1660), and a number of others who prepared textbooks to accompany their lectures. These books—many of them going through a large number of editions—followed a general pattern, beginning with a short section of theory that dealt with the elements and the relationship of chemistry to medicine, then proceeding to a discussion of laboratory equipment. The major part of all of them was reserved for descriptions of chemical preparations of medicinal application deriving from mineral, vegetable, and animal sources—a tripartite division in chemistry that can be traced to the work of al-Razi. The original format of the chemical textbook was never appreciably altered, but the contents were changed over the years to reflect current thought. Thus Nicholas Lemery's *Cours de chymie* (1675) maintained tradition with its familiar tripartite organization but offered

---

[4]Best described by Hélène Metzger, *Les doctrines chimiques en France du début XVIIe à la fin du XVIIIe siècle. Première partie* (Paris: P.U.F., 1923), pp. 35-99. See also J.R. Partington, *A History of Chemistry* (London: Macmillan, 1962), Vol. III, pp. 1-48.

[5]Metzger, pp. 45-50; Partington, pp. 4-7.

innovation as well with corpuscular explanations of chemical reactions. So successful was this work that it was repeatedly published in revised editions until the mid-eighteenth century.[6]

Lemery's textbook surely belongs to the iatrochemical tradition, but the impact of the Parcelso-Helmontian heritage is even more evident in the *Traicté de la chymie* (1660) of his predecessor, Nicholas Lefèvre (c. 1610-1669), who left his position at the Jardin du Roi to establish a chemistry laboratory for Charles II at the Court of St. James.[7] His work was widely appreciated in London, where he was soon elected to the Royal Society.

Lefèvre's textbook was a practical one, no less than others of this genre, but, it carries special interest because of its greater attention to theoretical chemistry. Lefèvre openly acknowledged his debt to Paracelsus, but above all van Helmont and Glauber were "as the two Beacons and Lights which we are to follow in the Theory of Chymistry, and the best practice of it."[8] The subject itself was divisible into three parts: the pharmaceutical preparation of medicines, medical chemistry (*Iatrochymy*), and philosophical chemistry.[9] The work of all great chemists from antiquity, the Middle Ages, and the current age attest to the fact that

> Chymistry is the true Key of Nature....[10]
> [Indeed] Chymistry is nothing else but the Art and Knowledge of Nature it self; that it is by her means we examine the Principles, out of which natural bodies do consist and are compounded; and by her are discovered unto us the causes and sources of their generations and corruptions, and of all the changes and alterations to which they are liable: ... [Further] it is known, that the ancient Sages have taken from Chymistry, the occasions and true motives of reasoning upon natural things, and that their monuments and writings do testifie this Art to be of no fresher date then Nature it self.[11]

[6]Metzger, pp. 281-340; Partington, pp. 28-41.
[7]Lefèvre offered a short autobiography in his prefatory piece "To the Apothecaries of England" in *A Compleat Body of Chymistry*, trans. P.D.C., 2 parts (London: O. Pulleyn, 1670), sig. A4^V-a^V. It is this English edition which will be cited here rather than the first French edition of 1660 or the first English edition of 1662.
[8]*Ibid.*, Part 1, p. 3.
[9]*Ibid.*, p. 7.
[10]*Ibid.*, p. 3.
[11]*Ibid.*, p. 1.

Above all for Lefèvre, chemistry emphasized the real "evidence and testimony of the senses" derived from the laboratory rather than the "bare and naked contemplation" characteristic of the Schoolmen.[12] Should the latter be asked,

> What doth make the compound of a body? He will answer you, that it is not yet well determined in the Schools: That, to be a body, it ought to have quantity, and consequently to be divisible; that a body ought to be composed of things divisible and indivisible, that is to say, of points and parts: but it cannot be composed of points, for a point is indivisible, and without quantity, and consequently cannot communicate any quantity to the body, since it hath none in it self, so that the answer should have concluded the body to be composed of divisible parts. But against this also will be objected, If it be so, let us know, whether the minutest part of the body is divisible or no, if it be answered, Divisible, then it is instanced again, that it is not the minutest, since there is yet a place left for division: but if this minutest part be affirmed to be indivisible, then the answer falleth again into the former difficulty, since it returns to affirm it a point, and consequently without quantity; of which being deprived, it is impossible it should communicate the same to the body, since divisibility is an essential property to quantity.[13]

How different for Lefèvre is chemistry, which rejected "such airy and notional Arguments, to stick close to visible and palpable things." He clearly thought that discussions of atoms or corpuscles could be no more firmly supported than the concepts of the Aristotelians or the Galenists. Only the evidence of the chemical laboratory would lead to a proper understanding of nature.

Lefèvre's chemical principles were the Paracelsian three plus phlegm and earth—a typical five-element system for the period.[14] There was no question in his mind but that a judicious analysis by the fire would yield these substances. He also believed in a universal spirit of life which was whirled about the earth in the air and was thus supplied as a necessity for all living things.[15] He rejected the belief that this spirit of life was common saltpeter and preferred to think of it rather as

[12]*Ibid.*, p. 9.
[13]*Ibid.*, pp. 9-10.
[14]*Ibid.*, pp. 17-29.
[15]*Ibid.*, p. 35.

... a Mysterious Salt, which is the soul of all Physical Generations, a Child and Son of Light, and the Father of all Germination and Vegetation; we confess that Salt in such a respect to be Universal: But we say at the same time, that it is more to be apprehended by the Intellect then the Senses, and that this Divine Salt cannot be comprehended nor hidden under any other covering or shape, then of the Sulphureous and Mercurial Volatile Salt of all natural Bodies, since this Salt is endowed of all the Essential and Centrick Vertues of Sublunary Mixts.[16]

The purpose of generation must then not be simply for the refreshment of bodies but rather to obtain this necessary vital spirit.[17]

The work of Lefèvre will be referred to again in this chapter. It is important to note here, however, that the textbook tradition, although basically a practical one, did transmit Paracelsian theory to the late-seventeenth- and even the eighteenth-century reading public. The work of Lefèvre went through nearly a dozen editions in French, German, and English by the close of the century.

Iatrochemistry was so well established by 1650 that numerous recent books were available for consultation then. An example is the work of the prolific German author Johannes Pharamundus Rhumelius (1597-1661), a spagyric physician who wrote on subjects ranging from the theory of disease and pharmaceutical chemistry to military fortification. His collected medical writings were published as the *Medicina spagÿrica tripartita* in 1648, the same year as van Helmont's *Ortus medicinae*. Rhumelius' *Medicina*, close to eight hundred pages, proved so useful that it required a second printing in 1662.[18]

Rhumelius represents orthodox Paracelsism in mid-century. The unity of nature and the universe was a fundamental theme as he expounded the macrocosm-microcosm concept, which he con-

[16]*Ibid.*, Part 2, p. 251.
[17]Nicasius le Febure, Royal Professor in Chymistry to his Majesty of England, and Apothecary in Ordinary to his Honorable Household, *A Copendious Body of Chymistry*, trans. P.D.C., 2 parts (London, 1664), Part 1, p. 40.
[18]The edition which will be cited is Johannes Pharamundus Rhumelius, *Medicina spagÿrica oder spagyrische Artzneykunst* . . . (Frankfurt: Christian Hermsdorffs, 1662). A French translation by Pierre Rabbe (*Médecine spagyrique*, Paris: Chacornac, 1932) is woefully incomplete and inaccurate. Rhumelius was born at Neumark in 1597 and died at Nuremberg in 1661. He is not referred to by Partington.

sidered the very essence of an understanding of both medicine and nature. The rays of the planets not only govern the course of the microcosm, they stamp their signature on man and at the same time impress all other earthly substances in a similar fashion. Thus it becomes the duty of the physician not only to study the complexions and the humors but also the "planets" of the microcosm. Only when the physician has learned the proper correspondence between the two worlds and the divine signatures will he know properly how to apply medicines.[19]

To accomplish this, he must know the five principles of disease—the *entia morborum* of Paracelsus[20]—but above all, the physician learns through work that is largely chemical in nature. Again we read that the whole process of Creation may best be understood in terms of a chemical separation. All nature, and therefore man himself, derives from salt, sulfur, and mercury.[21] And though the elements cannot be denied, it is the *tria prima* which form the essence of bodies and which the physician must be most concerned with.

Rhumelius clearly sought the cause of disease by means of chemistry. He wrote of the need for "spagyric anatomy," which separates the pure from the impure and thus makes possible those potent medicines which only chemical physicians could prepare.[22] Indeed, the bulk of the *Medicina spagÿrica tripartita* is composed of the ten sections of the "Antidotarium chymicum" and the eleven parts of the "Jatrium chymicum." These 650 pages represent one of the most extensive collections of chemical remedies to be published in the course of the seventeenth century.

As for method of cure, Rhumelius recognized two types in common use: the traditional cure by contraries and the "new Hermetic and Paracelsian medicine" based upon natural magic which asserted that like cures like.[23] It was the latter which was true, wrote Rhumelius, since only this method is founded on the universal law of sympathy and antipathy and observes the magnetic analogy and

---

[19] Rhumelius, *Compendium hermeticum* in *Medicina spagÿrica*, pp. 6, 7.
[20] *Ibid.*, p. 6. The various *entia* are discussed on the pages immediately following.
[21] *Ibid.*, p. 18.
[22] *Ibid.*, p. 46.
[23] *Ein ander Compendium Hermeticum* in *Medicina spagÿrica*, pp. 82-83.

concordance of the macrocosm and microcosm.[24] Again one must turn to the divine signatures, "the correct Alphabet of Nature through which the Book of Nature and of Medicine may be read."[25]

Many works similar to that of Rhumelius might be cited. There is relatively little new here, certainly nothing to compare with the more interesting contemporary thought of van Helmont or Glauber. However, the existence of these comprehensive texts on chemical medicine indicates that there was a continuing demand for works on Paracelsian medicine interpreted primarily through chemistry.

Geocosmic Considerations: F. M. van Helmont (1685) and John Webster (1671)

While medicine emerges as the primary theme in the chemical literature of the seventeenth century, a persistent secondary theme is the study of terrestrial phenomena. The conviction that valid analogies existed between the greater and lesser worlds continued to center attention on geocosmic events. Rhumelius, motivated essentially by a medical interest, wrote a tract on thermal springs, and he was thus led—as many of his predecessors had been—to the problem of accounting for the origin of spa waters and analyzing them chemically. While adding little new, his treatise stands as one of the most concise and clear examples of its genre in the pre-Boylean literature.[26]

Similarly, van Helmont's son, Franciscus Mercurius (1614-1699), was led to a consideration of the earth in his discussion of the macrocosm and the microcosm. Published in 1685, his *Paradoxal Discourses* reflect both current and traditional thought. Convinced of a perpetual circulation and harmony in the great world, the younger van Helmont felt that anything less in man would be unthinkable.

[24]*Ibid.*, p. 83.
[25]*Ibid.*, p. 86.
[26]Rhumelius, *Thermarum et acidularum desciiptio* . . . in *Medicina spagyrica*, pp. 305-336 (308-310).

The Sun, by his descension or influence, generates a fire in the Creatures, which in Man is to be likened to his bloud; but the influential descent of the Moon and Stars, generates a water: both of which are driven about with the self-same circulation in Man, the Microcosm, as they are in the Macrocosm or greater World.[27]

In this work we again note a traditional "chemical" insistence on a spiritual sulfur and saltpeter in the air and the use of these to explain thunder and lightning.[28]

No less than his father did, Franciscus Mercurius had an interest in the growth of vegetation, yet in contrast to his father, he insisted on the necessity of air for the nourishment of trees and plants.[29] He did not extend this postulate to metals in his explanation of their increased weight on calcination; in this case he argued that with a moderate heat lead will take up water and fire. He did note that the original metal could be recovered with a still greater heat.[30] He maintained a great interest in metals and, like his father, made the concept of the elemental sand a fundamental one for his interpretation of the earth. This sand, he wrote,

> ... is continually made in the water, is the foundation of the whole Earth, and of the highest and vastest Mountains, as being that on which they rest and are supported; and is indeed the very Root from whence the whole Earth, all Mountains, and other visible Bodies do arise and have their original ....[31]

With a microscope it is evident that each grain of sand has a different shape, leading to only one answer: each grain must have its own specific seed, since it is this that confers form on substance.[32]

The influence of the elder van Helmont is seldom more evident

---

[27] Franciscus Mercurius van Helmont, *The Paradoxal Discourses . . . Concerning the Macrocosm and Microcosm, or the Greater and Lesser World and their Union, Set down in Writing by J. B. and now Published* (London: J.C. and Freeman Collins, for Robert Kettlewel, 1685), in two parts separately paginated, Part 1, p. 18. See Partington, *History of Chemistry,* Vol. II, pp. 242-243 for references to the scanty secondary literature on F. M. van Helmont.

[28] F. M. van Helmont, *Paradoxal Discourses,* Part 1, p. 35.

[29] *Ibid.,* p. 52.

[30] *Ibid.,* pp. 108-109.

[31] *Ibid.,* p. 36.

[32] *Ibid.,* p. 43.

than in the *Metallographia* of John Webster (1671). Here Webster sought to make available in translation the views of the major authors who had written on this subject. The true knowledge of metals was traced through a series of sages extending from Tubal-Cain through Pythagoras and Plato to the alchemists of late antiquity and the Middle Ages.[33] Webster singled out Paracelsus for special praise, because he had described chemical processes in plain words, thus making alchemy available to a new audience. Before his time this divine subject had been "buried in the Cells of the Friers and Monks, and so came to no great improvement nor perfection."[34] Above all, notice was to be taken of van Helmont, "seeing his Writings are so much read, and studied, that now an *Helmontian* seems to overtop a common *Chymist, Paracelsian,* and *Galenist.*"[35] The works of these admirable authors, Webster added, stood in great contrast to the *Mundus subterraneus* of the Jesuit Athanasius Kircher (1665), which promised much but delivered little. This was a book "stuffed with Scandals and Lies against *Paracelsus, Arnoldus,* and *Lully,* whose Art of Transmutation or Maturation of Metals he laboureth to prove to be false; of which he knows no more then a blind man doth of Colours."[36]

Webster believed implicitly in the growth of metals, and he was quick to say that his own experience confirmed the statements of other authors regarding the existence of a metallic "Gur." He described this substance to a young miner with whom he was well acquainted, and

> ... according to my directions, providing himself of some woodden dishes to take with him, it was not very long ere he brought me a large quantity, found in a trench.... It was (as he most faithfully affirmed) when he first broke the hard stone in which it was inclosed, some of it especially very thin and liquid, so as he could hardly preserve it; and the other as soft as Butter, and the inmost part of that as soft as Butter to my touch and feeling, and the outside more hard; for the longer it lay to the air, the harder it grew. It was a greyish or whitish colour, and would spread with ones finger upon a table, or smooth piece of wood, as like Butter as could be, but not so fatty, or greasie: and as *Helmont* saith, was like unto soft soap, but most of it something harder,

[33] Webster, *Metallographia*, pp. 2-10.
[34] *Ibid.*, p. 16.
[35] *Ibid.*, p. 34.
[36] *Ibid.*, pp. 30-31.

for he had brought it near two miles to me, and though he had made haste, yet it had hardened by the air in the way.[37]

The generation of metals was to be considered a chemical process, and Webster felt that the subject had nowhere been treated with greater truth or detail than in the *Magnum oportet* of van Helmont.[38] After quoting van Helmont on the middle and the final lives of metals, he went on to summarize the process as he understood it:

> But to come to the true efficient cause of the generation of Metals ... the substance seems to be this. That the Solar particles, celestial spirit, or internal and incombustible Sulphur (which is the true fire of Nature) hid in the viscous matter or mercury, and excited and stirred up by the motion of the celestial bodies, central sun, or subterraneous fire or heat (which we shall not take upon us to determine, but leave it to the judgment of the learned Reader) doth generate, perfect, and ripen Metals ... And all do consent that Sulphur is the efficient cause, or father, and Mercury the passive or mother of all Metals.[39]

Webster did not claim to have added anything new to the subject, but his book was welcomed. Although the reviewer in the *Philosophical Transactions* was somewhat less than enthusiastic about Webster's espousal of transmutation, he paid tribute to the author's diligence in collecting and describing "the signs of Ores and Minerals both before and after digging."[40] In a chapter on the authors of books on minerals, Daniel Georg Morhof allotted more space to Webster than to anyone else. For him the *Metallographia* was of the utmost value for having collected the opinions of the ancient as well as the modern writers.[41]

J. J. Becher's *Physica subterranea* (1669)

The significance attached to essentially geological phenomena is nowhere more apparent than in the work of Johann Joachim

---

[37]*Ibid.*, pp. 51-52.
[38]*Ibid.*, pp. 80-81.
[39]*Ibid.*, p. 83.
[40]*Philosophical Transactions of the Royal Society of London,* **5** (1670), No. 66, 2034-2036 (2036).
[41]Daniel George Morhof, *Polyhistor, literarius, philosophicus et practicus* (4th ed.,

Becher. This is seen throughout the chemical works of this prolific author, but it will suffice to refer to the *Physica subterranea* (1669).[42] That Becher himself thought that this work was important is reflected in his preparation of three supplements (1671, 1675, 1680). The four parts were subsequently combined into a second complete edition (1681), and Becher also prepared his own translation, the *Chymisches laboratorium Oder Unter-erdische Naturkündigung* . . . (1680), which includes the first two supplements plus his *Oedipus chimicus* (first edition, 1664), a short piece which had already become a standard text on the elements, the principles, and chemical processes. The *Physica subterranea*—and the work of Becher as a whole—was, however, to become far more widely known after its "discovery" by Georg Ernst Stahl. Considering the *Physica subterranea* one of the most important chemical texts ever written, Stahl had it published in a new edition along with his own lengthy *Specimen Beccheriana* in 1703. Here, and in a later edition (1738), Becher's text with Stahl's commentary served as an introduction to phlogiston chemistry for many eighteenth-century chemists.

In singling out the *Physica subterranea* as a landmark volume, it would be misleading to divorce it from its sources. Examined in its historical context, it may be judged as a rather typical iatrochemical work. The method of the chemist is characterized as "realiter," while that of the Aristotelian is "verbaliter."[43] Again reminiscent of the Paracelsian literature, Becher's approach relied on the two-book concept. Becher went on to state that students must become thoroughly conversant with the two great macrocosmic laboratories, Nature and Art (the subterranean and the superterranean laboratories).[44] These two must be used together,

---

Lübeck: Peter Boeckmann, 1747); Book 2, Part 2, Ch. 29, sec. 4, pp. 402-404 (402).

[42]Again the reader is referred to the sketch on Becher by the present author in the *Dictionary of Scientific Biography*, Vol. I, pp. 548-551, where the major primary and secondary sources are discussed in a short descriptive bibliography. Two editions of the *Physica subterranea* have been consulted, that published by J. L. Gleditschium at Leipzig in 1703 and that published by Weidmann in the same city in 1738. Both contain Georg Ernst Stahl's *Specimen Beccherianum*.

[43]Becher, *Physica subterranea* (1703), p. 243.

[44]*Ibid.* (1738), p. 1.

since it is the duty of the natural philosopher to observe nature and then make use of his chemical laboratory in an effort to understand the world. There is in fact a fundamental need to use chemistry for the proper understanding of medicine and for the preparation of pharmaceuticals, but it should always be borne in mind that these fields are essentially practical. Theoretical studies are based fundamentally upon the study of subterranean phenomena.[45]

Once again with Becher we read that the study of nature properly begins with an explanation of the Mosaic account of the Creation.[46] By a gravitational effect the universe has been brought into being from the initial chaos into five principal regions of matter: sidereal, meteoric, and animal, vegetable, and mineral.[47] The creation of light has a special significance because this is nothing less than heat. The resultant duality of heat and cold made possible rarefaction and contraction—forces which are responsible for the origin of motion.[48]

As with all theoretical iatrochemists, the problem of the elements was a basic one for Becher.[49] He had little respect for the four Aristotelian elements as they were commonly taught, and like others before him, he rejected fire as a true element. Little better were the efforts of van Helmont—and by this date also Robert Boyle (in the *Sceptical Chymist*)—to show the elemental nature of water through the growth of vegetable substances. Becher rejected their belief that water could be changed into earth through vegetable growth, and he explained that the willow tree experiment could best be understood in terms of earth being carried by the water into the substance of the tree.[50] Similarly, he argued that observations show that the philosophical attributes of the Paracelsian triad have little in common with ordinary salt, sulfur, and mercury, so that they could not really be "principles."[51] However, he felt that on practical grounds, because of their familiarity, their

[45] *Ibid.*, p. 4.
[46] *Ibid.*, pp. 6 ff.
[47] *Ibid.*, p. 11.
[48] *Ibid.*, p. 13.
[49] *Ibid.*, pp. 19 ff.
[50] *Ibid.*, p. 87.
[51] *Ibid.*, p. 84.

use might be defended, and in the *Oedipus chimicus* (1664) he did suggest that sulfur was analogous to earth, and salt to water, while earth and water, in more subtle form, were mercurial in nature.[52]

Becher believed that air, water, and earth were the true elementary principles.[53] Since air was primarily an instrument for mixing, however, the last two formed the real basis of all material things. Thus, a mixture of water and earth (with air) will give rise to stars, animals, vegetables, and most minerals; similar mixtures of waters result in snow, frost, and meteors, while earthy mixtures form metals and stones. While the essential substance of subterranean bodies is earthy, there is a need for three types of earth in metals and minerals; one for substance, another for color and combustibility, and a third, more subtle, for form, odor, and weight. These are, respectively, the *terra vitrescible*, the *terra pinguis*, and the *terra fluida*, earths that have been improperly identified with the Paracelsian principles salt, sulfur, and mercury.[54]

The concept of *terra pinguis* as a fatty, oily, and combustible earth occurs in the older alchemical literature, and Becher also called it sulfur $\phi\lambda o\gamma\iota\sigma\tau \acute{o}\varsigma$. Here again he was following a long tradition, since as an adjective meaning "inflammable" $\phi\lambda o\gamma\iota\sigma\tau \acute{o}\varsigma$ is found in the works of Sophocles and Aristotle.[55] More recently $\phi\lambda o\gamma\iota\sigma\tau o\nu$ had been used in adjectival form by Hapelius (1609) and Poppius (1618), while Robert Boyle had cited Sennert's use of $\phi\lambda o\gamma\iota\sigma\tau \grave{o}\nu$ (1619) in *The Sceptical Chymist* (1661). Even van Helmont, in the *Tumulus pestis*, stated that "Sulphur, in the age of Hippocrates was called Phlogiston, that is inflameable . . . ."[56] Although Becher insisted that each combustible body must contain the cause of its combustibility within itself, he had no clearly defined position on the role to be played by this substance in the

[52]Becher, *Oedipus chimicus obscuriorum terminorum & principiorum chimicorum mysteria & resolvens* in J.J. Manget, ed., *Bibliotheca chemica curiosa*, 2 vols. (Geneva: Chouet, G. De Tournes, Cramer, Perachon, Ritter, & S. De Tournes, 1702), Vol. I, pp. 306-336 (315).

[53]Becher, *Physica subterranea* (1738), p. 19.

[54]*Ibid.*, pp. 60-76 (61, 66, 76).

[55]Becher's use of the word *phlogiston* is summarized by Partington in *History of Chemistry*, Vol. II, p. 646, and the earlier use of the Greek word is discussed on pp. 667-668.

[56]J. B. van Helmont, "Tumulus pestis" (Ch. 17), *Opuscula medica inaudita* (1648), p. 80; *Oriatrike* (1662), p. 1154.

burning process. Generally he spoke of the rarefaction of the burning substance through the dissolving power of flame, and he gave relatively little attention to the contribution of air. He was aware that metals grew heavier when calcined: like Boyle, he credited this to the addition of ponderable fire particles.[57]

Becher wrote in support of spontaneous generation,[58] and he believed firmly in metallic transmutation.[59] He wrote at great length on the solutive process in animals (putrefaction), vegetables (fermentation), and minerals (liquefaction).[60] Important though fermentation was to him, this process did not play the central role in his understanding of nature that it had for van Helmont. Becher defined fermentation as the rarefaction of dense bodies leading to perfection. It could not progress in closed vessels without a necessary supply of fresh air,[61] and it was distinct from putrefaction, in which a mixed body was completely broken down and destroyed.

Becher was a thoroughgoing vitalist who accepted the common belief that metals grow in the earth. He defended the macrocosm-microcosm analogy[62] and gave a lengthy comparison of animals, vegetables, and minerals.[63] Metals and minerals have their seed arising from the principles, they are nourished by a mineral vapor, and they have their own form of excrement. Yet, these are subterranean phenomena and require an understanding of the earth itself.

Turning then to the constitution of the earth, Becher was immediately faced with two problems long discussed by chemists: the origin of springs and the concept of the central fire. He rejected the central fire entirely, arguing that no air existed in the middle of the earth.[64] More recent opinions had suggested that water rather than fire resided there: for example, the mechanical model of Gaspar

---

[57] Partington, *History of Chemistry*, Vol. II, p. 648.

[58] *Ibid.*, p. 651.

[59] He persuaded the authorities in Haarlem to purchase a process from him for the transmutation of silver into gold (1678). This is discussed in the supplements to the *Physica subterranea*.

[60] These Becher discusses successively in the *Physica subterranea* (1703) starting on p. 261.

[61] He defined fermentation as "quod corporis densioris rarefactio, particularumque aërearum interpositio" (*ibid.*, p. 313).

[62] As an example, see *ibid.*, p. 212.

[63] *Ibid.*, p. 257.

[64] *Physica subterranea* (1738), p. 28.

Schott (1608-1666) postulated an internal terrestrial system of tubular channels, while van Helmont preferred to think of the internal waters containing a living spirit similar to human blood.[65]

Neither of these systems seemed entirely satisfactory to Becher. Nor could he accept the common analogy of mountains acting as terrestrial alembics perpetually condensing aqueous vapors rising from the center.[66] Rather, he suggested that the waters fall to the center of the earth through the gravitational force. But since laboratory experience shows that vapors arise when liquids are transferred from hot containers to cold ones, Becher reasoned that hot waters from the surface are rapidly cooled at the center of the earth, where they vaporize and return to the surface. There a portion is condensed and appears in the form of mountain streams. Elsewhere separate vapors from the interior reach higher temperatures through mixing before they erupt as hot springs or medicinal spas.[67] But in a fashion similar to this circulation of water other internal exhalations progress toward the formation of metals as their final product. Thus, to the extent there is a perpetual and universal circularity in nature one may properly compare such macrocosmic phenomena with chemical distillations.[68] This, indeed, is the proper meaning of the ascensions and descensions referred to in the *Emerald Table* of Hermes Trismegistus.[69]

G. E. Stahl and Chemical Tradition

Becher's title, the *Physica subterranea,* very aptly characterizes his volume, which consistently emphasizes themes long considered to be central to the chemical philosophy. However, if this is true for Becher, it would be an oversimplification to picture Georg Ernst Stahl (1660-1734) either as a Paracelsian or a typical late-seventeenth-century iatrochemist.[70] In the case of Stahl, historians

---

[65] *Ibid.*, p. 26.
[66] He cited Fromondus on this concept (*ibid.*, p. 35).
[67] *Ibid.*, pp. 34, 35.
[68] *Ibid.*, pp. 49, 50.
[69] *Ibid.*, p. 52.
[70] Stahl's chemistry is discussed at length by Partington, *History of Chemistry,* Vol. II, pp. 652-690. Still excellent is Hélène Metzger, *Newton, Stahl, Boerhaave et la doctrine chimique* (Paris: Félix Alcan, 1930), pp. 91-188. The reader is referred also

of chemistry have again chosen to emphasize the innovations which characterize his work rather than discuss his links with the past. To some extent this may be justified. Surely Stahl's rejection of chemical explanations in medicine represents an abrupt change in the iatrochemical tradition. The chemical medicine of an earlier period was suspect to this man, who postulated a deep gulf between living beings (possessed with a soul) and the inorganic world. It is thus understandable to find that he did not accept the belief that metals grow and perfect themselves in the earth over a period of time.

And yet it is possible to overemphasize what is new in the work of Stahl. For instance, his rejection of the Helmontian *archeus* in favor of the direct action of the sensitive soul on local chemical processes in the body[71] may be characterized more as a problem in semantics than a significant alteration in medical or scientific theory. As for his chemistry—and therefore much of eighteenth-century chemistry as well—it would be difficult to argue that Stahl's work was not strongly colored by his predecessors. It is evident that he knew well the books of Paracelsus and van Helmont, and he was clearly impressed by the work of Isaac Hollandus.[72] Among more recent authors he was particularly influenced by Kunckel and Becher. It was the former's *Nützliche Observationes* (1676) which first awakened his interest in chemistry.[73] And if in future years he was to persistently point out the errors to be found in this author, it is important to note that at all times Stahl hoped to develop, like Kunckel, a total philosophy of nature. In his *Laboratorium chymicum* (1716) Kunckel could still write:

> Chemistry is without contradiction one of the most useful arts and it would be no exaggeration to call it the mother or the instructress of

---

to David Oldroyd, "An Examination of G. E. Stahl's Principles of Universal Chemistry," *Ambix*, **20** (1973), 36-52. For Stahl's medical thought, see Lester S. King, "Stahl and Hoffmann: A Study in Eighteenth Century Animism," *Journal of the History of Medicine and Allied Sciences*, **19** (1964), 118-130; Walter Pagel, "Helmont. Leibniz. Stahl," *Sudhoffs Archiv für Geschichte der Medizin*, **24** (1931), 19-59; Albert Lemoine, *Le vitalisme et l'animisme de Stahl* (Paris: Germer Baillière, 1864).

[71]Partington, *History of Chemistry*, Vol. II, p. 656.

[72]Whose work was added as an appendix to Stahl's *Fundamenta chymiae dogmaticae* (1723).

[73]Partington, *History of Chemistry*, Vol. II, p. 652.

PLATE XXIX An allegorical representation of a vital universe having a valid relationship between its macrocosmic and geocosmic phenomena. Frontispiece, Joh. Joachim Becher, *Physica subterranea* ... with Georg Ernst Stahl, *Specimen Beccherianum* (Editio Novissima, Leipzig: Weidmann, 1738).

other arts; she alone can teach us to interpret the Sacred Scriptures; she alone teaches us the work of God; and it is thanks to her that we understand the Creation and the material world; Physic and medicine are her dependents; and again, she serves as the foundation for the science of animals and vegetables.[74]

Partington has noted that it was after reflecting upon Kunckel's discussion of the principles that Stahl was led to his doctrine of phlogiston.[75] Noting that Kunckel had only postualted mercurial, saline, and earthy principles in metals, Stahl felt that the omission of a sulfureous principle made this system unworkable. Turning then to the work of Becher he found an elemental system far more to his liking—one which he accepted with little question or criticism. The key to his conversion was Becher's system of three earths, including the *terra pinguis*, which Stahl readily identified as the sulfureous principle and renamed phlogiston.[76] Here we may point to a direct connection between the phlogiston theory and the traditional Paracelsian preoccupation with element theory, but it should be noted that phlogiston became the basis of a new theoretical system and method of explanation for Stahl. This conceptual scheme centered on the processes of oxidation and reduction, and it was supported by a considerable body of observational evidence that seemed to confirm the system as a whole.

Stahl sought a chemical philosophy of nature divorced from medicine, and the *Physica subterranea* served his purposes well. Here Becher had largely confined himself to inorganic processes,[77] that is, to a "subterranean" system. Furthermore, he had refrained from the all too frequent practice of preparing an alchemical cookbook, and he had not become embroiled either in the trivialities of pharmaceutical chemistry or medical theory.[78] The *Physica subterranea* had progressed from an interpretation of the

[74]Quoted in Metzger, *Newton, Stahl, Boerhaave*, p. 95.

[75]*History of Chemistry*, Vol. II, p. 666.

[76]G. E. Stahl, *Specimen Beccherianum* (appended to Becher's *Physica subterranea* (Leipzig: Joh. Ludov. Gleditschium, 1703), p. 94. "Ego *Phlogiston* appelari coepi; Nempe *primum Ignescible*, inflammabile, directe atque eminenter ad Calorem suscipiendum atque fovendum habile Principium: . . ." (p. 39).

[77]As noted by Stahl (*ibid.*, p. 3).

[78]*Ibid.*, p. 4.

divine Creation to a study of the earth, its structure and contents, and then on to the various types of mixed bodies and the theory of solution. The latter subjects were clearly of most interest to Stahl, whose chemistry always emphasized inorganic nature. Thus, although fermentation remained for Stahl a major chemical process,[79] he discussed it primarily in terms of chemical processes—in the Becherian tradition—rather than as a fundamental process for growth and change in all spheres of nature, as an Helmontian philosopher would have approached the subject. Ever convinced of the importance of theory in chemistry, Stahl lauded Becher for having distinguished between mixts, simple bodies, and compounds—indeed, for first having formulated a truly comprehensive chemical theory.[80]

It is of interest that Stahl criticized the Aristotelians, here in a fashion reminiscent of Lefèvre. Following their master, he wrote, the Schoolmen had freely discussed the infinite divisibility of bodies and had gone on to interpret nature in terms of mathematical distinctions and physical divisions. Little better were the discrete particles or atoms of Democritus. These had become a subject of central interest to many natural philosophers who had attempted to explain both physical and chemical phenomena by arguments based upon the supposed shapes of these corpuscles.[81] But surely this was all ridiculous, since nothing definite could be known about the actual shapes of bodies of this magnitude. These explanations could safely be rejected as "mathematical." Geometrical and mechanical demonstrations may often appear convincing, but ultimately the investigator is forced to go back to experimental chemistry to seek true demonstrations and real knowledge of substances.

We may properly look for signs of change and a gateway to

[79] Here see Stahl's *Zymotechnia fundamentalis, seu fermentationis theoria generalis, qua nobilissimae hujus artis, & partis chymiae, utilissimae atq̃: subtilissimae, causae & effectus in genere, ex ipsis Mechanico-physicis principiis, summo studio eruuntur, simulque experimentum novum sulphur verum arte producendi, et alia utilia experimenta atque observata, inseruntur (s.1.* [Halle: Christop. Salfeld, 1697]).

[80] Partington, *History of Chemistry*, Vol. II, pp. 656, 665.

[81] This argument is discussed by Partington (*ibid.*, p. 656) on the basis of the *De vera diversitate corporis mixtis et vivi* (1707). Metzger (*Newton, Stahl, Boerhaave*, pp. 101-103) discusses Stahl's attack on the mechanical philosophy on the basis of the relevant passages in the *Specimen Beccherianum*.

eighteenth-century chemistry and medicine in the work of Stahl. It would be impossible to simply equate phlogiston chemistry with the Paracelsian chemical philosophy, even though the concept of phlogiston itself was an old one. Yet at the same time even Stahl is best understood in relation to his background. Stahlian animism in medicine had its roots in Paracelso-Helmontian medical theory, and Stahlian chemistry derived openly from Becher's *Physica subterranea*, which was squarely in the mainstream of seventeenth-century chemical thought. Stahl's rejection of the "mathematical" mechanical philosophy with its special emphasis on unproven corpuscular explanations links him with his experimental and antimathematical chemical predecessors. His search was for a new chemical philosophy, but his solution was one that connects him closely with earlier chemists and iatrochemists.

The Chemical Corpuscularians: Walter Charleton and the Chemical Philosophy

The third quarter of the century was one of rapid change in the sciences. It was a period during which the mechanical philosophy became a major concern for those who sought a new philosophy of nature. And yet, among those chemists we note particularly as corpuscularians, there is much evidence to show that they were deeply versed in the earlier literature. This subject cannot be investigated exhaustively here, but reference to the work of several major authors should indicate the relationship.

The work of Walter Charleton (1620-1701) has received special attention in recent years, particularly because of his lengthy exposition of the mechanical philosophy in the *Physiologia Epicuro-Gassendo-Charltoniana* (1654).[82] And yet, if Charleton was a con-

---

[82]The reader is directed particularly to the works of Robert Kargon: "Walter Charleton, Robert Boyle, and the Acceptance of Epicurean Atomism in England," *Isis*, **55** (1964), 184-192; *Atomism in England From Hariot to Newton* (Oxford: Clarendon Press, 1966), pp. 77-92; Walter Charleton, *Physiologia Epicuro-Gassendo-Charltoniana* . . . Introduction and indexes by Robert Hugh Kargon (New York/London: Johnson Reprint, 1966). In addition to Kargon's treatment, I have leaned heavily on the following paper by one of my students: Nina Rattner Gelbart, "The Intellectual Development of Walter Charleton," *Ambix*, **18** (1971), 149-168.

vinced corpuscularian at that time, he had been an ardent Helmontian only four years earlier. Educated at Magdalen Hall, Oxford, Charleton had turned to medicine and was awarded the degree of Doctor of Physick in 1643. It was in the following year and in 1648 that the works of van Helmont appeared, and the youthful physician applied himself immediately to understanding these texts. Charleton's study of van Helmont resulted in an original work plus the first translations of this author to be made into English. Charleton's *Spiritus Gorgonicus* (1650) was thoroughly Helmontian in nature, a discussion of the universal stone-forming spirit which resulted in the growth of both macrocosmic and microcosmic concretions. Here is reflected van Helmont's "De lithiasi" and his views on tartar, but references are made throughout the text to Paracelsus, Severinus, and other iatrochemists.[83]

The *Ternary of Paradoxes*, published also in 1650, with a revised edition appearing later in the same year, included Charleton's translations of three Helmontian tracts.[84] His choice is a revealing one, for in addition to the "Imago Dei" he turned his pen again to the question of tartar to produce an English version of the "Tartari vini historia." Most interesting, however, is the fact that Charleton prepared a long introduction and an appendix to his translation of the highly controversial "De magnetica vulnerum curatione," which he placed first in this trilogy. Here he referred repeatedly to Digby, Severinus, Hartmann, Kircher, Cabeus, Porta, and to "that Torrent of Sympathetical Knowledge," Robert Fludd.[85] As for the "sagacious" van Helmont, Charleton felt that it would be improper to question even his most unexpected statements because

---

[83] Walter Charleton, *Spiritus Gorgonicus, vi sua axipara exutus; sive de causis, signis, & sanatione litheaseωs, diatriba* (Leiden, 1650), pp. 5, 6, 9, 43, 45, 55, 154-155, and passim.

[84] J. B. van Helmont, *Ternary of Paradoxes. The Magnetick Cure of Wounds. Nativity of Tartar in Wine. Image of God in Man,* translated, illustrated and ampliated by Walter Charleton (London: James Flesher for William Lee, (1650): to be cited as 1650a. J.B. van Helmont, *A Ternary of Paradoxes, Of the Magnetick Cure of Wounds, Nativity of Tartar in Wine. Image of God in Man,* second impression, more reformed, and enlarged with some Marginal Additions, trans. Walter Charleton (London: James Flesher for William Lee 1650): cited henceforth as 1650b.

[85] Charleton, *Ternary*, 1650a, sig. E3^V.

of the "poverty of our reason compared to the wealthy harvest of his."[86]

For Charleton, a new observational science founded upon chemistry and the truths of universal attraction would be essential for a proper understanding of nature. He found himself particularly attracted as well by van Helmont's investigation of the acid-base dichotomy.

> Now Pyrotechnicall Philosophy, and the Mechanick Experiments of Chymistry, have sufficiently instructed us: that every *Acidum* is, at first encounter, subdued by any *Alchahal*, or *Lixiviall Salt*; as is autoptically demonstrable, in the suddain Transformation of the *Spirit of vitrioll* into *Alumen*, by the inspresion of *Mercury* dulcified. The same effect may be also exemplified in the *Acid Spirit of Sulphur*, which for ever loseth its native *Acidity*, and in a moment degenerateth into an alumnious sweetnesse, at the Conjunction of the *Salt of Tartar*, resolved into an oyle by Deliquium.[87]

And here again it was van Helmont's glory to have understood the medicinal importance of this fundamental chemical reaction.

For all of Charleton's open acceptance of the chemical philosophy in the *Ternary*, there is also an interest in atomistic explanation. Sympathetic cures are real and true, but they are best explained in terms of corporeal atoms which contribute to a union of the blood of the wounded person with the blood on the cloth.[88] There is magnetic attraction—action at a distance—to be sure, but no cure could occur without actual physical contact. Charleton was convinced even at this time that "Every mixt Body, of an unctious composition, doth uncessantly vent, or expire a circumferentiall streame of invisible Atomes, homogeneous and consimilar, that is of the same identical nature with it selfe."[89] While these atoms were specific for any given substance, their properties were not dependent solely upon size, shape, and motion, as were those

---

[86]*Ibid.*, p. 95. And again, "I highly admire the sagacity of *Helmonts* wit . . ." (*Ternary*, 1650b, sig. E2^r).

[87]Charleton, *Ternary*, 1650a, sig. d3^v.

[88]*Ibid.*, 1650b, sig. D2^r.

[89]As quoted by Gelbart, "Intellectual Development," p. 157 from Charleton, *Ternary*, 1650a, sig. E3^v.

of Democritus and Epicurus; nor were they endowed with vital powers in the sense of the Helmontian semina.[90]

Charleton's ensuing works indicate an ever-increasing disenchantment with van Helmont, the Paracelsians, and the alchemists. His fourth Helmontian translation, the *Deliramenta catarrhi*, appeared late in 1650. By this time he lauded van Helmont only for having been "*stronger at* Demolishing *the Doctrines of the Antient Pillars of our Art, then* Erecting *a more substantial and durable* Structure *of his own, his* Witt *more acute and active at* Contradiction, *then his* judgement *profound and authentick at* Probation.*"*[91] It was evident that van Helmont was to be consulted primarily for his devastating attack on ancient medicine and philosophy rather than for the foundation of a new science. In his preface to this work Charleton cited a new source, the mechanist Thomas Hobbes,[92] and two years later, in his *Darkness of Atheism Dispelled by the Light of Nature*, he gave evidence of having turned to an entirely new set of authors: Copernicus, Gilbert, Galileo, Mersenne, Descartes, and Harvey.[93] Here Epicurean atoms are extolled as the proper basis of a new science of nature, since other philosophies evaporate in a cloud of metaphysical opinions. Yet, he added, the mysteries of life could not simply be reduced to a question of the size, shape, and motion of atoms; for the understanding of these problems the Helmontian *archei* were still essential. Indeed, Charleton continued to adhere to a life spirit inherent in the sun, in words that differed little from those of earlier Hermeticists.[94]

By 1654 Charleton had moved still further toward a mechanist position. Not only were Paracelsus and his arch-disciple Oswald Crollius to be condemned, even van Helmont was now termed "Hair-brain'd and Contentious."[95] As for the weapon-salve, this was of little value. Those who wrote of its effect through the universal *anima mundi* lived in a dream world, and even his own earlier

---

[90]Gelbart, p. 157.

[91]J. B. van Helmont, *Deliramenta catarrhi: or, The Incongruities, Impossibilities, and Absurdities Couched under the Vulgar Opinion of Defluxions,* Translator and Paraphrast, Dr. Charleton (London: E. G. for William Lee, 1650), sig. A4^r.

[92]*Ibid.*, sig. a1^v.

[93]Gelbart, "Intellectual Development," p. 160.

[94]*Ibid.*, p. 162.

[95]Charleton, *Physiologia Epicuro-Gassendo-Charltoniana*, p. 58.

atomistic explanation could not be defended, since experiments showed that the weapon-salve was a hoax.

> This Verdict, I praesume, was little expected from *Me*, who have, not many years past, publickly declared my self to be of a *Contrary* judgment; written profestly in Defence of the cure of wounds, at distance . . . And glad I am of this fair opportunity, to let the world know of my *Recantation* . . . and to endeavour the Eradication of any Unsound and Spurious Tenent, with so much more of readiness and sedulity, by how much more the unhappy influence of my Pen, or Tongue hath, at any time, contributed to the Growth and Authority thereof.[96]

In the *Physiologia*, as well as in other works later in the decade, Charleton rejected noncorporeal causes of phenomena and insisted on "mechanical" explanations based upon considerations of matter and motion. But the *Physiologia* was essentially limited to a discussion of physical problems. When confronted with the need for an explanation of digestion or the motion of the heart, Charleton still felt free to fall back upon inherent "faculties" as a cause for these effects.[97] In short, an understanding of Charleton the atomist requires an understanding of a younger Charleton who could be called a Helmontian chemical philosopher.

The "Helmontian" Robert Boyle

The influence of the chemical philosophy on the young Walter Charleton finds a parallel in the work of his contemporary Robert Boyle (1627-1691), who is also remembered primarily as a mechanical philosopher.[98] Boyle's early sojourn on the Continent followed by his intensive studies in England and Ireland had made

---

[96]*Ibid.*, p. 382.

[97]Gelbart discusses Charleton's later publications of the 1650s in "Intellectual Development," pp. 166-168.

[98]There is a voluminous literature on Robert Boyle. The most extensive recent "life" is that of R. E. W. Maddison, *The Life of the Honourable Robert Boyle* (London: Taylor and Francis, 1969). However, still of great value is Thomas Birch's *The Life of the Honourable Robert Boyle*. This has been used as it appears in Robert Boyle, *The Works of the Honourable Robert Boyle*, 6 vols. (London: J. and F. Rivington *et al.*, 1772), Vol. I, pp. v-ccxviii. The complete *Works* appeared first in 1744, but the 1772 edition will be cited here.

him fully conversant with all aspects of current scientific and medical thought. And while he was willing to place Gassendi and Mersenne among his favorite authors as early as 1647,[99] he was at the same time deeply interested in utopian thought and was urging Hartlib to prepare translations of Andreae's *Christianopolis* and Campanella's *Civitas solis*.[100]

In a letter to his brother, Lord Broghill, Boyle in 1652 confessed that "those excellent sciences, the mathematics," were "the first I addicted myself to, and was fond of," but that "experimental philosophy with its key, chemistry," had succeeded "them in my esteem and applications."[101] His rejection of mathematics in favor of experimental studies is a familiar approach among Paracelsian authors, and it indicates that other themes reflecting the influence of contemporary chemical thought might be found in his early works. This proves to be the case, especially in *The Sceptical Chymist* and *The Usefulness of Experimental Philosophy*. Although the latter was not published until 1663, Boyle tells us in his "Advertisement" to the reader that the first part of *Some Considerations Touching the Usefulness of Experimental Natural Philosophy* had been completed when he was "scarce above 21 or 22 years old" (1648 or 1649).[102] But while he could already speak of the human body as an "engine" and suggest that some illnesses occur strictly from mechanical causes (e.g., motion, in seasickness),[103] the bulk of the work is cast in a more familiar fashion.

The young Boyle tells us that the natural philosophy of the Schools is relatively easy to master since it is confined to rote learning from books.[104] How different this is, he addressed his listener, significantly named "Pyrophilus," to the experimental philosophy. Yet, no matter how difficult the latter may be, it is to this that one must have recourse, since the heathen Aristotelians continued to explain nature without allowing a place for God in his

---

[99] Boyle to Hartlib (May 8, 1647), cited by Birch in Boyle, *Works*, Vol. I, p. xli.
[100] Boyle to Hartlib (Apr. 8, 1647), cited by Birch, *ibid.*, p. xxxvi.
[101] Boyle to Lord Broghill (June 19, 1652), cited by Birch, *ibid.*, p. xlix.
[102] Robert Boyle, "Some Considerations Touching the Usefulness of Experimental Natural Philosophy" (Part I, "The Author's Advertisement about the following Essays") in *Works*, Vol. II, p. 4.
[103] *Ibid.* (Part II, Sect. 1), p. 180.
[104] *Ibid.* (Part I), p. 5.

PLATE XXX Robert Boyle (1627-1691). Line engraving by B. Baron from the painting by J. Kerseboom. From the frontispiece to the first volume of the *Works* of Robert Boyle (1744).

own Creation.[105] In reality, the study of nature instructs us in the knowledge of our Creator and thus becomes an act of devotion and praise.[106] Those who doubt this need only turn to their Bibles, where God has encouraged man to study nature by presenting an account of the Creation in the first chapters of Genesis. In simplest terms, the world is a temple and man its priest.[107] The postulation of atoms, for instance, is insufficient by itself to explain many natural phenomena, writes Boyle. How could one hope to explain the regular occurrence of crystals or the growth of minerals without a knowledge of the seminal principle regulating the appropriate life processes?[108] This alone negates the atheistic viewpoint of the Aristotelians.

Not only will man benefit from a greater knowledge of God through the study of nature, he will gain a new mastery over nature and learn to command its forces. Nowhere will this be more evident than in medicine, where the new knowledge will be quickly harvested as an improvement in the health of mankind.[109] Once again we are told to study both macrocosmic and microcosmic phenomena:

> ... that the naturalist's knowledge may assist the physician to discover the nature and causes of several diseases, may appear by the light of this consideration, that though divers Paracelsians (taught, as they tell us, by their master) do but erroneously suppose, that man is so properly a microcosm, that of all the sorts of creatures, whereof the macrocosm or universe is made up, he really consists; yet certain is it, that there are many productions, operations, and changes of things, which being as well to be met with in the great, as in the little world, and divers of them disclosing their natures more discernibly in the former, than in the latter; the knowledge of the nature of those things, as they are discoverable out of man's body, may well be supposed

---

[105] *Ibid.*, p. 36.

[106] *Ibid.*, pp. 6-8.

[107] *Ibid.*, pp. 19, 32.

[108] *Ibid.*, p. 44. Boyle's doubts concerning atomic theory follow on the succeeding pages: see esp. p. 49.

[109] *Ibid.* (Part II, Sect. 1, "... Of its Usefulness to Physick"), p. 64. Lester S. King has discussed the medical views expressed in these essays in his *The Road to Medical Enlightenment 1650-1695* (London: Macdonald; New York: American Elsevier, 1970), pp. 62-86. His treatment is oriented more toward the mechanical philosophy.

capable of illustrating many things in man's body, which receiving some modifications there from the nature of the subject they belong to, pass under the notion of the causes or the symptoms of diseases.[110]

The studies of van Helmont, continued Boyle, have shown that the growth of stones is the same in both worlds, and his examination of digestion has further proved the essential nature of chemistry in the examination of bodily processes.[111] In both cases, Boyle argued, the Helmontian concept of fermentation is essential. This is not to say, as some chemists do, that this art can perfectly imitate all bodily processes, yet few would deny that chemical investigations have aided our understanding of both digestion and disease.[112]

Boyle went on to criticize the chemists in general, and van Helmont in particular, for their general rejection of the study of anatomy,[113] but he tempered this judgment by praising them for their conviction that more cures would be found through the study of chemistry.[114] Thus, should the alkahest really exist, unlikely though this may be, it would indicate that the stone could be dissolved rather than removed through surgical methods.[115] A major purpose of the search for chemical medicines is the avoidance of pain:

> ... it is to be hoped, many patients may be rescued from a great deal of pain; and that is, by finding out medicaments, that may in several distempers, that are thought to belong peculiarly to the chirurgeon's hand, excuse the need of burning, cutting, trepanning, and other as well painful as terrible manual operations of chirurgery.[116]

These chemical medicines are more powerful than herbal medicines, and they are also welcome to the poor, since they are frequently less expensive than the traditional ones. Through their chemical knowledge true physicians have learned that gems, gold, and other precious metals are of no value when administered inter-

---

[110]Boyle, " ... Usefulness of Experimental Philosophy" (Part II, Sect. 1), *Works*, Vol. II, p. 76.
[111]*Ibid.*, pp. 76-77, 79.
[112]*Ibid.*, p. 83.
[113]*Ibid.*, p. 85. Boyle refers particularly to the need for vivesection.
[114]*Ibid.*, p. 92.
[115]*Ibid.*, p. 97.
[116]*Ibid.*, p. 115.

nally.¹¹⁷ Excluding such useless and expensive ingredients from medical use in the future should prove beneficial to the sick, be they rich or poor.

If chemistry were credited only with the separation of active principles, it would be a major factor in the improvement of pharmacy. However, if it is used properly in conjunction with the study of the body and disease, many new cures will be discovered.[118] On this point van Helmont should be read attentively:

> And, *Pyrophilus*, that you may not wonder, that I, who think much of *Helmont's* theory scarce intelligible, and take great exceptions at many things in his writings, should yet now and then commend medicines upon his authority; I must here confess to you once for all, that (always excepting his extragavant piece, *De magnetica vulnerum curatione*) I have not seen cause to disregard many things he delivers, as matters of fact, provided they be rightly understood; having not found him forward to praise remedies without cause, though he seem to do it sometimes without measure . . . . .[119]

As with Charleton, cure by sympathy proved to be a problem for Boyle. Fludd may have been an "odd chymist," but he correctly related the transplantation of a case of the gout from a man to a dog.

> And I cannot but commend the curiosity of Dr. *Harvey*, who, as rigid a naturalist as he is, scrupled not often to try the experiment mentioned by *Helmont*, of curing some tumours or excrescencies, by holding on them for a pretty while (that the cold might throughly penetrate) the hand of a man dead of a lingring disease; which experiment the doctor was not long since pleased to tell me, he had sometimes tried fruitlessly, but often with good success.[120]

Boyle openly admitted his doubt,[121] but this hardly stopped him from prescribing sympathetic medicine to his friends. Indeed, he wrote,

> I see not, why these remedies, that work, as it were, by emanation, may not deserve the name of medicines, if they sometimes un-

[117]*Ibid.*, pp. 123, 127.
[118]*Ibid.*, p. 150.
[119]*Ibid.*, p. 149.
[120]*Ibid.*, p. 167.
[121]*Ibid.*, p. 165.

questionably succeed, though they should not always prove successful ones; nor why they should, notwithstanding their sometimes not succeeding, be laid aside; especially since these sympathetical ways of cure are most of them so safe and innocent, that though, if they be real, they may do much good, if they prove fictions they can do no harm . . . .[122]

The sections of *The Usefulness of Experimental Philosophy* written in the late 1640s read much as a preamble to a typical Paracelso-Helmontian text. However, like Charleton's, Boyle's views changed in the following decade. There is an indication of this in the remaining essays that appeared under this title, published in 1671, but largely completed by 1658. His earlier statement (1652) that experimental chemistry had supplanted his interest in the mathematical sciences stands in contrast with an open admission here that he had once been overly influenced "by the great authority of those modern philosophers, who would have the use of mathematicks, as disciplines, that consider only abstracted quantity and figure, to be rather hurtful than advantageous to a naturalist, the object of whose studies ought to be matter."[123] Although we can see evidence of change in this passage, George Starkey could still write that van Helmont was "a deserved favourite" of his as well as Boyle's when he dedicated the *Pyrotechny Asserted and Illustrated* to his friend in 1658.[124]

We may turn to *The Sceptical Chymist* for further evidence of the influence of Paracelso-Helmontian chemistry on the young Robert Boyle. Although not published until 1661, this work was at least partially in first draft early in the preceding decade.[125] The purpose of the essay was to examine the elemental systems then in use—a time-honored chemical exercise. Boyle began by noting that "of

[122] *Ibid.*, p. 168.

[123] *Ibid.* (Part II, Sect. 2), *Works*, Vol. III, p. 426.

[124] George Starkey, *Pyrotechny Asserted and Illustrated, To be the surest and safest means For Arts Triumph over Natures Infirmities* (London, 1658), sig. B1^r. The best account of this Harvard graduate of 1636 is in Ronald Sterne Wilkinson, "George Starkey, Physician and Alchemist," *Ambix*, 11 (1963), 121-152.

[125] The present discussion of *The Sceptical Chymist* is taken from Allen G. Debus, "Fire Analysis and the Elements in the Sixteenth and the Seventeenth Centuries," *Annals of Science*, 23 (1967), 127-147 (139-143). A discussion of Boyle's early work on this subject (plus the text of the pertinent manuscript) will be found in Marie Boas, "An Early Version of Boyle's *Sceptical Chymist*," *Isis*, 45 (1954), 153-168.

late Chymistry begins, as indeed it deserves, to be cultivated by Learned Men who before despis'd it; and to be pretended to by many who never cultivated it, that they may be thought not to ignore it."[126] With this he found no fault, but he was clearly concerned about the quality of the authors being referred to. Already a convinced corpuscularian, Boyle felt that there were many phenomena that could not be explained satisfactorily by the *tria prima* "without taking much more Notice than they are wont to do ... of the Motions and Figures, of the small Parts of Matter."[127] However, he had no intention of alienating those chemical philosophers who based their philosophy on the Creation account in Genesis, and he carefully pointed out that the atomic hypothesis did not conflict with the Mosaic philosophy.[128]

By his criticism Boyle hoped to bring forth abler chemists, who "will be oblig'd to speak plainer than hitherto has been done, and maintain it by better Experiments and Arguments . . . ."[129] Yet Boyle still believed in the existence of an elite few who had a knowledge of great secrets; thus he distinguished between the "Cheats or but Laborants, and the true *Adepti*; By whom, could I enjoy their Conversation, I would both willingly and thankfully be instructed; especially concerning the Nature and Generation of Metals."[130] While Boyle attacked the mystical and occult nature of most alchemical texts, he carefully limited this attack to their unnecessary obscurity when describing the first principles of the art. He was quite willing to "Excuse the Chymists when they write Darkly, and Aenigmatically, about the Preparation of their *Elixir*, and Some few other grand *Arcana*, the divulging of which they may upon grounds Plausible enough esteem unfit."[131]

Boyle was aware that the most effective attack on the Aristotelians and the chemists would be through their favored sys-

---

[126] Robert Boyle, *The Sceptical Chymist or Chymico-Physical Doubts and Paradoxes, Touching the Spagyrist's Principles Commonly call'd Hypostatical: As they are wont to be Propos'd and Defended by the Generality of Alchymists. Wherunto is Premis'd Part of another Discourse relating to the same Subject* (London: J. Cadwell for J. Crooke, 1661; London: Dawson Reprint, 1965), sig. A2^r.

[127] *Ibid.*, sig. A2^v.

[128] *Ibid.*, p. 38.

[129] *Ibid.*, sig. A5^v.

[130] *Ibid.*, sig. A7^r.

[131] *Ibid.*, p. 203.

tems of elemental matter. He was able to deal with the four elements in a summary fashion. He complained that the Aristotelians generally had very little experimental knowledge of the elements and employed "Experiments rather to illustrate than to demonstrate their Doctrines."[132] The only example brought forth in support of the four elements—the burning twig—was quickly rejected as invalid, but in the ensuing discussion Boyle questioned whether it could be accepted that fire separates anything from a body that was not pre-existent in it. He then suggested that fire alters substances as well as separates them.[133]

The views of the chemical philosophers were of far more concern, since the "Doctrine about the Elements is more applauded by the Moderns, as pretending highly to be grounded upon Experience."[134] Boyle was fully aware that the basic observations supporting the chemical principles were dependent upon analysis by fire, and he immediately turned to this problem by asking "how far, and in what sence, Fire ought to be esteem'd the genuine and universal Instrument of analyzing mixt Bodies."[135] Here his arguments clearly reflect the literature of the previous ninety years. The analyst obtains far different results through burning and distillation, and different degrees of heat give quite different results in distillation as well.[136] In any case, it is certainly not true that all substances can be analyzed into three fractions:[137] there are some things—above all, the precious metals—from which one cannot separate salt, sulfur, or mercury. One might also ask how to classify glass, since it is not destroyed but is produced by the action of fire;[138] since nothing is separated from it, perhaps it should be considered a new element. Again, there are other substances which can be separated by the fire into more than three fractions.[139] Most organic substances, for instance, can be separated into five portions; but Boyle could not accept this as a true analysis, since it as-

[132] *Ibid.*, sig. A5^r, p. 20.
[133] *Ibid.*, pp. 27, 29.
[134] *Ibid.*, p. 349; see also p. 36.
[135] *Ibid.*, p. 48.
[136] *Ibid.*, pp. 49-51.
[137] *Ibid.*, pp. 55 ff. and *passim*.
[138] *Ibid.*, pp. 53, 224.
[139] *Ibid.*, p. 187.

sumed that fire actually separates the elements from a body.[140]

Boyle then proceeded to question the general belief that every "Distinct Substance that is separated from a Body by the Help of the Fire, was Pre existent in it as a Principle or Element of it."[141] Here too a wealth of arguments from the earlier literature were drawn upon. In particular, he turned to van Helmont's willow-tree experiment, which he believed had shown almost conclusively that vegetable matter is composed largely of water. Yet, if the results of a normal distillation analysis were accepted here, there would be found "Phlegme, a little Empyreumaticall Spirit, a small quantity of adust Oyl, and a *Caput mortuum,*" the latter consisting of salt and earth.[142] Therefore, "(as we have seen) out of fair Water alone, not only Spirit, but Oyle, and Salt, and Earth may be Produced; It will follow that Salt and Sulphur are not Primogeneal Bodies, and principles, since they are every Day made out of plain Water by the Texture which the Seed or Seminal principle of plants puts it into."[143] Van Helmont's soap experiment provided further information for Boyle. Water and a seminal principle result in vegetable matter, and then a reaction of the oil (or sulfur) of plants plus lixiviate salts yields soap, which in turn can be distilled back to water.[144] Even the widely accepted belief in the growth of metals in the earth presents a close parallel with the concept of a water-seminal principle.[145] It is true that Boyle had serious reservations about the water hypothesis,[146] but his high regard for van Helmont, as "an Author more considerable for his Experiments than many Learned men are pleas'd to think him,"[147] made it a concept that he felt required weighty consideration.

The Paracelsian attempts to base the three principles on qualities seemed tenuous to Boyle. Some said the cause of color is mercury, but Paracelsus stated that salt is the origin of color, while Sennert ascribed the source to sulfur. Boyle himself noted that

[140]*Ibid.,* pp. 286 f.
[141]*Ibid.,* p. 103.
[142]*Ibid.,* p. 111.
[143]*Ibid.,* p. 162.
[144]*Ibid.,* pp. 131, 382.
[145]*Ibid.,* p. 364.
[146]*Ibid.,* pp. 385 ff.
[147]*Ibid.,* p. 112.

colors result from the passage of light through a prism, and here the principles play no part at all.[148] The principles do not explain magnetism, the generation of chicks or plants, motion or gravity. How then do these chemists presume to call the principles the basis for a new and universal philosophy? At best they should be considered one useful concept among many others for the scientist.[149]

For a final summary Boyle returned to the problem of fire.[150] He felt that he had shown conclusively that fire does not analyze substances into their elements or into any determinate number of things. On the other hand, Eleutherius, the moderator of his discussion, urged Carneades to accept the fact that most mineral bodies can be separated into three parts and most vegetable substances into five. And, although these substances might be mixed, there should be no harm in calling them elements and letting them "bear the Names of these Substances which they most Resemble." Boyle's position here is indeed very close to that of van Helmont. Above all, Boyle stressed that in the future chemists should depend more on experiments and rely far less on the fire as the method for their analyses.[151] Through the final statement by Carneades it becomes clear that Boyle was not so determined to destroy the elements and the principles as he was to weaken the unwarranted confidence in the current analytical methods to demonstrate them. Indeed, the reader may feel somewhat shocked after surviving a four-hundred-page attack on these systems to find that the final opinion of Carneades is that either system "may be much more probably maintain'd than hitherto it seems to have been . . ."[152] In the end, although Boyle objected to analysis by fire, he did think that the three principles might be more firmly established through newly devised experiments.

The Analysis Problem

Boyle's denial of the validity of fire analysis was widely noted in

---

[148] *Ibid.*, pp. 327-328.
[149] *Ibid.*, pp. 301-313 (305).
[150] *Ibid.*, pp. 431-436.
[151] *Ibid.*, p. 434.
[152] *Ibid.*, p. 435-436.

the decades after the first appearance of *The Sceptical Chymist*.[153] Yet his arguments were far from conclusive to all of his contemporaries. A contrary position had been reached by Lefèvre, and this was retained in the 1670 English edition of his *Traicté de la chymie*. For him chemical resolutions by fire readily showed five fractions, but it was a question "of no small difficulty, *viz. Whether these five Substances, are Natural or Artificial Principles, and not rather Principles of Disunion and destructive, than of Composition and Mixture?*"[154] While admitting that these substances were obtained by artificial means, he insisted that they were natural, "since Art doth contribute nothing else but the Vessels to contain and receive them." The existence of the principles becomes quite clear if we examine the case of a normal corruption or putrefaction, where unpleasant odors betray the existence of saline and sulfureous spirits. This may only satisfactorily be explained by assuming that the salt has been dissolved by the internal phlegm. Salt is the bond that holds together mercury and sulfur, and when it is dissolved, these two are released. And since heat accompanies all putrefactions, all four of these principles are carried away in the air, leaving behind only the earthly residue. This evidence confirms Lefèvre's belief in the validity of analysis by fire, and he concluded that "these substances are not extracted from the Mixt by transmutation, but by a meer natural separation, assisted by the heat of the Vessels and the hand of the Artist: for all things cannot indifferently and immediately be transformed in the like and same things."[155]

Perhaps an even better example is found in Nicholas Lemery's *Cours de chymie* (1675). There are similarities between Boyle and Lemery, for the latter favored an atomic explanation of matter and attacked the current view of the meaning of "principle":

> The word *Principle* in *Chymistry* must not be understood in too nice a sense: for the substances which are so called, are only *Principles* in respect of us, and as we can advance no farther in the division of bodies; but we well know that they may be still divided into abundance of other parts which may more justly claim, in propriety of speech, the name of *Principles*: wherefore such substances are to be un-

---

[153]See Debus, "Fire Analysis," pp. 143-147.
[154]Lefèvre, *Compleat Body of Chymistry* (1670), Part 1, p. 20.
[155]*Ibid.*, p. 20-21.

derstood by *Chymical Principles*, as are separated, and divided, so far as we are capable of doing it by our weak imperfect powers.[156]

With this understanding, Lemery pointed out that a normal "Anatomy of a mixt body" yields the five principles water, spirit, oil, salt, and earth.[157]

Lemery was quite willing to accept the evidence obtained from distillation analyses, but he went on to refer to "some modern philosophers" who seemed uncertain whether the substances so separated actually existed earlier in the body or were changed by the fire during distillation.[158] It is true, he granted, that fire seems to change substances, but surely fire "does not make those *Principles*: for we see them and smell them in many bodies, before ever we bring them to undergo the *Fire*."[159] Thus we perceive oil to be in olives, almonds, and nuts, while the salt in plant juices is readily obtained by pressing and subsequent evaporation. The five principles are found in animal and vegetable matter; but, even Lemery was willing to admit, they were not obtained readily from minerals. Nevertheless, the reader was assured, these principles exist even in gold and silver, but here they are "so strictly involved" with one another that they are inseparable.[160] In Lemery we find a corpuscularian and an experimentalist arguing in much the same fashion as the Paracelsian Duchesne at the beginning of the century.[161]

---

[156]Nicholas Lemery, M.D., *A Course of Chymistry*, trans. Walter Harris, M.D. (2nd English ed. from the 5th French ed., London: R.N. for Walter Kettilby, 1686), p. 5. Lemery's position is similar to that taken by Mersenne in his questioning of the Paracelsian principles (*La verite des sciences*, 1625, p. 56; discussed above, Ch. 4).

[157]*Course of Chemistry*, p. 3.

[158]*Ibid.*, p. 6.

[159]*Ibid.*, p. 7.

[160]*Ibid.*, p. 9.

[161]An interesting comment on the views of Lemery is seen in the work of the "Pythagorean," Whitlocke Bulstrode (1692). He suggested that Lemery's argument for five principles was not conclusive, for "he honestly confesseth, that this is effected by the alteration the Fire makes on Bodies; not by a natural Analysis into their first Principles." In agreement with Lemery that fire may change the texture of substances but not destroy them, Bulstrode practically stated the law of conservation of matter: "Fire indeed may separate the Parts of a mix'd Body,

A final example of the value of the evidence given by fire might be taken from the writings of Daniel Georg Morhof, who specifically referred to *The Sceptical Chymist* in the second volume of his widely read *Polyhistor* (1692). Morhof stated that Boyle's arguments were quite clever and ingenious, but, he added, "I choose to dissent."[162] His line of argument is similar to that of Sennert. For Morhof the three principles should not be taken as prime matter but as a first stage of compound matter; in turn they form the base of more complex bodies.[163] As for the fire problem, Boyle was quite right in pointing out that different applications of heat give rise to apparently different analytical results. However, this may be simply explained. In the case of a violent or destructive fire, the most mobile parts of the substance fly away if the combustion takes place in the open. Even within a closed system such an analysis is spoiled by the force of propulsion which mixes the more fixed with the more volatile portions. On the other hand, with a gentle fire that is increased in temperature only gradually, there is a selective division of the substance in which "nothing new is generated, and only those things which were unknown before are discovered." Basing his opinion on such evidence as was available to him, he could only conclude that analysis by fire *does* result in a demonstration of the three principles in mixed bodies, and further, that fire does not seem to produce new species.[164]

In short, it appears that there was no real consensus among chemists either about analytical procedure or about systems of ele-

---

change the Figure, and so alter its Appearance, as to puzzle the best Mechanick to reduce it to its primitive state; yet this is no Annihilation, but Division. The burning of Wood or any Fuel, is a Destruction of it, I confess, as to the Proprietor; but not with respect to the Universe, no more than there is less Money in the World by the Profuseness of a Prodigal; as the one doth but change Hands, so the other alters only the situation of its Parts." Bulstrode argued that in truth "there are no other Principles, but the moist vapor impregnated with vital heat; for these two alone constitute all Bodies" (clearly a reference to earlier concepts of prime matter). See Whitlocke Bulstrode, *A Discourse of Natural Philosophy Wherein the Pythagorean Doctrine Is set in a true Light, and vindicated* (2nd ed., London: Jonas Browne, 1717), pp. 121, 137, 139.

[162]Daniel Georg Morhof, *Polyhistor* (4th ed., 2 vols., Lübeck: Peter Boeckmann, 1747), Vol. II, pp. 252-253, 333-334.

[163]*Ibid.*, p. 252.

[164]*Ibid.*, pp. 253, 252.

ments or principles in the period before and immediately after the publication of *The Sceptical Chymist*, even though the importance of the subject was widely appreciated. Boyle's approach shows a close acquaintance with the earlier literature and with the arguments of Erastus, Bacon, Sennert, and especially van Helmont. Indeed, his *Sceptical Chymist* may be considered the major summary of seventeenth-century chemical literature on this subject. Still both the defenders of the principles and their opponents were able to use the results of fire analysis to advance their views. In the closing years of the century, chemists continued to debate how the results of fire analysis might best be interpreted. In 1673, for example, the members of the French Academy discussed the problem and hopefully concluded that the earlier criticisms of distillation analysis might be overcome with due caution and effort on the part of chemists.[165] Indeed, there were few late-seventeenth-century chemists who would not have agreed with John Ray (1693) that this *was* the proper way to carry out the "Chymical Anatomy of mix'd Bodies."[166] Only gradually did the advantages of other methods of analysis convince chemists to change their procedures, and as late as 1778 Macquer thought it necessary to begin a discussion of analysis with an attack on the futility of the use of fire and heat.[167]

In the search for other methods of analysis Boyle's name is again one of great prominence. Indeed, his work forms a basis for Marie Boas Hall's attempt to define a seventeenth-century chemical revolution.[168] In her appraisal of Boyle's work she has

---

[165] *Histoire de l'Académie Royale des Sciences depuis son éstablissement en 1666 jusqu'a 1686* (Paris: Martin, Coignard and Guerin, 1733), Vol. I, pp. 161-167. For this reference I am indebted to F. L. Holmes, who has also discussed the problem of analysis in his "Analysis by Fire and Solvent Extractions: The Metamorphosis of a Tradition," *Isis*, **62** (1971), 129-148. Proponents of traditional fire and distillation analyses could find support in phlogiston chemistry. J. J. Becher argued that combustion is a destruction or a dissolution of the combustible substance into its components.

[166] John Ray, *Three Physico-Theological Discourses* (2nd ed., London: Sam. Smith, 1693), p. 325.

[167] P. J. Macquer, *Dictionnaire de chymie* (2nd ed., 4 vols., Paris: Didot, 1778), Vol. I, pp. 169-170. See also Allen G. Debus, "Solution Analyses Prior to Robert Boyle," *Chymia*, **8** (1962), 41-61.

[168] She has concluded that "seventeenth-century chemistry deserves a place in

cited three fundamental contributions to support her thesis: (1) Boyle's participation in the clarification of chemical classification, (2) his reform of chemical nomenclature, and (3) his study of reaction mechanisms.[169] She admits, however, that Boyle's attempt to study reaction mechanisms was not particularly successful,[170] and that such success as he actually obtained was due to his ability to use skillfully the chemical identification tests at his disposal. Similarly, she accepts the fact that Boyle's attempt to correlate the names of chemical reagents with their composition could only be a limited success in a period when high-powered analytical tests were not available to chemists.[171] And again, Boyle's advance in the field of chemical classification was due to his progress in chemical analysis.

The development of chemical analytical procedures thus assumes an extremely prominent place in any estimate of Boyle's achievements. In Boas Hall's investigation of Boyle's chemical analyses, she suggests that he was one of the first—if not the first—to see clearly the inadequacy of fire as the basic analytical agent. Boyle, she continues, then sought an alternative analytical method, and, guided by the belief that distinct chemical entities

---

the seventeenth-century scientific revolution" and that "in no sense was eighteenth-century chemistry a continuation of the old mystical science or the purely practical art." Marie Boas (Hall), *Robert Boyle and Seventeenth-Century Chemistry* (Cambridge: Cambridge University Press, 1958), pp. 231-232. The critique of this position presented here appeared in essentially the same form in Allen G. Debus, "The Significance of the History of Early Chemistry," *Cahiers d'Histoire Mondiale*, **9** (1965), 39-58 (52-55). See also by the present author "La philosophie chimique de la Renaissance et ses relations avec chimie de la fin du XVIIe siècle" in *Sciences de la Renaissance*. VIIIe Congrès International de Tours (1965), Jacques Roger, ed. (Paris: Vrin, 1973), pp. 273-283, and the review of Marie Boas Hall, *Robert Boyle on Natural Philosophy. An Essay with Selections from His Writings* (Bloomington: Indiana University Press, 1965) in *Isis*, **57** (1966), 125-126.

[169] Boas, *Robert Boyle*, Chs. 3-5.

[170] "The seventeenth-century chemists became profoundly interested in studying the course of a reaction. But they could predict very few; even Boyle, more interested in prediction than most other chemists, managed to predict the course of a reaction in very few cases, and to interpret the course of a reaction in not so very many more" (*ibid.*, p. 179).

[171] "The rational nomenclature of the later seventeenth-century was unsystematic, as the reformers a century later rightly claimed" (*ibid.*, p. 158).

should be sought rather than the various properties connected with the principles in distillation procedures, he turned to a series of important tests which form the basis of qualitative procedures. Among these were color tests (both the characteristic colors of salts in solution and color indicators), flame tests, the recognition of crystal form as a means of identification, and the regular use of specific gravity as a quantitative test for aqueous solutions. She has granted special significance to Boyle's work on *Mineral Waters*, which she implies included the first systematic development of aqueous analytical procedures.[172]

There can be no question that many of these tests became widely known among chemists because of the influence of Boyle's writings, but the works of earlier chemists indicate that Boyle was not as innovative as has been alleged. It is surely understandable that a belief in unchangeable chemical entities in the form of atoms or corpuscles would lead him to seek such entities rather than the elusive elements or principles; yet an examination of the work of iatrochemists who customarily used the principles as an explanatory device shows that they too were concerned with the search for distinct chemical entities. Here an example is Sir Thomas Browne's discussion in 1646 of the blackening which occurs in the well-known oak-galls test.[173] Starting from the Paracelsian position that color is dependent on the principles, Browne proceeded to investigate in detail not only the relation of the inflammable principle, sulfur, to blackness, but also the similar role played by salt. This he did through an intensive study of the best-known discoloration caused by a salt—the reaction of vitriol and oak galls. In the course of this investigation he showed that it was not oil of vitriol but the fixed part of the salt—and especially iron—which caused the blackening. Furthermore, he showed that this occurs not only with oak galls but also with a decoction or infusion from many other "astringent" plants. Clearly the concept of the

---

[172]*Ibid.*, pp. 108-141 *passim*.

[173]This example is discussed in considerable detail in the present author's "Sir Thomas Browne and the Study of Colour Indicators," *Ambix*, **10** (1962), 29-36. The problem of analysis referred to appears in Browne's *Pseudodoxia epidemica* (1646). The edition used is *The Works of Sir Thomas Browne*, ed. Charles Sayle, 3 vols. (London/Edinburgh: Grant Richards/John Grant, 1904, 1907), Vol. II, pp. 387-395.

three principles was being used as a basis for the search for chemical entities. One hardly needed to be a corpuscularian to conduct such a search.

It has already been shown that Boyle's arguments against the use of fire as an analytical agent were anticipated nearly a century prior to the publishing of *The Sceptical Chymist* and that he was fully aware of the literature on this subject through van Helmont's lengthy discussion. In regard to Boyle's mineral water procedure, this seems in essence to have been an extension—granted, an important one—of a development that had been in progress over the course of several hundred years.[174] The number of flame tests Boyle used was not significantly greater than Thurneisser had used a century earlier, although Boyle used them with greater regularity. Similarly, the emphasis placed on crystal forms by analysts after distillation residues began to be recrystallized clearly antedates Boyle's work in the field. Boyle's constant use of specific gravity was an important innovation. As shown above, prior to this time these comparisons were generally accepted in principle, but few felt that the small differences could be determined with the balances at their disposal.

The importance of Boyle's work on color indicators has always been recognized. The only color indicator in common use prior to his time was the oak-galls test—mentioned in relation to Browne—which was used occasionally for iron but more generally for vitriol or alum. Although Boyle admitted that others had observed the red color of acids with syrup of violets, on the basis of his own claim he is granted priority for the discovery that this indicator turns green in the presence of alkalis.[175] Here, however, Boyle has an important predecessor in Edward Jorden, who differentiated between "sowre juyces" and "Salts," two opposite classes of sub-

[174] See the sections on aqueous analysis above, Chs. 1, 2, and 5.

[175] Robert Boyle, *Experiments and Considerations Touching Colours* (London: Henry Herringham, 1664), pp. 245-246. Boyle wrote that others had known that "almost any Acid Salt will turn Syrrup of Violets Red. But to improve the Experiment, let me add what has not (that I know of) been hitherto observed, and has, when we first shew'd it them, appear'd something strange, even to those that have been inquisitive into the Nature of Colours; namely, that if instead of Spirit of Salt, or that of Vinegar, you drop upon the Syrrup of Violets a little Oyl of Tartar per deliquum, or the like quantity of solution of Potashes, and rubb them together with your finger, you shall find the Blew Colour of the Syrrup turn'd in a

stance, one of which turns a scarlet cloth red and the other blue. Boyle's influence was enormous, and there is no doubt that he added numerous new analytical tests to chemistry. Nevertheless, it may fairly be questioned whether he should be viewed as the architect of a seventeenth-century chemical revolution or as a transitional character in the history of chemistry on the basis of his analytical work.

The Nitro-Aerial Particles in Mid-Century

A topic of perpetual interest to historians of chemistry has been John Mayow's (1641-1679) well-known nitro-aerial particles. He discussed these in the *Tractatus duo* of 1668 and the *Tractatus quinque* of 1674. Because some of his statements suggest that he thought of these particles in much the same way we think of oxygen, he has frequently been referred to as a precursor of Lavoisier.[176] To be

---

moment into a perfect Green, and the like may be perform'd by divers other Liquors, as we may have occasion elsewhere to inform you." Thomas Willis, *Practice of Physick* . . ., trans. S. Pordage (London, 1684), p. 40 (in "Of Fermentation") observed that spirit of hart's horn will turn tincture of violets green, but he did not make this a general test, as Boyle was to do. (The *De fermentatione* was first published in 1659.)

[176]Mayow's work has been the subject of intensive research since the period of the chemical revolution. The earlier literature has been summarized by J. R. Partington in "The Life and Work of John Mayow (1641-1679)," *Isis*, **47** (1956), 217-230 and 405-417. This work has been expanded in his chapter on Mayow in his *History of Chemistry*, Vol. II, pp. 576-636. Here a discussion of the most pertinent literature will be found in the long footnote extending from p. 579 through 581. Note should also be taken of the recent papers by Walter Böhm: "John Mayow und Descartes," *Sudhoffs Archiv für Geschichte der Medizin und der Naturwissenschaften*, **46** (1962), 45-68; "John Mayow and His Contemporaries," *Ambix*, **11** (1964), 105-120; "John Mayow und die Geschichte des Verbrennungsexperiments," *Centaurus*, **11** (1965), 241-258. The background to Mayow has been discussed by Henry Guerlac, "John Mayow and the Aerial Nitre. Studies in the Chemistry of John Mayow. I, "*Actes du septième congrès international d'Histoire des sciences* (Jerusalem), 1953, pp. 332-349; "The Poets' Nitre. Studies in the Chemistry of John Mayow. II," *Isis*, **45** (1954), 243-255. See also Wlodzimierz Hubicki, "Michael Sendivogius' Theory, its Origin and Significance in the History of Chemistry," *Actes du dixième congrès international d'histoire des sciences* (Ithaca, 1962), 2 vols. (Paris: Hermann, 1964), Vol. II, pp. 829-833; Allen G. Debus, "The Aerial Niter in the 16th and Early 17th Centuries," *ibid.*, pp. 835-839; Allen G. Debus, "The Paracelsian Aerial Niter," *Isis*, **55** (1964), 43-61.

sure, Mayow's tracts are "experimental" in nature—typical of the period in which they were written. And yet, what is it that Mayow proposed? He suggested that there is a "nitrous" part of the air which is requisite for both combustion and respiration. In addition, he postulated that thunder and lightning result from aerial explosions of sal-nitric and sulfureous particles[177] and that some diseases are similarly caused. Warning of the introduction of sulfureous particles, he noted that

> ... if nitro-aërial particles constitute the animal spirits, it may sometimes happen that they in the brain follow the motion of the nitro-aërial particles forming the lightning in the air; so that the animal spirits would seem not so much struck by lightning as themselves to form lightning. And hence it is that they, being violently moved and as it were set on fire, are dissipated in a moment; and so on account of the flame kindled in its brain the animal, deprived of the common light and breath, is extinguished.[178]

These topics along with Mayow's discussion of muscular action[179] and his views of the vital nitrous salt of the air are all familiar to the student of the works of the earlier chemical philosophers.

Although Mayow did not refer to his sources with the regularity one could wish, they surely did exist. Douglas McKie has shown the relationship of Mayow's views to the work of Boyle and Hooke.[180] The former spoke of a volatile niter in the air, while the latter made an aerial saltpeter an integral part of his theory of combustion. Hooke was to expand his views of a "nitrous air" con-

---

[177]John Mayow, *Medico-Physical Works Being a Translation of Tractatus Quinque Medico-Physici* (re-issue edition for the Alembic Club, Edinburgh/London: E. and S. Livingstone, 1957), pp. 147-152. The "gunpowder" analogy is discussed in this citation.

[178]*Ibid.*, p. 257.

[179]Mayow's tract on muscular motion (*ibid.*, pp. 229-302) is based primarily on his theory of nitro-aerial particles and can be connected to the earlier statements of Paracelsus. The deleterious action of sulfureous particles on the nitro-aerial particles as a cause of illness or disease is also brought forth by Mayow and compared with the action of aerial niter and sulfur in thunder and lightning (*ibid.*, pp. 256-257, 259, 279).

[180]Douglas McKie, "Fire and the Flamma Vitalis," in E. A. Underwood, ed., *Science, Medicine and History*, 2 vols. (London: Oxford University Press, 1953), Vol. I, pp. 469 ff.

siderably in the period after 1682.[181] However, the founders and early members of the Royal Society derived inspiration not only from these authors; they in addition turned to the "vital salt spirit" of van Helmont[182] and the nitrous *spiritus universalis* of Glauber.[183] Coming closer to home, they looked to the writings of Francis Bacon.

In one of the most popular of his books, the *Sylva sylvarum* (first published posthumously in 1626), Bacon made some observations on the mixture of flame and air which may well have been read by Mayow, Boyle, and Hooke.[184] Stating that flame and air do not generally mix, he investigated the case of gunpowder and wrote that the motion of a bullet is due specifically to the niter or saltpeter which contains "a notable crude and windy spirit."[185] According to Bacon, the mechanism of gunpowder rests on (1) the rarefaction of the earthy substance into flame—a conversion to an element which requires more volume—thus resulting in (2) the heating of the niter, and the subsequent expansion of the enclosed air to expel the fire with which it cannot exist. However, Bacon added, "As for *living Creatures*, it is certain their *Vital Spirits* are a Substance compounded of an *Airy* and *Flamy* Matter; and though *Air* and *Flame*, being free, will not well mingle; yet bound in by a *Body* that hath some fixing, they will." It was to this vital spirit which is like and yet unlike the saltpeter in gunpowder that Bacon referred the motion of the muscles as had Paracelsus in the *Liber azoth,* and as Mayow would some fifty years later. For Bacon saltpeter was "the *Spirit* of the Earth,"[186] and he showed its remarkable effect on the growth of plant life.[187]

Mayow, Hooke, and Boyle need not have turned only to writers of major importance or to the older Paracelsians to become acquainted with this concept. John French (1651), Ralph Bathurst

---

[181] Partington, *History of Chemistry*, Vol. II, pp. 555-559.
[182] See above, Ch. 5.
[183] See above, Ch. 6.
[184] Debus, "Paracelsian Aerial Niter," p. 57.
[185] Francis Bacon, *Sylva sylvarum, or A Natuall History in Ten Centuries* (8th ed., London: J.F. and S.G. for William Lee, 1664), p. 8. Bacon went on to discuss the dual "aiery" and flamelike nature of the vital spirit in the separately paginated and appended *History of Life and Death* (1st ed., 1623), p. 64.
[186] Bacon, *History of Life and Death*, pp. 30-31.
[187] Bacon, *Sylva sylvarum*, p. 96.

(1654), Eugenius Philalethes (1657), Malachi Thruston (1664), and Sir Kenelm Digby (1661) all discussed the properties of the aerial niter.[188] Similarly, Lefèvre discoursed at length on the purpose of the respiration which he reported as necessary for drawing in the universal spirit in the air. "For it is this spirit contained in the Air we breath in, which subtilizeth, and maketh volatile, all the superfluities that are found both in the venal and arterial blood, the shop and matter of vital and animal spirits."[189] He went on to say that some (although not himself) considered saltpeter to be a "Universal Salt, believing that it possesses in it self the Soul of the World" and that this mysterious salt "is the soul of all Physical Generations, a Child and Son of Light, and the Father of all Germination and Vegetation."[190]

The influence on Mayow of earlier and contemporary authors may also be noted in his use of the candle-cupping glass experiment as a method of showing the diminution of air in the process of combustion. In a frequently quoted passage Mayow wrote,

> ... let a burning candle be placed in water so that the wick may stand about six finger-breadths above the water, and then let an inverted cupping-glass of sufficient height be put over the light and plunged immediately into the water surrounding the light.... Care, however, must be taken that the surface of the water enclosed within the glass be at the same level as the water without. But that this may be obtained in the present experiment, and also in those that follow, let one leg of an inverted syphon be enclosed within the cavity of the cupping-glass before it is put into the water while the other leg projects outside, yet so that the end of each leg may be above the surface of the water, as is seen in the said figure. The use of the syphon is to enable the air enclosed in the alembic, and compressed by the underlying water while the glass is being let down into the water, to pass out through the cavity of the syphon, so that the water within may not be depressed below the level of the water outside, as it would otherwise be. But when the air ceases to pass through the syphon (which will happen almost in an instant) the syphon should be at once withdrawn, that the air may not afterwards rush through it into the glass. When these arrangements are made let the cupping-glass be firmly fixed so that it may descend no further into the water, and you will presently

[188]See Guerlac, "Studies ... II," *Isis*, **45** (1954), *passim*.
[189]Lefèvre, *Compleat Body of Chymistry*, Part I, p. 38.
[190]*Ibid.*, Part II, p. 251.

see, while the light still burns, the water rising gradually into the cavity of the cupping-glass.[191]

Partington correctly praised this account for Mayow's insistence on the equalization of water levels by means of the siphon.[192]
Mayow's explanation of the phenomenon was dependent upon his belief in the reality of the nitro-aerial particles.

> I will not deny that the ascent of the water arises in part from the circumstance that when the light is about to expire, the air enclosed in the cupping-glass is less agitated and rarefied by the igneous particles than formerly.[193]
> ... [However] I think it should be maintained that the air mixed with the flame is, by the burning of the flame, quickly deprived of its nitro-aërial and elastic particles, so that this air not only becomes unfit for sustaining fire but also loses in part its elasticity. Hence when a flame enclosed in a glass vessel has exhausted the nitro-aërial particles of the air, it soon goes out and the space contained within is like a vacuum, not only on account of the diminished motion of the igneous particles, but also from the lack of elastic particles . . . .[194]

Mayow's observation may be compared with van Helmont's view that the aerial pores had collapsed during the combustion after their acceptance of smoke.[195]

In England Robert Hooke had carried out similar experiments at the Royal Society shortly before the publication of the *Tractatus quinque* (November 1672-March 1673),[196] but the candle experiment, originally described by Philo and Hero, had been a topic of considerable interest to natural philosophers through the course of the century. Not only had it been discussed by van Helmont and Guericke, it had been noted as a common observation both by Francis Bacon and Robert Fludd.[197]

---

[191] Mayow, *Tractatus quinque*, pp. 68-69.
[192] *History of Chemistry*, Vol. II, p. 595.
[193] Mayow, *Tractatus quinque*, p. 69.
[194] *Ibid.*, p. 68.
[195] See above, Ch. 5, pp. 40-43.
[196] Partington, *History of Chemistry*, Vol. II, p. 559.
[197] Although this experiment has been discussed here repeatedly in relation to a number of authors, a useful summary of seventeenth-century references will be found in *ibid.*, p. 595. A detailed discussion of the views of van Helmont and Fludd is in Ch. 5 above.

In this context it is of some interest to find that the candle experiment had been recently employed by William Simpson in the course of a current conflict involving the analysis of mineral waters. Such analyses by English physicians was surely practiced in the sixteenth century, and the description of the spa at Bath (1631) by Edward Jorden deservedly ranks as one of the more important English scientific treatises in the period prior to 1650. However, interest in this subject developed at a more rapid pace after the discovery of the mineral water springs at Scarborough and Knaresborough in 1626. Early accounts by Edmund Deane (1626) and Michael Stanhope (1626, 1632) were followed by a number of others, among them analyses by John French (1651) and Robert Witty (1660).[198] Witty's work on the spa at Scarborough was republished in 1667, initiating a debate over methods of aqueous analysis, in the course of which the results of the oak-gall test were questioned, and one of the disputants, William Simpson, referred the whole matter to the Royal Society. However, by this time (1669) the importance of Boyle's work on colors was well known, and soon these men were appealing to his authority.[199] The heat of the controversy may have been one reason why Boyle wrote that he purposely forebore to consult other authors on this matter when he wrote his *Short Memoirs for the Natural Experimental History of Mineral Waters* (1684-1685).[200]

Although Simpson's *Hydrologia chymica* (1669) was ostensibly devoted to the study of mineral waters and to the refutation of the

---

[198] These early English analyses are discussed by Debus in "Solution Analyses Prior to Robert Boyle," pp. 52-57.

[199] William Simpson, *Hydrologia chymica: or the Chymical Anatomy of the Scarbrough, and other Spaws in York-Shire* (London, 1669), p. 363. Good abstracts of the various works in this controversy will be found in Thomas Short, *The Natural, Experimental and Mechanical History of Mineral Waters* (London, 1734), pp. 112-172. On the appeal to the Royal Society, see Short quoting Simpson's *Hydrological Essays* (1670), p. 136; on the appeal to Boyle's authority, see Short quoting George Tonstall's *Scarbrough Spaw* (1670), pp. 139-143. Also see Short quoting Witty's *Scarbrough's Spagirical Anatomiser Dissected* (1672), p. 148. In the *Philosophical Transactions* these tracts were reviewed and articles appeared in reference to specific points of the controversy by Drs. Daniel Foote, Highmore, and Witty. See Short, *op. cit.*, pp. 152-156.

[200] Robert Boyle, *Short Memoirs for the Natural Experimental History of Mineral Waters Addressed by Way of a Letter to a Friend* (London: Samuel Smith, 1684/1685), sig. A2.

earlier work by Witty, there is much within it to connect him directly with the tradition of the chemical philosophers. Although Simpson hastened to point out that he disapproved of many chemical authorities (e.g. Crollius, Beguin, Libavius), he spoke highly of Lefèvre and he referred his readers above all to Paracelsus and to the "profound" van Helmont.[201] Here again is an author who found a valid analogy between the earthly circulation of water and the circulation of blood in man:

> ... a Circulation of water is as justly requisite ... for the upholding the Symmetry of parts, and intirenes of the whole terraqueous Globe, as the Circulation of blood is necessary for the preservation of life, and vital functions in the Microcosme or body of man: The earth can no more produce Vegetables, or Minerals without this connatural circulation, of water replenish'd with Celestial influences, than the blood in the body of man can produce Vital, or Animal Spirits, requisite for absolving the functions of life, without its inbred circulation ....[202]

Simpson was well aware of the aerial niter concept, but he made no use of this in his description of the candle experiment. Here he was concerned primarily with the qualitative change produced in air by the application of heat. Like Fludd, Simpson was interested in the difference between the candle experiment and the external application of heat in the case of the thermoscope.

> That the Ayr receives a considerable alteration by heat, is further confirm'd by the experiment of inverting a glass Cucurbit over a Candle, fastened with tallow upon the bottom of a glass or earthen Bason (wherein water is first poured, to the height of two or three fingers breadths) where the heat of the Candle doth so weaken the spring of the Ayr within the Glass, that it wanting the help of the circulating Ayr (always requisite to the perpetuating the motion of bodyes) which is intercepted by the body of water, that in stead thereof, the Water it self circulates, being forc'd thereto by the spring of the Air, that presseth upon it from without, and therefore it riseth up to a great height of the glass-body (as I have sometime seen upon tryal thereof and put out the Candle: which experiment seems somewhat to contradict the former of the Weatherglass, though in reality it doth not; for although there heat makes it descend, but here it makes it ascend: yet if we con-

---

[201] Simpson, *Hydrologia chymica*, from the prefatory "To the Reader."
[202] *Ibid.*, pp. 303-304.
[203] *Ibid.*, pp. 313-315.

sider that in that of the Weather-glass, the Air in it is first thinn'd by heat, before the glass be put into water, and therefore when its condensed by cold, it draws up water, or rather, the water is forced by the outward Spring of the Air, and follows it to an Aequilibrium; but in this last Experiment the glass is inverted into water, without any previous alteration of the Air therein: which being to supply the motion of a body, viz. the burning of the Candle, doth it for a while, but wanting a fresh supply from other Air without, to promote the Circulation thereof (always necessary for the motion of bodies) the want therof makes the strong spring of the air upon the surface of the water, to force up the water it self, into the glass body: ....[203]

Simpson's description of the candle experiment cannot be simply equated with Mayow's, any more than early speculations of the aerial niter can be termed the same as the views of Mayow. The latter's tracts were based far more upon experimental observations than those of earlier Paracelsians, and he regularly discussed his work in corpuscularian terms. Still, the concept of Mayow's nitro-aerial particles was not far removed from the earlier aerial niter of the chemical philosophers, and any discussion of his description of the candle experiment would be difficult without some knowledge of the abundant seventeenth-century literature on this then-perplexing problem. Mayow's interest in the principles, in fermentation in the earth as the cause of the heat of spa waters (here he cites and agrees with Edward Jorden), and specific ferments in the bodily organs mark him as deeply influenced by contemporary chemical medicine[204]—and once again, especially by the work of van Helmont.

The Acid-Alkali Theory after van Helmont

Van Helmont's investigation of the relationship of acids and alkalis in the digestive process had led him to recognize the phenomenon of neutralization. He further postulated that this process took place in the healing of wounds and that such chemical reactions resulted in the production of "neutral substances."[205] These views were rapidly adopted, and extended, by chemists and

[204]Mayow, *Tractatus quinque*, pp. 34-36, 170-182 (reference to Jorden on p. 175), 264-281.
[205]See above, Ch. 5.

physicians both. Interested in similar problems, Franciscus de le Boë, called Sylvius, independently developed a theory of disease that sought to explain illness in terms of either overacidity or alkalinity, and Otto Tachenius extended the medical implications of acids and alkalis. The tenets of late-seventeenth-century iatrochemistry will be discussed at greater length in the following sections of this chapter.

It has been suggested, however, that chemists rapidly accepted this theory as a convenient replacement for the "somewhat tarnished hypotheses of the four elements and the three chemical principles."[206] To be sure, François André had proposed to replace the older elemental systems with one in which the components of bodies were either acid or alkaline salts, in his *Entretiens sur l'acide et l'alcali* (1672).[207] It is also true that Nicolas Lemery considered salt to be one of the five chemical principles (in addition to water, spirit, oil, and earth),[208] and that he considered this principle to be composed of acid and alkaline elements. And, in a lengthy discussion, Lemery described the violent reaction of acids and bases in corpuscular terminology.[209]

According to this interpretation the popularity of the new doctrine spurred Boyle to write his *Reflections upon the Hypothesis of Alcali and Acidum* (1675), in which he argued against current methods of detection such as effervescence. Indeed, the suggestion that an alkali might be identified merely through any reaction of a substance with an acid was quite insufficient. Rather, Boyle argued, it was far better to use the new analytical tests he described, for they would lead to the development of a classification scheme based upon acids, bases, and neutral substances rather than acids and bases alone.[210] If understood in this fashion, Boyle's role is indeed significant.

When, in the *Sceptical Chymist*, Boyle destroyed utterly the experimental basis for the Aristotelian theory of the four elements and

---

[206]Marie Boas, "Acid and Alkali in Seventeenth Century Chemistry," *Archives Internationales d'Histoire des Sciences*, **9** (1956), 13-28 (16).

[207]*Ibid.*, p. 17. See also the section on the acid-base theory in Metzger, *Doctrines chimiques en France*, pp. 199-219 (206-210).

[208]Lemery, *A Course of Chymistry*, p. 3.

[209]Cited by Boas from *ibid.*, pp. 24-27.

[210]A paraphrase of Boas' discussion in "Acid and Alkali," pp. 18-21.

the chemists' theory of the tria prima he offered no real alternative theory and thereby left a void—abhorred by contemporary scientists if not by nature—never really filled by his corpuscular philosophy. With the destruction of the acid-alkali hypothesis he did offer a competing, if not all-embracing, unifying theory. He suggested a workable, demonstrably useful, experimentally based classification of three groups of simple substances: acid, alkali and neutral salts or solutions.[211]

Although by 1675 Boyle may have felt that the three principles were unworthy of serious consideration on the basis of the experiments brought forth on their behalf[212] and that "chemistry in gross" was "a discipline subordinate to physicks,"[213] this opinion was one that was far from universal. Both the elements and the principles were commonly referred to throughout the remainder of the seventeenth century, and surely Lemery himself had based his own theoretical views less upon the acid-base duality than he had upon a five element-principle system common to the period. And indeed, as late as 1778, Macquer felt it necessary to give a detailed discussion of the principles in his *Dictionnaire de chymie*.

The older elemental systems had not been destroyed with the publication of the *Sceptical Chymist*. Similarly, the concept of neutral substances, neither predominantly acidic nor basic, hardly originated with Boyle. This concept can be found in a strictly chemical sense in the work of van Helmont as well as in the publications of his followers. Note has been made of Walter Charleton's espousal of van Helmont's concept of neutralization (1650).[214] There was also Boyle's friend George Starkey, an avowed Helmontian who had written *Natures Explication and Helmont's Vindication* in 1657. The following year he dedicated his *Pyrotechny Asserted and Illustrated* to Boyle. In this text—surely read by Boyle—Starkey discussed the preparation of neutral salts from the reaction of acids and alkalis. "But as concerning many excellent prepara-

[211] *Ibid.*, pp. 20-21.

[212] This view has already been noted in Boyle's *Sceptical Chymist* (1661). His attitude was more critical in the discourse *Of the Imperfection of the Chemist's Doctrine of Qualities* (1675), but he still professed to be willing to be convinced of the truth of the three principles by sounder experiments. (Boyle, *Works*, Vol. IV, pp. 283-284.)

[213] Robert Boyle, *Reflections Upon the Hypothesis of Alcali and Acidum* (1675) in *Works*, Vol. IV, p. 291.

[214] See this chapter above.

tions that be made by Alcalies . . . it is sufficient that they be only made volatile, that is, imbibed with a Spirit, till between them and the alcaly, a neutral Salt be produced, an insipid Flegme being only rejected . . . ."[215] For him this chemical process could ensure the preparation of "pleasant sweet medicine" devoid of the sharpness which characterized the original substances.

In short, although Boyle clearly was disturbed about the current interest in the acid-base hypothesis, it would be only partially true to suggest that this concept had gained currency because of a void in element theory brought about by his attack in the *Sceptical Chymist*. Rather, this hypothesis may best be understood as auxiliary to existing chemical theory exemplified by the widespread appeal of the Helmontian chemical philosophy. Similarly the concept of a tripartite acid-base-neutral salt classification system stems ultimately from Helmontian sources rather than Boyle, while the aqueous analytical tests employed by him form part of a development that can be traced back nearly half a millennium.

Although our examples have been but few, they indicate that the chemically oriented mechanical philosophers cannot be well understood without some knowledge of the chemical philosophy. Thus, although Charleton was a militant mechanist by 1654, Mayow insistent upon corpuscular explanations not long after, and Boyle consciously placing "physicks" over chemistry while demanding that explanations of natural phenomena be based upon matter and motion, these men had all been thoroughly steeped in Paracelsian and Helmontian dogma not long before. And if it is possible to interpret them primarily in terms of the mechanical philosophy while ignoring their earlier views, it may be more profitable to look upon much of their work as a continuation of earlier thought which often appeared at a later date with little more than a corpuscular veneer.

Chemistry and Late-Seventeenth-Century Medicine: The Chemical Medicine of Noah Biggs (1651)

The results of the long coupling of chemistry and medicine by Paracelsians are readily apparent in the works of mid- and late-

[215] Starkey, *Pyrotechny Asserted*, p. 128.

seventeenth-century physicians. Johann Joachim Becher was a trained physician and at the same time the author of chemical works which were eventually among the most influential texts from this period. Similarly, Robert Boyle's interest in medicine betrayed his deep knowledge of the works of earlier iatrochemists.

If the influence of van Helmont was strong on both Becher and Boyle, it was even more so on lesser men. John Webster's attempt to reform the English universities (1654) and George Starkey's *Natures Explication or Helmont's Vindication* (1657) have already been noted. In addition, William Bacon wrote *A Key to Helmont. Or a Short Introduction To the better understanding of the Theory and Method of the most Profound Chymical Physicians* (1682). And for an interesting reflection of mid-seventeenth-century Helmontian medicine it is helpful to return again to *The Vanity of the Craft of Physick*, written by Noah Biggs in 1651.

We have already discussed Biggs' plea to Parliament for the reform of the universities and the strong religious bent of his medical interests.[216] His own allegiance is easy to detect, for he regularly paraphrased sections of van Helmont and proceeded to incorporate them into his own work. As one might expect, chemistry was essential for the study of nature and medicine:

> I praise God who hath been so bountiful to me as to call me to the practise of *Chymistry,* out of the dregs of other Professions: Since *Chymistry* hath *principles* not drawn from fallacious reasonings, but such as are known by nature, & conspicuous by fire; and she prepareth the *Intellect* to penetrate, not the upper deck or *surface* of things, but the deep hold, the *concentrick* and *hidden* things of nature; and maketh an investigation into the *America* of nature . . . .[217]

Like van Helmont, Biggs criticized Paracelsus; and here again we note van Helmont's rejection of Duchesne's experiment that supposedly established the doctrine of signatures. But although he felt that Duchesne had incorrectly interpreted his experiment on the nettle, Biggs did not question the fact that the exterior form

---

[216]See above, Ch. 6. The section on Biggs in the present chapter follows closely the relevant sections in Allen G. Debus, "Paracelsian Medicine: Noah Biggs and the Problem of Medical Reform," *Medicine in Seventeenth Century England,* ed. Allen G. Debus (Berkeley/Los Angeles: University of California Press, 1974).

[217]Noah Biggs, *Mataeotechnia medicinae praxews. The Vanity of the Craft of Physick* . . . (London: Giles Calvert, 1651), p. 57.

was but a hieroglyphic emblem of the interior essence—an essence which might be discovered through the correct use of the fire.[218] But how was the fire best applied? Again reflecting the view of van Helmont—and similar also to the argument offered later by Boyle in *The Sceptical Chymist*—Biggs claimed that too many chemists had used distillation primarily in a search for the principles, that is, for the salt, sulfur, and mercury of any given object. But, he noted, rather than obtaining the essences they sought, they brought about a destruction or transmutation as they proceeded to produce something entirely new.[219] Again, they misunderstood the value of the different fractions in this process, and in their search for essences they limited themselves too severely to distillates. Thus in their *Pharmacopoeia*, the fellows of the London College of Physicians recommended that after a distillation the *caput mortuum* should be discarded, but, Biggs advised, in many cases it is just this salt residue that is the substance that must be collected, since often the weaker exterior elements "may expire by violence of the fire."[220]

For Biggs the correct application of pyrotechny to medicine would result in the extinction of the Galenic doctrine of qualities and temperament as the essential virtues of substances were gradually uncovered.[221] The preparations of herbs as well as metals and minerals would be discovered, and all diseases would be vanquished with them. It was common knowledge that chemists had suffered from the charge that they employed poisons as remedies, but Biggs argued that this charge was most unfair, for chemical methods actually enabled the operator to change poisons to substances sweeter than sugar. The Creator had given the earth to the children of man and

> ... he created the same with all the contents thereof for man. At length taking a view of all things by *Chymistry*, and seeing them more clearly, we repented of our rashnesse, and former foolish ignorance. For in both we adored in suppliant wise with admiration the immense Clemency and wisedome of the *Architect*. For he would not have poisons be poisons, or prejudiciall to us. For he made not death; nor any exterminating medicine in the earth, but rather that by a little in-

[218]*Ibid.*, pp. 48-50.
[219]*Ibid.*, p. 219.
[220]*Ibid.*, p. 117.
[221]*Ibid.*, pp. 46-47.

dustry of ours they might be changed into great pledges of his love, for the use of mortalls, against the rage of future diseases.²²²

It was, of course, of fundamental importance to establish the correct methods of preparation for all medicinal substances, to learn their correct distillation procedures, and to determine the useful fractions to collect. But instead of discussing these subjects in detail, Biggs turned to the problem of the adulteration of chemical medicines by the use of dangerous metallic equipment. Specifically, he proposed to question

> ... whether that *product*, by the *vulgar* and rustick distillation of Apothecaries in the *common leaden stills*, be any other, then an *insipid*, *aqueous* humour, frighted out of the whole meerly by the violence of the *fire* ... without the *elementall, true, genuine, homogeneall* entity of the composition; ... ²²³

Not only do lead stills fail to retain the true force and virtue of the herbs, they result in products that "are noxious, evill, and pernitious, and destructive to the nature of Man in generall, nauseaous to the stomack and loathsome to the sick, wholly different from the nature of the herb of which they are distill'd . . . ."²²⁴ Even the College of Physicians forbade the preparation of acidic substances in brass equipment, because they "yeeld a *decoction* very *ungrateful*, and partaking of a canckerous and *aeruginous* quality." The official *Pharmacopoeia* thus recommended that barberries be boiled in an earthenware vessel to avoid acidic contamination from the brass. But, Biggs noted,

> If the decoctions may partake of a canckerous *aeruginous quality* from the brasen vessel, why also may not the distill'd waters in the *leaden stills* with peuter-heads partake of the *saturnine cerussal quality*, not to be digested by the most *struthiocameline Athanor* of the microcosmical *oeconomy*.²²⁵

If Galen had questioned the use of drinking water that had flowed through lead pipes, why was it that contemporary Galenists insisted that many herbs be distilled in lead equipment? Surely this

---

²²²*Ibid.*, p. 86-87.
²²³*Ibid.*, p. 113.
²²⁴*Ibid.*, p. 117.
²²⁵*Ibid.*, p. 120.

was a more dangerous practice, especially in distilling truly acidic substances such as vinegar; then "the force of the heat, or burntnesse ascending up with a vapour, many times also *acid* and tart, of a *vitriolated* nature and quality, doth infect and *tinct* the waters with a *saturnine cerussal evill quality*."[226] The danger was particularly great, Biggs added, with those herbs that partake of a "*vitriolated* quality, and an *acid* sharp spirit," for those are the substances which have a tendency to "draw *minerall* and *metallick* spirits unto them." Biggs could only conclude that the practice of distilling vinegar in leaden equipment was "childish, ignorant and unadvised.... Unto so low a pitch of stupid ignorance hath vulgar Physitians fallen that [they] so easily and implicitly entertain the customes and traditions of their predecessours, without any examination or due disquisition of the things."[227]

The poisonous qualities of lead could be avoided through the use of glass or glazed vessels, a practice that Biggs had taught to several apothecaries. Distillations performed in such equipment gave forth "the naturall *odour, savour* and *tast* of the *herbs* and *flowers* whereout they be taken, *absque impyreumate*, without any noisome smel or tast of smoake or burning, enjoing its *saline balsamick* conservatory of *vitality,* and from, *putrefaction* and *corruption*; which cannot be perform'd to a *moity* in the common leaden stills."[228]

An important corollary was that lead was also recognized as a danger in the home. Biggs noted that all "Kitchin wenches" knew that vinegar in a pewter saucer would produce a white substance like ceruse which could be scraped off and crumbled in the fingers. This substance easily upset weak stomachs, and as a result "*such vineger cannot be good for the stomach.* Besides it makes it more flat and dead, when it hath sated it self on its proportionate subject, the *peuter saucer*. It partakes then of a *sordid Saturnine evill quality*, pernitious to the tender tunicles of the stomack." Thus, as distilled waters should be prepared from glass vessels, Biggs entreated the "Ladies and Gentlewomen" to "lay aside your peuter saucers, and no more eat vinegar out of them, but instead thereof, you may use

---

[226]*Ibid.*, p. 121. On the history of lead poisoning see Ralph H. Major, "Some Landmarks in the History of Lead Poisoning," *Annals of Medical History*, N.S. 3 (1931), 218-227.

[227]Biggs, *Vanity of the Craft of Physick*, pp. 125, 126.

[228]*Ibid.*, p. 119.

saucers made of fine Earth, or silver plate."[229] Biggs' caveat against the use of lead in the home and in the preparation of medicines is very early—more than a century before Sir George Baker's *Essay Concerning the Cause of the Endemical Colic of Devonshire* (1767)[230]—but it should be noted that he did not describe in detail the symptoms of this affliction other than to refer to the "Saturnine evill quality" that ruined the stomach.

Although Biggs cannot be advanced as a major figure in medical or scientific history, he stands as an excellent example of a mid-seventeenth-century chemical philosopher. Here was an iatrochemist who had written an emotional appeal for educational reform, a man who had called for chemistry to be made the basis for a new science and new medicine. His desired reform would come, he claimed, after the works of the Galenists and the chemists had been reviewed and impartially judged. Then all physicians would clearly acknowledge that chemical methods and chemical theory offered new hope to a discipline that had been stagnant for centuries. However, the reformer Biggs was also a trained chemist, and if he closely adhered to current Paracelsian and Helmontian thought, he was also quite capable of investigating the dangerous medical effects of the use of lead.

Chemistry and the London College of Physicians

It was as a call for medical reform that Biggs' *Vanity of the Craft of Physick* attracted the attention of the London College of Physicians.[231] After 1618 the fellows of the College went beyond the pages of the *Pharmacopoeia* as they followed what would appear to be a genuine interest in chemistry. As early as 1597 Stephen Bredwell had proposed the establishment of a chemical lectureship at the College, since "the art of Chimistrie is in it selfe the most no-

---

[229]*Ibid.*, pp. 128-129.
[230]See R. M. S. McConaghey, "Sir George Baker and the Devonshire Colic," *Medical History*, 11 (1967), 345-360.
[231]See Allen G. Debus, " 'The Devil upon Dun,' a Seventeenth Century Attack on the Quack Chemist," *Die ganze Welt ein Apotheken. Festschrift für Otto Zekert*, ed. Sepp Domandl (Salzburger Beiträge zur Paracelsusforschung, Heft 8: Im Verlag Notring der wissenschaftlichen Verbände Österreichs, 1969), pp. 47-62.

ble instrument of naturall knowledges."[232] Bredwell's suggestion was not acted on at the time, but sometime between 1648 and 1651—when the College was preparing the second edition of its *Pharmacopoeia*—the distinguished chemist William Johnson was placed in charge of a newly created chemical laboratory at the College.[233] Yet, although Johnson himself edited new editions of a short Paracelsian tract and works by the Italian iatrochemist Leonardo Fioravanti and also prepared one of the most popular chemical lexicons of the century, he soon found himself forced to defend the College against other chemists.[234]

The monopolistic privileges of the College were no less hated by the dissident Paracelso-Helmontians of the civil war years than the traditional medical curricula of the universities.[235] Note has been made of the large number of new chemical translations appearing at this time, but no single item was more disturbing to the fellows than the translation of the *Pharmacopoeia* by Nicholas Culpeper (1649, 1653). To this work he added long explanatory

[232] Stephen Bredwell, "To the well affected Reader and peruser of this booke," in John Gerarde, *The Herball or Generall Historie of Plantes* (London: E. Bollifant for B. and J. Norton, 1597), sig. B4^r.

[233] Cecil Wall, H. Charles Cameron, and E. Ashworth Underwood, *A History of the Worshipful Society of Apothecaries of London*, Vol. I (London/New York/Toronto: Oxford University Press for the Wellcome Historical Medical Museum, 1963), p. 93.

[234] William Johnson, ed., *Three Exact Pieces of Leonard Phioravant Knight and Doctor in Physick . . . Whereunto is Annexed Paracelsus his One Hundred and fourteen Experiments: With certain Excellent Works of B.G. à Portu Aquitans, Also Isaac Hollandus his Secrets concerning his Vegetall and Animall Work With Quercetanus his Spagyrick Antidotary for Gun-shot* (London: G. Dawson for William Nealand, 1652). This collected edition is a reprint of John Hester's sixteenth-century translations. Here Johnson defends the position of the College against the publications of Noah Biggs and Nicholas Culpeper in the introductory pieces, sig. A1^r-C1^v. William Johnson, *Lexicon chymicum. Cum obscuriorum verborum, et rerum Hermeticarum, tum phrasium Paracelsicarum, In scriptis ejus: Et aliorum chymicorum . . .* (London: G. D[awson] for William Nealand, 1652). This work was considered of sufficient authority to be republished in the first volume of J. J. Manget's *Bibliotheca chemica curiosa* (1702), pp. 217-291.

[235] See C. Webster, "English Medical Reformers of the Puritan Revolution: A Background to the 'Society of Chymical Physicians,'" *Ambix*, **14** (1967), 16-41. On the significant work being carried out by the members of the College, see C. Webster, "The College of Physicians: 'Solomon's House' in Commonwealth England," *Bulletin of the History of Medicine*, **41** (1967), 393-412.

notes, and, ignoring the chemical interests of the fellows, he complained that most of them had "no more skil in Chemistry, than I have in building houses."[236] Rather than read the directions (which he faithfully translated), he suggested that true students should go to an alchemist to learn how to distill oils properly.

William Johnson took the opportunity to reply to this attack on the College in his edition of the chemical recipe books of Fioravanti (1652). Here he commented on the recent publications of Culpeper and Biggs in separate prefaces. The former was censured for his translation of the *Pharmacopoeia*,[237] while Johnson took it upon himself to defend both the College and its recommended preparations. Biggs, however, was subjected to harsher treatment: Johnson called him the "Helmontii Psittacum,"[238] a man capable of citing only that one author. The charge was a false one, but Johnson was correct in affirming that the *Pharmacopoeia* did list and recommend chemically prepared medicines.

The spirit of revolt that characterizes most of the works written by the chemists of this period was to become heightened in the ensuing years. The exchange between John Webster, Seth Ward, John Wilkins, and John Hall appeared in 1654, while George Starkey elaborated on van Helmont's challenge to the medical establishment in his *Vindication* in 1657. Starkey too suggested a comparison be made between his methods of cure and those of the Galenists, with equal numbers of sick chosen by lot. But he added,

> I will ingage on these grounds, That whatsoever they shall agree to give me for every Cure, I will forfeit twice as much for every one not

---

[236] I have consulted Culpeper's translation of the 2nd ed. of the *Pharmacopoeia*: Nicholas Culpeper, *Pharmacopoeia Londinensis: or the Dispensatory* ... (6th ed., London: Peter Cole, 1659), p. 328. On Culpeper see F. N. L. Poynter, "Nicholas Culpeper and His Books," *Journal of the History of Medicine and Allied Sciences*, **17** (1962), 152-167; F. N. L. Poynter, "Nicholas Culpeper and the Paracelsians," *Science, Medicine and Society in the Renaissance. Essays to Honor Walter Pagel*, ed. Allen G. Debus, 2 vols. (New York: Science History Publications; London: Heinemann, 1972), Vol. I, pp. 201-220; David L. Cowen, "The Boston Editions of Nicholas Culpeper," *Journal of the History of Medicine and Allied Sciences*, **11** (1956), 156-165.

[237] *Three Exact Pieces of Leonard Phioravant*, pp. 9-16, epistle to "Friend Culpeper" by William Johnson.

[238] *Ibid.*, pp. 1-7, "Short Animadversions upon the Book lately Published by one who stiles himselfe Noah Biggs, Helmontii Psittacum" by William Johnson.

cured in the time agreed on; that is, in all Feavers continual, Fluxes, and Pleurisies, in four daies; in Agues (not Hyemal quartanes) in four fits, in Hecticks and Chronical diseases in thirty (at most forty) daies, (now under continual Feavers I comprehend Calentures, small Pox, Measles, &c. which are of that head) provided they will be upon the same lay with me in as many Patients as I have for my share, which let them be divided by tens, they to divide one ten, and I another, and alway the divider to have the five Patients which the chooser leaves; I will engage to perform all my cures without bloud-letting, purging by any promiscuous Purge, or vomiting by any promiscuous Vomit . . . and let them perform their cures how they can . . . and if I wave the combate on these terms, let me be suspended from ever practicing as a vain-glorious boaster, and if they win of me, I will recant my opinion with the greatest both solemnity and ignominy they can devise to enjoyne me to.[239]

Starkey bitterly complained of the "hatred and opposition" he was subjected to "for performing cures so soon and cheap: yet I know that my reward will be a good name when I am gone, and from God hereafter . . . ."[240] He strongly recommended van Helmont's *Pharmacopolium ac dispensatorium modernum* to the reader, and so certain was he of his ability to heal the sick that he promised once more, "if I cure not six for one, I will recant what ever I have written publickly; let them do the same if they dare."[241]

By 1665 the Helmontian chemical physicians of London felt so alienated from the fellows of the College of Physicians that they prepared a declaration indicating their intention to organize their own society.[242] This proposal gained surprising support from influential members of the court, but the ignorance and illiteracy of some of the chemists surely contributed to the collapse of their project. Nor did the plague of 1665 aid their cause. Here, in fact, was

---

[239] George Starkey, *Natures Explication and Helmont's Vindication* (London, 1657), sig. a1 from the Epistle Dedicatory.

[240] *Ibid.*, pp. 225-226.

[241] *Ibid.*, p. 213 and sig. b1.

[242] Sir Henry Thomas was the first to call attention to this group in "The Society of Chymical Physitians" in Underwood, ed., *Science, Medicine and History*, Vol. I, pp. 56-71. The history and background have been discussed in greater detail by Webster, "English Medical Reformers," and P. M. Rattansi, "The Helmontian-Galenist Controversy in Restoration England," *Ambix*, **12** (1964), 1-23.

their long-sought test, but while most of the College physicians fled to the safety of the countryside with their rich patients, the Helmontians remained in London where their ranks were decimated by the dread disease. The failure of the Helmontians to halt the plague and their diminished numbers resulted in a correspondingly lessened influence after 1665. In their stead the apothecaries were to rise to new heights of influence. One senses this in the works of the Helmontian Everard Maynwaring, who, in the early 1670s, was as disturbed by the growing power of the apothecaries as he was by the privileges of the College.[243]

By this time, however, the College no longer prepared chemical medicines. William Johnson had died during the plague, as had his Helmontian adversary Starkey, and the laboratory had been destroyed the following year in the fire of London. As a result, the apothecaries resolved to manufacture and sell chemical medicines. A set of rules, dated January 4, 1672, includes the following preamble:

> Whereas the Compa[ny] of Apothecaries of London have bene publiquely traduced by the Pseudo Chimists of these tymes for their ignorance in the Spagirick part of Pharmacie to vindicate their reputations from those scandalous aspersions & alsoe to assure the Colledg of Phisitians, their patients & all others concerned that all Chimicall preparations shalbe skillfully faithfully & exactly made & sold by an Operator of their owne Fraternity at the Apothecaries hall: The Master Wardens & Court of Assistants of the Company of Apothecaries London are resolved to errect an Elaboratory at their owne hall which shalbe under the inspeccion & governem[en]t of themselves the Master Wardens Court of Assistants & a Com[mit]tee chosen for that purpose.[244]

A laboratory and shop were built which continued in operation until 1922, the first operator being one William Stringer (1672-1673).

The details of the continuing conflicts of privilege between the physicians, the apothecaries, and the distillers throughout the seventeenth and the eighteenth centuries often related to the

---

[243] Rattansi, "Helmontian-Galenist Controversy," pp. 21-23.
[244] Wall, Cameron, and Underwood, *Worshipful Society of Apothecaries*, pp. 149-150.

preparation of chemicals, but they need not concern us in further detail. Still, it is of interest to note that by the third quarter of the century chemical medicines were perhaps as well known to the London crowd as they were to the members of the medical profession. The early accounts of the laboratory include complaints from tenants living near Apothecaries' Hall who objected to the sulfur fumes spewed forth from the kitchen, which was being used by the chemists for their preparations.[245]

Much more indicative of the popular attitude toward chemicals is the contemporary ballad "The Devil upon Dun. Or The Downfall of the Upstart Chymist" (1672). Here a distinction is made between the true chemist and the Paracelsian quack. A sample stanza and the chorus of this popular song will indicate that it was the quack who was the victim of the author's anger:

> 'Mongst all Professions in the Town,
>   Held most in renown,
>   From th' Sword to the Gown,
> The upstart Chymist rules the Roast;
>   For He with his Pill
>   Does ev'n what he will,
>   Employing his skill,
>   Good Subjects to kill,
> That he of his dang'rous Art may boast:
>
> O 'tis the Chymist, that man of the fire,
>   Who by his Black Art
>   Does Soul and Body part:
>   He smoaks us, and choaks us,
> And leaves us like Dun in the mire.[246]

Chemistry and the Blood

The circulation of the blood proved to be a subject of persistent interest to seventeenth-century iatrochemists. We have seen that William Harvey was the friend of Robert Fludd and that he accompanied him in the inspection of the wares of apothecary shows on behalf of the College of Physicians. There can be little doubt

---

[245] *Ibid.*, p. 151.

[246] The complete text of this ballad is given in Debus, "Devil upon Dun," as cited above, n. 231.

that Harvey was aware of the chemical doctrines favored by his colleague. He appealed to the macrocosm-microcosm concept in the *De motu cordis*, and, as Boyle related, he investigated cure by sympathy on the basis of his reading of van Helmont. Fludd found in Harvey a kindred spirit, and it was he who first defended the doctrine of the circulation of the blood in print. But although he had praised Harvey for his anatomical skill, it is clear that Fludd considered this work to be simply ancillary to his own grand vision of mystical circulation common to both the macrocosm and the microcosm.

A more critical approach to the two-world system may well account for van Helmont's failure to refer to the circulation of the blood, but this was hardly the case with other chemists. We have already referred to pertinent passages in the work of his son as well as to others penned by Johann Rudolph Glauber and William Simpson. These men compared macrocosmic circulation (water and life spirit) with the similar circulation of the blood in man.

Here attention may also be drawn to Tobias Schütze, a surgeon of Brandenburg, whose *Harmonia macrocosmi cum microcosmo* (1654) shows the strong influence of the chemical philosophy. Schütze referred to both Sendivogius and Fludd with respect, the latter especially for having described nature in its relationship to God.[247] In many ways this is a typical work, emphasizing the correspondence between the macrocosm and microcosm, astral and polar (magnetic) influences, the theory of sympathy and antipathy, and the elements and the principles.[248] The chemical approach to nature is frequently praised.[249] The chapter on the "Astronomische Anatomie vom Primo Mobili" is of special interest.[250] Here, after having noted the Creator's authorship of the twenty-four-hour revolution of the stars and planets, Schütze discussed the inevitable cyclical influence of this revolution on man. The result was to be seen in the recent discovery of the circulation of the blood. Summarizing the system, Schütze repeated Harvey's

---

[247]Tobias Schütze, *Harmonia macrocosmi cum microcosmo. Das ist/Eine Ubereinstimung der grossen mit der kleinen Welt als dem Menschen/in zwei Theil abgetheilet* (Frankfurt am Main: Daniel Reicheln, 1654), sig.)??(iv^v.

[248]Summarized in *ibid.*, sig.)?(vi^{r-v}.

[249]As an example note the discussion on p. 4.

[250]*Ibid.*, pp. 21-27.

quantitative argument and traced the path of the blood from the heart through the arteries and the veins and then back again to the heart in a continuous circuit.[251] The discussion is both early and clear, but at the same time it is significant that the author, a mystical alchemist, was primarily concerned that this discovery helped to confirm the true harmony between the greater world of nature and the lesser world of man. Credit for this advance was given to Harvey (noted as the first discoverer) and also to Roger Drake and Johannes Walaeus.[252]

The approach to this problem by mid-seventeenth-century chemists can also be seen in John Webster's *Academiorum examen* of 1654. As was Schütze, Webster was strongly influenced by Fludd; however, Webster also noted that the field of anatomy "seems to be growing, and arising towards a *Zenith* of perfection; especially since our never-sufficiently honoured Countryman Doctor *Harvey* discovered that wonderful secret of the Bloods circularly motion." Yet, this was but vulgar anatomy, according to Webster, and had contributed little to the curing of disease. Above all, this type of study was

> ... defective as to that vive and *Mystical Anatomy* that discovers the true *Schematism* or signature of that invisible *Archeus* or *Spiritus mechanicus* that is the true opifex, and dispositor of all the salutary, and morbifick lineaments, both in the seminal *guttula*, the tender *Embrio*, and the formed Creature, of which *Paracelsus, Helmont,* and our learned Countryman, Dr. *Fludd,* have written most excellently.[253]

In short, for Webster, Harvey's work was splendid and justly renowned, but Fludd's work represented a sounder foundation for a new Christian interpretation of the universe. His mystical anatomy of the blood seemed to offer a far more profound understanding than did the more superficial anatomical work of Harvey.

We are fortunate in having Seth Ward's point-by-point reply to Webster. As a mechanist, Ward's reaction to this is somewhat predictable. The new experimental basis of medicine, he wrote, has resulted in the discovery of important anatomical and

---

[251] *Ibid.*, p. 24.
[252] *Ibid.*, pp. 26-27.
[253] John Webster, *Academiarum examen* (1654), p. 74.

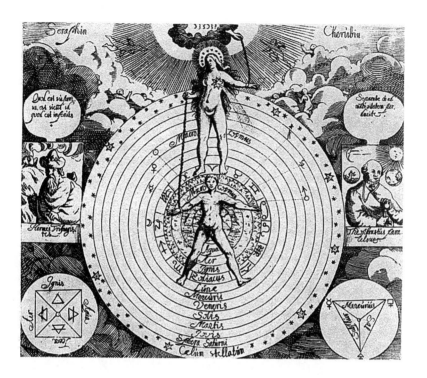

PLATE XXXI The macrocosm and the microcosm, derived from the title page and pages 4-5 of Robert Fludd, *Utriusque cosmi maioris scilicet minoris metaphysica, physica atque technica historia* (Oppenheim: J.T. De Bry, 1617). The portraits of Hermes Trismegistus and Paracelus are based upon those on the title page of the earliest editions of Oswald Crollius' *Basilica chymica*. From Tobias Schütze, *Harmonia macrocosmi cum microcosmo* (Frankfurt am Main: Daniel Reicheln, 1654).

physiological observations—especially the circulation of the blood—but these discoveries are in turn affecting medical practice, and thus it is that the College of Physicians is "the glory of this Nation, and indeed of Europe, for their Learning and felicity, in the cures of desperate Ulcers and diseases, even of the Cancer . . . ." But, he went on to complain, Webster preferred the mystical anatomy of Paracelsus and Fludd, and

> As for his *Postulatum* of discovering the signatures of the Invisible *Archeus* by Anatomy, it is one of his *Rosycrucian Rodomantados*; would he have us by dissection surprize the *anima mundi*, & shew him the impressions of a thing invisible? Yet the Schematismes of nature in matters of sensible bulk, have been observed amongst us, and collections made of them in our inquiries, and when the microscope shall be brought to the highest (whether it is apace arriving) we shall be able either to give the seminall figures of things, which regulates them in their production and growth, or envince them to lye in quantities insensible, and so to be in truth invisible.[254]

The diametrically opposed views of Webster and Ward in regard to just what constituted significant research on the blood is still reflected in Boyle's *Natural History of the Human Blood* thirty years later. Although this work has been lauded as the basis of modern physiological chemistry, A. R. Hall has correctly concluded such praise is overly extravagant.[255] Perhaps it is most interesting to observe that although this is a late composition, it retains a strong Helmontian influence. To be sure, Boyle was intimately aware of the recent natural analyses of blood made by Malpighi and Needham. And Hooke noted (January 24, 1678) that human blood consists of an undefined liquor and "an infinite number of exceedingly small parts, which were plainly perceived to be globular."[256] Nevertheless, Boyle remained more interested in the chemical process of analysis, as did most of his contemporaries. This method, first suggested by van Helmont, found avid sup-

---

[254]Seth Ward, *Vindiciae academiarum*, p. 35.

[255]Here reacting to John F. Fulton's appraisal (*A Bibliography of the Honourable Robert Boyle*, 2nd ed., Oxford: Clarendon Press, 1961, p. 99) in his "Medicine and the Royal Society," in *Medicine in Seventeenth Century England*, ed. Debus, p. 447.

[256]*Ibid.*, p. 444, here quoting Thomas Birch, *History of the Royal Society*, 4 vols. (London, 1756), Vol. III, pp. 346, 349, 374-375, 379-380.

porters among Boyle's friends and associates.[257] Like them Boyle had found five chemical principles in human blood through distillation,[258] but he felt that the chief of these was "spirit," to which he assigned marvelous properties. As a medicine it was to be administered internally as van Helmont had suggested in his "Spiritus vitae,"[259] yet, in the Paracelsian tradition of both Fludd and van Helmont it was important to Boyle to identify this spirit further. A series of tests were described, among them the "putting together [of] the volatile salt of human blood, and the spirit of nitre, with the more fugitive parts of which salt they conceive the air to be plentifully, and some of them to be vitally impregnated."[260] In this case there was observed the strong smell of niter after heating, but this volatile salt or spirit of blood was evidently very similar to the spirit of urine. Actually both are ammonia, but Boyle did not proceed to make this identification on the basis of his laboratory observations. Rather, after having raised this question, he felt confident in offering van Helmont's solution to the problem without further examination: that the spirit of the blood and the spirit of the urine are in fact different.

> ... yet, if we credit the famous *Helmont*, there is a considerable difference between the spirit of human blood and that of human urine, since he somewhere expressly notes (though I remember not the place, nor have his book at hand) that the spirit of human blood cures epilepsies, which is a thing the spirit of urine will not do.[261]

The recent research of Audrey Davis has added appreciably to our knowledge of the impact on disease theory of chemical studies of the blood in the late seventeenth century.[262] Noting Francis Glis-

---

[257] Thomas Willis, *De fermentatione* (1659); Walter Needham (1675) in Birch, *History*, Vol. III, pp. 233-241.

[258] Robert Boyle, "Memoirs for the Natural History of Human Blood" (1684) in *Works* (1772), Vol. IV, pp. 595-659 (incorrectly numbered 759) (610).

[259] *Ibid.*, p. 642.

[260] *Ibid.*, p. 607.

[261] *Ibid.*, p. 623.

[262] Audrey B. Davis, "Some Implications of the Circulation Theory for Disease Theory and Treatment in the Seventeenth Century," *Journal of the History of Medicine and Allied Sciences*, **26** (1971), 28-39; "The Circulation of the Blood and Chemical Anatomy" in *Science, Medicine and Society in the Renaissance*, ed. Debus, Vol. II, pp. 25-37.

son's search for the origin of the blood, she has examined the results of his discovery that the liver could not be its source. With the assumption that the blood must contain its own power of sanguifaction, it became theoretically respectable to suggest that the blood might be treated directly in cases involving the "blood diseases." However, such treatment took on chemical overtones with the rejection of the traditional humors and the increasingly acceptable belief that the blood might be discussed and analyzed as a chemical fluid. Here Boyle's analyses were influential, but it is evident that he had read van Helmont carefully on this subject; and his work was hardly unique in any case. He must also have been intimately aware of the studies of his colleague at Oxford, Thomas Willis. Willis had not only discussed the distillation of the blood in detail, he had gone further to assign pathological roles to the five chemical fractions he had isolated.

In short, late-seventeenth-century iatrochemists found the study of the blood a subject of special interest. They were to be among the staunchest defenders of the circulation of the blood. This concept seemed to them to have finally demolished the Galenic humors, since rapid circulation argued against the accumulation of a humoral excess anywhere in the body. In addition—at least for some—the discovery of this bodily circulation offered proof of long-sought geocosmic and macrocosmic influences. They were no less intrigued by van Helmont's chemical distillations of the blood, which suggested that it might be viewed as a chemical fluid. Here was a new approach to pathology in line with the contemporary interest in transfusion and infusion experiments, and one which offered a new method in the therapy of the diseases associated with the blood.

The Chemical Medicine of Willis and Sylvius

Thomas Willis (1621-1675) and Franciscus de le Boë, called Sylvius (1614-1672), were the two most influential iatrochemists of the late seventeenth century. It would be impossible, and unnecessary, to consider the totality of their work here. It is important, however, to note that both accepted the traditional search for a universal system of nature and that their work represented blends of philosophical and scientific work representing numerous schools. Here it will be necessary to indicate the relationship of their views to the Paracelso-Helmontian tradition. With both authors this connection is perhaps stronger than has been commonly admitted in the past.

In the case of Willis we have a young man born into humble circumstances whose ability was recognized at an early age.[263] Educated at Oxford, he graduated bachelor of medicine in 1646. A religious and political conservative, Willis was at the same time in the very center of English scientific life as a member of the Oxford *virtuosi*. Here he was both friend and colleague of John Wallis, Robert Boyle, and many others, including Webster's antagonists John Wilkins and Seth Ward. Appointed Sedleian Professor of Natural Philosophy at the restoration (1660), Willis later moved to London (1666), where he spent his remaining days.

The positive contributions of Willis have long been recognized and need not be catalogued again. He was a skilled anatomist, and his work on the brain includes a description of the main cerebral arteries indicating what is now called the "circle of Willis." His discussion of the nerves was the most detailed prepared to that date, while his descriptions of diabetes and hysteria stand as medical landmarks. It is of special interest to note that his work clearly indicates his awareness of corpuscular thought.

It has been possible to discuss Willis in terms of an often ill-defined new science as a "modern." Yet at the same time he has always been recognized for his chemical interests, and he may perhaps best be interpreted against a background of contemporary iatrochemical thought. His most recent biographer has noted that Willis was attracted at an early age by the chemists, and that this was to remain a dominant theme in his theoretical work.[264] His first publication, the *De fermentatione* (1659), was surely written with a

---

[263]On Willis see Dr. Hansruedi Isler, *Thomas Willis 1621-1675. Doctor and Scientist* (New York/London: Hafner, 1968). Also, although short, the discussion of José Maria López Piñero ("La iatroquimica de la segunda mitad del siglo XVII" in *Historia de la medicina*, ed. Pedro Laín Entralgo, Barcelona: Salvat Editores, 1973, Vol. IV, pp. 279-295) is one that the present author finds excellent both for Willis and Sylvius. Note should also be taken of the fact that a number of papers included in *Medicine in Seventeenth Century England*, ed. Debus, reassess the relationship of Willis to Sydenham.

[264]See esp. Isler, *Willis*, pp. 58-62.

PLATE XXXII Thomas Willis (1621-1675).

deep knowledge of the current chemical literature. Typically, Willis first discussed the "beginnings of natural things." Of the three recognizable systems the elements of the Aristotelians were of the least value in his eyes; indeed, it was little more helpful "to say an House consists of Wood and stone, as a Body of four Elements." Far more popular was the use of the atoms of Democritus and Epicurus. The proponents of this system suggested that

> ... all Natural effects ... depend upon the Conflux of Atoms diversly figured, so that in all Bodies, there be Particles Round, Sharp, Foursquare, Cylindrical, Chequer'd or Streaked, or of some other Figure; and from the divers changes of these, the Subject is of this or that Figure, Work, or Efficacy.[265]

But although the corpuscular philosophy offered a "mechanical" system by which to avoid the occult qualities of the Peripatetics, this approach "rather supposes, than demonstrates its principles," while it suggested notions which are "remote from sense."

In contrast to both the atomists and the Aristotelians, there was a third opinion, that of the chemists, which deserved the most weighty consideration. Willis thoroughly accepted the evidence of distillation analyses, which he felt proved the existence of both active (spirit, sulfur, salt) and passive (water, earth) principles. However, his definition was essentially a practical one, since "I mean by the name of Principles, not simple and wholly uncompounded Entities, but such kind of Substances only into which Physical things are resolved, as it were into parts, lastly sensible."[266] And, indeed, "because this hypothesis determinates Bodies into sensible parts, and cuts open things as it were to the life, it pleases us before the rest."[267]

For Willis the cause of change in natural phenomena could most frequently be ascribed to fermentation. This was "an intestine motion of Particles, or the Principles, of every Body, either tending to the perfection of the same Body, or because of its change

---

[265] Thomas Willis, "Of Fermentation or the Inorganical Motion of Natural Bodies" in *Practice of Physick...*, trans. S. Pordage (London: T. Dring, C. Harper and J. Leigh, 1681), p. 2. I am grateful to Lester S. King for the use of his copy of this rare volume.

[266] *Ibid.*, p. 3.

[267] *Ibid.*, p. 2.

into another."²⁶⁸ In his description of the effects of this process he proceeded from the mineral to the vegetable and the animal kingdoms. In traditional alchemical fashion he wrote of the earth as a "pregnant womb" in which metals and minerals grow.²⁶⁹ Like Edward Jorden and van Helmont, he ascribed this growth to local ferments. In the course of such fermentations saline particles are freed in the form of vapor. When this in turn is mixed with earthy matter or moistened with water, the result will be fountains and spa waters. The vapors resulting from subterranean fermentations thus "perpetually breath forth, and are diffused through the whole Region of Air."²⁷⁰

In vegetable life Willis identified the fermentation process with the growth of seeds, while his description of animal life centered on man. In the latter case he placed special emphasis on the life spirit, which resulted from a local fermentation in the heart. This spirit is distributed throughout the body as the blood circulates.²⁷¹ However, other local ferments exist in the body as well. Among them Willis paid special attention to the fermentation processes that exist in the bowels, the genital parts, and the spleen. Disease also results from a fermentation process—as does its cure.²⁷² Willis did not refer to local *archei* in the body, but there seems little doubt that his concept of the ferment was inspired by the work of van Helmont.²⁷³ His retention of the animal soul, his belief in metallic and mineral growth, and the evident influence of contemporary chemical theory all contributed to an approach far more favorable to vitalistic than iatromechanistic thought.

Willis constantly resorted to chemical analogies in his explanations. Disease was described as a fermentation process, as was its attendant fever, which could be seen as a boiling of the blood.²⁷⁴ In the Paracelsian tradition he explained the action of the muscles as the result of the reaction of nitrous and sulfureous spirits.²⁷⁵ Like

---

[268] *Ibid.* (1684 ed.), p. 8.
[269] *Ibid.* (1681 ed.), p. 10.
[270] *Ibid.*, p. 11.
[271] *Ibid.*, p. 13.
[272] *Ibid.*, pp. 14-16.
[273] Here I am completely in agreement with Isler, *Willis*, p. 61.
[274] Willis, "Of Fevers," in *Practice of Physick* (1681), p. 57.
[275] Willis, "Of Convulsive Diseases," *ibid.*, p. 2 (separate pagination).

most chemists he employed distillation frequently. Blood was shown to be composed of five principles through distillation analysis[276]—and Willis used the same "spagyric" means as an acceptable method of examining urine.[277] He employed the distillation analogy in his description of the separation of the animal spirits in the brain:

> ... the Brain with a Scull over it and the appending Nerves, represent the little Head or Glassie Alembic, with a Spunge laid upon it, as we use to do for the highly rectifying of the Spirit of Wine: for truly the Blood when Rarified by Heat, is carried from the Chimny of the Heart, to the Head even as the Spirit of Wine boyling in the Cucurbit, and being resolved into Vapour, is elevated into the Alembic; where the Spunge covering all the opening of the Hole, only transmits or suffers to pass through the more penetrating and very subtil Spirits, and carries them to the snout of the Alembic: in the mean time, the more thick Particles are stayed, and hindred from passing. Not unlike this manner, the blood being restrained within by the Skull, and its *menynges*, as by an Alembick, are drunk up by the spungy substance of the Brain, and there being made more noble or excellent, are derived into the Nerves, as so many snouts hanging to it. In the mean time the more crass or thick Particles of the blood, being hindred from entring, are carried back by Circulation . . . .[278]

In his more famous work on *The Anatomy of the Brain* Willis reiterated that "the business of extracting the animal Spirits is performed even as a Chymical Elixir."[279]

Willis broke with van Helmont in regard to bloodletting. This, he felt, was likely to prove beneficial if applied early; if the hoped-for result did not occur then, phlebotomy should not be performed later. With a well-advanced fever in the blood this practice would only weaken the patient,[280] and at such a time drugs would prove more effective, for they would help to produce a corrective fermentation in the blood which would cure the patient. Here, however, he may be seen as an eclectic. A glance at the *Pharmaceutice rationalis*

---

[276]Willis, "Of Fevers," *ibid.*, p. 59.
[277]Willis, "Of Urines," *ibid.*, p. 1 (separate pagination).
[278]Willis, "Of Fermentation," *ibid.*, pp. 14-15.
[279]Willis, "The Anatomy of the Brain," *ibid.*, p. 88.
[280]This subject is discussed in more detail by Davis, "Implications of Circulation Theory," pp. 35-36.

indicates that Willis neither adhered exclusively to chemicals nor to Galenicals. While strong chemicals (i.e., antimony compounds for purging) were recognized as important, he felt that more gentle herbs were often required; when neither worked, the physician might resort to sympathetic methods.[281]

Thomas Willis may be compared with his slightly older contemporary, Sylvius.[282] Educated at Sedan, Leiden, and Basel, Sylvius went on to practice medicine at Leiden (1638) and Amsterdam (1640-1641). In 1658 he was appointed Professor of Medicine at Leiden, where he remained until his death fourteen years later. As in the case of Willis, the career of Sylvius may be interpreted in terms of an official acceptance of iatrochemical doctrine by the university medical establishment. But again we find someone who was much more than a fervent partisan of Paracelsian medicine or of a mystical alchemical philosophy of nature. Sylvius had read widely in all areas of medical and scientific literature, and the thesis he presented for his bachelor of medicine degree at Leiden (1636) defended the pulmonary circulation within a Galenic context.[283] However, if his work displays a residual Galenism, it is also possible to find in it a keen awareness of contemporary anatomical thought, Cartesian mechanism, and, of course, iatrochemical theory.[284]

Sylvius was clearly Galenic in his adherence to the doctrine of innate heat and in his interest in the animal spirits.[285] At the same time he may be considered one of the earliest proponents of the circulation of the blood. During his first three years at Leiden (1638-1641), he had taught anatomy using as a text the *Institutiones anatomicae* of Casper Bartholinus (1611). His published lectures

---

[281]Willis, "Pharmaceutice rationalis" (Part I) in *Practice of Physick* (1681), pp. 24, 48 (separate pagination); (Part III), *ibid.*, pp. 126, 137. It will be noted that Willis' attitude toward sympathetic cure is very similar to that of Boyle.

[282]The most recent studies of Sylvius are those of López Piñero (see above, n. 263) and King, *Road to Medical Enlightenment*, pp. 93-112. The most detailed account will be found in E. D. Baumann, *François de la Boë Sylvius (1614-1672)* (Leiden: E. J. Brill, 1949). See also Partington, *History of Chemistry*, Vol. II, pp. 281-290.

[283]López Piñero, "La iatroquimica," p. 281.

[284]*Ibid.*, p. 284.

[285]*Ibid.*, p. 285.

PLATE XXXIII Franciscus de le Boë Sylvius (1614-1672). Line engraving by C. V. Dalen after himself.

show that he was teaching the circulation at this early date,[286] and it was Sylvius who converted the Galenist professor of medicine at Leiden, Johannes Walaeus, to the new system (1640).[287] It was also during the first period at Leiden that Sylvius and Descartes became close friends. Descartes' work influenced that of the young iatrochemist and, as López Piñero has noted, many of the followers of Sylvius can be counted as Cartesians.[288] Still, when Sylvius compared the interpretations of Harvey and Descartes on the pulse, he endorsed the view of Harvey rather than that of his friend.[289]

Above all, Sylvius insisted on the independence of his own thought. His willingness to correct Bartholinus on the blood flow can be cited as one example of this; another is his approach to chemistry. For him chemistry was essential for a proper understanding of nature and consequently was fundamental for medicine.[290] However, he rejected the charge of being dependent on the work of van Helmont and called attention to the fact that he had lectured on his system at Leiden prior to 1641 while the work of van Helmont had not appeared in print *in toto* until several years later.[291] Surely Sylvius made fermentation an integral part of his physiological system, but there was a real difference between his conception (chemical dissolution in the presence of water, salt, moderate heat, and air)[292] and the primary emphasis on force found in the texts of the Belgian systematist. Another distinction was Sylvius' relatively small interest in the concept of a universal chemical philosophy.

The use of chemical explanation in physiological phenomena by Sylvius is best seen in his discussion of digestion.[293] Here he argued that mastication prepares food for the necessary fermenta-

---

[286]Franciscus de le Boë, Sylvius, "Dictata ad Casparis Bartholini institutiones anatomicas," in *Opera medica* (Amsterdam: Daniel Elsevier and Abraham Wolfgang, 1680), p. 891. In his funeral oration Luca Schacht praised Sylvius for teaching the circulation of the blood at such an early age (*ibid.*, p. 928).

[287]López Piñero, "La iatroquimica," p. 282.

[288]*Ibid.*, p. 286.

[289]Sylvius, "Disputationes medicarum decas," in *Opera medica*, p. 42.

[290]*Ibid.*, p. 13.

[291]Sylvius, "Praefatio ad lectorem," *ibid.*, p. 10.

[292]Sylvius, "Disputationes medicarum decas," *ibid.*, pp. 11-12.

[293]Sylvius' views on digestion are well treated by King in his *Road to Medical Enlightment*, pp. 98-104. His discussion is the basis of the present account.

tion process in the stomach. A necessary preparatory agent is saliva, which contains the required fermentative forces (water, salt, and spirit). The stomach contributes a moderate heat (fire), which originates in the heart and is communicated through the blood. A chemical process of separation then follows in the intestines as a result of the bile and the pancreatic juices (alkaline and acid). The resultant chyle contains the volatile spirit of food plus small amounts of alkaline salt and acid spirit, and in this form it enters the blood stream. On reaching the left ventricle of the heart the still imperfect blood is further heated and rarefied before going on to the lungs and the right ventricle, where it is rarefied again and made ready for circulation. The circulatory process itself provides a continuous nourishment for the internal fire of the heart, at the same time maintaining vitality throughout the body.

Sylvius recognized the similarity of life and combustion, and he gave a significant role to the aerial niter in the respiratory process.[294] Convinced that fire could not exist without air, he still had to account for the maintenance of a constant heat level in the body—and therefore in the heart. This was the proper role of the niter, which exerted a moderating influence on the fire and contributed to the condensation of the rarefied blood in the lungs.

An admirer of the work of Glauber, Sylvius was further influenced by that chemist's belief in the dual nature of salt (acid-alkali).[295] Salt was to become an essential part of his concept of fermentation, and the result was an explanation of disease in terms of an excess of either acid or base. Acrimonious influences were ascribed to specific fluids (especially the lymph, the saliva, the pancreatic juice, and the bile) and characteristic diseases resulted from their acidic or alkaline natures. Whether or not Sylvius' concept of disease as a chemical imbalance of this sort originated in his reading of van Helmont, it is true that his views found many followers in the last third of the seventeenth century. The *Praxeos medicae idea nova* (1667) was reviewed in the *Philosophical Transactions* in 1668 and note was taken of the revised edition of 1671.[296] Boyle

---

[294] *Ibid.*, pp. 104-105.

[295] *Ibid.*, pp. 111-112; López Piñero, "La iatroquimica," pp. 282, 285.

[296] *Philosophical Transactions of the Royal Society of London from their Commencement in 1665 to the Year 1800, Abridged*, ed. Charles Hutton, George Shaw, and Richard Pearson, 18 vols. (London: C. and R. Baldwin, 1809), Vol. I, pp. 289, 595.

was to firmly reject the acid-base theory of disease, but other authors (most notably Raymond Vieussens in France and Otto Tachenius in Italy) did much to promote these ideas throughout Europe.[297] Lester King has further traced this concept in the medical systems of the mechanists Friedrich Hoffmann and Hermann Boerhaave,[298] but as yet we know far too little about the course of iatrochemistry in the eighteenth century.

A Newtonian Postscript

There has been no attempt in the course of the present chapter to define what is meant by the mechanical philosophy of the seventeenth century. Perhaps it is not necessary to do so, since most of the chemical philosophers themselves did not favor mechanical explanations. An additional problem is that historians of science have too often failed to define what they mean by the term. Most frequently the reader will find "mechanical philosophy" applied to a corpuscular-atomistic system freed of its original atheistic context and with an emphasis placed on quantitative features—that is, on matter and motion. The Epicurean revival of the seventeenth century influenced many authors to insist on the importance of the characteristic shape of the atoms of any given substance, and this was to become a useful explanatory tool for chemists such as Charleton and Boyle.

Still, the stark quantitative aspects of Epicurean atomism were less than satisfying for many steeped in the vitalistic chemical philosophy. The sizes and shapes of atoms did not in themselves seem to explain the most significant part of natural phenomena—the living world. Therefore it was to become important not only to make the atomic theory theologically respectable; it was necessary to accommodate it as well to the beliefs of those physician-scientists who sought a key to all nature in the study of chemistry. The works of early-seventeenth-century iatrochemists—men such as Sennert and Jungius—display a less starkly quantitative concept of atomism. Here we find atoms as *minima* associated with innate powers or properties which are not simply shape or motion. It

[297]López Piñero, "La iatroquímica," pp. 290-293.
[298]King, *Road to Medical Enlightment*, p. 111.

was in this form that atomic theory could, and did, appeal to chemical philosophers.

Recent research has indicated that as early as 1598, Arcerius, a Freisian philologist, identified Moses with the traditional founder of atomic philosophy, the Phoenician Moschus.[299] Here is a definite attempt to incorporate atomism within the Mosaic philosophy long before Gassendi attempted to Christianize the works of the Epicureans. We need not then be surprised to find Sennert referring to the derivation of Democritean atomism from the Hebrews,[300] or Fludd devoting one experiment in the *Philosophicall Key* to the probability that "all things wer made of Atoms as some Philosophers haue gessed."[301] Van Helmont referred frequently to atoms, and his concept of seeds with their inherent forces can be construed as a form of corpuscular theory. The young Boyle, already deeply influenced by Helmont, Gassendi, and Charleton, could take as a major proposition in the *Sceptical Chymist* that "*It seems not absurd to conceive that at the first Production of mixt Bodies, the Universal Matter whereof they among other Parts of the Universe consisted, was actually divided into little Particles of several sizes and shapes variously mov'd.*" Boyle went on to suggest the necessity of local motion associated with these particles "whether we chuse to grant the Origine of Concretions assign'd by *Epicurus*, or that related by *Moses.*"[302]

Late-seventeenth-century chemists varied considerably among themselves in their discussions of atomism. A wide spectrum of opinions could be introduced at this point. There were some Hermeticists who sought physical truth in this hypothesis, while others rejected it decisively, as we have noted earlier. For the most part, however, chemical philosophers found atomic theory a subject of lesser interest, since their attention was more likely to be focused on life processes. Accordingly, this subject did not general-

---

[299] J. E. McGuire and P. M. Rattansi, "Newton and the 'Pipes of Pan,' " *Notes and Records of the Royal Society of London,* **21** (1966), 108-143 (130). In the same paper these authors refer to Newton's reference to Moschus in the line of early atomists (115).

[300] *Ibid.* (here citing Sennert's *Hypomnemata physica,* 1636).

[301] Robert Fludd, *A Philosophicall Key* (c. 1619) (Trinity College, Cambridge, Western MS 1150 [0.2.46]), Book 2, Ch. 4, "The sixt experiment" (fol. 79$^{r, v}$).

[302] Boyle, *The Sceptical Chymist* (1661; Dawson reprint, 1965), pp. 37.

ly play a major role in their own works, and we have not emphasized it here. Nevertheless, because of the significance of Newton in the development of the mechanical philosophy, it is of real interest to note that he, too, was influenced by the chemical philosophy and well aware of the problems posed by contemporary atomic thought.

Newton's extensive alchemical papers are characterized by a fascination with traditional medieval alchemical authors.[303] In dealing with this complex individual we find that the mechanist of the *Principia* departs when we turn to the *De natura acidorum*, the Queries of the *Opticks*, the works on chronology and Biblical exegesis, and above all, the unpublished manuscripts. In the far-ranging discussion found in Query 31 of the *Opticks* Newton wrote of the elementary matter "that God in the beginning form'd . . . in solid, massy, hard, impenetrable, moveable Particles, of such Sizes and Figures, and with such other Properties, and in such Proportion to Space, as most conduced to the End for which he form'd them . . ."[304] Newton was to reject shape in favor of attractive power as an explanatory factor in the interaction of these particles. In addition it seemed to him

> . . . that these Particles have not only a *Vis inertiae*, accompanied with such passive Laws of Motion as naturally result from that Force, but also that they are moved by certain active Principles, such as that of Gravity, and that which causes Fermentation, and the Cohesion of Bodies. These principles, I consider, not as occult Qualities, supposed to result from the specifick Forms of Things, but as general Laws of Nature, by which the Things themselves are form'd; their Truth appearing to us by Phaenomena, though their Causes be not yet discover'd.[305]

In his research relating to the revision of the *Principia*, J. E. McGuire has discussed draft manuscripts referring to these active principles which seemed to be required to explain fermentation

---

[303] The present discussion is founded primarily in Allen G. Debus, "Motion in the Chemical Texts of the Renaissance," *Isis*, **64** (1973), 4-17 (15-16).

[304] Isaac Newton, *Opticks*, with a foreword by Albert Einstein, an introduction by Sir Edmund Whittaker, a preface by I. Bernard Cohen (New York: Dover, 1952), p. 400.

[305] *Ibid.*, p. 401.

and other effects involving motion.[306] Many of Newton's examples we would term strictly chemical, but in a draft passage related to Query 23 of the 1706 *Opticks* he emphasized the relationship of motion to life and he expressed his belief that "the laws of motion arising from life or will may be of universal extent."[307] These texts clearly indicate that Newton believed there were two kinds of motion in nature: one, based on a *Vis inertiae*, which can be described by the passive laws of motion in the *Principia;* and the other, a motion associated with life or will, the effects of which were normally seen in examples taken from chemistry or from life. However, the source of this second motion is clearly indicated by Newton's preoccupation with fermentation. The connection with van Helmont is confirmed by reference to his lengthy note taken from the *Opera omnia* of the earlier author. Summarizing the chapter on the *Blas meteoron* that we have discussed in an earlier chapter, Newton commented,

> As in man the passions arouse heat and cold, thus, in the air the notion and the light of the stars particularly arouse local and alterative motions, motions which I signify by the word Blas. Is Blas the operative principle of motion and change in the winds, the rain and the snow in the same fashion that Gas is the material principle?[308]

It is clear that the concept of the *Blas* was of interest to Newton primarily because of its universality as a principle of motion and change. His frequent reference to fermentation and other chemical

---

[306] J. E. McGuire, "Force, Active Principles, and Newton's Invisible Realm," *Ambix*, **15** (1968), 154-208.

[307] *Ibid.*, p. 205.

[308] "Sicut in homine passiones excitant calorem et frigus, sic motus et lumen stellarum praesertim vero motus excitant motus locales et alterativos in aere, quorum utrumque nomine Blas significo. Estque Blas operationem (?) principium motionum et mutationum in ventis pluvia nive et sicut Gas materialiter." King's College, Cambridge, Keynes, Newton MS 16, beginning "pag. 21. Causae et initia naturalium"; see fol. 2^V referring to "Blas Meteoron, p. 50." P. M. Rattansi has informed me that the page references are to the 1667 edition of the van Helmont *Opera omnia*. Additional evidence of Newton's interest in van Helmont may be found in *A Catalogue of the Portsmouth Collection of Books and Papers Written by or belonging to Sir Isaac Newton* ... (Cambridge: Cambridge University Press, 1888). Here, under Sec. II: Chemistry, see p. 13, No. 8, "Extracts apparently from van Helmont," and p. 16, No. 4, "De Peste. Van Helmont."

examples coupled with the specific notes in his Helmontian manuscript make it most likely that his "active principles" are, in effect, an application or a modification of the *Blas*.

Further evidence of the influence of van Helmont on Newton may perhaps be seen in the formulation of the third law of motion. Although this is frequently noted for its originality, it may be questioned whether Newton's serious consideration of van Helmont's discussion of action and reaction might not have contributed to his formulation of the law. As noted earlier, van Helmont had bitterly attacked the idea of action and equal and opposed reaction in his tract "Ignota actio regiminis."[309] He considered this a Galenic position which was the very root of the dictum that contraries cure. In a lengthy rebuttal—again based on the physics of the *Blas*—van Helmont proceeded to "prove" his own contention by extending the problem from medicine to motion, specifically discussing the subject of collisions. All of this was designed to show that under no circumstances could there be an equal and opposed reaction to any given action. Unfortunately, Newton's notes—if they were made—of this section of the *Opera omnia* of van Helmont are not included in the King's College collection of his alchemical papers. Nevertheless, it is not unlikely that his general interest in van Helmont's work coupled with his specific interest in the *Blas* would have lead him to ponder over, and eventually reject, this statement by van Helmont.

The work of Rattansi has shown that Newton's concept of an "aether" may be connected with his study of the alchemical *spiritus*,[310] while Westfall has emphasized Newton's union of the Hermetic and the mechanical traditions in his "Hypothesis of Light."[311] Newton's deep interest in the Helmontian work and especially the concept of the *Blas* can only reinforce the belief that in his search for a total physics of the earth and the heavens he felt free to consider the views of the chemical philosophers as well as those of the mechanists.

---

[309]See above, Ch. 5.

[310]P. M. Rattansi, "Newton's Alchemical Studies," in *Science, Medicine and Society in the Renaissance*, ed. Debus, Vol. II, pp. 167-182.

[311]Richard S. Westfall, "Newton and the Hermetic Tradition," *ibid.*, pp. 183-198.

Conclusion

It is customary to interpret the last half of the seventeenth century in terms of a rapid acceptance of the mechanical philosophy accompanied by a decline of the influence of the ancient philosophers and the mysticism of the Renaissance. It has not been our task here to discuss the work of Bacon, Descartes, the iatromechanists, or other figures who normally play a major role in this drama. Rather, by ranging widely in this chapter, an attempt has been made to show that the normal "script" of the scientific revolution is at least partially unsatisfactory.

It is evident that there was no gradual decline of interest in the chemical philosophy in the middle years of the seventeenth century. Both traditional alchemy and chemical medicine reached new peaks of influence at that time. Pansophic and utopian schemes were in tune with the works of earlier mystical alchemists, Paracelsians, and Rosicrucians, while the Helmontian corpus offered an updated approach to the chemical philosophy. Then it was still possible to write lengthy and popular works favoring Paracelsian medicine—as did Rhumelius. Respected authors could also sharply criticize the atomic revival of the period. Those chemists who became converts of the mechanical philosophy frequently did so in such a way that their views were little more than a thin veneer to cover their earlier beliefs. Walter Charleton's "rejection" of his earlier ideas was not a total one, John Mayow's interest in nitro-aerial particles was in fact little removed from the aerial niter of the chemical philosophers before him, and the strong influence of earlier chemists permeates the work of Robert Boyle. With Becher and Stahl a different chemical tradition developed. Yet, if the phlogiston theory was to be less dependent upon corpuscular explanations than that of the chemical mechanists, its fundamental tenets were no less tied to Paracelsian origins.

The influence of chemistry on the medical world did not diminish during this period. Seemingly minor figures such as Noah Biggs and John Webster demanded that the medical curriculum be fundamentally changed to conform with the views of the chemical philosophers, while other dissident chemists gained surprisingly strong support in their challenge of the monopolistic practices of the Royal College of Physicians. Yet, those Helmontians who wished to destroy the medical establishment failed. Far more successful were Sylvius and Willis, who occupied major chairs of medicine at Leiden and Oxford. Less partisan than Starkey, Biggs, or Webster, these men found relatively little difficulty in being accepted by their colleagues. In this they were aided by the developments of the preceding century, when chemistry had made very real progress in medicine. Surely by the time of Sylvius and Willis iatrochemistry had become a recognized force in the medical world.

To be sure, the period was one of synthesis; perhaps as many syntheses existed as there were authors. Yet, as far as the chemical influence is concerned, the major figure of the period was certainly van Helmont. This is evident both in the writings of physicians and in the work of those whose interest might properly be titled "terrestrial astronomy." Even Newton had read the *Opera omnia* in detail and had considered deeply van Helmont's views on the origin of motion and its relationship to the life force. If Sylvius alone denied the influence of van Helmont, it should be remembered that his statement was made in sharp reaction to a charge of near plagiarism.

In short, the chemical philosophy remained influential in the late seventeenth century. With some authors it persisted in a form little removed from that seen early in the century. With a far larger number it became very much a "Helmontian" chemical philosophy. However, with the shifting patterns of the period one finds most often the incorporation of fundamental concepts and themes of the chemical philosophy in new contexts and new syntheses. In almost all of these cases Jean Baptiste van Helmont emerges as a major influence.

8
POSTSCRIPT

[The mind of the chemical philosopher] should always be awake to devotional feeling, and in contemplating the variety and the beauty of the external world, and developing its scientific wonders, he will always refer to that infinite wisdom, through whose beneficence he is permitted to enjoy knowledge; and, in becoming wiser, he will become better, he will rise at once in the scale of intellectual and moral existence, his increased sagacity will be subservient to a more exalted faith, and in proportion as the veil becomes thinner through which he sees the causes of things, he will admire more the brightness of the divine light by which they are rendered visible.

Sir Humphry Davy (1829)*

IT WOULD BE PREMATURE to attempt a definitive assessment of the influence of the chemical philosophy. To date we do not know enough about either the texts or their authors. That this is so is largely a result of the development of the history of science as a field. Perhaps far too many historical studies of sixteenth- and seventeenth-century science and medicine have emphasized "positive" achievements to the exclusion of other factors; surely far too few studies have sought the totality of scientific and medical themes evidenced in the literature of this period. Only when that task has been accomplished will it be possible to present a more balanced account of the scientific revolution. Nevertheless, even the limited survey of the literature we have presented here indicates that the chemical philosophy was the subject of considerable discussion and debate in both the sixteenth and the seventeenth centuries. On this basis alone its role should be assessed in future studies of the scientific revolution.

---

*Sir Humphry Davy, Bart., *Consolations in Travel, or The Last Days of a Philosopher* (Philadelphia: John Grigg, 1830), p. 233 (from "Dialogue the Fifth, The Chemical Philosopher").

The Chemical Philosophy in Retrospect

The study of early modern chemical texts quickly reveals that a fundamental change occurred in the century and a half that separated the student years of Paracelsus from the earliest publications of Robert Boyle. Chemical works dating from the early sixteenth century for the most part can be categorized as metallurgical, pharmaceutical, or alchemical (transmutational). An Agricola, a Brunschwig, or a Norton might on occasion think of chemistry as the "key to nature" in the sense of the pseudo-Lull, but it was not his goal to establish a broad chemical philosophy which could be used as a basis for a new understanding of the universe. Nor did these authors revel in the thought of revolt. Rather, they openly utilized the works of the ancients as auxiliary to their own, and they sought no serious confrontation with the medical or the educational establishments. Indeed, they travelled in paths already well worn by their medieval predecessors.

It was the destiny of Paracelsus to think consciously of a unified chemical approach to nature. And yet, had this been his sole accomplishment his work might well have had far less impact than it did. To be sure, his thought was deeply influenced by the alchemical, metallurgical, and pharmaceutical information he had come in contact with as a youth, but Paracelsus was also deeply imbued with Germanic folk medicine, with the religious concepts of the early years of the Reformation, and with a heady mixture of Renaissance Hermetic, Gnostic, and Cabalistic thought. It is little wonder that he reacted sharply against the traditional learning he was exposed to in the universities. Rather than embracing Aristotle and Galen, Paracelsus turned to chemistry and natural magic. In a deliberate spirit of reform he sought a new medicine and a new natural philosophy based upon the Holy Scriptures and nature rather than Greek philosophy.

The diffuse and scattered texts of Paracelsus were gathered and systematized by his followers in the third quarter of the sixteenth century. It is with these writers that we see a conscious effort to develop a true chemical philosophy of nature and medicine. Through their work runs a deep undercurrent of Christianity. They argued that the ancients had been atheists who had improperly turned to mathematics as a guide in their search for knowledge when they should have sought truth in nature and medicine. Here they found inspiration in the words of Ecclesiasticus 38. Certainty for them was only to be found in the book of God's revelation, the Bible, and in the book of God's Creation, nature. Since in both cases illumination came only through the grace of the Lord, students could safely ignore the natural philosophy and medicine of antiquity. Indeed, the universities were either to be condemned or thoroughly reformed. Youth was told to reject a formal education in preference for the firsthand study of nature. It is little wonder that we find a high proportion of sixteenth- and seventeenth-century Paracelsians who were wandering scholars in their student years. While encouraged to study the world through observation, they were also told to read and reread the early chapters of Genesis as a proper basis for their new knowledge. Here the story of the Creation was to lead them quickly to the problem of the elements. They were agreed that fire could be rejected as a true element, but this seemed to bring about the collapse of the entire system of the ancients: if there were no four elements, there could be no four humors nor four temperaments. Rather, laboratory experience indicated the existence of a triad: salt, sulfur, and mercury. The consequences of tampering with the traditional elemental system account for the many folio pages devoted to the theory of the elements in Renaissance medical and scientific texts.

The Paracelsian chemical philosopher, however, did not confine his study of nature to the problem of the elements. A deeply rooted belief in the macrocosm-microcosm analogy led to the conviction that all nature is alive and interconnected. The study of astral emanations, the growth of metals, and the origin of mountain streams must lead to information about man's own body. In this fashion medicine was considered an integral part of the study of nature; a universe understood in chemical terms required analogous processes in the microcosm of man. But here the rejection of the four elements had required the rejection of the mainstay of ancient medicine—the four humors. And if humoral pathology could not be accepted, an additional problem was to develop from the new interest expressed in folk medicine, where an emphasis on cure by similitude clashed directly with the Galenic dictum that contraries cure. The Paracelsian was indeed encouraged to seek out poisons, alter them by his chemical art, and then prescribe them to his patient. In short, one finds repeatedly that the Paracelsian chemical philosophers differed significantly from the medieval medical alchemists. Not only did they seek a new philosophy that united chemistry and medicine, they were revolutionaries who consciously opposed the establishment on many crucial points.

And yet, if we find a number of common themes in the late sixteenth and the early seventeenth centuries, it would be far from the truth to indicate that all of these chemical physicians were in agreement. Some were led to a mystical alchemical understanding of the universe, while others flatly rejected Paracelsian mysticism in a search for the practical benefits of the new chemical preparations. Though these benefits were frequently seen as medical, they were not exclusively so: the sixteenth-century texts of Palissy and Plat emphasized the application of chemical knowledge to the reform of agriculture, while the later work of Glauber described the chemical production of food concentrates and the use of chemical warfare as a basis for the prosperity of Germany.

Interesting as the chemical philosophers are in their own right, it would still be possible to relegate them to a lesser place in history if they had been ignored or summarily dismissed by their contemporaries. Such was not the case. In his own day Paracelsus did not loom as a major figure, but after his death a rapidly growing interest in his writings made the debate over his work one of the crucial scientific and medical problems of the late sixteenth and seventeenth centuries. Both his theology and his rejection of ancient philosophy and medicine were questioned at an early date, and both enemies and supporters were forced to digest the weighty critique of the conservative Thomas Erastus. Others sought a more moderate course, and a number of eclectics—best typified by Guinter von Andernach—attempted to graft the best of the chemical innovations to the most revered tenets of ancient theory. But if a middle ground was eventually to prove the most appealing to savants who had no desire to engage in polemics, this did not prevent a series of bitter debates that centered on the medical and chemical views of the reformers. The Parisian conflict of the opening years of the seventeenth century attracted the attention of scholars in all parts of Europe, while the mid-century debate in England between Webster and Ward involved a rethinking of the curricula of the English universities. Great names as well as lesser figures took note of the Paracelsians. Daniel Sennert, clearly one of the most distinguished men of seventeenth-century medicine, was accused of founding his own chemical sect, and Francis Bacon read the works of the Paracelsians with great care.

The publication of the Rosicrucian documents in the second decade of the new century resulted in the many folio volumes on the "chemical physics" of the macrocosm and the microcosm by Robert Fludd. These in turn brought forth another series of books by Mersenne, Kepler, Gassendi, and others which vividly expressed the alarm felt by mathematicians and the early mechanical philosophers who viewed a "new philosophy" defined in Paracelsian terms as a potent threat to their own goals. The complexity of the period is nowhere more evident than in Mersenne's acceptance of the chemical philosopher van Helmont and in his curt dismissal of Harvey as a disciple of Fludd. To be sure, van Helmont may properly be contrasted with Fludd, but his own plea for a universal chemical philosophy is perhaps the most eloquent of any penned by members of this school.

Acceptance and Rejection: The Question of Influence

It may be argued with conviction that the work of the chemical philosophers must be understood if only because of the sheer volume of their work. But one can go further to state that the reaction against the chemical universe by those recognized as key figures in the scientific revolution requires an assessment of the Paracelsians. Still, a recognition of the chemical philosophers on either of these terms—and I believe both are justifiable—ignores the question of direct influence.

The goal of the present volume has centered on the description and discussion of the chemical philosophy itself rather than its influence. Nevertheless, I think that the question of whether it exerted a direct influence may be answered mainly in the affirmative. Any discussion of the relationship of the chemical philosophers to the mechanists of the seventeenth century requires a careful examination of specific concepts in the work of both schools. It is unfortunate that relatively little of this type of research has been accomplished to date. As a result one finds in the recent literature some who would trace the most significant aspects of the scientific revolution to the work of the Hermetic philosophers of the fifteenth and sixteenth centuries, while others see little if any value in the work of pre-Boylean chemists.[1] As we have noted, the evidence would indicate that neither position is entirely correct.

The influence of the chemical philosophy is most dramatic in medicine. By the early years of the seventeenth century many Galenists who were highly critical of the Paracelsians were quite willing to accept many chemical remedies. The inclusion of chemical preparations in the *Pharmacopoeia Londinensis* of 1618 was to formalize this acceptance by the medical establishment. No less significant was the impact of the chemical philosophy on medical theory. Although the concept of a fluid imbalance in disease has never been totally eradicated from medicine, the validity of the traditional humoral theory was under heavy attack during the sixteenth and seventeenth centuries. The Paracelsian views of macrocosmic-microcosmic growth processes combined with views of localization and specificity favored by chemists were to prove major factors in the development of seventeenth-century medical theory. In addition, chemical explanations of physiological processes developed by van Helmont and Sylvius found numerous adherents in the latter part of the century. There is no better example than Thomas Willis, who saw no difficulty in blending elements of contemporary corpuscular thought with Helmontian chemical philosophy. Another case is revealed in the concept of irritability, where the classic works of Haller (1752) and Francis Glisson (1677) have been shown to have roots in the Galenic corpus,[2] but also, as Pagel has indicated, Glisson's work may properly be compared with that of Harvey, who considered motion and sensibility intrinsic to the blood.[3] It is clear that the involuntary

[1] The best example of the first viewpoint is found in Frances A. Yates, *The Rosicrucian Enlightenment* (London/Boston: Routledge and Kegan Paul, 1972), while the most recent papers and reviews by A. R. Hall and Marie Boas Hall are illustrative of the latter opinion.

[2] Owsei Temkin, "The Classical Roots of Glisson's Doctrine of Irritation," *Bulletin of the History of Medicine,* **38** (1964), 297-328 (325).

[3] Walter Pagel, "Harvey and Glisson on Irritability: With a Note on van Helmont," *Bulletin of the History of Medicine,* **41** (1967), 497-514 (504-509); see also Walter Pagel, "William Harvey Revisited—Part 1," *History of Science,* **8** (1969), 1-31 (11-13).

natural movements described by Harvey have a similarity to the concept of the Helmontian *archeus*.

There are a number of specific cases which reflect influence in areas other than medicine. The chemical analysis of spa waters is one of special significance, since this forms the basis of modern aqueous analysis. Here it can be shown that sixteenth-century Paracelsians adopted a medieval tradition which they proceeded to develop and expand, and then Boyle in turn integrated their work into his own. Another case of direct influence involves the concept of the aerial niter. Already discussed in the *Liber azoth* (1591) in a Paracelsian context, the belief in a vital aerial saltpeter can be traced through the works of numerous seventeenth-century authors before it appeared in corpuscular form in the work of John Mayow (1668, 1674). Indeed, we meet with Paracelsian and Helmontian concepts only slightly altered in the works of the late-seventeenth-century natural philosophers who formed the membership of the Royal Society of London. The problem of the elements, the origin of mountain streams and metals, the improvement of agricultural yields through the use of salts—all these topics exercised the minds of the Society's *virtuosi* no less than they had their chemical predecessors. And if an understanding of the phlogiston theory is dependent upon the writings of Stahl, surely his work cannot be understood without Becher, whose *Physica subterranea* is clearly within the tradition of the chemical philosophy.

The corpuscular framework of late-seventeenth-century science was also dependent upon an earlier worldview. We have seen a distinct sixteenth-century connection between atomism and the Mosaic philosophy, and even Boyle insisted that the adoption of a corpuscular philosophy did not rule out the acceptance of a natural philosophy based on Genesis. It is interesting to note that the new interest in Democritean atomism in the earliest decades of the seventeenth century was associated with chemical philosophers such as Fludd, Sennert, and van Helmont. At that time it was the mechanist Mersenne who viewed this revival as an incipient danger to a new mathematically oriented science. The favoring of innate powers within the atoms, corpuscles, or seeds of substances came to distinguish this chemical influence from the work of physically oriented natural philosophers later in the century, who emphasized purely physical properties such as size, shape, and motion. Yet, even here, the distinction is not easily drawn. The one-time Helmontian Walter Charleton continued to refer to the "life" properties of corpuscles when discussing living organisms in his later works, even though he ignored them when concerned primarily with less complex phenomena associated with the physics of motion.

It is difficult to avoid Isaac Newton in any discussion of the influence of the chemical philosophy. On the surface nothing seems more opposed to the severe mathematical approach of the *Principia mathematica* than the conscious effort in the Helmontian works to avoid a mathematically inspired "Philosophia nova." But going beyond the published scientific texts of Newton, one is forced to deal with a great mass of chemical manuscripts. Here, in addition to an interest in traditional alchemy, it is seen that Newton had read van Helmont with considerable care. Indeed, his postulation of a motion associated with living organisms—and distinguished from the motion of corpuscles in the study of local motion—can most easily be understood as a reflection of the Helmontian *Blas* or principle of motion. This acceptance of a concept based in the chemical literature stands in contrast to Newton's rejection of other significant views expressed in the work of van Helmont. It is interesting to find that van Helmont discussed—and denied—the concept of action with a corresponding equal and opposed reaction, a position (originally based upon an opposition to Galen) that Newton would have found impossible to uphold in the physics of motion. Newton's possible familiarity with this Helmontian text may well prove to be an important key to the development of his third law of motion.

The case of Newton is a difficult one, but it is also somewhat typical among the problems of dealing with the influence of the chemical philosophy. Many authors who are most frequently referred to as mechanical philosophers deal with key themes inherent in the writings of the iatrochemists. And if their judgment of the earlier chemists is frequently harsh, their works no less reflect the very real influence of these chemists. To a surprising degree one finds the rejection of the chemical philosophy of the Paracelsians and the Helmontians coupled with an acceptance of many of its key concepts. Major figures of late-seventeenth-century science, men such as Boyle, Willis, and Newton, become neither "ancients" nor "moderns," but rather eclectics who consciously relied on their chemical heritage when they found it useful.

Aftermath

If it is a complicated process to assess the impact of the chemical philosophy on the seventeenth-century scientific revolution, it is far more difficult to grasp its later influence. In large measure eighteenth-century science has been judged in terms of the triumph of Newtonian science. There have been numerous studies of the Leibniz-Newton debate on the origin of the calculus, the work of Euler and the Bernoullis, and the defeat of the Cartesians in the French Academy. But while the *Principia* has been a subject of frequent papers and monographs, no less attention has been centered on the "experimental" Newton. Cohen has connected eighteenth-century optical and electrical research as well as the events leading to the chemical revolution with the deep and long-lasting influence of Newton's *Opticks* (1704).[4]

The eighteenth-century scientific mainstream can be characterized as "Newtonian" provided one understands the adjective to refer to a science divorced from Newton's chemical and alchemical speculations—other than those published in the Queries of the *Opticks* or the *De natura acidorum*. But to find evidence of the continued interest in the earlier world picture of the chemical philosophers, one need not dig deeply. The phlogiston theory is one example, but there is additional evidence of a strong interest in traditional chemistry throughout the century. Van Helmont's *Opera* was reprinted as late as 1707, and an abstracted version appeared two years later in Wallachia. The most extensive edition of alchemical texts ever prepared is the two-volume folio edition of J. J. Manget printed at Geneva in 1702. Another large collection, the *Musaeum Hermeticum*, was republished in 1738, while the chemical writings of Basil Valentine appeared in five eighteenth-century editions as late as 1775. The massive three-volume *Deutsches theatrum chemicum* (1728-1732), edited by Friedrich Roth-Scholz, made available to the German reader a collection that rivaled the

[4] I. Bernard Cohen, *Franklin and Newton: An Inquiry into Speculative Newtonian Experimental Science and Franklin's Work in Electricity as an Example Thereof* (Philadelphia: The American Philosophical Society, Memoirs, Vol. 43, 1956).

earlier Latin works of Zetzner and Manget. In France the four-volume alchemical collection of Richebourg was published between 1751 and 1754, while Pernety's alchemical *Dictionnaire Mytho-Hermétique* is separated by only a few years from the first edition of Macquer's *Dictionnaire de chymie* (1776).[5]

The interest in a chemically inspired vitalistic approach to nature can be supported by means other than listing the vast number of alchemical texts printed in the eighteenth century. The triumph of Newtonian science over the proponents of Descartes in the Academy in the fourth decade of the century was paralleled by a strong popular reaction against this mechanized universe seen in the many utopian novels published after that date.[6] This same dissatisfaction is reflected in the article on "Chymie" published in the *Encyclopédie* (1753),[7] while Robinet wrote at length in opposition to a purely mechanistic universe which he felt was insufficient as a means of describing life processes.[8] Editions of the works of Willis (Venice, 1720) and Sylvius (Paris, 1771) attest to the continued appeal of a vitalistic seventeenth-century iatrochemistry on the Continent until well into the new century. Knight has properly

[5] Most of the works referred to here have been noted earlier. Exceptions are the last four: Friedrich Roth-Scholz, ed., *Deutsches theatrum chemicum*, 3 vols. (Nuremberg: Adam Jonathon Felssteckern, 1728, 1730, 1732); Jean Maughin de Richebourg (attributed editorship), *Bibliothèque des philosophes chimiques*, 4 vols. (Paris: Cailleau, 1741, 1754; this edition is a considerably augmented version of two parts that had appeared in 1672 and 1678); Dom Antoine-Joseph Pernety, *Dictionnaire Mytho-Hermétique, dan lequel on trouve les allégories fabuleauses des poetes, les métaphores, les énigmes et les termes barbares des philosophes Hermétiques expliqués* (Paris: Bauche, 1758); Pierre-Joseph Macquer, *Dictionnaire de chymie . . .*, 2 vols. (Paris: Lacombe, 1766).

[6] Here I refer to the unpublished doctoral dissertation by Nina Rattner Gelbart, "Science in Enlightenment Utopias: Power and Purpose in Eighteenth Century French Voyages Imaginaires" (The University of Chicago, 1973).

[7] D. Diderot, ed., *Encyclopédie, ou dictionnaire raisonné, des arts et les métiers . . .*, 17 vols. (Paris: Briasson, David, Le Breton & Durant, 1751-1765). The article "Chymie," unsigned, is in Vol. III (1753), pp. 408-437. Diderot's antimechanistic views have been noted by Charles Coulston Gillispie in his "The *Encyclopédie* and the Jacobin Philosophy of Science: A Study in Ideas and Consequences," *Critical Problems in the History of Science*, Marshall Clagett, ed. (Madison: University of Wisconsin Press, 1962), pp. 255-289.

[8] This forms a central theme of the doctoral dissertion of Terence Murphy at the University of Chicago, "Jean Baptiste Robinet and Eighteenth-Century Materialism" (1975).

called our attention to the anti-mechanistic side of eighteenth-century science, while noting that for many savants chemistry remained "an exciting discipline promising to reveal the unity of matter—a sphere in which mechanics had merely scratched the surface."[9]

In addition to serving as the center for the publication of collected alchemical editions, eighteenth-century German publishers reflected the interest of their public also in the numerous printings of single works. Sudhoff lists over fifty eighteenth-century editions of various works of Paracelsus, nearly all from Central Europe.[10] Leibniz, around 1709, could still attack the "fanatical ... *Philosophia Mosaica*" of Robert Fludd, "which saves all phenomena by attributing them to God immediately and by miracle."[11] But all judgments were not so harsh. A revived interest in the Rosicrucian texts resulted in an annotated translation of Fludd's *Apologia* (1782),[12] while a major defense of alchemy was penned by Johann Christian Wiegleb in 1777.[13] Early in his own intellectual development Johann Wolfgang Goethe was influenced by alchemical and Paracelsian works, and it has been suggested that he modeled Faust's description of his father on the life of van Helmont.[14] His late *Farbenlehre* (1805-1810) exhibits his alchemical interests in the

---

[9]D. M. Knight, "The Physical Sciences and the Romantic Movement," *History of Science*, 9 (1970), 54-75 (63).

[10]Karl Sudhoff, *Bibliographia Paracelsica* (Berlin, 1894; reprint Graz: Akademische Druck — u. Verlagsanstalt, 1958), pp. 631-669.

[11]Gottfried Wilhelm Leibnitz, *New Essays Concerning Human Understanding*, trans. Alfred Gideon Langley (2nd ed., Chicago: Open Court, 1916), p. 63. Professor Garlon Treash of Mount Allison University has pointed out to me that Leibniz criticized Fludd a number of times on the same point [see his *Critical Thoughts on the General Part of the Principles of Descartes* (1692) (Art. 64), *Specimen dynamicum* (1695; Part 1), *On Nature Itself* (1698; Sect. 10), Letter to Louis Bourquet (August 1715)]. See also the important early paper by Walter Pagel, "Helmont, Leibniz, Stahl," *Sudhoffs Archiv für Geschichte der Medizin*, 24 (1931), 19-59.

[12]Robert Fludd, *Schutzschrift für Aechtheit der Rosenkreutzergesellschaft . . . übersetzt von Ada Mah Booz* (Leipzig: Adam Friedrich Böhme, 1782).

[13]Dietlinde Goltz, "Alchemie und Aufklärung: Ein Beitrag zur Naturwissenschaftsgeschichtsschreibung der Aufklärung," *Medizin historisches Journal*, 7 (1972), 31-48.

[14]Alice Raphael, *Goethe and the Philosopher's Stone: Symbolical Patterns in 'The Parable' and the Second Part of 'Faust'* (London: Routledge and Kegan Paul, 1965), pp. 11-13.

content of a work that has been described as a "frontal assault upon Newton."[15] And if the Scottish physician John Brown (1735-1788) found little support in England in his quest for certitude in a philosophical system of medicine, this was hardly the case in Germany, where many scholars after 1795 found Brown's system a revelation.[16] Among them was Friedrich von Schelling, for whom medicine and science were based upon a knowledge of the natural world first determined through philosophical reflection. Schelling, like earlier Paracelsians, thought of medicine as a "supreme" science which rested on irreducible vital principles.[17] Again, in Hans Christian Oersted's *Soul in Nature*, the scientist is pictured as having been given the power to govern nature so that mankind might progress in spiritual development. Here too is pictured an all-embracing natural philosophy with religious overtones.[18]

Eighteenth-century England seems to have been less interested in the chemical tradition of the Paracelsians. One can point to a few scattered publications of individual texts in the early years, and also to a few members of the Royal Society who professed a belief in transmutation, but for the most part one senses a real dominance of the Newtonian experimental tradition in a form that was relatively untouched by its Helmontian background. In the later years of the century it is possible to refer to Thomas Taylor's (1758-1835) translations of neo-Platonic texts or to early figures in the Romantic movement such as Samuel Taylor Coleridge (1772-1834) for themes common to the chemical philosophy, but here we are outside the normal bounds of the sciences. Of more interest is the very influential *Natural Theology* (1802) of William Paley (1743-1805), which includes an insistence on the "correspondency" of all inhabitants of land and sea.[19] For Paley the true theist is willing to argue by analogy, and his veneration of God "will incline him to attend with the utmost seriousness, not only to

[15]Knight, "Physical Sciences," p. 57.

[16]Guenther B. Risse, "Kant, Schelling, and the Early Search for a Philosophical 'Science' of Medicine in Germany," *Journal of the History of Medicine and Allied Sciences*, **27** (1972), 145-158 (146).

[17]*Ibid.*, pp. 154-155.

[18]David M. Knight, "The Scientist as Sage," *Studies in Romanticism*, **6** (1967), 65-88 (82-88).

[19]William Paley, D.D., *The Complete Works . . . and a Life of the Author, by the Rev. Robert Lynam, A.M.*, 4 vols. (London: George Cowie et al., 1825), Vol. III, p. 175.

all that can be discovered concerning him by researches into nature, but to all that is taught by a revelation, which gives reasonable proof of having proceeded from him."[20] Here we meet once more with the familiar two-book doctrine of knowledge.

Perhaps the most interesting English work of the early nineteenth century in this connection is Sir Humphry Davy's (1778-1829) *Consolations in Travel* (1830). This was a work of great popularity that was composed during Davy's final illness. Although it may be dramatically contrasted with Davy's earlier published experimental work, much of it is similar to lecture notes and other material written over the previous thirty years.[21] Davy included a chapter on "The Chemical Philosopher" in the *Consolations*. Here he emphasized that religious certainty was intimately connected with a knowledge of nature.

> The true chemical philosopher sees good in all the diversified forms of the external world. Whilst he investigates the operations of infinite power guided by infinite wisdom, all low prejudices, all mean superstitions disappear from his mind. He sees man, an atom amidst atoms, fixed upon a point in space; and yet modifying the laws that are around him by understanding them; and gaining, as it were, a dominion over time, and an empire in material space, and exerting on a scale infinitely small, a power seeming a sort of shadow or reflection of a creative energy, and which entitles him to the distinction of being made in the image of God, and animated by a spark of the divine mind. Whilst chemical pursuits exalt the understanding, they do not depress the imagination or weaken genuine feeling; whilst they give the mind habits of accuracy, by obliging it to attend to facts, they likewise extend its analogies; and, though conversant with the minute forms of things, they have for their ultimate end the great and magnificent objects of nature. They regard the formation of a crystal, the structure of a pebble, the nature of a clay or earth; and they apply to the causes of the diversity of our mountain chains, the appearances of the winds, thunder storms, meteors, the earthquake, the volcano, and all those phenomena which offer the most striking images to the poet and the painter. They keep alive that inextinguishable thirst after

---

[20] *Ibid.*, p. 319.

[21] In a private communication (June 3, 1974) Dr. Knight adds that "it seems probable that Davy's audiences at the Royal Institution and those who heard his conversation would have heard most of the arguments put forward in the *Consolation*."

knowledge, which is one of the greatest characteristics of our nature; for every discovery opens a new field for investigation of facts, shows us the imperfection of our theories. It has justly been said, that the greater the circle of light, the greater the boundary of darkness by which it is surrounded. This strictly applies to chemical inquiries; and hence they are wonderfully suited to the progressive nature of the human intellect, which, by its increasing efforts to acquire a higher kind of wisdom, and a state in which truth is fully and brightly revealed, seems as it were to demonstrate its birthright to immortality.[22]

These words might easily have been endorsed by Davy's seventeenth-century predecessors who also had called themselves chemical philosophers.

With Davy, Paley, and the German scientists of the nineteenth century we seem to have ranged far from the direct influence of Paracelsus and his immediate followers. Now we find ourselves referring to the Romantic movement, to *Naturphilosophie* and to natural theology, terms that are normally confined to the late eighteenth and nineteenth centuries. And yet, these categories include within them themes and concepts we have found integral to the works of the chemical philosophers of the Renaissance. Common to both groups is the quest for unity in a nature created by divinity—and an ever-recurrent emphasis on the importance of chemistry and medicine in the search for that understanding. Surely it would be a mistake to attempt to equate the seventeenth-century chemical philosophy with nineteenth-century views of nature, but there are enough concepts in common and enough evidence of the survival of a non-Newtonian vitalistic natural philosophy in the eighteenth century to encourage us to seek the existence of any valid relationships that might exist.[23] Here we will only note the great foresight of George Sarton, who commented in one of his last essays on the importance of the Paracelsians. For him an understanding of Paracelsism was requisite for "medicine,

[22]Davy, *Consolations in Travel*, pp. 225-226.

[23]Walter Pagel was acutely aware of the parallels that might be drawn between seventeenth- and nineteenth-century biological thought in his "Religious Motives in the Medical Biology of the Seventeenth Century," *Bulletin of the Institute of the History of Medicine*, 3 (1935), 97-128, 213-231, 265-312. Here see especially pp. 286-297.

chemistry, and philosophy not only in the Renaissance, but also throughout the seventeenth and eighteenth centuries."[24] With this statement we can fully concur.

[24]George Sarton, *Appreciation of Ancient and Medieval Science During the Renaissance* (1st ed., 1955; New York: Perpetua edition, 1961), p. 5.

BIBLIOGRAPHY

THIS BIBLIOGRAPHY has been arranged alphabetically by primary and secondary sources and then chronologically when an author is represented by more than one work. As mentioned earlier, no attempt has been made to list all sixteenth- and seventeenth-century texts in this field; only those that have actually been consulted will be found here. It should also be noted that a number of English translations have been cited. The present research originated with a paper on the English Paracelsians prepared for Professor W.K. Jordan's seminar on Tudor England at Harvard University in 1956. In the course of that work the author soon became aware of the rich, and largely untapped, mine of sixteenth-and seventeenth-century translations of continental chemical and medical works. These early translations are delightful to read—far more so than modern renditions. Thus, although many new translations have been made for this book, early versions are quoted whenever possible.

It seems unnecessary to comment on all of the translations listed in the bibliography. A few examples may suffice. As one would expect, there are real problems in using such works, since the translators were not all of the same caliber. There are, indeed, some very poor translations. Paracelsus and the alchemists have excited the interest of occultists for a long period of time, and translations made by such enthusiasts are to be used with caution. A prime example is the French translation of the *Medicina spagÿrica tripartita* (1648) of Johannes Pharamundus Rhumelius [*Médecine spagyrique (1648)*, trans. Pierre Rabbe (Paris: Bibliothèque Chacornac, 1932)]. The work is highly condensed, and through the elimination of essential words and phrases the translator has actually altered the meaning of the original text. This twentieth-century effort may be compared with the rather remarkable translation of Daniel Sennert's *De chymicorum cum Aristotelicis et Galenicis consensu ac dissensu liber* (1619) made by Nicholas Culpeper and Abdiah Cole [*Chymistry Made Easie and Useful, Or, The Agreement and Disagreement of the Chymists and Galenists* (London: Peter Cole, 1662)]. In this case key sentences have been translated in their entirety, and generally with considerable accuracy. But if long passages are missing, the finished work still maintains the thrust of Sennert's argument and leads the reader accurately and swiftly to the subjects of interest in the Latin original.

A final example is John Chandler's translation of van Helmont's *Ortus medicinae* and *Opuscula* (1648; English translation, 1662). Here we have the case of the translation of the *opera omnia* of a major author. Chandler's work has been frequently attacked for its cumbersome English and the use of words which today might be misinterpreted. Although specific cases can be raised to support this view, the same charges could be levelled against most of the works originally written in English from that period. In any case, it would be unfortunate if such criticism resulted in the neglect of Chandler's work, for it is surely a most useful one. Chandler was successful in preserving van Helmont's general meaning, and his translation may be used with relative safety in its nontechnical sections. Even in the more technical chapters the version offered is frequently acceptable. And if Chandler's translation may be faulted for not being entirely accurate, it remains significant for vividly preserving the flavor of seventeenth-century science and medicine.

Little need be said about the secondary sources. An attempt has been made to use all recent studies of value. Still, it seems inevitable that some published research will be overlooked and that other papers and books will be found too late for their effective use. I was fortunate in being able to work from the galley proofs of Walter Pagel's "Van Helmont's Concept of Disease—To Be or Not To Be? The Influence of Paracelsus" [*Bulletin of the History of Medicine*, **46** (1972), 419-454], but other important works bearing on problems discussed in these pages appeared in print later. Among these may be cited Owsei Temkin's *Galenism: Rise and Decline of a Medical Philosophy* (Ithaca/London: Cornell University Press, 1973), Charles B. Schmitt's "Toward a Reassessment of Renaissance Aristotelianism" [*History of Science*, **9** (1973), 159-193], and Audrey B. Davis' *Circulation Physiology and Medical Chemistry in England 1650-1680* (Lawrence, Kansas: Coronado Press, 1973).

In the final stages—after most of the type for this volume had been set—there appeared several new studies that would have been discussed in the text had they appeared earlier. Among these are two parts of Joseph Needham's long-awaited volume on chemistry: *Science and Civilisation in China*, Volume V: *Chemistry and Chemical Technology*, Part II, *Spagyrical Discovery and Invention: Magisteries of Gold and Immortality* (with Lu Gwei-Djen) (Cambridge: Cambridge University Press, 1974); and Part III, *Spagyrical Discovery and Invention: Historical Survey, from Cinnabar Elixirs to Synthetic Insulin* (with Ho Ping-Yü and Lu Gwei-Djen) (Cambridge: Cambridge University Press, 1976). In another new work Owen Hannaway has explored the conflicting opinions of Andreas Libavius and Oswald Crollius in his *The Chemists and the Word. The Didactic Origins of Chemistry* (Baltimore: Johns Hopkins Press, 1975). My thoughts on this work have just appeared in a review, "Paracelsians and Pragmatists," *Times Literary Supplement*, June 18, 1976, p. 746. And finally, Betty Jo Teeter Dobbs has presented the results of her study of Isaac Newton's alchemical manuscripts in *The Foundations of Newton's Alchemy or "The Hunting of the Green Lyon"* (Cambridge: Cambridge University Press, 1976). In the case of this work my reflections are to be found in a review that is yet to be printed in *Centaurus*. Although other recent titles might well have been mentioned, these are representative of the ever-increasing volume of truly important research in this area of Renaissance science and medicine.

Primary Sources

Académie Royale des Sciences, Paris. *Histoire de l'Académie Royale des Sciences depuis son établissement en 1666 jusqu'a 1686*. Paris: Martin, Coignard and Guerin, 1733.

Agricola, Georgius. *De re metallica*. Trans. H. C. and L. H. Hoover. 1912. Reprint New York: Dover, 1950.

―――. *De natura eorum quae effluunt ex terra libri IV*. 1545. In *Schriften zur Geologie und Mineralogie* (from the *Ausgewählte Werke*, ed. Hans Prescher. Berlin: VEB Deutscher Verlag der Wissenschaften, 1956).

Agrippa, Henry Cornelius. *De occulta philosophia. Libri tres*. [Cologne]: J. Seter, 1533.

―――. *Three Books of Occult Philosophy or Magic. Book One—Natural Magic* . . . . Ed. Willis F. Whitehead. 1897. London: Aquarian Press, 1971.

Albertus Magnus. *Book of Minerals*. Trans. Dorothy Wyckoff. Oxford: Clarendon Press, 1967.

Andreae, Johann Valentin. *Christianopolis. An Ideal State of the Seventeenth Century.* Trans. with an historical introduction by Felix Emil Held. New York: Oxford University Press 1916.

Anthony, Francis. *Medicinae chymiae, et veri potabilis auri assertio.* Cambridge: C. Legge, 1610.

———. *Apologia veritatis illucescentis pro auro potabile.* London: J. Legatt, 1616.

———. *The Apologie, or, defence of a verity heretofore published concerning a medicine called Aurum Potabile.* London: J. Legatt, 1616.

Aristotle. *The Metaphysics.* Books I-IX. Trans. Hugh Tredennick. London: Heinemann; New York: Putnam's, 1933.

———. *Metaphysics.* Newly translated as a postscript to natural science with an analytical index of technical terms. Trans. Richard Hope. New York: Columbia University Press, 1952.

———. *Meteorologica.* Trans. H. D. P. Lee. London: Heinemann; Cambridge, Mass.: Harvard University Press, 1952.

Arnald of Villanova. *Opera nuperrime reuisa.* Lyon: Apud Scipionem de gabiano, 1532.

———. *Here is a Newe Boke, Called the Defence of Age and Recovery of Youth.* Trans. J. Drummond. London: R. Wyer, 1540.

Ashmole, Elias. *Elias Ashmole (1617-1692). His Autobiographical and Historical Notes, His Correspondence, and Other Contemporary Sources Relating to his Life and Work.* Ed. with a biographical introduction, by C. H. Josten. 5 vols. Oxford: Clarendon Press, 1966.

Aubertus Vindonis, Iacobus. *De metallorum ortu & causis contra chemistas brevis & dilucida explicatio.* Lyon: I. Berion, 1575.

———. *Duae apologeticae responsiones ad Iosephum Quercetanum.* (Lyon: I. Ausulti, 1576).

Baccius, Andrea. *De thermis.* Venice: Felix Valgrisius, 1588.

Bacon, Francis. *Works.* Ed. Basil Montagu. 3 vols. Philadelphia: Carey and Hart, 1842.

———. *Works* (Philosophical). Ed. James Spedding, Robert L. Ellis, and D.D. Heath. 5 vols. London: Longmans *et al.*, 1870.

———. *The Philosophical Works of Francis Bacon . . . Reprinted from the Texts and Translations, with the Notes and Prefaces of Ellis and Spedding.* Ed. with an introduction by John M. Robertson. London: Routledge and Sons; New York: Dutton, 1905.

———. *Sylva Sylvarum, or A Naturall History in Ten Centuries.* 8th ed. London: J.F. and S.G. for William Lee, 1664.

Bacon, Roger. *Opera quaedam hactenus inedita.* Vol. I. Ed. J. S. Brewer. London: Longman *et al.*, 1859.

Baillif, Roch le. Edelphe medecin spagiric. *Le demosterion . . . auquel sont contenuz trois cens aphorisemes Latins & Francois. Sommaire veritable de la medecine Paracelsique, extraicte de luy en la plus part, par le dict Baillif. Le sommaire duquel se trouvera a fueillet suyvant.* Renaes: Pierre le Bret, 1578.

———. *Premier traicte de l'homme, et son essentielle anatomie, avec les elemens, & ce qui est an eux: De ses maladies, medecine, & absoluts remedes ès tainctures d'or, corail, & antimoine: & magistere des perles: & de leur extraction.* Paris: Abel l'Angelier Libraire, 1580.

Baker, George. *The composition or making of the moste excellent and pretious Oil called Oleum Magistrale.* London: J. Alde, 1574.

Basil Valentine. *The Last Will and Testament of Basil Valentine, Monke of the Order of St. Bennet . . . To which is added Two Treatises the First declaring his Manual Operations. The Second shewing things Natural and Supernatural. Never before Published in English.* Trans. J[ohn] W[ebster]. London: S. G. and B. G. for Edward Brewster, 1671.

———. *Chymische Schrifften.* 3rd ed. 2 vols. Hamburg: Gottfried Leibezeits, 1700.

———. *The Triumphal Chariot of Antimony with the Commentary of Theodore Kerckringius.* Trans. Arthur Edward Waite. London: James Elliott, 1893.

Becher, J. J. *Oedipus chymicus, obscuriorum terminorum & principiorum mysteria aperiens & resolvens.* In *Bibliotheca chemica curiosa*, Vol. I, pp. 306-336.

———. *Physica subterranea.* 1669. Leipzig: Joh. Ludov. Gleditschium, 1703.

———. *Physica subterranea.* 1669. 3rd ed., Leipzig; Weidmann, 1738.

Beguin, Jean. *Les elemens de chymie.* Lyon: Pierre & Claude Rigaud, 1666.

———. *Tyrocinium chymicum: or, Chymical Essays, Acquired from the Fountain of Nature, and Manual Experience.* Trans. Richard Russell. London: Thomas Passenger, 1669.

Berthelot, M. P. E. *Collection des anciens alchimistes Grecs.* 3 vols. Paris, 1887-1888. Reprint London: Holland Press, 1963.

———. *La chimie au moyen âge.* 3 vols. Paris, 1893. Reprint Osnabruck: Otto Zeller; Amsterdam: Philo Press, 1967.

Bibliotheca chemica curiosa. ed. Jean Jacques Manget. 2 vols. Geneva: Chouet, G. De Tournes, *et al.* 1702.

Bibliotheca scriptorum medicorum. ed. J. J. Manget. Geneva, 1731.

Bibliothèque des philosophes chimiques. Jean Maughin de Richebourg (attributed ed.). 4 vols. Paris: Cailleau, 1741, 1754.

Biggs, Noah, Chymiatrophilos. *Mataeotechnia medicinae praxeωs. The Vanity of the Craft of Physick. Or, A New Dispensatory. Wherein is dissected the Errors, Ignorance, Impostures and Supinities of the SCHOOLS, in their main Pillars of Purges, Blood-letting, Fontanels, or Issues, and Diet, &c. and the particular*

Medicines of the Shops. With an humble Motion for the Reformation of the Universities. And the whole Landscap of Physick, And discovering the Terra incognita of Chymistrie. To the Parliament of England. London: Giles Calvert, 1651.

Birch, Thomas. *The History of the Royal Society of London* .... 4 vols. London: A Millar, 1756.

Biringuccio, Vannoccio. *The Pirotechnia.* Trans. C. S. Smith and M. T. Gnudi. New York: American Institute of Mining and Metallurgical Engineers, 1942. Reissued New York: Dover, 1959.

Borrichius, Olaus. *De ortu & progressu chemiae dissertatio.* In *Bibliotheca chemica curiosa,* Vol. I, pp. 1-37.

R. B. Esquire [R. Bostocke]. *The difference betwene the auncient Phisicke, first taught by the godly forefathers, consisting in vnitie peace and concord: and the latter Phisicke proceeding from Idolaters, Ethnikes, and Heathen: as Gallen, and such other consisting in dualitie, discorde, and contrarietie* .... London: Robert Walley, 1585.

Boyle, Robert. *The Works of the Honourable Robert Boyle.* 6 vols. London: J. and F. Rivington *et al.,* 1772.

———. *The Sceptical Chymist or Chymico-Physical Doubts and Paradoxes, Touching the Spagyrist's Principles Commonly call'd Hypostatical: As they are wont to be Propos'd and Defended by the Generality of Alchymists. Wherunto is Praemis'd Part of another Discourse relating to the same Subject.* London: J. Cadwell for J. Crooke, 1661. Reprint London: Dawson, 1965.

———. *Experiments and Considerations Touching Colours.* London: Henry Herringham, 1664.

———. *Short Memoirs for the Natural Experimental History of Mineral Waters Addressed by Way of a Letter to a Friend.* London: Samuel Smith, 1684/1685.

Browne, Thomas. *The Works of Sir Thomas Browne.* Ed. Charles Sayle. 3 vols. Vols. I and II, London: Grant Richard, 1904. Vol. III, Edinburgh: John Grant, 1907.

Brunschwig, Hieronymus. *Liber de arte distillandi de compositis.* 1512. Reprint Leipzig: Zentralantiquariat der deutschen demokratischen Republik, 1972.

———. *The vertuose boke of distyllacyon* .... Trans. L. Andrewe. London: L. Andrewe, 1527.

———. *Book of Distillation.* Intro. Harold J. Abrahams. Sources of Science, No. 79. New York/London: Johnson Reprint, 1971.

Bullein, William. *Bulleins Bulwarke of defēce against all Sicknes, Sornes and woundes* ... *(Here after insueth a little Dialogue, betwene twoo men, the one called Sorenes, and the other Chyrurgi.* ...) London: J. Kyngston, 1562.

Bulstrode, Whitlocke. *A Discourse of Natural Philosophy Wherein the*

Pythagorean Doctrine Is set in a true Light, and vindicated. 1692. 2nd ed. London: Jonas Browne, 1717.

Cardan, Jerome. *De subtilitate.* Nuremberg: I. Petreium, 1550.

———. *De varietate libri XVII.* Basel: Henricus Petrus, 1557.

———. *The First Book of Jerome Cardan's De Subtilitate.* Latin text, commentary, and translation by Myrtle Marquerite Cass. Williamsport, Pa.: Bayard Press, 1934.

Catalogue des manuscrits alchimiques Grecs. Ed. J. Bidez, F. Cumont, J. L. Heiberg, O. Lagercrantz, *et al.* 8 vols. Brussels: Union Académique International, 1924-1932.

Catalogue des manuscrits alchimiques Latin. Ed. J. Bidez, F. Cumont, A. Delatte, F. Kenyon, V. de Falco. 2 pts. Brussels: Union Académique International, 1939, 1951.

Charleton, Walter. *Spiritus Gorgonicus, vi sua axipara exutus; sive de causis, signis, & sanatione litheaseωs, diatriba.* Lieden: Elsevir, 1650.

———. *Physiologia Epicuro-Gassendo-Charltoniana* .... Intro. and indexes by Robert Hugh Kargon. New York/London: Johnson Reprint, 1966.

Clowes, William. *A Briefe and necessarie Treatise touching the cure of the disease called Morbus Gallicus.* London: T. Cadman, 1585.

———. *A Right Frutefull and Approoved Treatise for the Artficiall Cure of that Malady called in Latin Struma.* London: E. Allde, 1602.

Comenius, John Amos. *A Reformation of Schooles* .... Trans. Samuel Hartlib. London: Michael Sparke, 1642.

———. *The Great Didactic.* Trans. and intro. M. W. Keatinge. London: Adam and Charles Black, 1896.

Cotta, John. *A Short Discoverie of the Unobserved Dangers of severall sorts of ignorant and unconsiderate Practisers of Physicke in England.* London: [R. Field] for W. Jones and R. Boyle, 1612.

———. *Cotta Contra Antonium: or an Ant-Antony* . . . Oxford: J. Lichfield and J. Short for H. Cripps, 1623.

Courtin, Germani. *Adversus Paracelsi de tribus principiis, auro potabile totâque pyrotechniâ, portentosas opiniones, disputatio.* Paris: Ex Officina Petri L'Hillier, 1579.

Coxe, Daniel. "Enquiries concerning Agriculture." *Philosophical Transactions of the Royal Society of London,* **1** (1665), 91-94.

———. "The improvement of Cornwall of Sea Sand, communicated by an Intelligent Gentleman well acquainted in those parts to Daniel Coxe." *Philosophical Transactions of the Royal Society of London,* **9** (1675), 293-296.

Crollius, Oswald. *Basilica chymica.* Frankfurt: Godfrid Tampach, [1623].

———. *Discovering the Great and Deep Mysteries of Nature.* In *Philosophy Reformed and Improved.* Trans. H. Pinnell. London: M.S. for Lodowick Lloyd, 1657.

——— and John Hartman, *Bazilica chymica, & praxis chymiatricae or Royal and Practical Chymistry in Three Treatises.* Trans. Richard Russell. London: John Starkey and Thomas Passinger, 1669, 1670.

Cusanus, Nicholas. *The Idiot in Four Books. The first and second Wisdome. The third of the Minde. The fourth of statick Experiments, Or Experiments of the Ballance.* London: William Leake, 1650.

———. *Of Learned Ignorance.* Trans. Fr. Germain Heron. London: Routledge and Kegan Paul, 1954.

Dariot, Claude. *De praeparatione medicamentorum,* 1582, and *Discours de la goutte. Auquel les causes d'icelle sont amplement declarees, avec sa guerison et precaution,* 1588. In Paracelsus, *La grand chirurgie* .... Montbeliart: Iaques Foillet, 1608. With separate pagination.

Davy, Humphry. *Consolations in Travel, or The Last Days of a Philosopher.* Philadelphia: John Grigg, 1830.

Dee, John. "Mathematicall Praeface" to *The Elements of Geometrie of the most auncient Philosopher Euclide of Megara.* Trans. H. Billingsley. London: John Daye, 1570.

———. *The Mathematicall Praeface.* Intro. Allen G. Debus. New York: Science History Publications, 1975.

Descartes, René. *Discours de la méthode.* Intro. and notes by Etienne Gilson. Paris: J. Vrin, 1964.

Deutsches Theatrum Chemicum. Ed. Friedrich Roth-Scholz. 3 vols. Nuremberg: Adam Jonathon Felssteckern, 1728, 1730, 1732.

Diderot, D., ed. *Encyclopédie, ou dictionnaire raissoné, des arts et les métiers* .... 17 vols. Paris: Braisson, Dávid, Le Breton & Durant, 1751-1765.

Digby, Kenelm. *A late discourse made in a solemne assembly of nobles and learned men at Montpellier in France touching the cure of wounds by the powder of sympathy.* 2nd ed. London: R. Lowndes, 1658.

Dondi, Giacomo de. *Tractatus de causa salsedinis aquarum, et modo.* In *De balneis omnia quae extant apud Graecos, Latinos conficiendi salis ex eis et Arabas.* Venice: Apud Juntas, 1553.

Donne, John. *Complete Poetry and Selected Prose.* London: Nonesuch Press, 1929.

Dorn, Gerhard. *Clavis totius philosophiae chemisticae per quam potissima philosophorum dicta reserantur.* 1567. In *Theatrum chemicum* (1659), Vol. I, pp. 192-361.

———. *Monarchia triadis, in vnitate, soli deo sacra.* In Paracelsus, *Aurora thesaurusque philosophorum.* Basel: Palma Guarini, 1577.

Bibliography

―――. *Liber de natura luce physica, ex Genesi desumpta.* 1583. In *Theatrum chemicum* (1659), Vol. I, pp. 326-361.

Dryander, Johann. *Vom Eymser Bade.* Coblenz, 1538.

Duchesne, Joseph (Quercetanus). *Ad Iacobi Avberti Vindonis de ortv et cavsis metallorvm contra chymicos explicationem . . . eivsdem de exqvisita mineralium, animalium, & vegetibilium medicamentorum spagyrica praeparatione & usu, perspicua tractatio.* Lyon: Joannem Lertotium, 1575.

―――. *The Sclopotarie of Iosephus Quercetanus, Phisition, or His booke containing the cure of wounds received by shot of Gunne or such like Engines of warre, Whereunto is Added his Spagericke antidotary of medicines against the aforesayd wounds.* Trans. John Hester, practitioner in the said spagiricall Arte. London: R. Ward for J. Sheldrake, 1590.

―――. *A Breefe Aunswere . . . to the exposition of Iacobus Aubertus Vindonis, concerning the original, and causes of Mettalles. Set foorth against Chimistes. Another exquisite and plaine Treatise of the same Josephus, concerning the Spagericall preparations, and use of minerall, animall, and vegitable secretes, not heeretofore knowne of many.* By Iohn Hester, practitioner in the Spagericall Arte. London, 1591.

―――. *Ad veritatem Hermeticae medicinae ex Hippocratis veterumq́ue decretis ac therapeusi, nec non viuae rerum anatomiae exegesi, ipsiusq́ue naturae luce stabiliendam, adversus cuiusdam anonymi phantasmata responsio.* 1604. Frankfurt: Wolffgang Richter and Conrad Nebenius, 1605.

―――. *The Practise of Chymicall, and Hermeticall Physicke, for the preseruation of health.* Trans. Thomas Tymme. London: Thomas Creede, 1605.

―――. *Liber de priscorum philosophorum verae medicinae materia, praeparationis modo, atque in curandis morbis, praestantia. Déque simplicium, & rerum signaturis, tum externis, tum internis, seu specificis, à priscis & Hermeticis philosophis multa cura, singularíque industria comparatis, atque introductis, duo tractatus. His accesserunt ejusdem Ios. Quercetani de dogmaticorum medicorum legitima, & restituta, medicamentorum praeparatione, libri duo. Itémque selecta quaedam consilia medica, clarissimis medicis Europaeis dicata.* 1603. Leipzig: Thom. Schürer and Barthol. Voight, 1613.

―――. *Traicté de la matiere, preparation et excellente vertu de la medecine balsamique des anciens philosophes. Avquel sont adiovstez deux traictez, l'un des signatures externes des choses, l'autre des internes & specifiques conformément à la doctrine & pratique des Hermetiques.* Paris: C. Morel, 1626.

Duclo, Gaston. *Apologia chrysopoeiae & argyropoeiae adversus Thomam Erastum.* In *Theatrum chemicum* (1613), Vol. II, pp. 1-83.

Erastus, Thomas. *Disputationum de medicina nova P. Paracelsi. Pars prima, in qua quae de remediis superstitiosis et magicis curationibus prodidit praecipue examinantur.* Basel, [1572].

―――. *Disputationum de nova P. Paracelsi medicina. Pars altera in qua philosophiae Paracelsicae principia et elementa explorantur.* [Basel], 1572.

Ercker, Lazarus. *Treatise on Ores and Assaying.* Trans. from the German ed. of 1580 by A. G. Sisco and Cyril Stanley Smith. Chicago: University of Chicago Press, 1951.

Espagnet, Jean d'. *Enchyridion physicae restitutae; or, the Summary of Physicks Recovered.* London: W. Bentley, to be sold by W. Sheares, and Robert Tulchein, 1651.

Fallopius Mutinensis, Gabriel. *De medicatis aquis atque de fossilibus.* 1564. Venice, 1569.

The Fame and Confession of the Fraternity of R:C: Commonly, of the Rosie Cross. With a Praeface annexed thereto, and a short Declaration of their Physicall Work, by Eugenius Philalethes (Thomas Vaughan). London: J.M. for Giles Calvert, 1652.

Fenot, John Antony. *Alexipharmacum, sive antidotus apologetica, ad virulantius Josephi cuiusdam Quercetani armeniaci, eomitas in libellum Jacobi Auberti, de ortu & causis metallorum contra chymistas ... In quo ... omnia argumēta refelluntur, quibus chymistae probare conantur, aurum argentumq; arte fieri posse ....* Basel, [1575].

(Fioravanti) Leonard Phioravant. *Three Exact Pieces of Leonard Phioravant Knight and Doctor in Physick ... Whereunto is Annexed Paracelsus his One hundred and fourteen Experiments: With certain Excellent Works of B.G. à Portu Aquitans. Also Isaac Hollandus his Secrets concerning his Vegetall and Animall Work With Quercetanus his Spagyrick Antidotary for Gunshot.* London: G. Dawson for William Nealand, 1652.

Fludd, Robert. *Apologia compendiaris fraternitatem de Rosea Cruce suspicionis et infamiae maculis aspersam, veritatis quasi fluctibus abluens et abstergens.* Leiden: Godfrid Basson, 1616.

———. *Tractatus apologeticus integritatem societatis de Rosea Cruce defendens.* Leiden: Godfrid Basson, 1617.

———. *Utriusque cosmi maioris scilicet et minoris metaphysica, physica atque technica historia ... tomus primus de macrocosmi historia in duas tractatus diuisa.* 1617.

———. *Tomas secundus de supernaturali, praeternaturali et contranaturali microcosmi historia, in tractatus tres distributa.* Oppenheim: J. T. De Bry, 1619.

———. "A Philosophicall Key." Trinity College, Cambridge. Western MS 1150 (0.2.46).

———. *Tomus secundus de supernaturali, naturali praeternaturali et contranaturali microcosmi historia, in tractatus tres distributa.* Oppenheim: J. T. De Bry, 1619.

———. *Anatomiae amphitheatrum.* Frankfurt: J.T. De Bry, 1623.

———. *Tractatus secundus de naturae simia seu technica macrocosmi historia.* 2nd ed. Frankfurt: J.T. De Bry, 1624.

―――. *Medicina catholica seu mysticum artis medicandi sacrarium in tomos divisum duos.* Frankfurt: Wilhelm Fitzer, 1629.

―――. *Sophiae cum moria certamen, in quo, lapis lydius a falso structore, Fr. Marino Mersenno, monacho, reprobatus, celeberrima voluminis sui Babylonici (in Genesin) figmenta accurate examinat.* [Frankfurt], 1629.

―――. (Joachim Frizius). *Summum bonum, quod est verum [(Magiae, Cabalae, Alchymiae: Verae); Fratrum Roseae Crucis verorum] subjectum....* [Frankfurt?], 1629.

―――. *Pulsus seu nova arcana pulsuum historia, e sacro fonte radicaliter extracta, nec non medicorum ethnicorum dictis & authoritate comprobata.* [Frankfurt?], n.d.

―――. *Doctor Fludds answer unto M. Foster, Or, The squeesing of Parson Fosters sponge, ordained by him for the wiping away of the weapon-salve.* London: Nathaniel Butter, 1631.

―――. *Integrum morborum mysterium: sive medicinae catholicae tomi primi tractatus secundus.* Frankfurt: Wilhelm Fitzer, 1631.

―――. *Clavis philosophiae et alchymiae Fluddanae, sive Roberti Fluddi Armigeri, et medicinae doctoris, ad epistolicam Petri Gassendi theologi exercitationem responsum.* Frankfurt: Wilhelm Fitzer, 1633.

―――. *Philosophia Moysaica.* Gouda: Petrus Rammazenius, 1638.

―――. *Mosaicall Philosophy: Grounded upon the Essentiall Truth or Eternal Sapience.* London: Humphrey Moseley, 1659.

―――. *Schutzschrift für die aechtheit der Rosenkreutzergesellschaft . . . übersetzt von Ada Mah Booz.* Leipzig: Adam Friedrich Böhme, 1782.

Forestus, P. *The arraignment of urines . . . epitomized and translated . . . by I. Hart.* London: G. Eld for Robert Mylbourne, 1623.

Fortunatus. *Decas elementorum mysticae geometrae quibus praecipua divinitatis arcana explicantur.* Padua: Peter Paul Tozzi, 1617.

Foster, William. *Hoplocrisma-spongus: or, A sponge to wipe away the weapon-salve.* London: Thomas Cotes, 1631.

French, John. *Art of Distillation.* 1650. 4th ed., London: E. Cotes for T. Williams, 1674.

Fuller, Thomas. *The Holy State.* 2nd ed. Cambridge: John Williams, 1648.

―――. *The Worthies of England,* ed. John Freeman. London: Geo. Allen & Unwin. 1952.

Gaffarel, James. *Vnheard-of Curiosities: Concerning the Talismanical Sculpture of the Persians; The Horoscope of the Patriarkes; And the Reading of the Stars.* Trans. Edmund Chilmead. London: Humphrey Moseley, 1650.

Gassendi, Pierre. *Opera omnia.* 6 vols. Florence: Typis apud Joannem Cajetanum Tartini, & Sanctem Franchi, 1727.

―――. *Epistolica exercitatio in qua principia philosophiae Roberti Fluddi, medici,* reteguntur, et ad recentes illius libros adversus R.P.F. Marinum Mersennum . . . respondetur. Paris: S. Cramoisy, 1630.

Geber. *The Works of Geber.* Trans. Richard Russell. London: William Cooper, 1686.

———. *The Works of Geber.* Trans. Richard Russell (1678). Intro. E. J. Holmyard. London/Toronto: J.M. Dent; New York: E.P. Dutton, 1928.

Gell, Robert. *Stella Nova A New Starre, Leading wisemen unto Christ* . . . . London: Samuel Satterthwaite, 1649.

———. ΑΓΓΕΛΟΚΡΑΤΙΑ ΘΕΟΥ *Or a Sermon Touching Gods government of the World by Angels* . . . . London: John Legatt for Nath. Webb and William Grantham, 1650.

A General Collection of Discourses of the Virtuosi of France, upon Questions of all Sorts of Philosophy, and other Natural Knowledg. Made in the Assembly of the Beaux Esprits at Paris, by the most Ingenious Persons of that Nation. Trans. G. Havers, Gent. and (Vol. II) J. Davies, Gent. 2 vols. London: Thomas Dring and John Starkey, 1664, 1665.

Gerarde, John. *The Herball or Generall Historie of Plantes.* London: E. Bollifant for B. and J. Norton, 1597.

Gesner, Conrad. *Bibliotheca universalis.* Zurich 1545.

———. *Euonymus . . . Liber secundus.* Zurich, [1569].

———. *The Treasure of Euonymus: conteyninge the wonderfull hid secretes of nature.* Trans. P. Morwyng. London: John Daie, 1559.

———. *The New Jewell of Health.* London: H. Denham, 1576.

Gilbert, William. *De magnete.* London: P. Short, 1600.

———. *On the Loadstone and Magnetic Bodies, and on The Great Magnet the Earth.* Trans. P. Fleury Mottelay. London: Bernard Quaritch, 1893.

Glauber, Johann Rudolph. *The Works . . . containing, Great Variety of Choice Secrets in Medicine and Alchymy* . . . . Trans. Christopher Packe. London: Thomas Milbourn, 1689.

———. *Opera chymica, Bücher und Schrifften* . . . . 2 parts. Frankfurt am Main: Thomae-Matthiae Götzens, 1658-1659.

———. *A Description Of New Philosophical Furnaces, Or A New Art of Distilling* . . . . Trans. John French. London: Richard Coats, for Tho. Williams, 1651.

Goodall, Charles. *The Royal College of Physicians of London.* London: M. Flesher for Walter Kettilby, 1684.

[Gott, Samuel]. *Nova Solyma The Ideal City: Or Jerusalem Regained. An anonymous romance written in the time of Charles I. Now first drawn from obscurity, and attributed to the illustrious John Milton.* Intro., trans., essays, and bib. by the Rev. Walter Bayley. 2 vols. London: John Murray, 1900.

Grévin, Jacques. *Discours . . . sur les vertus et facultez de l'antimoine, contre de qu'en a escrit maistre Loys de Launay.* Paris: A. Wechel, 1566.

———. *La second discours . . . sur les vertus et facultez de l'antimoine . . . pour la confirmation de l'advis des médecins de Paris et pour servir d'apologie contre ce qu'a escrit M. Loïs de Launoy . . . .* Paris: J. Du Puys, n.d.

(Guinter) J. Guintherius von Andernach. *De medicina veteri et nova tum cognoscenda, tum faciunda commentarij duo.* 2 vols. Basel: Henricpetrina, 1571.

Gwinne, Matthew. *In Assertorem chymicae, sed verae medicinae desertorem, Fra. Anthonium.* London: R. Field, 1611.

Hakewill, George. *An Apologie or Declaration of the Power and Providence of God in the Government of the World.* 3rd ed. Oxford: William Turner, 1635.

H[all], J[ohn]. *An Humble Motion To the PARLIAMENT of ENGLAND Concerning the ADVANCEMENT of Learning and Reformation of the Universities.* London: John Walker, 1649.

Hall, Joseph. *The Discovery of a New World (Mundus alter et idem) c. 1605.* Ed. Huntington Brown with a foreword by Richard E. Byrd. Cambridge, Mass.: Harvard University Press, 1937.

Hall, Thomas. *Vindiciae literarum, The Schools Guarded: Or the Excellency and Usefulnesse of Humane Learning in Subordination to Divinity, and preparation to the Ministry . . . Whereunto is added an Examination of John Websters delusive Examen of Academies.* London: W.H. for Nathaniel Webb and William Grantham, 1655.

Hart, James. *The Anatomie of Urines.* London: James Field for Robert Mylbourne, 1625.

———. *KAINIKH, or The Diet of the Diseased.* London: John Beale, 1633.

Hartlib, Samuel. *A Description of the Famous Kingdom of Macaria: shewing the excellent Government, wherein the Inhabitants live in great Prosperity, Health and Happiness; the King obeyed, the Nobles honored, and all good men respected; Vice punished, and Virtue rewarded. An Example to other Nations: In a Dialogue between a Scholar and a Traveller printed 1641.* In *The Harleian Miscellany: A Collection of Scarce, Curious, and Entertaining Pamphlets and Tracts, as well in Manuscript as in Print. Selected from the Library of Edward Harley, Second Earl of Oxford.* Vol. I. London: John White, John Murray and John Harding, 1808.

———. *His Legacy of Husbandry . . .* 3rd ed. London: J.M. for Richard Wodnothe, 1655.

Harvey, William. *The Works of William Harvey, M.D.* Trans. and with a life by Robert Willis. London: Sydenham Society, 1847.

Heer, Henricus van. *Spadacrene, hoc est Fons Spadanus; ejus singularia, bibendi modus, medicamina bibentibus necessaria.* Liége: Arn. de Corswaremia, 1614.

Helmont, Franciscus Mercurius van. *The Paradoxal Discourses . . . Concerning the Macrocosm and Microcosm, or the Greater and Lesser World And Their Union, Set down in Writing by J.B. and now Published.* London: J.C. and Freeman Collins, for Robert Kettlewel, 1685.

Helmont, Jan Baptiste van. *Ortus medicinae. Id est initia physicae inaudita. Progressus medicinae novus, in morborum ultionem, ad vitam longam.* Amsterdam: Ludovicus Elzevir, 1648. Reprinted Brussels: Culture et Civilisation, 1966.

———. *Opuscula medica inaudita* (including the "De lithiasi" the "De febribus," the "De humoribus Galeni," and the "De peste." Amsterdam: Ludovicus Elzevir, 1648. Reprinted Brussels: Culture et Civilisation, 1966.

———. *Ternary of Paradoxes. The Magnetick Cure of Wounds. Nativity of Tartar in Wine. Image of God in Man.* Trans., illustrated, and ampliated by Walter Charleton. London: James Flesher for William Lee, 1650.

———. *A Ternary of Paradoxes. Of the Magnetick Cure of Wounds, Nativity of Tartar in Wine. Image of God in Man.* Second impression, more reformed, and enlarged with some marginal additions, trans. Walter Charleton. London: James Flesher for William Lee, 1650.

———. *Deliramenta catarrhi: or, The Incongruities, Impossibilities, and Absurdities Couched under the Vulgar Opinion of Defluxions.* Translator and Paraphrast, Dr. Charleton. London: E. G. for William Lee, 1650.

———. *Oriatrike or Physick Refined. The Common Errors therein Refuted, and the whole Art Reformed & Rectified. Being A New Rise and Progress of Phylosophy and Medicine, for the Destruction of Diseases and Prolongation of Life.* Trans. J[ohn] C[handler]. London: Lodowick Loyd, 1662.

———. *Opuscula medica inaudita.* Frankfurt: John. Just. Erythropili, 1682.

Henry, William. *The Elements of Experimental Chemistry.* 1st American ed. from the 8th London ed. 2 vols. Philadelphia: R. Desilve, 1819.

Herman, Phillip. *An Excellent Treatise.* Trans. John Hester. London: J. Charlwood, 1590.

Hermes Trismegistus. *Hermetica.* Ed., trans. Walter Scott (Vol. IV with A. S. Ferguson). 4 vols. Oxford: Clarendon Press, 1924, 1925, 1926, 1936.

Heydon, Christopher. *An Astrological Discourse, Manifestly proving the Powerful Influence of Planets and Fixed Stars upon Elementary Bodies, in Justification of the verity of Astrology, Together with an Astrological Judgement upon the great Conjunction of Saturn and Jupiter 1603.* London: John Macock, for Nathaniel Brooke, 1650.

Johnson, William. *Lexicon chymicum. Cum obscuriorum verborum, et rerum Hermeticarum, tum phrasum Paracelsicarum, in scriptis ejus: et aliorum chymicorum . . . .* London: G. D[awson] for William Nealand, 1652.

Jones, John. *Galens Bookes of Elementes.* London: W. Jones, 1574.

Jorden, Edward. *A briefe discourse of a disease called the Suffocation of the Mother, written upon occasion which hath beene of late taken thereby, to suspect possession of an evill spirit or some such like supernaturall power. Wherein is declared that divers strange actions and passions of the body of man. which . . . are imputed to the Divell, have their true natural causes, and do accompanie this disease.* London, 1603.

———. *A Discourse of Naturall Bathes, and Minerall Waters.* 1631. 2nd ed. ("in many points enlarged") London: Thomas Harper, 1632.

Kepler, Johannes. *Gesammelte Werke.* 18 vols. Munich: C.H. Beck, 1937-1949.

Kircher, Athanasius. *Magnes sive de arte magnetica.* Rome: Ludovici Grignani, 1641.

Ko Hung. *Alchemy, Medicine and Religion in the China of A.D. 320. The Nei P'ien of Ko Hung.* Trans. and ed. James R. Ware. Cambridge, Mass.: M.I.T. Press, 1966.

(Lefèvre, Nicholas) Nicasius le Febure, Royal Professor in Chymistry to his Majesty of England, and Apothecary in Ordinary to his Honorable Household. *A Compendious Body of Chymistry.* Trans. P. D. C. 2 parts. London, 1664.

———. *A Compleat Body of Chymistry.* Trans. P.D.C. 2 parts. London: O. Pulleyn, 1670.

Leibniz, Gottfried Wilhelm. *New Essays Concerning Human Understanding.* Trans. Alfred Gideon Langley. 2nd ed. Chicago: Open Court, 1916.

Lemery, Nicholas. *A Course of Chymistry.* 1675 (French). Trans. Walter Harris. 2nd English ed. from the 5th French ed. London: R.N. for Walter Kittilby, 1686.

Libavius, Andreas. *Alchemia.* Frankfurt: Saurius, 1597.

———. *Alchymia, recognita, emendata, et aucta, tum dogmatibus & experimentis nonullis . . . .* Frankfurt: Excudebat Joannes Saurius, impensis Petri Kopffii, 1606.

———. *Commentariorum alchymiae.* Frankfurt: Saurius/Kopffii, 1606.

———. *Syntagma selectorum undiquaque et perspicue traditorum alchymiae arcanorum.* Frankfurt: Excudebat Nicolaus Hoffmannus, impensis Petri Kopffii, 1611.

Macquer, Pierre-Joseph. *Dictionnaire de chymie.* 2 vols. Paris: Lacombe, 1766.

———. *Dictionnaire de chymie* 2nd ed. 4 vols. Paris: Didot, 1778.

Matthioli, Petrus Andreas. *Opera quae extant omnia . . .* Frankfurt: N. Bassaei, 1598.

―――. *Les commentaires de M.P. Andre Matthiole, medecin Sienois, sur les six livres de la matiere medecinale de Pedacius Dioscoride, Anazarbéen.* Trans. M. Antoine du Pinet. Lyon: Jean Baptiste de Ville, 1680.

Mayow, John. *Tractatus quinque.* Trans. as *Medico-Physical Works* by A. C. B. and L. D. Edinburgh: Alembic Club, 1907.

―――. *Medico-Physical Works Being a Translation of Tractatus Quinque Medico-Physici.* Re-issue for the Alembic Club. Edinburgh/London: E. and S. Livingstone, 1957.

Mersenne, Marin. *Correspondance du P. Marin Mersenne. Religieux Minime.* Pub. Mme. Paul Tannery, ed. Cornelis de Waard with the collaboration of René Pintard. 10 vols. (*1617-1641*). Paris: P.U.F. Centre National de la Recherche Scientifique, 1932-1967.

―――. *La verite des sciences. Contre les septiques ou Pyrrhoniens.* Paris: Toussainct Du Bray, 1625. Reprint Stuttgart/Bad Cannstatt: Friedrich Fromman Verlag (Günther Holzboog), 1969.

Moffett, Thomas. *De iure et praestantia chemicorum medicamentorum.* In *Theatrum chemicum* (1613), Vol. I, pp. 63-112; (1659), Vol. I, pp. 64-108.

Montagnana, Bartolomeo. *Tractatus tres de balneis patavinis.* In his *Consilia CCCV.* Ed. J. de Vitalibus. Venice: G. L. Per Bonetū Locatellū, 1564.

Morhof, Daniel George. *Polyhistor, literarius, philosophicus et practicus.* 1692. 4th ed. Lübeck: Peter Boeckmann, 1747.

Naudé, Gabriel. *Instruction à la France sur la verité de l'histoire des freres de la Roze-Croix.* Paris: Francois Iulliot, 1623.

Newton, Isaac. *Philosophiae naturalis principia mathematica.* London: Jussu Societatis Regiae ac Typis Josephi Streater, 1687.

―――. *Opticks.* Foreword by Albert Einstein, intro. by Sir Edmund Whittaker, a preface by I. Bernard Cohen. New York: Dover, 1952.

Paley, William. *The Complete Works . . . and a Life of the Author, by the Rev. Robert Lynam, A.M.* 4 vols. London: George Cowie *et al.*, 1825.

Pantheus, J. A. *Ars & theoria transmutationis metallicae cum voarchadumia, proportionibus, numeris, & iconibus rei accomodis illustrata.* In *Theatrum chemicum* (1613), Vol. II, pp. 499-598.

Paracelsus. *La grande, vraye, et parfaicte chirurgie.* Trans. M. Pierre Hassard d'Armentières. Anvers: Guillaume Silvius, 1568.

―――. *La grand chirurgie de Philippe Aoreole Paracelse . . . traduite en Francois de la version Latin de Iosquin d'Alhem . . . par M. Claude Dariot plus un discours de la goutte . . . item III. Traittez de la preparation des medicaments.* 3rd ed. Montbeliart: Iaques Foillet, 1608.

―――. *Opera Bücher und Schrifften.* Ed. Johann Huser. 2 vols. Strasbourg: L. Zetzner, 1616.

―――. *Chirurgische Bücher und Schrifften.* ed. J. Huser. Strasbourg: Zetzner, 1618.

―――. *Of the Supreme Mysteries of Nature.* Trans. R. Turner. London: J.C. for N. Brook and J. Harison, 1656.

―――.*Three Books of Philosophy Written to the Athenians* in *Philosophy Reformed and Improved.* Trans. H. Pinnell. London: M.S. for L. Lloyd, 1657.

―――. *Opera omnia medico-chemico-chirurgica.* Trans. Fridericus Bitiskius. 3 vols. Geneva: Ioan. Anton. & Samuel. de Tournes, 1658.

―――. *His Aurora & Treasure of the Philosophers. As Also The Water-Stone of the Wise Men; Describing the Matter of, and manner how to attain the universal Tincture.* Trans. J[ames] H[owell]. London: Giles Calvert, 1659.

―――. *The Hermetic and Alchemical Writings of Paracelsus.* Trans. Arthur Edward Waite. 2 vols. London: James Elliott, 1894.

―――. *Sämtliche Werke.* Ed. Karl Sudhoff and Wilhelm Matthiessen. 15 vols. Munich/Berlin: R. Oldenbourg (Vols. VI-IX: O.W. Barth) 1922-1933.

―――. *Volumen medicinae paramirum* Trans. and preface by Kurt F. Leidecker. *Bulletin of the History of Medicine.* Supplement No. 11. Baltimore: Johns Hopkins Press, 1949.

―――. *Selected Writings.* Ed. Jolande Jacobi. Trans. Norbert Guterman. Bollingen Series XXVIII. New York: Pantheon Books, 1951.

(Paré) Parey, Ambroise. *The Workes of that famous Chirurgion . . . .* Trans. Thomas Johnson. London: Richard Cotes and Willi Du-gard, 1649.

―――. *Oeuvres complètes d'Ambroise Paré.* Notes and intro. J. F. Malgaine. 3 vols. Paris: J.B. Baillière, 1840-1841.

Peiresc, Nicolas-Claude Fabri de. *Lettres de Peiresc publiées par Philippe Tamizey de Larroque . . . Tome quatrième. Lettres de Peiresc à Borilly, à Bouchard et à Gassendi. Lettres de Gassendi à Peiresc 1626-1637.* Paris: Imprimerie national, 1893.

Pernety, Dom Antoine-Joseph. *Dictionnaire Mytho-Hermétique, dan lequel on trouve les allégories fabuleuses des poetes, les métaphores, les énigmes et les termes barbares des philosophes hermétiques expliqués.* Paris: Bauche, 1758.

Petrus Bonus of Ferrara. *The New Pearl of Great Price.* Trans. and abridged by Arthur Edward Waite. London: James Elliott. 1894. Reprint London: Vincent Stuart, 1963.

Philiatros [pseud.]. *Philiatros, or the copie of an Epistle, wherein sundry fitting Considerations are propounded to a young Student of Physicke.* London: W. White, 1615.

The Philosophical Epitaph of W.C. Esquire . . . . London: William Cooper, 1673.

Pico della Mirandola, Count Giovanni. *Opera quae extant omnia.* 2 vols. Basel: Sebastianum Henricpetri, 1601.

Pietro da Eboli. *I bagni di Pozzuoli (1195).* Naples, 1887.

Planis Campy, David de. *L'hydre morbifiqve exterminee par L'Hercvle chymiqve ou les sept maladies tenues pour incurables iusques à present rendues guerissables par l'art chymique medical.* Paris: Herué du Mesnil, 1628.

Plat, Hugh. *Diuerse new Sorts of Soyle not yet brought into any publique use, for manuring both of pasture and arable ground, with sundrie concepted practises belonging therunto.* London: Peter Short, 1594.

———. *The Jewell House of Art and Nature. Conteining diuers rare and profitable Inuentions, together with sundry new Experiments in the Art of Husbandry, Distillation and Moulding.* London: Peter Short, 1594.

———. *The Jewell House of Art and Nature: Containing Divers Rare and Profitable Inventions, together with sundry new Experiments in the Art of Husbandry. With Divers Chymical Conclusions concerning the Art of Distillation, and the rare practices and uses thereof. Faithfully and familiarly set down, according to the Authors own Experience . . . Whereunto is added, A rare and excellent Discourse of Minerals, Stones, Gums and Rosins; with the vertues and use thereof, By D.B. Gent.* London: Elizabeth Alsop, 1653.

———. Review of *The Garden of Eden, or an Accompt of the Culture of Flowers and Fruits now growing in England . . .* in 2 parts in 8^0, written by Sir Hugh Plat Kt., newly reprinted. In *Philosophical Transactions of the Royal Society of London,* 9 (1675), 302-304.

Plattes, Gabriel. *A Discovery of Subterraneall Treasure.* London, 1639.

Porta, John Baptista. *Natural Magick.* London: Thomas Young and Samuel Speed, 1658.

Rabelais. *Les oeuvres.* Ed. Ch. Marty-Laveaux. 6 vols. Paris: Lemerre, 1870-1903.

Rawlin, Thomas. *Admonitio pseudo-chymicis seu alphabetarium philosophicum.* London: E. Allde, c. 1610.

Ray, John. *Three Physico-Theological Discourses.* 2nd ed. London: Sam. Smith, 1693.

Raynalde, Thomas. *A compendious declaration of the . . . vertues of a certain lateli inventid oile . . . .* Venice: J. Gryphius, 1551.

Recueil général des questions traictées ès conférences du Bureau d'Adresse. Ed. E. Renaudot. 4 vols. Paris: L. Chamhoudry, 1655-1656.

Reeve, Edmund. *The New Jerusalem: The Perfection of Beauty: The joy of the whole earth.* London: J.G. for Nath. Brooks, 1652.

Reyher, Samuel. *Dissertatio de nummis quibusdam ex chymico metallo factis.* Kiel: Joachim Reumann, 1692.

Rhumelius, Johannes Pharamundus. *Médecine spagyrique.* Trans. Pierre Rabbe. Paris: Chacornac, 1932.

———. *Medicina spagÿrica oder Spagyrische Artzneykunst* .... Frankfurt: Christian Hermsdorffs, 1662.

Ridley, Mark. *A short Treatise of Magneticall Bodies and Motions.* London: Nicolas Okes, 1613.

Royal College of Physicians. *Pharmacopoeia Londinensis.* 2nd ed. London: J. Marriott, 1618 (British Museum copy with the notes of Theodore Turquet de Mayerne).

Royal College of Physicians. *Pharmacopoeia Londinensis of 1618 Reproduced in Facsimile with a Historical Introduction by George Urdang.* Madison: University of Wisconsin Press, 1944.

Royal College of Physicians. *Pharmacopoeia Londinensis: or the Dispensatory* .... Trans. and annotated by Nicholas Culpeper. 6th ed. London: Peter Cole, 1659.

(Royal Society). *Philosophical Transactions of the Royal Society of London from their Commencement in 1665 to the Year 1800, Abridged.* Ed. Charles Hutton, George Shaw, and Richard Pearson. 18 vols. London: C. and R. Baldwin, 1809.

Ruland, Martin. *Lexicon alchemiae.* 1612. Reprint Hildesheim: Georg Olms, 1964.

———. *A Lexicon of Alchemy.* Trans. A.E. Waite. 1893. Reprint London: John M. Watkins, 1964.

Savonarola, Johannes Michael. *De balneis et termis naturalibus omnibus ytalie.* Ferrara: Nov. 10, 1485.

———. *Practica canonica de febribus.* Venice, 1552.

Schütze, Tobias. *Harmonia macrocosmi cum microcosmo. Das ist eine Übereinstimmung der grossen mit der kleinen Welt als dem Menschen/in zwei Theil abgetheilet.* Frankfurt am Main: Daniel Reicheln, 1654.

Scot, Patrick. *The Tillage of Light, or a true Discoverie of the Philosophical Elixir commonly called the Philosophers Stone Serving to enrich all true, noble and generous Spirits, as will adventure some few labours in the tillage of such a light, as is worthy the best observance of the most Wise.* London: William Lee, 1623.

Scot, Reginald. *The Discoverie of Witchcraft.* Intro. Rev. Montague Summers. 1930. Reprint New York; Dover, 1972.

Sendivogius, Michael. *A New Light of Alchymy* .... Trans. J[ohn] F[rench]. London: A. Clark for Tho. Williams, 1674.

Sennert, Daniel. *De chymicorum cum Aristotelicis et Galenicis consensu ac dissensu liber 1., controversias plurimas tam philosophis quam medicis cognitu utiles continens.* Wittenberg: Apud Zachariam Schurerum, 1619.

———. *De chymicorum cum Aristotelicis et Galenicis consensu ac dissensu.* 3rd ed. Paris: Apud Societatem, 1633.

———. *Epitome naturalis scientiae.* Paris: Apud Societatem, 1633.

——— . *The weapon-salves maladie; or, A declaration of its insufficiencie to performe what is attributed to it,* trans. out of his 5th booke Pt. 4 Chap. 10 *Practicae medicinae.* London: John Clark, 1637.

———, Nich. Culpeper, and Abdiah Cole. *Chymistry Made Easie and Useful. Or, The Agreement and Disagreement Of the Chymists and Galenists.* London: Peter Cole, 1662.

Severinus, Petrus. *Idea medicinae philosophicae.* 1571. 3rd ed., The Hague: Adrian Clacq, 1660.

Short, Thomas. *The Natural, Experimental and Mechanical History of Mineral Waters.* 2 vols. London: F. Gyles, 1734; Sheffield: J. Garnett, 1740.

Simpson, William. *Hydrologia chymica: or the Chymical Anatomy of the Scarbrough, and other Spaws in York-Shire.* London: W.G. for Richard Chiswel, 1669.

Stahl, G. E. *Zymotechnia fundamentalis, seu fermentationis theoria generalis, qua nobilissimae hujus artis, & Partis chymiae, utilissimae atq: subtilissimae, causae & effectus in genere, ex ipsis mechanico-physicis principiis, summo studio eruuntur, simulque experimentum novum sulphur verum arte producendi, et alia utilia experimenta atque observatur, inseruntur.* [Halle: Christop. Salfeld, 1697].

——— . *Specimen Beccherianum . . . .* Appended to Becher's *Physica subterranea.* Leipzig: Joh. Ludov. Gleditschium, 1703.

Starkey, George. *Natures Explication and Helmont's Vindication.* London: E. Cotes for Thomas Alsop, 1657.

——— . *Pyrotechny Asserted and Illustrated, To be the surest and safest means For Arts Triumph over Natures Infirmities.* London: R. Daniel, for Samuel Thomson, 1658.

The Statutes at Large . . . . Ed. Danby Pickering. Cambridge, 1763.

Stavenhagen, Lee, ed. and trans. *A Testament of Alchemy: Being the Revelations of Morienus to Khālid ibn Yazīd.* Hanover, N.H.: Published for the Brandeis University Press by the University Press of New England, 1974.

Suavius, Leo. (Jacques Gohory). *Theophrasti Paracelsi philosophiae et medicinae utriusque universae, compendium, ex optimi quibusque eius libris: Cum scholijs in libros IIII eiusdem de vita longa, plenos mysteriorum, parabolorum, aenigmatum.* Basel, 1568.

Sylvius, Franciscus de la Boë. *Opera medica.* Amsterdam: Daniel Elsevier and Abraham Wolfgang, 1680.

Tagliacozzi, Gaspare. *Cheirurgia nova . . . de narium, aurium labiorumque defectu, per insitionem cutis ex humero, arte hactenus omnibus ignota, sarciendo.* Frankfurt, 1598. Reedited M. Troschel, Berlin: Reimer, 1831.

Theatrum chemicum. Ed. Lazarus Zetzner. 2nd ed., 4 vols., Strasbourg:

Zetzner, 1613. 5th vol., Strasbourg: Zetzner, 1622. 3rd ed., 6 vols., Strasbourg: Zetzner, 1659-1661.

Theatrum chemicum Britannicum. Comp. and annotated by Elias Ashmole. London, 1652. Reprinted with intro. by Allen G. Debus. New York: Johnson Reprint, 1967.

Theatrum sympatheticum auctum. Nuremburg: J.A. Endterum, & Wolfgang; Juniores Haeredes 1662.

Thurneisser zum Thurn, Leonhart. *Pison.* Frankfurt on der Oder, 1572.

――――. *Historia und Beschreibung Elementischer und Natürlicher Wirckungen aller Erdgewechssen.* Berlin: M. Hentzske, 1578.

A Transcript of the Registers of the Company of Stationers of London 1554-1640 A.D. London, Privately printed, May 1, 1877.

The Turba Philosophorum or Assembly of the Sages. Trans. Arthur Edward Waite. London: George Redway, 1896.

Turner, William. *A Booke of the natures and properties as well of the bathes in England as of other bathes in Germanye and Italye.* Cologne: A Birckmann, 1562.

Tymme, Thomas. *A Dialogue Philosophicall.* London: T. S[nodham] for C. Knight, 1612.

Vigenère, Blaise de. *A Discourse of Fire and Salt, Discovering Many Secret Mysteries, as well Philosophicall as Theologicall.* Trans. Edward Stephens. London: Richard Cotes, 1649.

Violet, Fabius. *La parfaicte et entiere cognoissance de toutes maladies du corps humain, causées par obstruction.* Paris: Pierre Billaine, 1635.

Vives, Juan Luis. *On Education. A Translation of the De tradendis Disciplinis.* 1531. Intro. Foster Watson. Cambridge: Cambridge University Press, 1913.

[War]D, [Set]H. *Vindiciae academiarum containing, Some briefe Animadversions upon Mr. Websters Book, Stiled The Examination of Academies. Together with an Appendix concerning what M. Hobbs, and M. Dell have published on this Argument.* Intro. [Joh]N [Wilkin]S. Oxford: Leonard Lichfield Printer to the University for Thomas Robinson, 1654.

Webster, John. *The Saints Guide, or Christ the Rule and Ruler of Saints. Manifested by way of Positions, Consectaries, and Queries . . . .* London: Giles Calvert, 1654.

――――. *Academiarum examen, or the Examination of Academies. Wherein is discussed and examined the Matter, Method and Customes of Academick and Scholastick Learning, and the insufficiency thereof discovered and laid open; As also some Expedients proposed for the Reforming of Schools, and the perfecting and promoting of all kind of Science. Offered to the judgments of all those that love the proficiencie of Arts and Sciences, and the advancement of learning.* London: Giles Calvert, 1654.

———. Practitioner in Physick and Chirurgery, *Metallographia; or an History of Metals. Wherein is declared the signs of Ores and Minerals both before and after digging, the causes and manner of their generations, their kinds, sorts, and differences; with the description of sundry new Metals, or Semi Metals, and many other things pertaining to Mineral knowledge. As also, The handling and shewing of their Vegetability, and the discussion of the most difficult Questions belonging to Mystical Chymistry, as of the Philosophers Gold, their Mercury, the Liquor Alkahest, Aurum potabile, and such like* . . . . London: A. C. for Walter Kettilby, 1671.

———. *The Displaying of Supposed Witchcraft.* London: J.M., 1677.

Widman (dicti Mechinger), Johannes. *Tractatus de balneis thermarum ferinarum (vulgo Uuildbaden) perutilis balneari volentibus ibidem.* Tübingen, 1513.

Willis, Thomas. *Practice of Physick* . . . . Trans. S. Pordage. London: T. Dring, C. Harper and J. Leigh, 1681, 1684 (type reset).

Wimpenaeus, Johannes Albertus. *De concordia Hippocraticorum et Paracelsistarum libri magni excursiones defensiuae, cum appendice, quid medico sit faciundum.* Munich: Adamus Berg, 1569.

Woodall, John. *The Surgions Mate.* London: E. Griffen for L. Lisle, 1617.

Woodward, John. "Thoughts and Experiments on Vegetation." 1699. In *The Philosophical Transactions of the Royal Society of London from their Commencement, in 1665, to the Year 1800; Abridged* . . . . Vol. IV, pp. 382-398.

Secondary Sources

Adams, Frank Dawson. *The Birth and Development of the Geological Sciences.* New York: Dover, 1954.

Ahonen, Kathleen. "Johann Rudolph Glauber." In *Dictionary of Scientific Biography*, ed. Charles C. Gillispie, Vol. V, pp. 419-423. New York: Scribner's, 1972.

Allen, Phyllis. "Medical Education in 17th Century England." *Journal of the History of Medicine and Allied Sciences,* **1** (1946), 115-143.

———. "Scientific Studies in the English Universities of the Seventeenth Century." *Journal of the History of Ideas,* **10** (1949), 219-253.

Armytage, W. H. G. "The Early Utopists and Science in England." *Annals of Science,* **12** (1956), 247-254.

Arnold, Paul. *Histoire des Rose-Croix et les origines de la Franc-Maçonnerie.* Paris: Mercure de France, 1955.

Bădărău, Dan. *Filozofia lui Dimitrie Cantemir*. Bucharest: Editura Academiei Republicii Populare Romîne, 1964.

Barrett, C. R. B. *The History of the Society of Apothecaries of London*. London: Elliot Stock, 1905.

Bastholm, Eyvind. "Petrus Severinus (1542-1602). A Danish Paracelsist." *Proceedings of the XXI International Congress of the History of Medicine (Sienna, 22-28 September, 1968)*, pp. 1080-1085.

———. "Petrus Severinus (1542-1602). En dansk paracelsist." *Sydveska ur Medicinhistoriska Sallskapets Årsskrift* (1970), 53-72.

Baumann, E. D. *François de la Boë Sylvius (1614-1672)*. Leiden: E. J. Brill, 1949.

Birch, Thomas. *The Life of the Honourable Robert Boyle*. In Robert Boyle. *The Works of the Honourable Robert Boyle*. 5 vols. London: A. Millar, 1744.

Boas, Marie. "The Establishment of the Mechanical Philosophy." *Osiris*, **10** (1952), 412-541.

———. "An Early Version of Boyle's *Sceptical Chymist*." *Isis*, **45** (1954), 153-168.

———. "Acid and Alkali in Seventeenth Century Chemistry." *Archives Internationales d'Histoire des Sciences*, **9** (1956), 13-28.

———. *Robert Boyle and Seventeenth-Century Chemistry*. Cambridge: Cambridge University Press, 1958.

Böhm, Walter. "John Mayow und Descartes." *Sudhoffs Archiv für Geschichte der Medizin und der Naturwissenschaften*, **46** (1962), 45-68.

———. "John Mayow and His Contemporaries." *Ambix*, **11** (1964), 105-120.

———. "John Mayow und die Geschichte des Verbrennungsexperiments." *Centaurus*, **11** (1965), 241-258.

Bolton, Henry C. *Evolution of the Thermometer, 1592-1743*. Eaton, Pa.: Chemical Publishing Co., 1900.

Broeckx, C. *Commentaire de J. B. van Helmont sur le premier livre du Régime d'Hippocratei Peri diaites*. Antwerp, 1849.

———. *Interrogatoires du J. B. van Helmont sur le magnétisme animal*. Anvers: Buschmann, 1856.

Cafiero, Luca. "Robert Fludd e la polemica con Gassendi." *Revista Critica di Storia della Filosofia*, **19** (1964), 367-410; **20** (1965), 3-15.

Calder, I. R. F. "John Dee Studied as an English Neo-Platonist." Ph.D. dissertation, University of London, 1952.

A Catalogue of the Portsmouth Collection of Books and Papers Written by or belonging to Sir Isaac Newton . . . . Cambridge: Cambridge University Press, 1888.

Chevalier, A.G. "The 'Antimony-War'—A Dispute Between Montpellier and Paris." *Ciba Symposia*, **2** (1940), 418-423.

Clulee, Nicholas. "The Glas of Creation: Renaissance Mathematicism and Natural Philosophy in the Work of John Dee." Ph.D. dissertation, University of Chicago, 1972.

Cohen, I. Bernard. *Franklin and Newton: An Inquiry into Speculative Newtonian Experimental Science and Franklin's Work in Electricity as an Example Thereof.* Philadelphia: The American Philosophical Society, Memoirs, Vol. 43, 1956.

Copeman, W. S. C. *Doctors and Disease in Tudor Times.* London: Wm. Dawson and Sons, 1960.

Cornford, Francis M. *Plato's Cosmology.* New York: Liberal Arts Press, 1957.

Costello, William T. *The Scholastic Curriculum at Early Seventeenth-Century Cambridge.* Cambridge, Mass.: Harvard University Press, 1958.

Cowen, David L. "The Boston Editions of Nicholas Culpeper." *Journal of the History of Medicine and Allied Sciences*, **11** (1956), 156-165.

Craven, J. B. *Doctor Robert Fludd (Robertus de Fluctibus). The English Rosicrucian. Life and Writings.* Kirkwall, 1902. Reprinted New York: Occult Research Press, n.d.

———. *Count Michael Maier.* Kirkwall, 1910. Reprint London: Dawsons, 1968.

Curtis, Mark H. *Oxford and Cambridge in Transition (1558-1642).* Oxford: Clarendon Press, 1959.

Davis, Audrey B. "The Circulation of the Blood and Chemical Anatomy." In *Science, Medicine and Society in the Renaissance*, ed. Allen G. Debus, Vol. II, pp. 25-37. New York: Science History Publications, 1972.

———. "Some Implications of the Circulation Theory for Disease Theory and Treatment in the Seventeenth Century." *Journal of the History of Medicine and Allied Sciences*, **26** (1971), 28-39.

Debus, Allen G. "The Paracelsian Compromise in Elizabethan Medicine." *Ambix*, **8** (1960), 71-97.

———. "Gabriel Plattes and his Chemical Theory of the Formation of the Earth's Crust." *Ambix*, **9** (1961), 162-165.

———. "Robert Fludd and the Circulation of the Blood." *Journal of the History of Medicine and Allied Sciences*, **16** (1961), 374-393.

———. "Sir Thomas Browne and the Study of Colour Indicators." *Ambix*, **10** (1962), 29-36.

———. "Solution Analyses Prior to Robert Boyle." *Chymia* **8** (1962), 41-61.

———. "John Woodall—Paracelsian Surgeon." *Ambix*, **10** (1962), pp. 108-118.

———. "Paracelsian Doctrine in English Medicine." In *Chemistry in the Service of Medicine*, ed. F. N. L. Poynter, pp. 5-26. London: Pitman, 1963.

———. "Robert Fludd and the Use of Gilbert's *De magnete* in the Weapon-Salve Controversy." *Journal of the History of Medicine and Allied Sciences*, **19** (1964), 389-417.

———. "The Paracelsian Aerial Niter." *Isis*, **55** (1964), 43-61.

———. "The Aerial Niter in the Sixteenth and Early Seventeenth Centuries." *Actes du Dixième Congrès International d'Histoire des Sciences*, Vol. II, pp. 835-839. Paris, 1964.

———. "An Elizabethan History of Medical Chemistry." *Annals of Science*, **18** (1962, published 1964), 1-29.

———. "The Significance of the History of Early Chemistry." *Cahiers d'Histoire Mondiale*, **9** (1965), 39-58.

———. "The Sun in the Universe of Robert Fludd." In *Le soleil à la Renaissance—sciences et mythes*, Travaux de l'Institut pour l'étude de la Renaissance et de l'Humanisme, Vol. II, pp. 257-278. Brussels: P.U.B./P.U.F., 1965.

———. *The English Paracelsians*. London: Oldbourne Press, 1965; New York: Franklin Watts, 1966.

———. Review of Marie Boas Hall, *Robert Boyle on Natural Philosophy, An Essay with Selections from His Writings* (Bloomington: Indiana University Press, 1965). In *Isis*, **57** (1966), 125-126.

———. "Fire Analysis and the Elements in the Sixteenth and the Seventeenth Centuries." *Annals of Science*, **23** (1967), 127-147.

———. "Palissy, Plat and English Agricultural Chemistry in the 16th and 17th Centuries." *Archives Internationales d'Histoire des Sciences*, **21** (1968), 67-88.

———. "Mathematics and Nature in the Chemical Texts of the Renaissance." *Ambix*, **15** (1968), 1-28, 211.

———. *The Chemical Dream of the Renaissance*. Cambridge: Heffer, 1968; Indianapolis: Bobbs Merrill, 1972.

———. "Edward Jorden and the Fermentation of the Metals: An Iatrochemical Study of Terrestrial Phenomena." In *Toward a History of Geology: Proceedings of the New Hampshire Interdisciplinary Conference on the History of Geology, September 7-12, 1967*, ed. Cecil J. Schneer, pp. 100-121. Cambridge, Mass.: M.I.T. Press, 1969.

———. " 'The Devil upon Dun,' a Seventeenth Century Attack on the Quack Chemist." In *Die ganze Welt ein Apotheken. Festschrift für Otto*

Zekert, ed. Sepp Domandl, Salzburger Beiträge zur Paracelsusforschung, Heft 8: Im Verlag Notring der wissenschaftlichen Verbände Österreichs, 1969, pp. 47-62.

———. "The Webster-Ward Debate of 1654: The New Philosophy and the Problem of Educational Reform." *Proceedings of the Institute of Medicine of Chicago*, **27** (1969), 248-249.

———. "Harvey and Fludd: The Irrational Factor in the Rational Science of the Seventeenth Century." *Journal of the History of Biology*, **3** (1970), 81-105.

———. *Science and Education in the Seventeenth Century. The Webster-Ward Debate.* London: Macdonald; New York: American Elsevier, 1970.

———. "John Webster and the Educational Dilemma of the Seventeenth Century." *Actes du XIIe Congrès International d'Histoire des Sciences*, Vol. III B, pp. 15-23. Paris: A. Blanchard, 1970.

———. "The Webster-Ward Debate of 1654: The New Philosophy and the Problem of Educational Reform." *L'univers à la Renaissance: Microcosme et macrocosme,* Travaux de l'Institut pour l'étude de la Renaissance et de l'Humanisme, Vol. IV, pp. 33-51. Brussels: P.U.B./P.U.F., 1970.

———. "Johann Joachim Becher." In *Dictionary of Scientific Biography*, Vol. I, pp. 548-551. New York: Scribner's, 1970.

———. "Joseph Duchesne." In *Dictionary of Scientific Biography*, Vol. IV, pp. 208-210. New York: Scribner's, 1971.

———. "The Paracelsians and the Chemists: The Chemical Dilemma in Renaissance Medicine." *Clio Medica*, **7** (1972), 185-199.

———. "Some Comments on the Contemporary Helmontian Renaissance." *Ambix*, **19** (1972), pp. 145-150.

———, ed. *Science, Medicine and Society in the Renaissance. Essays to Honor Walter Pagel.* 2 vols. New York: Science History Publications; London: Heinemann, 1972.

———. "Robert Fludd." In *Dictionary of Scientific Biography*, Vol. V, pp. 47-49. New York: Scribner's, 1972.

———. "Guintherius—Libavius—Sennert: The Chemical Compromise in Early Modern Medicine." In *Science, Medicine and Society in the Renaissance*, ed. Allen G. Debus, Vol. I, pp. 151-165. New York: Science History Publications, 1972.

———. "La philosophie chimique de la Renaissance et ses relations avec la chimie de la fin du XVIIe siècle," *Sciences de la Renaissance.* VIIIe Congrès International de Tours (1965), ed. Jacques Roger, pp. 273-283. Paris: Vrin, 1973.

———. "A Further Note on Palingenesis: The Account of Ebenezer Sibley in the *Illustration of Astrology* (1792)." *Isis*, **64** (1973), 226-230.

―――. "Motion in the Chemical Texts of the Renaissance." *Isis,* **64** (1973), 4-17.

―――. "Alchemy." In the *Dictionary of the History of Ideas,* ed. Philip P. Weiner, Vol. I, pp. 27-34. New York: Scribner's 1973.

―――. "Paracelsian Medicine: Noah Biggs and the Problem of Medical Reform." In *Medicine in Seventeenth Century England,* ed. Allen G. Debus, pp. 33-48. Los Angeles/Berkeley: University of California Press, 1974.

―――. "Peter Severinus." In *Dictionary of Scientific Biography,* Vol. XII, pp. 334-336. New York: Scribner's, 1975.

―――. "Basil Valentine." In *Dictionary of Scientific Biography,* Vol. XIII, pp. 558-560. New York: Scribner's, 1976.

Dulieu, Louis. *La pharmacie à Montpellier.* Avignon: Les Presses Universelles, 1973.

Duveen, Denis I. *Bibliotheca alchemica et chemica.* London: Dawsons, 1949.

Eis, Gerhard. *Vor und nach Paracelsus. Untersuchungen über Hohenheims Traditionsverbundenheit und Nachrichten über seine Anhänger.* Medizin in Geschichte und Kultur, Band 8. Stuttgart: Gustav Fisher Verlag, 1965.

Eliade, Mircea. *The Forge and the Crucible.* New York: Harper, 1962.

―――. "The Forge and the Crucible: A Postscript." *History of Religion,* **8** (1968), 74-88.

Enselme, Jean. "L'étrange 'vingt-sixième livre' des oeuvres complétes d'Ambroise Paré." *Lyon Médicale,* **227** (1972), 39-47.

Eurich, Nell. *Science in Utopia. A Mighty Design.* Cambridge, Mass.: Harvard University Press, 1967.

Evans, George H. "The Contribution of Francis Anthony to Medicine." *Annals of Medical History,* 3rd Ser., **2** (1940), 171-173.

Ferguson, J. *Bibliotheca chemica.* 2 vols. Glasgow, 1906; London: Academic and Bibliographical Publications, 1954.

Figurovskij, N. A. "Die Chemie in Russland im Zeitalter der Iatrochemie." *Nova Acta Leopoldina,* **27** (1963), 351-366.

Forbes R. J., *Studies in Ancient Technology.* Vol. 7. Leiden: E. J. Brill, 1963.

Frank, Robert G. "Early Modern English Universities." *History of Science,* **11** (1973), 194-216; 239-269.

French, Peter J. *John Dee. The World of an Elizabethan Magus.* London: Routledge and Kegan Paul, 1972.

Fulton, John F. *A Bibliography of the Honourable Robert Boyle.* 2nd ed. Oxford: Clarendon Press, 1961.

Fussell, G. E. "Low Countries Influence on English Farming." *English Historical Review,* **74** (1959), 612-622.

———. *Crop Nutrition: Science and Practice before Liebig.* Lawrence, Kansas: Coronado Press, 1971.

Gardner, F. Leigh. *A Catalogue Raisonné of Works of the Occult Sciences.* Vol. I: *Rosicrucian Books.* Intro. William Wynn Westcott. 2nd ed. Privately printed, 1923.

Garin, Eugenio. *L'educazione in Europa 1400-1600. Problemi e programmi.* Bari: Editori Laterza, 1957.

Gelbart, Nina Rattner. "The Intellectual Development of Walter Charleton." *Ambix,* **18** (1971), 149-168.

———. "Science in Enlightenment Utopias: Power and Purpose in Eighteenth Century French Voyages Imaginaires." Ph.D. dissertation, University of Chicago, 1973.

Gibson, Thomas. "A Sketch of the Career of Theodore Turquet de Mayerne." *Annals of Medical History,* N.S. **5** (1933), 315-326.

Gillispie, Charles Coulston. "The Encyclopédie and the Jacobin Philosophy of Science: A Study in Ideas and Consequences." In *Critical Problems in the History of Science,* ed. Marshall Clagett, pp. 255-289. Madison: University of Wisconsin Press, 1962.

Gnudi, M. T. and J. P. Webster. *The Life and Times of Gaspare Tagliacozzi, Surgeon of Bologna (1545-1599).* New York: Herbert Reichner, 1950.

Goldammer, Kurt. "Bemerkungen zur Struktur des Kosmos und der Materie bei Paracelsus." In *Medizingeschichte in unserer Zeit. Festgabe für Edith Heischkel und Walter Artelt zum 65. Geburtstag,* ed. Hans-Heinz Eulner, *et al.,* pp. 121-144. Stuttgart: Ferdinand Enke, 1971.

———. "Die Paracelsische Kosmologie und Materietheorie in ihrer wissenschaftsgeschichtlichen Stellung und Eigenart." *Medizin historisches Journal,* **6** (1971), 5-35.

Goltz, Dietlinde. "Die Paracelsisten und die Sprache." *Sudhoffs Archiv für Geschichte der Medizin und der Naturwissenschaften,* **56**, (1972) 337-352.

———. "Alchemie und Aufklärung: Ein Beitrag zur Naturwissenschaftsgeschichtsschreibung der Aufklärung." *Medizin historisches Journal,* **7** (1972), 31-48.

Graves, Frank Pierrepont. *Peter Ramus and the Educational Reformation of the Sixteenth Century.* New York: Macmillan, 1912.

Gregory, T. *Scetticismo ed empirismo: Studio su Gassendi.* Bari: Editori Laterza, 1961.

———. "Studi sull'atomismo del seicento." *Giornale critico della Filosofia Italiana,* **43** (1964), 38-65; **45** (1966), 43-63; **46** (1967) 528-541.

Gruman, Gerald J. *A History of Ideas about the Prolongation of Life. The Evolution of Prolongevity Hypotheses to 1800.* Philadelphia: American Philosophical Society, Transactions, N.S. **56** (Pt. 9), 1966.

Guerlac, Henry. "John Mayow and the Aerial Nitre. Studies in the Chemistry of John Mayow. I." *Actes du Septième Congrès International d'Histoire des Sciences*, Jerusalem, 1953, pp. 332-349.

———. "The Poets' Nitre. Studies in the Chemistry of John Mayow. II." *Isis*, **45** (1954), 243-255.

———. "Guy De La Brosse and the French Paracelsians." In *Science, Medicine and Society in the Renaissance*, ed. Allen G. Debus, Vol. I, pp. 177-199. New York: Science History Publications, 1972.

Gugel, Kurt F. *Johann Rudolph Glauber 1604-1670, Leben und Werk*. Würzburg: Freunde Mainfränkischer Kunst und Geschichte, 1955.

Hall, A. Rupert. "Medicine and the Royal Society." In *Medicine in Seventeenth Century England*. ed. Allen G. Debus, pp. 421-452. Berkeley/Los Angeles: University of California Press, 1974.

Hamy, E.T. "Un précurseur de Guy de la Brosse. Jacques Gohory et le Lycium Philosophal de Saint-Marceau-les-Paris (1571-1576)." *Nouvelles Archives du Museum d'histoire naturelle*, 4th Ser., **1** (1899), 1-26.

Hannaway, Owen. "Guy de la Brosse." In *Dictionary of Scientific Biography*, Vol. V, pp. 447-448. New York: Scribner's, 1972.

Hassinger, Herbert. *Johann Joachim Becher 1635-1682. Ein Beitrag zur Geschichte des Merkantilismus*. Veröffentlichungen der Kommision für Neuere Geschichte Österreichs, No. 38. Vienna: A. Holzhausens, 1951.

Hirst, Desirée. *Hidden Riches*. London: Eyre & Spottiswoode, 1963.

Holmes, F. L. "Analysis by Fire and Solvent Extractions: The Metamorphosis of a Tradition." *Isis*, **62** (1971), 129-148.

Holmyard, E. J. *Alchemy*. Harmondsworth: Penguin, 1957.

Hooykaas, R. "Die Elementenlehre des Paracelsus." *Janus*, **39** (1935), 175-188.

———. "Die Elementenlehre der Iatrochemiker." *Janus*, **41** (1937), 26-28.

———. "Chemical Trichotomy before Paracelsus?" *Archives Internationale d'Historic des Sciences*, **28** (1949), 1063-1074.

———. *Humanisme, science et reforme: P. de la Ramée, 1515-72*. Leiden: E.J. Brill, 1958.

Hopkins, Arthur John. *Alchemy, Child of Greek Philosophy*. New York: Columbia University Press, 1934.

Howe, Herbert M. "A Root of van Helmont's Tree." *Isis*, **56** (1965), 408-419.

Hubicki, Wlodzimierz. "Chemie und Alchemie des 16. Jahrhunderts in Polen." *Annalen Universitatis Mariae Curie-Sklodowska Polonia-Lublin*, **10** (1955), 61-100.

―――. "Chemistry and Alchemy in Sixteenth-Century Cracow." *Endeavour*, **17** (1958), 204-207.

―――. "Alexander von Suchten." *Sudhoffs Archiv für Geschichte der Medizin und der Naturwissenschaften*, **44** (1960), 54-63.

―――. "Michael Sendivogius' Theory, Its Origin and Significance in the History of Chemistry." *Actes du Dixième Congrès International d'Histoire des Sciences* (Ithaca, 1962), Vol. II, pp. 829-833. Paris: Hermann, 1964.

Hutin, Serge. *Robert Fludd (1574-1637): Alchimiste et philosophe Rosicrucien.* Paris: Omnium Litteraire, 1971.

Isler, Hansruedi. *Thomas Willis 1621-1675. Doctor and Scientist.* New York/London: Hefner, 1968.

Jones, R. F. "The Humanistic Defence of Learning in the Mid-Seventeenth Century." In *Reason and the Imagination: Studies in the History of Ideas, 1600-1800*, ed. J. A. Mazzeo, pp. 71-92. New York: Columbia University Press; London: Routledge and Kegan Paul, 1962.

de Jong, H. M. E. "Glauber und die Weltanschauung der Rosenkreuzer." *Janus*, **56** (1969), 278-304.

Josten, C. H. "Truth's Golden Harrow. An Unpublished Alchemical Treatise of Robert Fludd in the Bodleian Library." *Ambix*, **3** (1949), 91-150.

―――. "Robert Fludd's 'Philosophicall Key' and his Alchemical Experiment on Wheat." *Ambix*, **11** (1963), 1-23.

―――. "A Translation of John Dee's 'Monas Hieroglyphica' (Antwerp, 1564), With an Introduction and Annotations." *Ambix*, **12** (1964), 84-220.

―――. "Robert Fludd's Theory of Geomancy and his Experiences at Avignon in the Winter of 1601 to 1602." *Journal of the Warburg and Courtauld Institutes*, **27** (1964), 327-335.

Kangro, Hans. "Erklärungswert und Schwierkeiten der Atomhypothese und ihrer Anwendung auf chemische Probleme in der ersten Hälfte des 17. Jahrhunderts." *Technikgeschichte*, **35** (1968), 14-36.

―――. *Joachim Jungius' Experimente und Gedanken zur Begründung der Chemie als Wissenschaft.* Wiesbaden: Franz Steiner: 1968.

Kargon, Robert. "Walter Charleton, Robert Boyle and the Acceptance of Epicurean Atomism in England." *Isis*, **55** (1964), 184-192.

―――. *Atomism in England From Hariot to Newton.* Oxford: Clarendon Press, 1966.

Kearney, Hugh. *Scholars and Gentlemen. Universities and Society in Pre-Industrial Britain.* Ithaca: Cornell University Press, 1970.

Keller, Alex. "Mathematical Technologies and the Growth of the Idea of

Technical Progress in the Sixteenth Century." In *Science, Medicine and Society in the Renaissance,* ed. Allen G. Debus, Vol. I, pp. 11-27. New York: Science History Publications; London: Heinemann, 1972.

Kent, A. and O. Hannaway. "Some New Considerations on Beguin and Libavius." *Annals of Science,* **16** (1960), 241-250.

King, Lester S. *The Growth of Medical Thought.* Chicago: University of Chicago Press, 1963.

———. "Stahl and Hoffman: A Study in Eighteenth Century Animism." *Journal of the History of Medicine and Allied Sciences,* **19** (1964), 118-130.

———. "The Road to Scientific Therapy: 'Signatures,' 'Sympathy,' and Controlled Experiment." *Journal of the American Medical Association,* **197** (1966), 250-256.

———. *The Road to Medical Enlightenment 1650-1695.* London: Macdonald; New York: American Elsevier, 1970.

———. "The Transformation of Galenism." In *Medicine in Seventeenth-Century England,* ed. Allen G. Debus, pp. 7-31 Berkeley/Los Angeles: University of California Press, 1974.

Klibansky, R. *The Continuity of the Platonic Tradition during the Middle Ages: Outlines of a Corpus Platonicum Medii Aevi.* London: Warburg Institute, 1939.

Knight, David M. "The Scientist as Sage." *Studies in Romanticism,* **6** (1967), 65-88.

———. "The Physical Sciences and the Romantic Movement." *History of Science,* **9** (1970), 54-75.

Kocher, P.H. "Paracelsan Medicine in England (ca. 1570-1600)." *Journal of the History of Medicine and Allied Sciences,* **2** (1947), 451-480.

Kopp, Hermann. *Geschichte der Chemie.* 4 vols. Braunschweig, 1843, 1844, 1845, 1847. Reprint Hildesheim: Georg Olms, 1966.

———. *Beiträge zur Geschichte der Chemie.* 3 pts. in 2 vols. Braunschweig: F. Vieweg, 1869-1875.

Koyré, Alexandre. *Mystiques, spirituels, alchimistes: Schwenckfeld, Seb. Franck, Weigel, Paracelse.* Paris: A. Colin, 1955.

Kraus, Paul. *Jābir ibn Hayyān. Contribution à l'histoire des idées scientifiques dans l'Islam.* 2 vols. Cairo: Impr. de l'Institut français d'archéologie orientale, 1942-1943.

Kremers, Edward, and George Urdang. *History of Pharmacy.* Philadelphia: Lippincott, 1951.

Kristeller, Paul Oskar. *The Philosophy of Marsilio Ficino.* Trans. Virginia Conant. New York: Columbia University Press, 1943.

———. *Renaissance Thought. The Classic, Scholastic and Humanist Strains.* New York: Harper and Row, Harper Torchbooks, 1961.

———. *Eight Philosophers of the Italian Renaissance.* Stanford: Stanford University Press, 1964.

Krivatsky, Peter. "Erasmus' Medical Milieu." *Bulletin of the History of Medicine,* **47** (1973), 113-154.

Kuhn, Thomas S. *The Copernican Revolution.* Cambridge, Mass.: Harvard University Press, 1957.

Kvačala, J. J.V. *Andreä's Antheil an geheimen Gesellschaften.* Dorpat: C. Mattiesen, 1899.

Lasswitz, Kurd. *Geschichte der Atomistik vom Mittelalter bis Newton.* 2 vols. Hamburg/Leipzig: Leopold Voss, 1890.

Leclerc, Ivor. "Atomism, Substance, and the Concept of Body in Seventeenth Century Thought." *Filosofia della Scienza,* **27**, 1-16.

Leicester, Henry M. *The Historical Background of Chemistry.* New York: John Wiley, 1956.

Lemoine, Albert. *Le vitalisme et l'animisme de Stahl.* Paris: Germer Baillière, 1864.

Lenoble, Robert. *Mersenne ou la naissance du mécanisme.* Paris: Vrin, 1943.

———. *La géologie au milieu du XVIIe siècle.* Les Conférences du Palais de la Découverte, Serie D, No. 27. Paris: University of Paris, 1954.

Lindroth, Sten. *Paracelsismen i Sverige till 1600—talets mitt.* Uppsala, 1943.

Lindsay, Jack. *The Origins of Alchemy in Graeco-Roman Egypt.* London: Muller, 1970.

Lloyd, O. M. "The Royal College of Physicians of London and Some City Livery Companies." *Journal of the History of Medicine and Allied Sciences,* **11** (1956), 412-421.

López Piñero, José María. "Juan de Cabriada y las primeras etapas de la iatroquímica y de la medicina moderna en España." *Cuadernos de História de la Medicina Española,* **2** (1962), 129-154.

———. "La iatroquímica de la segunda mitad del siglo XVII." In *Historia de la medicina,* ed. Pedro Laín Entralgo, Vol. IV, pp. 279-295. Barcelona: Salvat Editores, 1973.

Lu Gwei-Djen, Joseph Needham, and Dorothy Needham. "The Coming of Ardent Water." *Ambix,* **19** (1972), 69-112.

McConaghey, R. M. S. "Sir George Baker and the Devonshire Colic." *Medical History,* **11** (1967), 345-360.

McGuire, J. E. "Force, Active Principles, and Newton's Invisible Realm." *Ambix,* **15** (1968), 154-208.

———, and P. M. Rattansi. "Newton and the 'Pipes of Pan.' " *Notes and Records of the Royal Society of London,* **21** (1966), 108-143.

McKie, Douglas. "Fire and the Flamma Vitalis." In *Science, Medicine and*

History, ed. E.A. Underwood, Vol. I, pp. 469 ff. London: Oxford, University Press, 1953.

McVaugh, Michael. "Arnald of Villanova and Bradwardine's Law." *Isis*, **58** (1967), 56-64.

Maddison, R.E.W. *The Life of the Honourable Robert Boyle*. London: Taylor and Francis, 1969.

Major, Ralph H. "Some Landmarks in the History of Lead Poisoning." *Annals of Medical History*, N.S. **3** (1931), 218-227.

Marx, Jacques. "Alchimie et palingénésie." *Isis*, **62** (1971), 274-289.

Merck Luengo, José Guillermo. *La quimiatría en España*. Madrid: Instituto Arnaldo de Vilanova, 1959).

Metzger, Hélène. *Les doctrines chimiques en France du début XVIIe à la fin du XVIIIe siècle. Première partie*. Paris: P.U.F., 1923.

———. *Newton, Stahl, Boerhaave et la doctrine chimique*. Paris: Felix Alcan, 1930.

Mévergnies, Paul Nève. *Jean-Baptiste van Helmont. Philosophe par le feu*. Paris: Libraire E. Droz, 1935.

———. "Sur les lettres de J. B. Van Helmont au P. Marin Mersenne." *Revue Belge de Philosophie et d'Histoire*, **26** (1948), 61-83.

Milt, Bernhard. "Conrad Gesner als Balneologe." *Gesnerus*, **2** (1945), 1-16.

Morwitz, E. *Geschichte der Medicin*. Wiesbaden: Dr. Martin Sändig, n.d.

Multhauf, Robert P. "John of Rupescissa and the Origin of Medical Chemistry." *Isis*, **45** (1954), 359-367.

———. "Medical Chemistry and 'The Paracelsians.' " *Bulletin of the History of Medicine*, **28** (1954), 101-126.

———. "J.B. Van Helmont's Reformation of the Galenic Doctrine of Digestion." *Bulletin of the History of Medicine*, **29** (1955), 154-163.

———. "The Significance of Distillation in Renaissance Medical Chemistry." *Bulletin of the History of Medicine*, **30** (1956), 329-346.

———. *The Origins of Chemistry*. London: Oldbourne, 1966.

Munk, William. *The Roll of the Royal College of Physicians of London* .... 3 vols. 2nd ed. London: Published for the College, 1878.

Murphy, Terence. "The Historicization of the Great Chain of Being. An Intellectual Biography of J. B. R. Robinet." Ph.D. dissertation, University of Chicago, 1975.

Murray, John J. "The Cultural Impact of the Flemish Low Countries on Sixteenth- and Seventeenth-Century England." *American Historical Review*, **62** (1957), 837-854.

Nauert, Charles G., Jr. *Agrippa and the Crisis of Renaissance Thought*. Urbana: University of Illinois Press, 1965.

Nebe, August. *Vives, Alsted, Comenius in ihrem Verhältnis zu einande.* Elberfeld, 1891.

Needham, Joseph. "Artisans et Alchimistes en Chine et dans le Monde Hellénistique." *La Pensée,* No. 152 (1970), 2-24.

Neuburger, Max. *Geschichte der Medizin.* 2 vols. Stuttgart: Ferdinand Enke, 1911.

Niebyl, Peter H. "Sennert, Van Helmont, and Medical Ontology." *Bulletin of the History of Medicine,* **45** (1971), 115-137.

———. "Galen, Van Helmont and Blood Letting." In *Science, Medicine and Society in the Renaissance,* ed. Allen G. Debus, Vol. II, pp. 13-23. New York: Science History Publications; London: Heinemann, 1972.

Norpoth, Leo. "Kölner Paracelsismus in der 2. Hälfte des 16. Jahrhunderts." *Jahrbuch des Kölnischen Geschichtsvereins,* **27** (1953), 133-146.

———. "Paracelsismus und Antiparacelsismus in Köln in der 2. Hälfte des 16. Jahrhunderts." In *Medicinae et artibus. Festschrift für Professor Dr. phil. Dr. med. Wilhelm Katner zu seinem 65. Geburtstag, Düsseldorfer Arbeiten zur Geschichte der Medizin,* Beiheft 1 pp. 91-102. Düsseldorf: Michael Triltsch, 1968.

Oldroyd, David. "An Examination of G.E. Stahl's Principles of Universal Chemistry." *Ambix,* **20** (1973), 36-52.

O'Malley, C. D. *Andreas Vesalius of Brussels.* Berkeley/Los Angeles: University of California Press, 1964.

———. "Joannes Guinter." In the *Dictionary of Scientific Biography,* Vol. V, pp. 585-586. New York: Scribner's, 1972.

Ong, W. J. *Ramus: Method and the Decay of Dialogue.* Cambridge, Mass.: Harvard University Press, 1958.

Oppenheim, A. Leo. "Mesopotamia in the Early History of Alchemy." *Revue d'Assyriologie et d'Archéologie Orientale,* **60** (1966), 29-45.

Pachter, Henry M. *Paracelsus, Magic into Science.* New York: Schuman, 1951.

Pagel, Walter. *Jo. Bapt. van Helmont, Einführung in die philosophische Medizin des Barock.* Berlin: Julius Springer, 1930.

———. "Helmont, Leibniz, Stahl." *Sudhoffs Archiv für Geschichte der Medizin,* **24** (1931), 19-59.

———. "Religious Motives in the Medical Biology of the XVIIth Century." *Bulletin of the Institute of the History of Medicine,* **3** (1935), 97-128, 213-231, 265-312.

———. "The Debt of Science and Medicine to a Devout Belief in God. Illustrated by the Work of J.B. Van Helmont." *Journal of the Transactions of the Victoria Institute,* **74** (1942), 99-115.

———. *The Religious and Philosophical Aspects of van Helmont's Science and Medicine*. Bulletin of the History of Medicine. Supp. No. 2. Baltimore: Johns Hopkins Press, 1944.

———. "J. B. van Helmont, De Tempore, and Biological Time." *Osiris*, **8** (1949), 346-417.

———. "Harvey and the Purpose of the Circulation." *Isis*, **42** (1951), 25-26.

———. "The Reaction to Aristotle in Seventeenth-Century Biological Thought." In *Science, Medicine and History. Essays in Honour of Charles Singer*, ed. E. A. Underwood, Vol. I, pp. 489-509. London: Oxford University Press, 1953.

———. "J. B. Van Helmont's Reformation of the Galenic Doctrine of Digestion—And Paracelsus." *Bulletin of the History of Medicine*, **29** (1955), 563-568.

———. "Van Helmont's Ideas on Gastric Digestion and the Gastric Acid." *Bulletin of the History of Medicine*, **30** (1956), 524-536.

———. "The Position of Harvey and Van Helmont in the History of European Thought." *Journal of the History of Medicine and Allied Sciences*, **13** (1958), 186-198.

———. *Paracelsus. An Introduction to Philosophical Medicine in the Era of the Renaissance*. Basel: S. Karger, 1958.

———. "Paracelsus and the Neoplatonic and Gnostic Tradition." *Ambix*, **8** (1960), 125-166.

———. "The Prime Matter of Paracelsus." *Ambix*, **9** (1961), 117-135.

———. *Das Medizinische Weltbild des Paracelsus: Seine Zusammenhänge Neuplatonismus und Gnosis*. Wiesbaden: Franz Steiner, 1962.

———. "Paracelsus' ätherähnliche Substanzen und ihre pharmakologische Auswertung an Hühnern. Spachgebrauch (henbane) und Konrad von Megenbergs "Buch der Natur" also mögliche Quellen." *Gesnerus*, **21** (1964), 113-125.

———. *William Harvey's Biological Ideas. Selected Aspects and Historical Background*. Basel/New York: S. Karger, 1967.

———. "Harvey and Glisson on Irritability: With a Note on van Helmont." *Bulletin of the History of Medicine*, **41** (1967), 497-514.

———. "Paracelsus: Traditionalism and Medieval Sources." In *Medicine, Science and Culture: Historical Essays in Honour of Owsei Temkin*, ed. Lloyd G. Stevenson and Robert P. Multhauf, pp. 51-76. Baltimore: Johns Hopkins Press, 1968.

———. "Chemistry at the Cross-Roads: The Ideas of Joachim Jungius." *Ambix*, **16** (1969), 100-108.

———. "William Harvey Revisited." *History of Science*, **8** (1969), 1-31 and **9** (1970), 1-41.

———. "Thomas Erastus." In the *Dictionary of Scientific Biography*, Vol. IV, pp. 386-388. New York: Scribner's, 1971.

———. "Van Helmont's Concept of Disease—To Be Or Not To Be? The Influence of Paracelsus." *Bulletin of the History of Medicine*, **46** (1972), 419-454.

———. "Johannes (Joan) Baptista van Helmont." In *Dictionary of Scientific Biography*, Vol. VI, pp. 253-259. New York: Scribner's, 1972.

———, and P. Rattansi. "Vesalius and Paracelsus." *Medical History*, **8** (1964), 309-328.

———, and Marianne Winder. "The Eightness of Adam and Related 'Gnostic' Ideas in the Paracelsian Corpus." *Ambix*, **16** (1969), 119-139.

———, ———. "The Higher Elements and Prime Matter in Renaissance Naturalism and in Paracelsus." *Ambix*, **21** (1974), 93-127.

Partington, J. R. "Joan Baptist van Helmont." *Annals of Science*, **1** (1936), 359-384.

———. "The Life and Work of John Mayow (1641-1679)." *Isis*, **47** (1956), 217-230 and 405-417.

———. *A History of Greek Fire and Gunpowder*. Cambridge: Heffer, 1960.

———. *A History of Chemistry*. Vols. II, III. London: Macmillan, 1961, 1962.

Patterson, T. S. "Jean Beguin and His *Tyrocinium Chymicum*." *Annals of Science*, **2** (1937), 243-298.

Pauli, Wolfgang. "The Influence of Archetypal Ideas on the Scientific Theories of Kepler." In C.G. Jung and W. Pauli, *The Interpretation of Nature and the Psyche*, trans. Priscilla Silz. Bollingen Series 51, pp. 147-240. New York: Pantheon Books, 1955.

Peuckert, Will-Erich. *Die Rosenkreutzer. Zur Geschichte einer Reformation*. Jena: E. Diederichs, 1928.

Pietsch, Erich. *Johann Rudolph Glauber. Der Mensch, sein Werk und seine Zeit*. Deutsches Museum Abhandlungen und Berichte. Munich: R. Oldenbourg, 1956.

Plessner, Martin. "The Place of the *Turba philosophorum* in the Development of Alchemy." *Isis*, **45** (1954), 331-338.

———. "*The Turba philosophorum*." *Ambix*, **7** (1959), 159-163.

Poynter, F.N.L. "Nicholas Culpeper and His Books." *Journal of the History of Medicine and Allied Sciences*, **17** (1962), 152-167.

———. "Nicholas Culpeper and the Paracelsians." In *Science, Medicine and Society in the Renaissance. Essays to Honor Walter Pagel*, ed. Allen G. Debus, Vol. I, pp. 201-220. New York: Science History Publications, 1972.

———, ed. *Chemistry in the Service of Medicine*. London: Pitman, 1963.

Purver, Margery. *The Royal Society: Concept and Creation.* London: Routledge and Kegan Paul, 1967.

Randall, John Herman, Jr. "The Development of Scientific Method in the School of Padua." *Journal of the History of Ideas,* 1 (1940), 177-206.

———. "Scientific Method in the School of Padua." In *Roots of Scientific Thought. A Cultural Perspective,* ed. Philip P. Wiener and Aaron Noland, pp. 139-146. New York: Basic Books, 1957.

Raphael, Alice. *Goethe and the Philosopher's Stone: Symbolical Patterns in 'The Parable' and the Second Part of 'Faust.'* London: Routledge and Kegan Paul, 1965.

Rath, Gernot. "Die Anfänge der Mineralquellenanalyse." *Medizinischen Monatsschrift,* 3 (1949), 539-541.

———. "Die Mineralquellenanalyse im 17. Jahrhundert." *Sudhoff's Archiv für Geschichte der Medizin und der Naturwissenschaften,* 41 (1957), 1-9.

Rattansi, P.M. "The Helmontian-Galenist Controversy in Restoration England." *Ambix,* 12 (1964), 1-23.

———. "Jean Beguin." In *Dictionary of Scientific Biography,* Vol. I, pp. 571-572. New York: Scribner's, 1970.

———. "Newton's Alchemical Studies." In *Science, Medicine and Society in the Renaissance,* ed. Allen G. Debus, Vol. II, pp. 167-182. New York: Science History Publications; London: Heineman, 1972.

Ray, A. P. *History of Chemistry in Ancient and Medieval India, incorporating the History of Hindu Chemistry by Achaiya Profulla Chandra Ray.* Calcutta: Indian Chemical Society, 1956.

Risse, Guenther B. "Kant, Schelling, and the Early Search for a Philosophical 'Science' of Medicine in Germany." *Journal of the History of Medicine and Allied Sciences,* 27 (1972), 145-158.

Rommelaere, Willem. *Études sur J.-B. Van Helmont.* Brussels: Manceaux, 1868.

Rossi, Paolo. *Francis Bacon: From Magic to Science.* Trans. Sacha Rabinovitch. Chicago: University of Chicago Press, 1968.

Santillana, Giorgio de. *The Age of Adventure: The Renaissance Philosophers.* New York: Mentor, 1956.

Sarton, George. *Appreciation of Ancient and Medieval Science During the Renaissance.* 1955. New York: Perpetua edition, 1961.

Schmidt, Albert-Marie. *Paracelse ou la force qui va.* Paris: Plon, 1967

Schmitt, Charles B. "Towards a Reassessment of Renaissance Aristotelianism." *History of Science,* 11 (1973), 159-193.

———. *Gianfrancesco Pico della Mirandola (1469-1533) and his Critique of Aristotle.* The Hague: Martinus Nijhoff, 1967.

Schneider, Wolfgang. "A Bibliographical Review of the History of Pharmaceutical Chemistry (with particular reference to German Literature). *American Journal of Pharmaceutical Education*, **23** (1959), 161-172.

————. "Der Wandel des Arzneischatzes im 17. Jahrhundert und Paracelsus." *Sudhoffs Archiv für Geschichte der Medizin und der Naturwissenschaften*, **45** (1961), 201-215.

————. "Die deutschen Pharmakopöen des 16. Jahrhunderts und Paracelsus." *Pharmazeutische Zeitung*, **106** (1961), 3-16.

————. "Probleme und neuere Ansichten in der Alchemiegeschichte." *Chemiker-Zeitung - Chem. Apparatur*, **85** (1961), 643-652.

————. *Geschichte der pharmazeutische Chemie*. Weinheim: Verlag Chemie, 1972.

Schröder, G. "Oswald Crollius." *Pharmaceutical Industry*, **21** (1959), 405-408.

————. "Studien zur Geschichte der Chemiatrie." *Pharmazeutische Zeitung*, **111** No. 35 (1966), 1246 ff.

————. "Oswald Crollius." In *Dictionary of Scientific Biography*. Vol. III, pp. 471-472. New York: Scribner's, 1971.

Secret, F. *L'astrologie et les Kabbalistes Chrétiens à la Renaissance*. La Tour St. Jacques, 1956.

Shapiro, Barbara. *John Wilkins 1614-1672. An Intellectual Biography*. Berkeley/Los Angeles: University of California Press, 1969.

Sivin, Nathan. *Chinese Alchemy: Preliminary Studies*. Cambridge, Mass.: Harvard University Press, 1968.

Solomon, Howard. *Public Welfare, Science, and Propaganda in Seventeenth Century France*. Princeton: Princeton University Press, 1972.

Spronson, J.W. van. "Glauber Grondlegger van Chemishe Industrie." *Nederlandse Chemische Industrie* (Mar. 3, 1970), 3-11.

Stapleton, H.E. "The Antiquity of Alchemy." *Ambix*, **5** (1953), 1-43.

Stavenhagen, Lee. "The Original Text of the Latin *Morienus*." *Ambix*, **17** (1970), 1-12.

Steinhüser. *Johann Joachim Becher und die Einzelwirtschaft: Ein Beitrag zur Geschichte der Einzelwirtschaftslehre und Kameralismus*. Nuremburg: Nürnberger Beiträge zu den Wirtschaftswissenschaften, 1931.

Stoddart, A.M. *The Life of Paracelsus: Theophrastus von Hohenheim 1493-1541*. London: William Rider, 1915.

Sudhoff, Karl. "Ein Beitrag zur Bibliographie der Paracelsisten im 16. Jahrhundert." *Centralblatt fur Bibliothekswesen*, **10** (1893), 316-326, 386-407.

————. *Bibliographia Paracelsica*. Berlin: Verlag Georg Reimer, 1894. Reprinted Graz: Akademische Druck- u. Verlagsanstalt, 1958.

―――. *Iatromathematiker vor nehmlich im 15. und 16. Jahrhundert.* Breslau: J.U. Kern, 1902.

―――. "The Literary Remains of Paracelsus." In *Essays in the History of Medicine.* New York: Medical File Press, 1926.

Taylor, F. Sherwood, and C. H. Josten. "Johannes Banfi Hunyades 1576-1650." *Ambix,* **5** (1953), 44-52.

―――, ―――. "Johannes Banfi Hunyades. A Supplementary Note." *Ambix,* **5** (1956), 115.

Telepnef, Basilio de. *Paracelsus: A Genius amidst a Troubled World.* St. Gall: Zollikofer, 1945.

Temkin, Owsei. *The Falling Sickness.* Baltimore: Johns Hopkins Press, 1945.

―――. "The Classical Roots of Glisson's Doctrine of Irritation." *Bulletin of the History of Medicine,* **38** (1964), 297-328.

―――. *Galenism. Rise and Decline of a Medical Philosophy.* Ithaca/London: Cornell University Press, 1973.

Thomas, Henry. "The Society of Chymical Physitians." In *Science, Medicine and History,* ed. E. A. Underwood, Vol. I, pp. 56-71. London: Oxford University Press, 1953.

Thorndike, Lynn. *A History of Magic and Experimental Science.* 8 vols. New York: Columbia University Press, 1923-1958.

―――. "Alchemy During the First Half of the Sixteenth Century." *Ambix,* **2** (1938), 26-37.

Trevor-Roper, H.R. "Three Foreigners and the Philosophy of the English Revolution." *Encounter,* **14** (1960), 3-20.

―――. "The Sieur de la Rivière, Paracelsian Physician of Henry IV." In *Science, Medicine and Society in the Renaissance,* ed. Allen G. Debus, Vol. II, pp. 227-250.

Turnbull, G.H. *Hartlib, Dury and Comenius. Gleanings from Hartlib's Papers.* Liverpool: University of Liverpool Press, 1947

Urdang, George. "The Mystery About the First English (London) Pharmacopoeia (1618)." *Bulletin of the History of Medicine,* **12** (1942), 304-313.

―――. "How Chemicals Entered the Official Pharmacopoeias." *Archives Internationales d'Histoire des Sciences,* **7** (1954), 303-314.

Vandevelde, A.J.J. "Helmontiana." 5 parts, in *Verslagen en Mededeelingen. K. Vlaamsche Academie voor Taal-en Letterkunde,* Pt. 1 (1929), 453-476; Pt. 2 (1929), 715-737; Pt. 3 (1929), 857-879; Pt. 4 (1932), 109-122, Pt. 5 (1936), 339-387.

Vannier, L. "L'oeuvre de O. Crollius (1580-1609)." *Bulletin de la Société Française d'Histoire de la Médecine,* **31** (1937), 91-108.

Waite, A.E. *The Real History of the Rosicrucians.* London: G. Redway, 1887.

———. *The Brotherhood of the Rosy Cross.* London: W. Rider and Sons, 1924.

Walden, P. "Glauber." In *Das Buch der grossen Chemiker,* ed. Günther Bugge, Vol. I, pp. 151-172. Berlin, 1929.

Walker, D. P. *Spiritual and Demonic Magic from Ficino to Campanella.* London: Warburg Institute, 1958.

Wall, Cecil, H. Charles Cameron, and E. Ashworth Underwood. *A History of the Worshipful Society of Apothecaries of London.* Vol. I. London/New York/Toronto: Oxford University Press for the Wellcome Historical Medical Museum, 1963.

Webster, Charles. "The Origins of the Royal Society." *History of Science,* **6** (1967), 106-127.

———. "The College of Physicians: 'Solomon's House' in Commonwealth England." *Bulletin of the History of Medicine,* **41** (1967), 393-412.

———. "English Medical Reformers of the Puritan Revolution: A Background to the 'Society of Chymical Physicians.' " *Ambix,* **14** (1967), 16-41.

———. "Science and the Challenge to the Scholastic Curriculum 1640-1660." In *The Changing Curriculum,* published by the History of Education Society. London: Methuen, 1971.

Weiss, Hélène. "Notes on the Greek Ideas Referred to in Van Helmont, De Tempore." *Osiris,* **8** (1949), 418-449.

Westfall, Richard S. "Newton and the Hermetic Tradition." In *Science, Medicine and Society in the Renaissance,* ed. Allen G. Debus, Vol. II, pp. 183-198. New York: Science History Publications; London: Heinemann, 1972.

Wightman, W.P.D. *Science and the Renaissance. An Introduction to the Study of the Emergence of the Sciences in the Sixteenth Century.* 2 vols. Edinburgh/London: Oliver and Boyd, 1962.

———. "Myth and Method in Seventeenth-century Biological Thought." *Journal of the History of Biology,* **2** (1969), 321-336.

Wilkinson, Ronald Sterne. "George Starkey, Physician and Alchemist." *Ambix,* **11** (1963), 121-152.

Woodward, William Harrison. *Vittorino da Feltre, and Other Humanist Educators. Essays and Versions.* Cambridge: Cambridge University Press, 1904.

———. *Desiderius Erasmus Concerning the Aim and Method of Education.* Cambridge: Cambridge University Press, 1904.

―――. *Studies in Education during the Age of the Renaissance 1400-1600.* Cambridge: Cambridge University Press, 1906.

Yates, Frances A. "The Art of Ramon Lull: An Approach to it through Lull's Theory of the Elements." *Journal of the Warburg Institute,* **17** (1954), pp. 115-173.

―――. *Giordano Bruno and the Hermetic Tradition.* Chicago: University of Chicago Press, 1964.

―――. "The Hermetic Tradition in Renaissance Science." In *Art, Science, and History in the Renaissance,* ed. Charles Singleton, pp. 255-274. Baltimore: Johns Hopkins Press, 1968.

―――. *Theatre of the World.* London: Routledge & Kegan Paul, 1969.

―――. *The Rosicrucian Englightenment.* London/Boston: Routledge and Kegan Paul, 1972.

Zekert, Otto. *Die grosse Wanderung des Paracelsus. De Peregrinatione Paracelsi Magna. Von Einsiedeln nach Salzburg.* Ingelheim am Rhein: C.H. Boehringer Sohn, 1965.

INDEX

NUMBERS IN BOLDFACE type indicate pages on which illustrations appear.

acid-alkali theory, *see* neutralization, acid-base
action at a distance, *see* magnetism; sympathetic action
aerial niter, or saltpeter (aerial spirit, vital spirit, celestial fire, *flamma vitalis*), 53, 87-88, 107-109, 130, 220-223, 220 n43, 224, 231-236, 232 n77, 248, 248 n106, 333, 414 n115, 456, 494-499, 530, 536, 545
  and Glauber, 432-433, 437, 442
  and van Helmont, 365-368
  and Lefèvre, 452-453
  and Mayow (nitro-aerial particles), 108, 492-499, 536
  and Paracelsus, 53, 87-88, 107-109, 109 n103
Agricola, Georgius, 17, 17 n28, 540
agriculture, chemical, 411-425, 434-438, 441, 442, 542, 545
Agrippa, Henry Cornelius, 34, 38, 43
air, as an element, *see* elements
Albertus Magnus, 11-12
alchemy and the alchemical tradition, 2, 3 n2, 6, 11 n10, 14, 19-25 *passim*, 32, 43-45, 61, 81, 158-159, 224, 261-270 *passim*, 391, 536, 549, and *passim*
  in antiquity, 3-6, 5 n4, 9
  Chinese, 9-10, 9 n9
  Indian, 10
  Islamic, 7-10
  Latin, 11-14
  and Boyle, 481
  and Fludd, 249, 255, 276
  and van Helmont, 322-327
  and Libavius, 169
  and Newton, 533, 546
  analytical tests, 14-18, 17 n28, 37, 59, 110-112, 110 n111, 306 n32, 484-492; *see also* color tests; crystalline form in analysis; distillation; fire; spa waters, analysis of; specific gravity; weight, analysis by
  ancients, authority of

incorporation with Paracelsian works, 135-139
influence of, 125, 131
rejection of
  by Bostocke, 180-182
  by Fludd, 215-223, 225
  by Paracelsians, 66, 117, 125-126, 127 n*
  by Paracelsus, 46, 48, 51-52, 54-55, 58, 61
  see also specific names, e.g., Aristotle
André, François, 500
Andreae, Johann Valentin, 214 n15 and n18, 384-386, 389, 390
Andrews, Dr., 253 n116, 254
Anthony, Francis, 184-185
antimony, as a medicine, 59, 115, 116, 123, 135-136, 144, 147, 148, 148 n64, 150, 190, 374-375
Apollonios of Tyana, 7
archeus (van Helmont), 107, 158, 192, 340-344, 360-362, 464, 545
Aquinas, Thomas, 11, 269
Aristotle and Aristotelianism, 5, 8, 16, 25, 26-30, 38, 60, 81, 81 n44, 103, 131, 173, 205, 220 n43, 291, and *passim*
  defense of
    Foster, 280-281
    T. Hall, 406-409
    Libavius, 169, 173
    Sennert, 195-196
  rejection of, 66-69, 75, 261, 390-391, and *passim*
    by Boyle, 474, 477, 481-482
    by Fludd, 216, 223, 225
    by van Helmont, 305, 312-314, 317, 330, 333
    by Mersenne, 265
    by Paracelsus, 51
    by Ramus, 29-30
    by Stahl, 468
    by Webster, 395-397, 403-404
    by Willis, 523
  see also elements

Arnald of Villanova, 20-21, 20 n37, 22, 135, 176, 217
Ashmole, Elias, 387 n14, 409, 448
astrology, 33, 38, 43, 219, 386-387, 408
Atkins, Henry, 186, 187
atomism and corpuscularianism, 69, 69 n10, 191 n202, 469, 531-533, 545-546
  and Boyle, 481, 490
  and Charleton, 469-473, 531
  and Fludd, 220, 270
  and Lemery, 485-486
  and Sennert, 191-192, 531, 532
  and Willis, 520, 523
Aubert, Jacques, 148-154
Aumay, Loys d l', 148
aurum potabile, *see* gold, as a medicine
Avicenna (Ibn Sina), 10, 11
  attacked by Paracelsus, 48, 51, 52

Baccius, Andrea, 111
Bacon, Francis, 83 n50, 212, 281, 292, 383-384, 389, 398-399, 404, 405, 410, 420, 423, 448, 494, 496, 543
Bacon, Roger, 19, 19 n33, 33, 33 n78, 43, 148, 148 n62, 269
Bacon, William, 503
Baillif, Roch le, sieur de la Rivière, 155-157, 159, 160 n97, 201
Baker, George, 18, 177, 178, 507
Banister, John, 178
Basil Valentine, 94-95, 94 n70, 99, 123, 261, 320, 547
Bathurst, Ralph, 494
Bauchinet, Guillaume, 167
Baugy, Nicolas de, 277
Becher, Johann Joachim, 80, 89 n62, 415 n119, **443**, 445-446, 447, 458-463, 464, 467-469, 488 n165, 503, 545
Beguin, Jean, 82, 83, 167-168, 167 n129
Bible, *see* scriptural authority
Biggs, Noah, 103 n89, 367-368, 392-393, 503-507, 509, 536
Biringuccio, Vannoccio, 17
Blas (van Helmont), 314-317, 332, 337, 534-535, 546
blood
  circulation of, 165, 165 n121, 208, 235-236, 236 n81, 271-276, 498, 512-519, 526
  and the vital spirit, 235, 248, 366-368, 518
    bloodletting, 109, 367, 525

Boehme, Jacob, 397
Boerhaave, Hermann, 531
Bolos Democritos, 6, 7
Bolos of Mendes, 5
Boot, Arnold de, 420, 420 n141, 421, 422 n148
Boot, Gerard de, 420, 421
Bostocke, R., 64-65, 69, 69 n11, 73-74, 75 n24, 80-81, 116, 179-182, 202, 212
Boyle, Robert, 112, 134, 198, 321, 389, 447, 460, 473-485, **475**, 488-492, 493, 494, 497, 500-502, 503, 517-519, 536
Bredwell, Stephen, 23, 507-508
Browne, Thomas, 490
Bruno, Giordano, 258 n138
Brunschwig, Hieronymus, 21-22, 59, 115, 176, 178, 202, 540
Bullein, William, 177
Bulstrode, Whitlocke, 486 n161
Burgravius, Johann Ernst, 281

Cabala and Cabalistic number mysticism, 35, 35 n83, 38, 45, 255, 258 n138, 267, 268, 305, 405
candle experiment (diminution of air during combustion), 330-334, **335**, 495-499
Cardan, Jerome, 135, 136, 161, 161 n103, 281
Chandler, John, 290
Charleton, Walter, 290, 469-473, 501, 502, 536, 546
chemical warfare (Glauber), 438-441, 542
Child, Robert, 421
Clapham, Henoch, 184 n187
Clave, Estienne de, 261, 265
Clowes, William, 178
Coleridge, Samuel Taylor, 550
color tests, in analysis, 6, 6 n5, 8, 16, 17-18, 112, 306 n32, 351, 490-492, 491 n175, 497
  oak galls as a color indicator, 15, 16-17, 18, 111, 112 n114, 490, 491, 497
Comenius, Jan Amos, 389-390, 390 n24
Copernicus and Copernican theory, 230, 230 n71, 240, 243-244, 243 n90, 270, 291, 334
Cotta, John, 185
Courtin, Germain, 154
Coxe, Daniel, 423-424
Creation, and elementary matter

Index

in antiquity, 5, 6
described in Genesis
  interpreted as a chemical process, 86-87, 125, 132 n14, 196
    by Dorn, 77-78
    by Duchesne, 161
    by Ficino, 32
    by Fludd, 226-229, 291
    by van Helmont, 318, 361
    by Paracelsus, 55-56, 76
    by Rhumelius, 454
    by Tymme, 77
  interpreted as a mathematical process, 36, 42-43
Crollius, Oswald, 63, 66, 82-83, 100, 116, 117-124, 126, 139, 196, 200, 246, 261, 281, 284, 292, 472, **515**
crystalline form, in analysis, 110 n111, 306 n32, 350-351, 350 n180, 490, 491
Culpeper, Nicholas, 249-250, 508-509
cure by contraries, 60, 117, 143, 161, 371, 454
cure by similitude, 60, 117, 117 n128, 121, 130, 143, 148, 157, 166, 181, 373, 454, 541
Cusanus, Nicholas, 36-38, 319 n80, 327 n105, 328 n108

Dariot, Claude, 146, 157-159
Davy, Humphry, 539, 551-552
Deane, Edmund, 422, 497
De Bry, John Theodore, 224, 224 n47
Dee, Arthur, 184
Dee John, 41-45, 41 n97, 46, 257, 257 n138, 292
Dell, William, 401
Democritus and Democritean atomism, 220, 468, 523, 545; *see also* atomism
Descartes, René, 212, 262, 383, 405, 448, 529
Dessenius, Bernhard, 131
Digby, Kenelm, 284 n232, 290, 495
digestion, 368-371, 529-530
Dioscorides, 22, 23, 190
disease, concept of
  caused by acid-base imbalance, 500, 530-531
  caused by external factors, 104-107
    Bostocke (seeds), 181
    Crollius (seeds or *astrum*), 120-121
    Fludd (*spiritus mali*), 250, **251**
    van Helmont (seeds), 359-363
  Paracelsus, 58-59
  caused by humoral imbalance (Galen), 58
  caused by nitroso-sulfureous particles, 108
distillation, 17, 21, 24, 80, 84, 107, 109, 113-116, **113**, 486, 488, 488 n165
  in Alexandrian texts, 6
  medieval, 12, 15, 116, 123
  and Gesner, 176 n152
  and Guinter, 143
  and Paracelsus, 59
  in the *Pharmacopoeia Londinensis*, 190
  and Sennert, 197-198
  and Willis, 107, 525
  "distillation books," 21-24
  to explain origin of mountain streams, rivers, minerals, 89, 89 n62, 90, 107, 463
Dondi, Giacomo de, 15, 16
Donne, John, 281 n220
Dorn, Gerhard, 54 n129, 74, 77-78, **113**, 147, 148
Drake, Roger, 514
Dryander, Johann, 16
Duchesne, Joseph (Quercetanus), 100-109, 115, 148-153, **151**, 160-168, 173, 178, 179, 184, 186, 201, 202, 235, 236 n81, 292, 486, 503
Duclo, Gaston, 159
Durelle, Jean, 278

educational reform, 29, 378-379, 381-400, 441
  Scholastic self-criticism, 26-30
  see also Schools and Scholasticism
elements (air, earth, fire, water), 5, 8, 78-84, 541
  and Boyle, 481-484
  and Dariot, 158
  and Duchesne, 161-162
  and Erastus, 134
  and Fludd, 227-229, 229 n65
  and van Helmont, 317-320, 339, 359
  and Paracelsus, 56-57, 78-79
  and Sennert, 197
  and Wimpanaeus, 136
  correlation with four evangelists, 84, 84 n51
  rejection of fire as an element, 161, 161 n103, 318, 353, 460-461
Engelhard, Matthias, 253 n116

Epicurus and Epicurean atomism, 472, 523, 531; *see also* atomism
epigenesis, 130-131
Erasmus, 28
Erastus (Liebler), Thomas, 99 n82, 131-134, 147, 159, 159 n95, 180, 200-201, 542
Ercker, Lazarus, 345, 347-348, 348 n171
d'Espagnet, Jean, 88
Euclid, 219

Fabre, Pierre Jean, 261
Fallopius, Gabriel, 17-18, 22, 110, 110 n109
Feltre, Vittorino da, 27
Fenot, John Antony, 153-154
fermentation, 341-343, 344, 344 n161 and n162, 352-357, 468, 478, 523-525, 529-530, 533-534
fertilizer, *see* agriculture, chemical
Ficino, Marsilio, 31-33, 148, 148 n62, 280, 397
Fioravanti, Leonardo, 177-178, 508, 509
fire
  as a means of analysis (separation), 81-83, 100, 121, 122, 134, 136, 154, 164, 181, 182, 320-326 *passim*, 378, 484-490
  and Biggs, 504
  and Boyle, 482-485, 489
  and van Helmont, 320-326, 398
  and Lefèvre, 452, 485
  and Lemery, 486, 486 n161
  and Morhof, 487
  and Severinus, 83
  as a means of dissolution, 134, 197-198, 462
  as one of the four elements, *see* elements
  celestial fire, *see* aerial niter
  central fire (or sun) in the earth, 88-93, **91**, 338, 356, 431-432, 462-463, 504
Fludd, Robert, 44, 202, 205-293, **209, 221, 233, 237, 241**, 295, 296, 330, 333-334, 366, 398, 399, 405, 409, 496, 513, 543, 549; *see also* Frizius, Joachim
Fortunatus, 218 n31
Foster, William, 184 n187, 208, 274 n197, 276, 279-283, 279 n214, 286, 286 n241
French, John, 112, 391, 494, 497

Frizius, Joachim (perhaps a pseudonym of Robert Fludd), 267, 268-269, 268 n179
Frobenius, Johannes, 48
Fuller, Thomas, 208

Gaffarel, Jacques, 267, 308
Galen and Galenism, 24-25, 26, 60, 124, 156-157, 174, 203, 205, 235, and *passim*
  and Biggs, 504-507
  and Duchesne, 166
  and Fludd, 250
  and Libavius, 170
  and Sennert, 192, 195, 197, 198-199, 203
  rejection of
    by Bostocke, 66 n9, 80-81
    by Fludd, 216, 223
    by van Helmont, 312-314, 377
    by Paracelsians, 66-69, 75, 117
    by Paracelsus, 51, 52, 58, 61
    by Penotus, 81, 159
    by Severinus, 75, 129
  see also humors
Galileo, 216, 277
Gassendi, Pierre, 236, 243 n90, 269-278, 296, 532, 543
Gaultier, Joseph, 277
Geber (Latin, 12th-13th centuries), 12; *see also* Jābir ibn Hayyān
Genesis, *see under* Creation
geological phenomena, *see* metals, generation of; minerals, generation of; waters
Gerarde, John, 23
Gesner, Conrad, 18, 22-23, 24, 25, 51, 81-82, 110 n109, 115, 176, 177, 178, 202
Geynes, John, 174
Gilbert, William, 104, 208, 212, 240-245, 243 n90, 246 n99 and n102, 248, 270, 287-290, 304, 349
Glauber, Johann Rudolph, 90, 99, 381, 422, 425, **427**, 441, 442, 449, 450, 451, 494, 513, 530, 542
Glisson, Francis, 518-519, 544
Goclenius (Göckel), Rudolph, 246, 281, 287, 303
Goethe, Johann Wolfgang, 549-550
Gohory, Jacques (Leo Suavius), 128, 146-148, 146 n56

Index

gold, as a medicine, 123, 144, 147, 154, 164, 170, 176, 185-186, 285, 325, 374
Gott, Samuel, 388
Grévin, Jacques, 148
Guinter of Andernach, Joannes, 139-145, 139 n36, **141**, 157, 196, 201, 542
Gwinne, Matthew, 185

Haack, Theodore, 278
Hakewill, George, 387
Hall, John, 392
Hall, Joseph, 388
Hall, Thomas, 406-409
Hart, James, 283-285, 286 n241, 289
Hartlib, Samuel, 388-390, 410-411, 420-425
Hartmann, Johann, 117-118, 125 n155
Harvet, Israel, 167
Harvey, William, 26, 26 n58, 37 n89, 208, 269, 271-276, 512-514, 529, 543, 544-545
Hassard, Pierre, 145-146
Heer, Henricus van, 306-307, 306 n32, 309, 339, 350
Helmont, Franciscus Mecurius van, 311, 312, 455-456
Helmont, Jean Baptiste van, 14, 30, 103 n89, 246, 247 n103, 277, 281, 290, 295-345, **299**, 357-379, 397, 398, 422, 447-451 *passim*, 456-457, 470-484 *passim*, 496, 499-503 *passim*, 517-519, 524, 525, 530, 534-537, 543, 546, 547
Helt, Justus, 253 n116, 254
Henricus de Rochas, 108
Heraclitus, 220
Hermes Trismegistus and Hermeticism, 6, 7, 31-35, 45, 69, 70, 77-78, 225, 226, 227, 253-254, 253 n116, 262, 280, 291, 432, 463, **515**, and *passim*
Hester, John, 177-178, 411, 448
Heydon, Christopher, 386-387, 387 n15
Heydon, John, 387, 387 n15
Hill, Thomas, 177
Hippocrates, 150, 217, 226, 261, 312, 353
Hobbes, Thomas, 401, 472
Hoffman, Friedrich, 531
Hollandus, Isaac, 464
Hooke, Robert, 496, 517
Horstius, Gregor, 253 n116, 254
humors (blood, phlegm, yellow bile, black bile)
 acceptance of
  by Erastus, 133
  by Fludd, 250, 250 n115, 291
  by van Helmont, 371
  by Paré, 25
  by Sennert, 192, 198-199, 203
 rejection of, 519, 544
  by Bostocke, 181
  by Crollius, 120
  by Duchesne, 164-165
  by Severinus, 130
 theory of three humors, 104
Hunyades, John, 394

Inquisition, 307-308

Jābir ibn Hayyān (Geber, c. 721-815), 8, 12
John of Rupescissa, 21, 22, 59, 217
Johnson, Christopher, 182
Johnson, William, 508, 509, 511
Jones, John, 177
Jorden, Edward, 112, 344-357, 491-492, 497
Jungius, Joachim, 389, 531

Kepler, Johannes, 44, 206
 versus Fludd, 219, 256-260, 277, 292, 543
Khālid ibn Yazid, 7
Khunrath, Heinrich, **67**, 270
Kircher, Athanasius, 277, 290, 457
Kunckel, Johann, 442-445, 464-467

Langton, Thomas, 182
Lanovius, Franciscus, 270
lead (dangers of), 505-507
Lefèvre, Nicholas (Le Febure, Nicasius), 109 n102, 413, 425-426, 450, 451-453, 468, 485, 495
Leibniz, Gottfried Wilhelm, 549, 549 n11
Lemery, Nicholas, 350 n180, 450-451, 485-486, 500, 501
Libavius, Andreas, 111-112, 117, 167, 168, 169-173, **171**, 190, 192, 195, 199, 201-203, 215, 217, 253, 346, 348 n171
Liébaut, Jean, 18, 24
light, divine, 229-230
 light/darkness dichotomy, 226, 226 n55, 228, 236

"like-like" principle, *see* cure by similitude
Linacre, Thomas, 174
loadstone, *see* magnetism
London College of Physicians, *see* Royal College of Physicians of London
longevity (prolongation of life), 4, 9-10, 19-20, 148
Louvain, Faculty of Medicine, 306, 308
Lull, Raymond, 13, 22, 35 n83, 217, 269, 540

macrocosm-microcosm analogy, 4, 69, 87, 96-109, **97**, **465**, **515**, 541, 551
  and Agrippa, 38
  and Andreae, 386
  and le Baillif, 156
  and Becher, 459-460, 462, **465**
  and Bostocke, 181
  and Boyle, 477
  and Comenius, 390
  and Crollius, 119-124, 139
  and Dariot, 159
  and Dee, 42-43
  and Duchesne, 165-166
  and Ficino, 32
  and Fludd, 207, 211, 218-220, 231, 235, 253-254, 291-292, 543
  and Glauber, 431, 433
  and Gohory, 147
  and Guinter, 143
  and Harvey, 513
  and F.M. van Helmont, 455-456
  and J.B. van Helmont, 306-308, 315
  and John of Rupescissa, 21
  and Libavius, 170, 173
  and Paracelsus, 52-53, 57, 96, 99 n82
  and Plato, 5
  and Rhumelius, 453-455, 536
  and Schütze, 513-515, **515**
  and Sennert, 196
  and Severinus, 129-130
  and Webster, 396
  and Wimpanaeus, 136, 139
  rejection of, 205
magic, natural, 33-34, 35, 38, 400, 410
  and Crollius, 122, 123
  and Dee, 41-45
  and Gohory, 147
  and van Helmont, 305, 308, 310
  and Paracelsus, 54, 55, 96
  and Sennert, 195
  and Ward, 404

  and Webster, 398-400, 407
magnal (van Helmont), 332-333, 337
magnetism, 104, 239-240, 244-249, 246 n99, 248 n108, 284-290, 291, 304
Maier, Michael, **71**, 215, 442
marl, 154, 411, 414-419, 420, 421, 423-425
Martinus del Rio, 298
mathematics, in the study of nature, 35-43
  and Agrippa, 38
  and Andreae, 386
  and Cusanus, 36-38
  and Dee, 41-45, 74
  and Dorn, 74-75
  and Fludd, 217-219, 257, 259-260
  and van Helmont, 312-313
  and Kepler, 259-260
  and Paracelsus, 55, 73-76
  and Ward, 402-403
  and Webster, 395-396
  sidereal mathematics, 55, 73-74
Matthioli, Petrus Andreas, 23
Mayerne, Theodore Turquet de, 167, 173, 186-190, 186 n193, **187**, 202
Maynwaring, Everard, 511
Mayow, John, 108, 447, 492-496, 499, 502, 536
mechanical philosophy and mechanism, 260-279 *passim*, 291, 469, 472-473, 514, 526, 531-535 *passim*, 536, 546, 548
medicines, chemically prepared, 19-25, 51, 59-60, 112-117, 123, 133, 140, 143-145, 153, 154-155, 166, 170, 177, 178, 183, 200, 374-376, 430, 454, 503-507, 544
mercury
  as a medicine, 51, 59, 116, 123-124, 133, 144, 190, 374-375
  as one of the Paracelsian three principles, *see* principles
  sulfur-mercury theory, *see under* sulfur
Mersenne, Marin, 206, 262, 262 n158, 265-270, 276, 277-279, 295-296, 308-310, 543
metals
  diseases of, 96, 125,
  generation of, 93, 107
    and alchemy, 4, 8
    and Basil Valentine, 94-96
    and Glauber, 430-432
    and van Helmont, 342

and Jorden, 347, 352-357
and Paracelsus, 47,
and Sendivogius, 90
and Stahl, 464
and Webster, 457-458
in medicine, 23, 93, 135-136, 144, 176, 430
minerals
generation of, 90, 93, 340-343, 352-357, 458, 477
mineral waters, see spa waters, analysis of
Moffett, Thomas, 79-80, 183-184, 202
monad, 44, **44**, 45, 257, **257**, 257 n138
Montagnana, Bartolmeo, 15, 16, 17
Morhof, Daniel Georg, 458, 487
Morienos, 11, 11 n11
Morin, Jean-Baptiste, 265

Naudé, Gabriel, 262, 277
neutralization, acid-base, 351, 368-371, 471, 499-502
Newton, Isaac, 447, 532-35, 537, 546, 547
niter, see aerial niter
Norton, Thomas, 14, 215, 540
Nuisement, Jacques Sieur de, 267

oak galls, see under color tests
observation and experiment in the study of nature, 4, 13, 14, 30-31, 61, 70, 212-213, 385, 471, 541
and Boyle, 474-477
and Cusanus, 37-38
and Severinus, 30-31, 83
and Webster, 396-400, 406
Oxford University, 400-401, 520

Paddy, William, 208
Padua, University of, 15, 25, 27
Paley, William, 550-551
palingenesis, 100-103, **101**, 207
Palissy, Bernard, 89, 154, 411-419, 413 n110, 425
Palmarius, Petrus, 168
Palmieri, Matteo, 28
Pantheus, J. A., 35 n83
Paracelsus (Philippus Areolus Theophrastus Bombastus von Hohenheim), 45-61, **49**, 88-89, 261, 412-414, **515**, 540-544, and *passim*
life, 45-51
work, 51-61

astral emanations and aerial niter, 53, 87-88, 107-109, 109 n103
chemically prepared medicines, 51, 58, n141, 59-60, 59, 59 n146, 116
Creation as a chemical process, 56
cure by similitude, 60, 117
disease theory, 58-59, 108
elements, 56-57
macrocosm-microcosm analogy, 52-53, 57, 96, 99 n82
mathematics, 55, 73-76
principles, 57-58, 79, 79 n37
rejection of ancient authority, 46, 48, 51-52, 54-55, 58, 60-61
signatures, 53, 55
weapon-salve, 246, 246 n100
defense of
by van Helmont, 302, 304-305, 309
by Severinus, 129-130
attempts to integrate with ancient authority, 135-145, 203
by Duchesne, 150-153
by Guinter, 140-145, 201, 203
by Sennert, 192-200 *passim*
by Wimpanaeus, 135-139, 201
rejection of
by Charleton, 472
by Dessenius, 131
by Erastus, 131-134, 200-201
by Foster, 281
by the French, 148-157, 166-173
by Glauber, 430-431
by Hart, 285
by Libavius, 170-171, 195, 199
by Sennert, 195-200 *passim*
mentioned in *Pharmacopoeia Londinensis*, 190
Paracelsians, 63-126, 64 n2, 128, 134, 145-158, and *passim*
Paré, Ambroise, 23-24, 24 n50
Paris, Medical Faculty, 156-157, 166-173, 186, 201, 206
Payanus, 273, 274
Peiresc, Nicolas-Claude Fabri de, 271, 278
Pell, John, 278
Penotus, Bernard G., 81, 154-155, 159, 184
Peter of Eboli, 15
Petrarch, 28 n63
Petrus Bonus of Ferrara, 13, 344
pharmacopoeias, 116, 117
Pharmacopoeia Augustana, 190

Pharmacopoeia Londinensis, 116, 182-183, 186-190, 189 n197, 202, 504, 505, 507-509, 544
Philalethes, Eugenius (Thomas Vaughan), 387, 495
Philo of Byzantium, 330, 496
philosopher's stone, 8, 170, 255; *see also* transmutation
phlogiston theory, 446, 447, 459, 467, 469, 488 n165, 536, 545
Pico della Mirandola, 35-36, 270, 398
Planis Campy, David de, 261-262, **263**
Plat, Hugh, 89, 411-419, 420, 421 n142, 423-425, 441
Plato, Platonism, neo-Platonism, 5, 28 n63, 31, 32, 35, 37 n88, 42, 69, 96, 226, 227, 230, 280, 397
Plattes, Gabriel, 93, 350 n180, 421
Pliny, 15, 217
Porta, John Baptista, 34, 43, 281
prime matter
  Aristotle (*prima materia*), 5
  Fludd, 226-230, 226 n55
  Paracelsus (*mysterium magnum*), 55-57, 55 n134
  see also Creation and elementary matter
principles (Paracelsian *tria prima*: salt, sulfur, mercury), 78-84, 201, 413, 488-491, 500-501, 541
  and Becher, 460-461
  and Biggs, 504
  and Bostocke, 180-181
  and Boyle, 481-484, 490
  and Crollius, 82, 121
  and Dariot, 158-159
  and Duchesne, 161-164
  and Erastus, 133-134, 201
  and Fludd, 229
  and Gohory, 147
  and Guinter, 140-143
  and van Helmont, 305, 320-321
  and Jorden, 353
  and Kepler, 259
  and Lefèvre, 452, 485
  and Moffet, 183
  and Paracelsus, 57-58, 79
  and Rhumelius, 454
  and Sendivogius, 84
  and Sennert, 197-199
  and Severinus, 83, 130
  and Tymme, 77
  and Wimpanaeus, 136
  correlated with the Trinity, 166, 180
pyrotechny, *see* fire
Pythagorus and Pythagorean tradition, 35, 44-45, 90 n63, 386
  and Cusanus, 38
  and Fludd, 217-218, 226, 257, 259

qualities (humid, dry, hot, cold)
  and Aristotle, 5
  and Fludd, 227
  and Paracelsus, 56
quantification, 36, 37, **39**, 55, 318-319, 327-329; *see also* mathematics; weight, analysis by
quellem (van Helmont), 337-338, 340

Rabelais, 30, 212
Ramus, Peter, 29-30, 212
Rawlin, Thomas, 185
Ray, John, 488
Raynalde, Thomas, 176
reaction products, 124, 190
Rhazes (al-Razi), 10, 450
Rhumelius, Johannes Pharmundus, 453-455, 536
Ribit, Jean, sieur de la Rivière, 159-160, 160 n97, 167, 186, 186 n193
Ridley, Mark, 208, 240, 287
Riolan, Jean (the elder), 166-167, 186, 198
Riolan, Jean (the younger), 167
Ripley, George, 269
Roberti, Jean, 246, 279, 303-304
Rosenkreutz, Christian, 213
Rosicrucian texts, 211-215, 211 n5, 223, 224, 253-254, 258 n138, 262, 265, 267, 291-292, 387, 397, 543
Royal College of Physicians of London, 23, 174-175, 174 n147, 175 n149, 182-191, 202, 206, 207, 208, 249, 260, 391, 504, 505, 507-512, 508 n235, 517
Royal Society of London, 385, 385 n7, 400, 409, 410, 423-424, 494, 496, 497, 545, 550
Ruland, Martin, the Elder, 169 n136, 344, 415

salt
  in chemical agriculture, 154, 412-419, 420, 422-434, 437, 442

Index

as one of the Paracelsian three principles, *see* principles
saltpeter, *see* aerial niter
Savanarola, Johannes Michael, 15-16, 17
Schegkius, 198
Schelling, Friedrich von, 550
Schools and Scholasticism
　attacks on, 14, 25-30, 28 n63, 212, 213, 215-217, 223, 297-298, 313, 372-373, 376-379, 429-430, 452
　defense of, 400-410
　see also educational reform; *specific names* (e.g., Aristotle)
Schott, Gaspar, 462-463
Schröder, Johann, 168
Schütze, Tobias, 523-515
Scot, Patrick, 208, 255
Scot, Reginald, 34
scriptural authority (in the study of nature), 61, 75, 93, 205, 540-541
　and Cusanus, 36-37
　and Ficino, 33
　and Fludd, 225-226, 230, 244, 248, 255, 270, 287, 287 n245, 291, 292
　and van Helmont, 314, 318, 334, 337, 358, 361, 366
　and Jorden, 355
　and Mersenne, 265
　and Paracelsus, 52, 53, 55
seeds and the seminal concept
　in generation, 130-131, 340-341, 344, 351, 353, 355, 357, 483
　in the origin of disease, 107, 120-121, 181, 360-363
　in the origin of metals, 107, 352, 431, 456
　see also archeus
Selden, John, 253 n116, 254
Sendivogius, Michael, 84, **85**, 88-90, 235, 513
Sennert, Daniel, 76, 191-200, **193**, 203, 246, 246 n100, 285-286, 531, 532
Servetus, 220 n43
Seton, Alexander, 84 n52
Severinus (Sørenson), Peter, 30, 70, 75, 83, 83 n50, 108, 108 n99, 128-131, 157, 183, 200, 212
signatures, doctrine of divine, 100-103, 125, 199
　and Andreae, 386
　and Crollius, 100, 122-124

and Duchesne, 100-103, 403-504
and van Helmont, 374, 503
and Kepler, 258
and Paracelsus, 53, 55
and Rhumelius, 454, 455
and Webster, 396
Simpson, William, 497-499, 513
Soucy, François de, sieur de Gerson, 262
spa waters (mineral and spring waters), analysis^of, 109-112, 422, 497, 499, 545
　medieval, 14-18, 25
　and Jorden, 346, 347-352, 356-357
　and Paracelsus, 59
　and Rhumelius, 455
specific gravity, in analysis, 327, 329, 490, 491
Stahl, Georg Ernst, 447, 459, 463-469
Starkey, George, 368, 480, 501, 502, 503, 509-510, 511
Stephanos, 6, 7
Stoicism, 220, 298
Stringer, William, 511
Suavius, Leo, pseudonym of Gohory, Jacques, q.v.
sulfur
　aerial or vital sulfur, 107-108, 130, 235, 456, 493, 493 n179; *see also* aerial niter
　as one of the Paracelsian three principles, *see* principles
　sulfur-mercury theory, in describing growth of metals, 8, 8 n8, 12, 57, 353
sun, as the font of life, 88, 224, 230-231, 235, 259-260, 432
Sylvius, Franciscus de le Boë, 448, 500, 526-530, **527**, 537, 544, 548
sympathetic action, 103, **105**, 247, 280-281, 284 n232, 287-289, 303, 308, 348, 471, 479, 526 n281; *see also* magnetism; weapon-salve

Tachenius, Otto, 500, 531
tartaric disease (Paracelsus), 107, 199, 362-365
Taylor, Thomas, 550
Telesio, 212
thermoscope, **237**, 239, 239 n82, 290 n259, 328, 333, 498
Thölde, Johann, 94 n70
Thruston, Malachi, 495

Thurneisser, Leonard, 110-111, 110 n111, 178, 328, 491
transmutation, 3, 6, 550
 in Alexandrian texts, 6
 in Islamic texts, 8
 in medieval texts, 12, 13
 and R. Bacon, 19
 and Becher, 462
 and Fludd, 255
 and van Helmont, 325-326
 and Jorden, 347
 and Libavius, 170
 and Sennert, 195
tria prima, see principles
Trithemius, Johannes, 46, 74, 147
Turner, William, 177, 184 n187
Tymme, Thomas, 69-70, 76-77, 160 n98, 167, 179

Ulstadius, Philip, 115
urine, analysis of, 59, 59 n143, 113, 328 n108

vacuum, 329-334, **335**
Valentianus, Ionnes, 147
Valentine, *see* Basil Valentine
Valetius, Franciscus, 417, 417 n125, 418
Vesalius, Andreas, 136, 139
Vieussens, Raymond, 531
Vigenère, Blaise de, 159
Violet, Fabius, 369
vital spirit, *see* aerial niter
Vitruvius, 15
Vives, Juan Luis, 28-29

Walaeus, Johannes, 514, 529
Ward, Seth, 400-406, 409, 514-517, 520
water
 generative, 414-416, 415 n120, 419
 origin of springs, streams, rivers, 89, 89 n62, 337-339, 430, 462, 463, 545
 as one of the four elements, *see* elements
 see also spa waters, analysis of
weapon-salve, 104, 104 n90, 246-248, 246 n100
 and Charleton, 472-473
 and Fludd, 279-290
 and van Helmont, 303-306, 307
Webster, John, 393-400, 402-409, 457-458, 503, 514-517, 536
weight, analysis by
 and the balance, **39**, 75-76, 349
 and Cusanus, 37, 319 n80, 327 n105, 328 n108
 and van Helmont, 318-319, 327-329
 and Jorden, 349
 and Libavius, 111
 and Thurneisser, 110 n111
 see also spa waters, analysis of
Weston, Richard, 421
wheat, in Fludd's experiment, 231-232, 232 n76, 253-254, 270
Widman, Johann, 16
Wiegleb, Johann Christian, 549
Wilkins, John, 400-402, 406, 409, 520
Willis, Thomas, 107, 448, 519-526, **521**, 537, 544, 546, 548
willow tree, in van Helmont's experiment, 75, 319, 420, 460, 483
Wimpanaeus, Johannes Albertus, 135-139, **137**, 144, 201
Witty, Robert, 497-498
Woodall, John, 161 n103, 179
Woodward, John, 420
Wright, Robert, junior, 249-250

Zoroaster, 220
Zosimos, 7
Zorzi, Francesco, 267

ERRATA

The first figure in the list is the page number. The second is the line of text, counting from the top of the page or, if asterisked, from the foot of the page (not including footnotes). If the page number is followed by "n," the line indicated is in a footnote on that page. The faulty word is not repeated but the place of the correct reading will be obvious.

PAGE, LINE	CORRECTION
vi, 17*	Lefèvre
viii, 9	phenomena (1682)
21n, 7	"John
22, 1	known up to
24, 13	aboundance
26, 4*	whom Walter
31, 2	text,
34, 3*	related to Christianity
35n, 17	*delete* The
36, 6	is found in
38n, 9	99–101.
39, 4-5	Early illustrations of enclosed analytical balances occur in Georg Agricola's *De re metallica* (1556) and in the *Theatrum chemicum Britannicum*
43, 14	importantly
43, 16	accomplished
45, 3–2*	mystical-mathematical
47n, 7	this effect by
53, 7-6*	saltpeter which is directly traceable
54n, 7	1656)
58n, 3	stated that the
70n, 9	Vlacq,
75, 4	mystical-mathematical
79, 2–3	nature, and the objects we call by these names are only crude approximations of them.[35]
97n, 3	*scilicet et minoris*
104, 3	Paracelsian
107, 13	they would no longer be able to act
107, 10*	phlegm in the nose

118, 1	Michaelis,
127n, 15	presidents
128n, 1	impatience of
145, 5	had been
148, 10*	Loys de l'Aunay and Jacques Grévin
150, 3*	salt peter
150, 3*	cruelly.
150, 2*	shot,
150, 1*	preuaile
153, 3	many thẽ loose
153, 3	maimd
153, 9	Yea
153, 11	yeelds to such as
153, 11*	houes
153, 8*	vse
159n, 1	p. 32.
160n, 22	notes.
162, 5	matter composed of the
169, 3	1540–1616),
176, 2*	Protestants
197, 15	question of
206, 11*	was no advocate
206, 9*	century more avid than
216, 7	teaching
216, 11*	important than the
221, 2*	*scilicet et minoris*
233, 2*	*scilicet et minoris*
235n, 2	writes,
241, 3	*scilicet et minoris*
257, 17*	hieroglyphic
281, 11	Paracelsus
295n, 4	Loyd
308, 8	College
309, 7	il ne peult
311, 3*	French (1670–1671)
318, 9	Galen
327, 1*	distillation
345, 4*	philosophy;
358, 3*	Almighty
358n, 16–17	Ecclesiasticus
358n, 17	See
368, 10	attributed digestion to

Errata

371, 5†	(sects. 11 *seq.*),
376, 1	truly,
390, 10*	problem and to
402, 6*	possible
421, 1*	Here was (delete there)
434, 10	benefit the common
437, 12*	subsequently
442, 11	reconstituted into liquid
448, 7*	there was a
464, 14*	problem of semantics
467, 8	postulated
467n, 2	[75]Partington, *History*
474, 4*	from the experimental
478n, 5	vivisection
483, 5	literature was
497, 4	physicians were surely
502, 7	disturbed by the
515, 3	*scilicet et minoris*
513, 5	Paracelsus
519, 10	subject,
519, 14	detail, but had
520n, 3	Piñero, "La
520n, 3	in
520n, 5	279–295,
546, 11	But in going
548, 3	1741 and 1754
548, 5	(1766)
549n, 1	David M. Knighti
558, 3	Press, 1916.
558, 15–14*	*Quercetanum.* Lyon: I. Ausulti, 1576.
559, 18*	3rd ed. Leipzig: Weidmann,
561, 6	Marguerite
568, 10	"De peste").
569, 3-2*	insert Maier, Michael. *Atalanta fugiens, hoc est, Emblemata nova de secretis chymica.* Oppenheim: J. T. De Bry, 1618.
574, 8	Vlacq
583, 17	Owen. "Jacques Gohory."
584, 11	Hafner,
589, 6–7	22–38.
592, 4*	für

A CATALOG OF SELECTED
DOVER BOOKS
IN SCIENCE AND MATHEMATICS

A CATALOG OF SELECTED

DOVER BOOKS

IN SCIENCE AND MATHEMATICS

Astronomy

BURNHAM'S CELESTIAL HANDBOOK, Robert Burnham, Jr. Thorough guide to the stars beyond our solar system. Exhaustive treatment. Alphabetical by constellation: Andromeda to Cetus in Vol. 1; Chamaeleon to Orion in Vol. 2; and Pavo to Vulpecula in Vol. 3. Hundreds of illustrations. Index in Vol. 3. 2,000pp. 6⅛ x 9¼.
23567-X, 23568-8, 23673-0 Three-vol. set

THE EXTRATERRESTRIAL LIFE DEBATE, 1750–1900, Michael J. Crowe. First detailed, scholarly study in English of the many ideas that developed from 1750 to 1900 regarding the existence of intelligent extraterrestrial life. Examines ideas of Kant, Herschel, Voltaire, Percival Lowell, many other scientists and thinkers. 16 illustrations. 704pp. 5⅜ x 8½. 40675-X

A HISTORY OF ASTRONOMY, A. Pannekoek. Well-balanced, carefully reasoned study covers such topics as Ptolemaic theory, work of Copernicus, Kepler, Newton, Eddington's work on stars, much more. Illustrated. References. 521pp. 5⅜ x 8½.
65994-1

AMATEUR ASTRONOMER'S HANDBOOK, J. B. Sidgwick. Timeless, comprehensive coverage of telescopes, mirrors, lenses, mountings, telescope drives, micrometers, spectroscopes, more. 189 illustrations. 576pp. 5⅝ x 8¼. (Available in U.S. only.)
24034-7

STARS AND RELATIVITY, Ya. B. Zel'dovich and I. D. Novikov. Vol. 1 of *Relativistic Astrophysics* by famed Russian scientists. General relativity, properties of matter under astrophysical conditions, stars, and stellar systems. Deep physical insights, clear presentation. 1971 edition. References. 544pp. 5⅝ x 8¼. 69424-0

Chemistry

CHEMICAL MAGIC, Leonard A. Ford. Second Edition, Revised by E. Winston Grundmeier. Over 100 unusual stunts demonstrating cold fire, dust explosions, much more. Text explains scientific principles and stresses safety precautions. 128pp. 5⅜ x 8½. 67628-5

THE DEVELOPMENT OF MODERN CHEMISTRY, Aaron J. Ihde. Authoritative history of chemistry from ancient Greek theory to 20th-century innovation. Covers major chemists and their discoveries. 209 illustrations. 14 tables. Bibliographies. Indices. Appendices. 851pp. 5⅜ x 8½. 64235-6

CATALYSIS IN CHEMISTRY AND ENZYMOLOGY, William P. Jencks. Exceptionally clear coverage of mechanisms for catalysis, forces in aqueous solution, carbonyl- and acyl-group reactions, practical kinetics, more. 864pp. 5⅜ x 8½.
65460-5

CATALOG OF DOVER BOOKS

THE HISTORICAL BACKGROUND OF CHEMISTRY, Henry M. Leicester. Evolution of ideas, not individual biography. Concentrates on formulation of a coherent set of chemical laws. 260pp. 5⅜ x 8½. 61053-5

A SHORT HISTORY OF CHEMISTRY, J. R. Partington. Classic exposition explores origins of chemistry, alchemy, early medical chemistry, nature of atmosphere, theory of valency, laws and structure of atomic theory, much more. 428pp. 5⅜ x 8½. (Available in U.S. only.) 65977-1

GENERAL CHEMISTRY, Linus Pauling. Revised 3rd edition of classic first-year text by Nobel laureate. Atomic and molecular structure, quantum mechanics, statistical mechanics, thermodynamics correlated with descriptive chemistry. Problems. 992pp. 5⅜ x 8½. 65622-5

Engineering

DE RE METALLICA, Georgius Agricola. The famous Hoover translation of greatest treatise on technological chemistry, engineering, geology, mining of early modern times (1556). All 289 original woodcuts. 638pp. 6¾ x 11. 60006-8

FUNDAMENTALS OF ASTRODYNAMICS, Roger Bate et al. Modern approach developed by U.S. Air Force Academy. Designed as a first course. Problems, exercises. Numerous illustrations. 455pp. 5⅜ x 8½. 60061-0

DYNAMICS OF FLUIDS IN POROUS MEDIA, Jacob Bear. For advanced students of ground water hydrology, soil mechanics and physics, drainage and irrigation engineering and more. 335 illustrations. Exercises, with answers. 784pp. 6⅛ x 9¼. 65675-6

ANALYTICAL MECHANICS OF GEARS, Earle Buckingham. Indispensable reference for modern gear manufacture covers conjugate gear-tooth action, gear-tooth profiles of various gears, many other topics. 263 figures. 102 tables. 546pp. 5⅜ x 8½. 65712-4

MECHANICS, J. P. Den Hartog. A classic introductory text or refresher. Hundreds of applications and design problems illuminate fundamentals of trusses, loaded beams and cables, etc. 334 answered problems. 462pp. 5⅜ x 8½. 60754-2

MECHANICAL VIBRATIONS, J. P. Den Hartog. Classic textbook offers lucid explanations and illustrative models, applying theories of vibrations to a variety of practical industrial engineering problems. Numerous figures. 233 problems, solutions. Appendix. Index. Preface. 436pp. 5⅜ x 8½. 64785-4

STRENGTH OF MATERIALS, J. P. Den Hartog. Full, clear treatment of basic material (tension, torsion, bending, etc.) plus advanced material on engineering methods, applications. 350 answered problems. 323pp. 5⅜ x 8½. 60755-0

A HISTORY OF MECHANICS, René Dugas. Monumental study of mechanical principles from antiquity to quantum mechanics. Contributions of ancient Greeks, Galileo, Leonardo, Kepler, Lagrange, many others. 671pp. 5⅜ x 8½. 65632-2

CATALOG OF DOVER BOOKS

METAL FATIGUE, N. E. Frost, K. J. Marsh, and L. P. Pook. Definitive, clearly written, and well-illustrated volume addresses all aspects of the subject, from the historical development of understanding metal fatigue to vital concepts of the cyclic stress that causes a crack to grow. Includes 7 appendixes. 544pp. 5⅜ x 8½. 40927-9

STATISTICAL MECHANICS: Principles and Applications, Terrell L. Hill. Standard text covers fundamentals of statistical mechanics, applications to fluctuation theory, imperfect gases, distribution functions, more. 448pp. 5⅜ x 8½. 65390-0

THE VARIATIONAL PRINCIPLES OF MECHANICS, Cornelius Lanczos. Graduate level coverage of calculus of variations, equations of motion, relativistic mechanics, more. First inexpensive paperbound edition of classic treatise. Index. Bibliography. 418pp. 5⅜ x 8½. 65067-7

THE VARIOUS AND INGENIOUS MACHINES OF AGOSTINO RAMELLI: A Classic Sixteenth-Century Illustrated Treatise on Technology, Agostino Ramelli. One of the most widely known and copied works on machinery in the 16th century. 194 detailed plates of water pumps, grain mills, cranes, more. 608pp. 9 x 12. 28180-9

ORDINARY DIFFERENTIAL EQUATIONS AND STABILITY THEORY: An Introduction, David A. Sánchez. Brief, modern treatment. Linear equation, stability theory for autonomous and nonautonomous systems, etc. 164pp. 5⅜ x 8¼. 63828-6

ROTARY WING AERODYNAMICS, W. Z. Stepniewski. Clear, concise text covers aerodynamic phenomena of the rotor and offers guidelines for helicopter performance evaluation. Originally prepared for NASA. 537 figures. 640pp. 6⅛ x 9¼. 64647-5

INTRODUCTION TO SPACE DYNAMICS, William Tyrrell Thomson. Comprehensive, classic introduction to space-flight engineering for advanced undergraduate and graduate students. Includes vector algebra, kinematics, transformation of coordinates. Bibliography. Index. 352pp. 5⅜ x 8½. 65113-4

HISTORY OF STRENGTH OF MATERIALS, Stephen P. Timoshenko. Excellent historical survey of the strength of materials with many references to the theories of elasticity and structure. 245 figures. 452pp. 5⅜ x 8½. 61187-6

ANALYTICAL FRACTURE MECHANICS, David J. Unger. Self-contained text supplements standard fracture mechanics texts by focusing on analytical methods for determining crack-tip stress and strain fields. 336pp. 6⅛ x 9¼. 41737-9

Mathematics

HANDBOOK OF MATHEMATICAL FUNCTIONS WITH FORMULAS, GRAPHS, AND MATHEMATICAL TABLES, edited by Milton Abramowitz and Irene A. Stegun. Vast compendium: 29 sets of tables, some to as high as 20 places. 1,046pp. 8 x 10½. 61272-4

FUNCTIONAL ANALYSIS (Second Corrected Edition), George Bachman and Lawrence Narici. Excellent treatment of subject geared toward students with background in linear algebra, advanced calculus, physics and engineering. Text covers introduction to inner-product spaces, normed, metric spaces, and topological spaces; complete orthonormal sets, the Hahn-Banach Theorem and its consequences, and many other related subjects. 1966 ed. 544pp. 6⅛ x 9¼. 40251-7

ASYMPTOTIC EXPANSIONS OF INTEGRALS, Norman Bleistein & Richard A. Handelsman. Best introduction to important field with applications in a variety of scientific disciplines. New preface. Problems. Diagrams. Tables. Bibliography. Index. 448pp. 5⅜ x 8½. 65082-0

FAMOUS PROBLEMS OF GEOMETRY AND HOW TO SOLVE THEM, Benjamin Bold. Squaring the circle, trisecting the angle, duplicating the cube: learn their history, why they are impossible to solve, then solve them yourself. 128pp. 5⅜ x 8½. 24297-8

VECTOR AND TENSOR ANALYSIS WITH APPLICATIONS, A. I. Borisenko and I. E. Tarapov. Concise introduction. Worked-out problems, solutions, exercises. 257pp. 5⅝ x 8¼. 63833-2

THE ABSOLUTE DIFFERENTIAL CALCULUS (CALCULUS OF TENSORS), Tullio Levi-Civita. Great 20th-century mathematician's classic work on material necessary for mathematical grasp of theory of relativity. 452pp. 5⅝ x 8¼. 63401-9

AN INTRODUCTION TO ORDINARY DIFFERENTIAL EQUATIONS, Earl A. Coddington. A thorough and systematic first course in elementary differential equations for undergraduates in mathematics and science, with many exercises and problems (with answers). Index. 304pp. 5⅜ x 8½. 65942-9

FOURIER SERIES AND ORTHOGONAL FUNCTIONS, Harry F. Davis. An incisive text combining theory and practical example to introduce Fourier series, orthogonal functions and applications of the Fourier method to boundary-value problems. 570 exercises. Answers and notes. 416pp. 5⅜ x 8½. 65973-9

COMPUTABILITY AND UNSOLVABILITY, Martin Davis. Classic graduate-level introduction to theory of computability, usually referred to as theory of recurrent functions. New preface and appendix. 288pp. 5⅜ x 8½. 61471-9

ASYMPTOTIC METHODS IN ANALYSIS, N. G. de Bruijn. An inexpensive, comprehensive guide to asymptotic methods–the pioneering work that teaches by explaining worked examples in detail. Index. 224pp. 5⅜ x 8½ 64221-6

ESSAYS ON THE THEORY OF NUMBERS, Richard Dedekind. Two classic essays by great German mathematician: on the theory of irrational numbers; and on transfinite numbers and properties of natural numbers. 115pp. 5⅜ x 8½. 21010-3

CATALOG OF DOVER BOOKS

APPLIED COMPLEX VARIABLES, John W. Dettman. Step-by-step coverage of fundamentals of analytic function theory—plus lucid exposition of five important applications: Potential Theory; Ordinary Differential Equations; Fourier Transforms; Laplace Transforms; Asymptotic Expansions. 66 figures. Exercises at chapter ends. 512pp. 5⅜ x 8½. 64670-X

INTRODUCTION TO LINEAR ALGEBRA AND DIFFERENTIAL EQUATIONS, John W. Dettman. Excellent text covers complex numbers, determinants, orthonormal bases, Laplace transforms, much more. Exercises with solutions. Undergraduate level. 416pp. 5⅜ x 8½. 65191-6

MATHEMATICAL METHODS IN PHYSICS AND ENGINEERING, John W. Dettman. Algebraically based approach to vectors, mapping, diffraction, other topics in applied math. Also generalized functions, analytic function theory, more. Exercises. 448pp. 5⅜ x 8¼. 65649-7

CALCULUS OF VARIATIONS WITH APPLICATIONS, George M. Ewing. Applications-oriented introduction to variational theory develops insight and promotes understanding of specialized books, research papers. Suitable for advanced undergraduate/graduate students as primary, supplementary text. 352pp. 5⅜ x 8½. 64856-7

COMPLEX VARIABLES, Francis J. Flanigan. Unusual approach, delaying complex algebra till harmonic functions have been analyzed from real variable viewpoint. Includes problems with answers. 364pp. 5⅜ x 8½. 61388-7

AN INTRODUCTION TO THE CALCULUS OF VARIATIONS, Charles Fox. Graduate-level text covers variations of an integral, isoperimetrical problems, least action, special relativity, approximations, more. References. 279pp. 5⅜ x 8½. 65499-0

CATASTROPHE THEORY FOR SCIENTISTS AND ENGINEERS, Robert Gilmore. Advanced-level treatment describes mathematics of theory grounded in the work of Poincaré, R. Thom, other mathematicians. Also important applications to problems in mathematics, physics, chemistry and engineering. 1981 edition. References. 28 tables. 397 black-and-white illustrations. xvii + 666pp. 6⅛ x 9¼. 67539-4

INTRODUCTION TO DIFFERENCE EQUATIONS, Samuel Goldberg. Exceptionally clear exposition of important discipline with applications to sociology, psychology, economics. Many illustrative examples; over 250 problems. 260pp. 5⅜ x 8½. 65084-7

NUMERICAL METHODS FOR SCIENTISTS AND ENGINEERS, Richard Hamming. Classic text stresses frequency approach in coverage of algorithms, polynomial approximation, Fourier approximation, exponential approximation, other topics. Revised and enlarged 2nd edition. 721pp. 5⅜ x 8½. 65241-6

INTRODUCTION TO NUMERICAL ANALYSIS (2nd Edition), F. B. Hildebrand. Classic, fundamental treatment covers computation, approximation, interpolation, numerical differentiation and integration, other topics. 150 new problems. 669pp. 5⅜ x 8½. 65363-3

CATALOG OF DOVER BOOKS

THE FUNCTIONS OF MATHEMATICAL PHYSICS, Harry Hochstadt. Comprehensive treatment of orthogonal polynomials, hypergeometric functions, Hill's equation, much more. Bibliography. Index. 322pp. 5⅜ x 8½. 65214-9

THREE PEARLS OF NUMBER THEORY, A. Y. Khinchin. Three compelling puzzles require proof of a basic law governing the world of numbers. Challenges concern van der Waerden's theorem, the Landau-Schnirelmann hypothesis and Mann's theorem, and a solution to Waring's problem. Solutions included. 64pp. 5¾ x 8¼. 40026-3

CALCULUS REFRESHER FOR TECHNICAL PEOPLE, A. Albert Klaf. Covers important aspects of integral and differential calculus via 756 questions. 566 problems, most answered. 431pp. 5⅜ x 8½. 20370-0

THE PHILOSOPHY OF MATHEMATICS: An Introductory Essay, Stephan Körner. Surveys the views of Plato, Aristotle, Leibniz & Kant concerning propositions and theories of applied and pure mathematics. Introduction. Two appendices. Index. 198pp. 5⅜ x 8½. 25048-2

INTRODUCTORY REAL ANALYSIS, A.N. Kolmogorov, S. V. Fomin. Translated by Richard A. Silverman. Self-contained, evenly paced introduction to real and functional analysis. Some 350 problems. 403pp. 5⅜ x 8½. 61226-0

APPLIED ANALYSIS, Cornelius Lanczos. Classic work on analysis and design of finite processes for approximating solution of analytical problems. Algebraic equations, matrices, harmonic analysis, quadrature methods, much more. 559pp. 5⅜ x 8½. 65656-X

AN INTRODUCTION TO ALGEBRAIC STRUCTURES, Joseph Landin. Superb self-contained text covers "abstract algebra": sets and numbers, theory of groups, theory of rings, much more. Numerous well-chosen examples, exercises. 247pp. 5⅜ x 8½. 65940-2

SPECIAL FUNCTIONS, N. N. Lebedev. Translated by Richard Silverman. Famous Russian work treating more important special functions, with applications to specific problems of physics and engineering. 38 figures. 308pp. 5⅜ x 8½. 60624-4

QUALITATIVE THEORY OF DIFFERENTIAL EQUATIONS, V. V. Nemytskii and V.V. Stepanov. Classic graduate-level text by two prominent Soviet mathematicians covers classical differential equations as well as topological dynamics and ergodic theory. Bibliographies. 523pp. 5⅜ x 8½. 65954-2

NUMBER THEORY AND ITS HISTORY, Oystein Ore. Unusually clear, accessible introduction covers counting, properties of numbers, prime numbers, much more. Bibliography. 380pp. 5⅜ x 8½. 65620-9

THEORY OF MATRICES, Sam Perlis. Outstanding text covering rank, nonsingularity and inverses in connection with the development of canonical matrices under the relation of equivalence, and without the intervention of determinants. Includes exercises. 237pp. 5⅜ x 8½. 66810-X

CATALOG OF DOVER BOOKS

INTRODUCTION TO ANALYSIS, Maxwell Rosenlicht. Unusually clear, accessible coverage of set theory, real number system, metric spaces, continuous functions, Riemann integration, multiple integrals, more. Wide range of problems. Undergraduate level. Bibliography. 254pp. 5⅜ x 8½. 65038-3

MODERN NONLINEAR EQUATIONS, Thomas L. Saaty. Emphasizes practical solution of problems; covers seven types of equations. ". . . a welcome contribution to the existing literature...."–*Math Reviews*. 490pp. 5⅜ x 8½. 64232-1

MATRICES AND LINEAR ALGEBRA, Hans Schneider and George Phillip Barker. Basic textbook covers theory of matrices and its applications to systems of linear equations and related topics such as determinants, eigenvalues and differential equations. Numerous exercises. 432pp. 5⅜ x 8½. 66014-1

MATHEMATICS APPLIED TO CONTINUUM MECHANICS, Lee A. Segel. Analyzes models of fluid flow and solid deformation. For upper-level math, science and engineering students. 608pp. 5⅜ x 8½. 65369-2

ELEMENTS OF REAL ANALYSIS, David A. Sprecher. Classic text covers fundamental concepts, real number system, point sets, functions of a real variable, Fourier series, much more. Over 500 exercises. 352pp. 5⅜ x 8½. 65385-4

AN INTRODUCTION TO MATRICES, SETS AND GROUPS FOR SCIENCE STUDENTS, G. Stephenson. Concise, readable text introduces sets, groups, and most importantly, matrices to undergraduate students of physics, chemistry, and engineering. Problems. 164pp. 5⅜ x 8½. 65077-4

SET THEORY AND LOGIC, Robert R. Stoll. Lucid introduction to unified theory of mathematical concepts. Set theory and logic seen as tools for conceptual understanding of real number system. 496pp. 5⅜ x 8¼. 63829-4

TENSOR CALCULUS, J.L. Synge and A. Schild. Widely used introductory text covers spaces and tensors, basic operations in Riemannian space, non-Riemannian spaces, etc. 324pp. 5⅜ x 8¼. 63612-7

ORDINARY DIFFERENTIAL EQUATIONS, Morris Tenenbaum and Harry Pollard. Exhaustive survey of ordinary differential equations for undergraduates in mathematics, engineering, science. Thorough analysis of theorems. Diagrams. Bibliography. Index. 818pp. 5⅜ x 8½. 64940-7

INTEGRAL EQUATIONS, F. G. Tricomi. Authoritative, well-written treatment of extremely useful mathematical tool with wide applications. Volterra Equations, Fredholm Equations, much more. Advanced undergraduate to graduate level. Exercises. Bibliography. 238pp. 5⅜ x 8½. 64828-1

FOURIER SERIES, Georgi P. Tolstov. Translated by Richard A. Silverman. A valuable addition to the literature on the subject, moving clearly from subject to subject and theorem to theorem. 107 problems, answers. 336pp. 5⅜ x 8½. 63317-9

CATALOG OF DOVER BOOKS

POPULAR LECTURES ON MATHEMATICAL LOGIC, Hao Wang. Noted logician's lucid treatment of historical developments, set theory, model theory, recursion theory and constructivism, proof theory, more. 3 appendixes. Bibliography. 1981 edition. ix + 283pp. 5⅜ x 8½. 67632-3

CALCULUS OF VARIATIONS, Robert Weinstock. Basic introduction covering isoperimetric problems, theory of elasticity, quantum mechanics, electrostatics, etc. Exercises throughout. 326pp. 5⅜ x 8½. 63069-2

THE CONTINUUM: A Critical Examination of the Foundation of Analysis, Hermann Weyl. Classic of 20th-century foundational research deals with the conceptual problem posed by the continuum. 156pp. 5⅜ x 8½. 67982-9

CHALLENGING MATHEMATICAL PROBLEMS WITH ELEMENTARY SOLUTIONS, A. M. Yaglom and I. M. Yaglom. Over 170 challenging problems on probability theory, combinatorial analysis, points and lines, topology, convex polygons, many other topics. Solutions. Total of 445pp. 5⅜ x 8½. Two-vol. set.
Vol. I: 65536-9 Vol. II: 65537-7

A SURVEY OF NUMERICAL MATHEMATICS, David M. Young and Robert Todd Gregory. Broad self-contained coverage of computer-oriented numerical algorithms for solving various types of mathematical problems in linear algebra, ordinary and partial, differential equations, much more. Exercises. Total of 1,248pp. 5⅜ x 8½. Two volumes.
Vol. I: 65691-8 Vol. II: 65692-6

INTRODUCTION TO PARTIAL DIFFERENTIAL EQUATIONS WITH APPLICATIONS, E. C. Zachmanoglou and Dale W. Thoe. Essentials of partial differential equations applied to common problems in engineering and the physical sciences. Problems and answers. 416pp. 5⅜ x 8½. 65251-3

THE THEORY OF GROUPS, Hans J. Zassenhaus. Well-written graduate-level text acquaints reader with group-theoretic methods and demonstrates their usefulness in mathematics. Axioms, the calculus of complexes, homomorphic mapping, p-group theory, more. Many proofs shorter and more transparent than older ones. 276pp. 5⅜ x 8½. 40922-8

DISTRIBUTION THEORY AND TRANSFORM ANALYSIS: An Introduction to Generalized Functions, with Applications, A. H. Zemanian. Provides basics of distribution theory, describes generalized Fourier and Laplace transformations. Numerous problems. 384pp. 5⅜ x 8½. 65479-6

Math–Decision Theory, Statistics, Probability

ELEMENTARY DECISION THEORY, Herman Chernoff and Lincoln E. Moses. Clear introduction to statistics and statistical theory covers data processing, probability and random variables, testing hypotheses, much more. Exercises. 364pp. 5⅜ x 8½. 65218-1

CATALOG OF DOVER BOOKS

STATISTICS MANUAL, Edwin L. Crow et al. Comprehensive, practical collection of classical and modern methods prepared by U.S. Naval Ordnance Test Station. Stress on use. Basics of statistics assumed. 288pp. 5⅜ x 8½. 60599-X

SOME THEORY OF SAMPLING, William Edwards Deming. Analysis of the problems, theory and design of sampling techniques for social scientists, industrial managers and others who find statistics important at work. 61 tables. 90 figures. xvii +602pp. 5⅜ x 8½. 64684-X

STATISTICAL ADJUSTMENT OF DATA, W. Edwards Deming. Introduction to basic concepts of statistics, curve fitting, least squares solution, conditions without parameter, conditions containing parameters. 26 exercises worked out. 271pp. 5⅜ x 8½. 64685-8

LINEAR PROGRAMMING AND ECONOMIC ANALYSIS, Robert Dorfman, Paul A. Samuelson and Robert M. Solow. First comprehensive treatment of linear programming in standard economic analysis. Game theory, modern welfare economics, Leontief input-output, more. 525pp. 5⅜ x 8½. 65491-5

DICTIONARY/OUTLINE OF BASIC STATISTICS, John E. Freund and Frank J. Williams. A clear concise dictionary of over 1,000 statistical terms and an outline of statistical formulas covering probability, nonparametric tests, much more. 208pp. 5⅜ x 8½. 66796-0

PROBABILITY: An Introduction, Samuel Goldberg. Excellent basic text covers set theory, probability theory for finite sample spaces, binomial theorem, much more. 360 problems. Bibliographies. 322pp. 5⅜ x 8½. 65252-1

GAMES AND DECISIONS: Introduction and Critical Survey, R. Duncan Luce and Howard Raiffa. Superb nontechnical introduction to game theory, primarily applied to social sciences. Utility theory, zero-sum games, n-person games, decision-making, much more. Bibliography. 509pp. 5⅜ x 8½. 65943-7

FIFTY CHALLENGING PROBLEMS IN PROBABILITY WITH SOLUTIONS, Frederick Mosteller. Remarkable puzzlers, graded in difficulty, illustrate elementary and advanced aspects of probability. Detailed solutions. 88pp. 5⅜ x 8½. 65355-2

PROBABILITY THEORY: A Concise Course, Y. A. Rozanov. Highly readable, self-contained introduction covers combination of events, dependent events, Bernoulli trials, etc. 148pp. 5⅜ x 8¼. 63544-9

STATISTICAL METHOD FROM THE VIEWPOINT OF QUALITY CONTROL, Walter A. Shewhart. Important text explains regulation of variables, uses of statistical control to achieve quality control in industry, agriculture, other areas. 192pp. 5⅜ x 8½. 65232-7

THE COMPLEAT STRATEGYST: Being a Primer on the Theory of Games of Strategy, J. D. Williams. Highly entertaining classic describes, with many illustrated examples, how to select best strategies in conflict situations. Prefaces. Appendices. 268pp. 5⅜ x 8½. 25101-2

Math–Geometry and Topology

ELEMENTARY CONCEPTS OF TOPOLOGY, Paul Alexandroff. Elegant, intuitive approach to topology from set-theoretic topology to Betti groups; how concepts of topology are useful in math and physics. 25 figures. 57pp. 5⅜ x 8½. 60747-X

COMBINATORIAL TOPOLOGY, P. S. Alexandrov. Clearly written, well-organized, three-part text begins by dealing with certain classic problems without using the formal techniques of homology theory and advances to the central concept, the Betti groups. Numerous detailed examples. 654pp. 5⅜ x 8½. 40179-0

EXPERIMENTS IN TOPOLOGY, Stephen Barr. Classic, lively explanation of one of the byways of mathematics. Klein bottles, Moebius strips, projective planes, map coloring, problem of the Koenigsberg bridges, much more, described with clarity and wit. 43 figures. 210pp. 5⅜ x 8½. 25933-1

CONFORMAL MAPPING ON RIEMANN SURFACES, Harvey Cohn. Lucid, insightful book presents ideal coverage of subject. 334 exercises make book perfect for self-study. 55 figures. 352pp. 5⅝ x 8¼. 64025-6

THE GEOMETRY OF RENÉ DESCARTES, René Descartes. The great work founded analytical geometry. Original French text, Descartes's own diagrams, together with definitive Smith-Latham translation. 244pp. 5⅜ x 8½. 60068-8

THE THIRTEEN BOOKS OF EUCLID'S ELEMENTS, translated with introduction and commentary by Sir Thomas L. Heath. Definitive edition. Textual and linguistic notes, mathematical analysis. 2,500 years of critical commentary. Unabridged. 1,414pp. 5⅜ x 8½. Three-vol. set.
Vol. I: 60088-2 Vol. II: 60089-0 Vol. III: 60090-4

GEOMETRY OF COMPLEX NUMBERS, Hans Schwerdtfeger. Illuminating, widely praised book on analytic geometry of circles, the Moebius transformation, and two-dimensional non-Euclidean geometries. 200pp. 5⅝ x 8¼. 63830-8

DIFFERENTIAL GEOMETRY, Heinrich W. Guggenheimer. Local differential geometry as an application of advanced calculus and linear algebra. Curvature, transformation groups, surfaces, more. Exercises. 62 figures. 378pp. 5⅜ x 8½. 63433-7

CURVATURE AND HOMOLOGY: Enlarged Edition, Samuel I. Goldberg. Revised edition examines topology of differentiable manifolds; curvature, homology of Riemannian manifolds; compact Lie groups; complex manifolds; curvature, homology of Kaehler manifolds. New Preface. Four new appendixes. 416pp. 5⅜ x 8½. 40207-X

TOPOLOGY, John G. Hocking and Gail S. Young. Superb one-year course in classical topology. Topological spaces and functions, point-set topology, much more. Examples and problems. Bibliography. Index. 384pp. 5⅝ x 8¼. 65676-4

CATALOG OF DOVER BOOKS

LECTURES ON CLASSICAL DIFFERENTIAL GEOMETRY, Second Edition, Dirk J. Struik. Excellent brief introduction covers curves, theory of surfaces, fundamental equations, geometry on a surface, conformal mapping, other topics. Problems. 240pp. 5⅜ x 8½. 65609-8

Math–History of

A SHORT ACCOUNT OF THE HISTORY OF MATHEMATICS, W. W. Rouse Ball. One of clearest, most authoritative surveys from the Egyptians and Phoenicians through 19th-century figures such as Grassman, Galois, Riemann. Fourth edition. 522pp. 5⅜ x 8½. 20630-0

THE HISTORY OF THE CALCULUS AND ITS CONCEPTUAL DEVELOPMENT, Carl B. Boyer. Origins in antiquity, medieval contributions, work of Newton, Leibniz, rigorous formulation. Treatment is verbal. 346pp. 5⅜ x 8½. 60509-4

THE HISTORICAL ROOTS OF ELEMENTARY MATHEMATICS, Lucas N. H. Bunt, Phillip S. Jones, and Jack D. Bedient. Fundamental underpinnings of modern arithmetic, algebra, geometry and number systems derived from ancient civilizations. 320pp. 5⅜ x 8½. 25563-8

A HISTORY OF MATHEMATICAL NOTATIONS, Florian Cajori. This classic study notes the first appearance of a mathematical symbol and its origin, the competition it encountered, its spread among writers in different countries, its rise to popularity, its eventual decline or ultimate survival. Original 1929 two-volume edition presented here in one volume. xxviii+820pp. 5⅜ x 8½. 67766-4

GAMES, GODS & GAMBLING: A History of Probability and Statistical Ideas, F. N. David. Episodes from the lives of Galileo, Fermat, Pascal, and others illustrate this fascinating account of the roots of mathematics. Features thought-provoking references to classics, archaeology, biography, poetry. 1962 edition. 304pp. 5⅜ x 8½. (Available in U.S. only.) 40023-9

OF MEN AND NUMBERS: The Story of the Great Mathematicians, Jane Muir. Fascinating accounts of the lives and accomplishments of history's greatest mathematical minds–Pythagoras, Descartes, Euler, Pascal, Cantor, many more. Anecdotal, illuminating. 30 diagrams. Bibliography. 256pp. 5⅜ x 8½. 28973-7

HISTORY OF MATHEMATICS, David E. Smith. Nontechnical survey from ancient Greece and Orient to late 19th century; evolution of arithmetic, geometry, trigonometry, calculating devices, algebra, the calculus. 362 illustrations. 1,355pp. 5⅜ x 8½. Two-vol. set. Vol. I: 20429-4 Vol. II: 20430-8

A CONCISE HISTORY OF MATHEMATICS, Dirk J. Struik. The best brief history of mathematics. Stresses origins and covers every major figure from ancient Near East to 19th century. 41 illustrations. 195pp. 5⅜ x 8½. 60255-9

CATALOG OF DOVER BOOKS

Physics

OPTICAL RESONANCE AND TWO-LEVEL ATOMS, L. Allen and J. H. Eberly. Clear, comprehensive introduction to basic principles behind all quantum optical resonance phenomena. 53 illustrations. Preface. Index. 256pp. 5⅜ x 8½. 65533-4

ULTRASONIC ABSORPTION: An Introduction to the Theory of Sound Absorption and Dispersion in Gases, Liquids and Solids, A. B. Bhatia. Standard reference in the field provides a clear, systematically organized introductory review of fundamental concepts for advanced graduate students, research workers. Numerous diagrams. Bibliography. 440pp. 5⅜ x 8½. 64917-2

QUANTUM THEORY, David Bohm. This advanced undergraduate-level text presents the quantum theory in terms of qualitative and imaginative concepts, followed by specific applications worked out in mathematical detail. Preface. Index. 655pp. 5⅜ x 8½. 65969-0

ATOMIC PHYSICS (8th edition), Max Born. Nobel laureate's lucid treatment of kinetic theory of gases, elementary particles, nuclear atom, wave-corpuscles, atomic structure and spectral lines, much more. Over 40 appendices, bibliography. 495pp. 5⅜ x 8½. 65984-4

AN INTRODUCTION TO HAMILTONIAN OPTICS, H. A. Buchdahl. Detailed account of the Hamiltonian treatment of aberration theory in geometrical optics. Many classes of optical systems defined in terms of the symmetries they possess. Problems with detailed solutions. 1970 edition. xv + 360pp. 5⅜ x 8½. 67597-1

THIRTY YEARS THAT SHOOK PHYSICS: The Story of Quantum Theory, George Gamow. Lucid, accessible introduction to influential theory of energy and matter. Careful explanations of Dirac's anti-particles, Bohr's model of the atom, much more. 12 plates. Numerous drawings. 240pp. 5⅜ x 8½. 24895-X

ELECTRONIC STRUCTURE AND THE PROPERTIES OF SOLIDS: The Physics of the Chemical Bond, Walter A. Harrison. Innovative text offers basic understanding of the electronic structure of covalent and ionic solids, simple metals, transition metals and their compounds. Problems. 1980 edition. 582pp. 6⅛ x 9¼.
66021-4

HYDRODYNAMIC AND HYDROMAGNETIC STABILITY, S. Chandrasekhar. Lucid examination of the Rayleigh-Benard problem; clear coverage of the theory of instabilities causing convection. 704pp. 5⅜ x 8¼. 64071-X

INVESTIGATIONS ON THE THEORY OF THE BROWNIAN MOVEMENT, Albert Einstein. Five papers (1905–8) investigating dynamics of Brownian motion and evolving elementary theory. Notes by R. Fürth. 122pp. 5⅜ x 8½. 60304-0

THE PHYSICS OF WAVES, William C. Elmore and Mark A. Heald. Unique overview of classical wave theory. Acoustics, optics, electromagnetic radiation, more. Ideal as classroom text or for self-study. Problems. 477pp. 5⅜ x 8½. 64926-1

CATALOG OF DOVER BOOKS

PHYSICAL PRINCIPLES OF THE QUANTUM THEORY, Werner Heisenberg. Nobel Laureate discusses quantum theory, uncertainty, wave mechanics, work of Dirac, Schroedinger, Compton, Wilson, Einstein, etc. 184pp. 5⅜ x 8½. 60113-7

ATOMIC SPECTRA AND ATOMIC STRUCTURE, Gerhard Herzberg. One of best introductions; especially for specialist in other fields. Treatment is physical rather than mathematical. 80 illustrations. 257pp. 5⅜ x 8½. 60115-3

AN INTRODUCTION TO STATISTICAL THERMODYNAMICS, Terrell L. Hill. Excellent basic text offers wide-ranging coverage of quantum statistical mechanics, systems of interacting molecules, quantum statistics, more. 523pp. 5⅜ x 8½.
 65242-4

THEORETICAL PHYSICS, Georg Joos, with Ira M. Freeman. Classic overview covers essential math, mechanics, electromagnetic theory, thermodynamics, quantum mechanics, nuclear physics, other topics. First paperback edition. xxiii + 885pp. 5⅜ x 8½. 65227-0

PROBLEMS AND SOLUTIONS IN QUANTUM CHEMISTRY AND PHYSICS, Charles S. Johnson, Jr. and Lee G. Pedersen. Unusually varied problems, detailed solutions in coverage of quantum mechanics, wave mechanics, angular momentum, molecular spectroscopy, more. 280 problems plus 139 supplementary exercises. 430pp. 6½ x 9¼. 65236-X

THEORETICAL SOLID STATE PHYSICS, Vol. 1: Perfect Lattices in Equilibrium; Vol. II: Non-Equilibrium and Disorder, William Jones and Norman H. March. Monumental reference work covers fundamental theory of equilibrium properties of perfect crystalline solids, non-equilibrium properties, defects and disordered systems. Appendices. Problems. Preface. Diagrams. Index. Bibliography. Total of 1,301pp. 5⅜ x 8½. Two volumes. Vol. I: 65015-4 Vol. II: 65016-2

A TREATISE ON ELECTRICITY AND MAGNETISM, James Clerk Maxwell. Important foundation work of modern physics. Brings to final form Maxwell's theory of electromagnetism and rigorously derives his general equations of field theory. 1,084pp. 5⅜ x 8½. Two-vol. set. Vol. I: 60636-8 Vol. II: 60637-6

OPTICKS, Sir Isaac Newton. Newton's own experiments with spectroscopy, colors, lenses, reflection, refraction, etc., in language the layman can follow. Foreword by Albert Einstein. 532pp. 5⅜ x 8½. 60205-2

THEORY OF ELECTROMAGNETIC WAVE PROPAGATION, Charles Herach Papas. Graduate-level study discusses the Maxwell field equations, radiation from wire antennas, the Doppler effect and more. xiii + 244pp. 5⅜ x 8½. 65678-5

INTRODUCTION TO QUANTUM MECHANICS With Applications to Chemistry, Linus Pauling & E. Bright Wilson, Jr. Classic undergraduate text by Nobel Prize winner applies quantum mechanics to chemical and physical problems. Numerous tables and figures enhance the text. Chapter bibliographies. Appendices. Index. 468pp. 5⅜ x 8½. 64871-0

CATALOG OF DOVER BOOKS

METHODS OF THERMODYNAMICS, Howard Reiss. Outstanding text focuses on physical technique of thermodynamics, typical problem areas of understanding, and significance and use of thermodynamic potential. 1965 edition. 238pp. 5⅜ x 8½. 69445-3

TENSOR ANALYSIS FOR PHYSICISTS, J. A. Schouten. Concise exposition of the mathematical basis of tensor analysis, integrated with well-chosen physical examples of the theory. Exercises. Index. Bibliography. 289pp. 5⅜ x 8½. 65582-2

RELATIVITY IN ILLUSTRATIONS, Jacob T. Schwartz. Clear nontechnical treatment makes relativity more accessible than ever before. Over 60 drawings illustrate concepts more clearly than text alone. Only high school geometry needed. Bibliography. 128pp. 6⅛ x 9¼. 25965-X

THE ELECTROMAGNETIC FIELD, Albert Shadowitz. Comprehensive undergraduate text covers basics of electric and magnetic fields, builds up to electromagnetic theory. Also related topics, including relativity. Over 900 problems. 768pp. 5⅝ x 8¼. 65660-8

GREAT EXPERIMENTS IN PHYSICS: Firsthand Accounts from Galileo to Einstein, edited by Morris H. Shamos. 25 crucial discoveries: Newton's laws of motion, Chadwick's study of the neutron, Hertz on electromagnetic waves, more. Original accounts clearly annotated. 370pp. 5⅜ x 8½. 25346-5

RELATIVITY, THERMODYNAMICS AND COSMOLOGY, Richard C. Tolman. Landmark study extends thermodynamics to special, general relativity; also applications of relativistic mechanics, thermodynamics to cosmological models. 501pp. 5⅜ x 8½. 65383-8

LIGHT SCATTERING BY SMALL PARTICLES, H. C. van de Hulst. Comprehensive treatment including full range of useful approximation methods for researchers in chemistry, meteorology and astronomy. 44 illustrations. 470pp. 5⅜ x 8½. 64228-3

STATISTICAL PHYSICS, Gregory H. Wannier. Classic text combines thermodynamics, statistical mechanics and kinetic theory in one unified presentation of thermal physics. Problems with solutions. Bibliography. 532pp. 5⅜ x 8½. 65401-X

Paperbound unless otherwise indicated. Available at your book dealer, online at **www.doverpublications.com**, or by writing to Dept. GI, Dover Publications, Inc., 31 East 2nd Street, Mineola, NY 11501. For current price information or for free catalogues (please indicate field of interest), write to Dover Publications or log on to **www.doverpublications.com** and see every Dover book in print. Dover publishes more than 500 books each year on science, elementary and advanced mathematics, biology, music, art, literary history, social sciences, and other areas.